T0205865

Infosys Science Foundation Series

Infosys Science Foundation Series in Mathematical Sciences

The *Infosys Science Foundation Series in Mathematical Sciences* is a sub-series of The *Infosys Science Foundation Series*. This sub-series focuses on high quality content in the domain of mathematical sciences and various disciplines of mathematics, statistics, bio-mathematics, financial mathematics, applied mathematics, operations research, applies statistics and computer science. All content published in the sub-series are written, edited, or vetted by the laureates or jury members of the Infosys Prize. With the Series, Springer and the Infosys Science Foundation hope to provide readers with monographs, handbooks, professional books and textbooks of the highest academic quality on current topics in relevant disciplines. Literature in this sub-series will appeal to a wide audience of researchers, students, educators, and professionals across mathematics, applied mathematics, statistics and computer science disciplines.

More information about this series at http://www.springer.com/series/13817

Ramji Lal

Algebra 1

Groups, Rings, Fields and Arithmetic

 Springer

Ramji Lal
Harish Chandra Research Institute (HRI)
Allahabad, Uttar Pradesh
India

ISSN 2363-6149   ISSN 2363-6157   (electronic)
Infosys Science Foundation Series
ISSN 2364-4036   ISSN 2364-4044   (electronic)
Infosys Science Foundation Series in Mathematical Sciences
ISBN 978-981-13-5088-7   ISBN 978-981-10-4253-9   (eBook)
DOI 10.1007/978-981-10-4253-9

This Springer imprint is published by Springer Nature
The registered company is Springer Nature Singapore Pte Ltd.
The registered company address is: 152 Beach Road, #21-01/04 Gateway East, Singapore 189721, Singapore

*Dedicated to the memory of
my mother
(Late) Smt Murti Devi,
my father
(Late) Sri Sankatha Prasad Lal, and
my father-like brother
(Late) Sri Gopal Lal*

Preface

Algebra has played a central and decisive role in all branches of mathematics and, in turn, in all branches of science and engineering. It is not possible for a lecturer to cover, physically in a classroom, the amount of algebra which a graduate student (irrespective of the branch of science, engineering, or mathematics in which he prefers to specialize) needs to master. In addition, there are a variety of students in a class. Some of them grasp the material very fast and do not need much of assistance. At the same time, there are serious students who can do equally well by putting a little more effort. They need some more illustrations and also more exercises to develop their skill and confidence in the subject by solving problems on their own. Again, it is not possible for a lecturer to do sufficiently many illustrations and exercises in the classroom for the purpose. This is one of the considerations which prompted me to write a series of three volumes on the subject starting from the undergraduate level to the advance postgraduate level. Each volume is sufficiently rich with illustrations and examples together with numerous exercises. These volumes also cater for the need of the talented students with difficult, challenging, and motivating exercises which were responsible for the further developments in mathematics. Occasionally, the exercises demonstrating the applications in different disciplines are also included. The books may also act as a guide to teachers giving the courses. The researchers working in the field may also find it useful.

The present (first) volume consists of 11 chapters which starts with language of mathematics (logic and set theory) and centers around the introduction to basic algebraic structures, viz. group, rings, polynomial rings, and fields, together with fundamentals in arithmetic. At the end of this volume, there is an appendix on the basics of category theory. This volume serves as a basic text for the first-year course in algebra at the undergraduate level. Since this is the first introduction to the abstract-algebraic structures, we proceed rather leisurely in this volume as compared with the other volumes.

The second volume contains ten chapters which includes the fundamentals of linear algebra, structure theory of fields and Galois theory, representation theory of finite groups, and the theory of group extensions. It is needless to say that linear algebra is the most applicable branch of mathematics and it is essential for students of any discipline to develop expertise in the same. As such, linear algebra is an integral part of the syllabus at the undergraduate level. General linear algebra, Galois theory, representation theory of groups, and the theory of group extensions follow linear algebra which is a part, and indeed, these are parts of syllabus for the second- and third-year students of most of the universities. As such, this volume may serve as a basic text for second- and third-year courses in algebra.

The third volume of the book also contains 10 chapters, and it can act as a text for graduate and advanced postgraduate students specializing in mathematics. This includes commutative algebra, basics in algebraic geometry, homological methods, semisimple Lie algebra, and Chevalley groups. The table of contents gives an idea of the subject matter covered in the book.

There is no prerequisite essential for the book except, occasionally, in some illustrations and starred exercises, some amount of calculus, geometry, or topology may be needed. An attempt to follow the logical ordering has been made throughout the book.

My teacher (Late) Prof. B.L. Sharma, my colleague at the University of Allahabad, my friend Dr. H.S. Tripathi, my students Prof. R.P. Shukla, Prof. Shivdatt, Dr. Brajesh Kumar Sharma, Mr. Swapnil Srivastava, Dr. Akhilesh Yadav, Dr. Vivek Jain, Dr. Vipul Kakkar, and above all the mathematics students of the University of Allahabad had always been the motivating force for me to write a book. Without their continuous insistence, it would have not come in the present form. I wish to express my warmest thanks to all of them.

Harish-Chandra Research Institute (HRI), Allahabad, has always been a great source for me to learn more and more mathematics. I wish to express my deep sense of appreciation and thanks to HRI for providing me all the infrastructural facilities to write these volumes.

Last but not least, I wish to express my thanks to my wife Veena Srivastava who had always been helpful in this endeavor.

In spite of all care, some mistakes and misprint might have crept in and escaped my attention. I shall be grateful to any such attention. Criticisms and suggestions for the improvement of the book will be appreciated and gratefully acknowledged.

Allahabad, India   Ramji Lal
April 2017

Contents

About the Author

Ramji Lal is Adjunct Professor at the Harish-Chandra Research Institute (HRI), Allahabad, Uttar Pradesh. He started his research career at the Tata Institute of Fundamental Research (TIFR), Mumbai, and served at the University of Allahabad in different capacities for over 43 years: as a Professor, Head of the Department, and the Coordinator of the DSA program. He was associated with HRI, where he initiated a postgraduate (PG) program in mathematics and Coordinated the Nurture Program of National Board for Higher Mathematics (NBHM) from 1996 to 2000. After his retirement from the University of Allahabad, he was an Advisor cum Adjunct Professor at the Indian Institute of Information Technology (IIIT), Allahabad, for over 3 years. His areas of interest include group theory, algebraic K-theory, and representation theory.

Notations from Algebra 1

$\langle a \rangle$	Cyclic subgroup generated by a, p. 122
a/b	a divides b, p. 57
$a \sim b$	a is an associate of b, p. 57
A^t	The transpose of a matrix A, p. 201
$A^\star$	The hermitian conjugate of a matrix A, p. 215
$Aut(G)$	The automorphism group of G, p. 103
A_n	The alternating group of degree n, p. 175
$B(n, \mathbb{R})$	Borel subgroup, p. 189
$C_G(H)$	The centralizer of H in G, p. 160
$\mathbb{C}$	The field of complex numbers, p. 78
D_n	The dihedral group of order $2n$, p. 90
det	Determinant map, p. 193
$End(G)$	Semigroup of endomorphisms of G, p. 103
$f(A)$	Image of A under the map f, p. 33
$f^{-1}(B)$	Inverse image of B under the map f, p. 33
$f|Y$	Restriction of the map f to Y, p. 29
E_{ij}^λ	Transvections, p. 201
$Fit(G)$	Fitting subgroup, p. 357
$g.c.d.$	Greatest common divisor, p. 58
$g.l.b.$	Greatest lower bound, or inf, p. 39
$G/^l H (G/^r H)$	The set of left(right) cosets of G mod H, p. 133
G/H	The quotient group of G modulo H, p. 150
$[G : H]$	The index of H in G, p. 133
$|G|$	Order of G
$G' = [G, G]$	Commutator subgroup of G
G^n	nth term of the derived series of G, p. 348
$GL(n, \mathbb{R})$	General linear group, p. 187
I_X	Identity map on X, p. 29
i_Y	Inclusion map from Y, p. 30
$Inn(G)$	The group of inner automorphisms

$ker\,f$	The kernel of the map f, p. 35
$L_n(G)$	nth term of the lower central series of G
$l.c.m.$	Least common multiple, p. 58
$l.u.b.$	Least upper bound, or sup, p. 39
$M_n(R)$	The ring of $n \times n$ matrices with entries in R
$\mathbb{N}$	Natural number system, p. 22
$N_G(H)$	Normalizer of H in G, p. 160
$O(n)$	Orthogonal group, p. 198
$O(1, n)$	Lorentz orthogonal group, p. 202
$PSO(1, n)$	Positive special Lorentz orthogonal group, p. 203
$\mathbb{Q}$	The field of rational numbers, p. 73
Q_8	The Quaternion group, p. 88
$\mathbb{R}$	The field of real numbers, p. 75
$R(G)$	Radical of G, p. 349
S_n	Symmetric group of degree n, p. 88
$Sym(X)$	Symmetric group on X, p. 88
S^3	The group of unit Quaternions, p. 91
$\langle S \rangle$	Subgroup generated by a subset S, p. 116
$SL(n, \mathbb{R})$	Special linear group, p. 196
$SO(n)$	Special orthogonal group, p. 199
$SO(1, n)$	Special Lorentz orthogonal group, p. 203
$SP(2n, \mathbb{R})$	Symplectic group, p. 202
$SU(n)$	Special unitary group, p. 204
$U(n)$	Unitary group, p. 204
U_m	Group of prime residue classes modulo m, p. 99
V_4	Klein's four group, p. 101
X/R	The quotient set of X modulo R, p. 36
R_x	Equivalence class modulo R determined by x, p. 27
X^+	Successor of X, p. 20
X^Y	The set of maps from Y to X, p. 33
$\subset$	Proper subset, p. 15
$\wp(X)$	Power set of X, p. 19
$\prod_{k=1}^{n} G_k$	Direct product of groups $G_k, 1 \le k \le n$, p. 142
$\trianglelefteq$	Normal subgroup, p. 148
$\trianglelefteq\trianglelefteq$	Subnormal subgroup, p. 335
$Z(G)$	Center of G, p. 112
$\mathbb{Z}_m$	The ring of residue classes modulo m, p. 80
$p(n)$	The number of partition of n, p. 209
$H \prec K$	Semidirect product of H with K, p. 206
$\sqrt{A}$	Radical of an ideal A, p. 233
$R(G)$	Semigroup ring of a ring R over a semigroup G, p. 239
$R[X]$	Polynomial ring over the ring R in one variable, p. 241
$R[X_1, X_2, \cdots, X_n]$	Polynomial ring in several variables, p. 249
μ	The Mobius function, p. 257

σ	Sum of divisor function, p. 257
$\left(\frac{a}{p}\right)$	Legendre symbol, p. 282
$Stab(G, X)$	Stabilizer of an action of G on X, p. 298
G_x	Isotropy subgroup of an action of G at x, p. 298
X^G	Fixed point of an action of G on X
$Z_n(G)$	nth term of the upper central series of G, p. 354
$\Phi(G)$	The Frattini subgroup of G, p. 358

Chapter 1
Language of Mathematics 1 (Logic)

The principal aim of this small and brief chapter is to provide a logical foundation to sound mathematical reasoning, and also to understand adequately the notion of a mathematical proof. Indeed, the incidence of paradoxes (Russell's and Cantor's paradoxes) during the turn of the 19th century led to a strong desire among mathematicians to have a rigorous foundation to all disciplines in mathematics. In logic, the interest is in the form rather than the content of the statements.

1.1 Statements, Propositional Connectives

In mathematics, we are concerned about the truth or the falsity of the statements involving mathematical objects. Yet, one need not take the trouble to define a statement. It is a primitive notion which everyone inherits. Following are some examples of statements.

1. Man is the most intelligent creature on the earth.
2. Charu is a brave girl, and Garima is an honest girl.
3. Sun rises from the east or sun rises from the west.
4. Shipra will not go to school.
5. If Gaurav works hard, then he will pass.
6. Gunjan can be honest if and only if she is brave.
7. 'Kishore has a wife' implies 'he is married.'
8. 'Indira Gandhi died martyr' implies and implied by 'she was brave.'
9. For every river, there is an origin.
10. There exists a man who is immortal.

   The sentences 'Who is the present President of India?', 'When did you come?', and 'Bring me a glass of water' are not statements. A statement asserts something(true or false).

© Springer Nature Singapore Pte Ltd. 2017

R. Lal, *Algebra 1*, Infosys Science Foundation Series in Mathematical Sciences,

DOI 10.1007/978-981-10-4253-9_1

We have some operations on the class of statements, namely 'and', 'or', 'If ...,
then ...', 'if and only if' (briefly iff), 'implies', and 'not'. In fact, we consider a
suitable class of statements (called the valid statements) which is closed under the
above operations. These operations are called the **propositional connectives**.

The rules which govern the formation of valid statements are in very much use (like
those of english grammar) without being conscious of the fact, and it forms the content
of the **propositional calculus**. For the formal development of the language, one is
referred to an excellent book entitled 'Set Theory and Continuum Hypothesis' by
P.J. Cohen. Here, in this text, we shall adopt rather the traditional informal language.

Conjunction

The propositional connective '*and*' is used to conjoin two statements. The conjunction of a statement P and a statement Q is written as 'P and Q'. The symbol '$\wedge$'
is also used for '*and*'. Thus, '$P \wedge Q$' also denotes the conjunction of P and Q. The
example 2 above is an example of a conjunction.

Disjunction

The propositional connective '*or*' is used to obtain the disjunction of two statements.
The disjunction of a statement 'P' and a statement 'Q' is written as 'P or Q'. The
symbol '$\vee$' is also used for '*or*'. The disjunction of a statement 'P' and a statement
'Q' is also written as '$P \vee Q$'. The example 3 above is an example of a disjunction.

Negation

Usually '*not*' is used at a suitable place in a statement to obtain the negation of the
statement. The negation of a statement 'P' is denoted by '$-P$'. The example 4 above
is the negation of 'Shipra will go to school'. The negation of this statement can also
be expressed by '-(Shipra will go to school)'.

Conditional statement

A statement of the form '*If P, then Q*' is called a *conditional statement*. The statement 'P' is called the *antecedent* or the *hypothesis*, and 'Q' is called the *consequent*
or the *conclusion*. The example 5 above is a conditional statement. 'If P, then Q' is
also expressed by saying that 'Q is a necessary condition for P'. An other way to
express it is to say that 'P is a sufficient condition for Q'.

Implication

A statement of the form 'P implies Q' (in symbol '$P \implies Q$') is called an *implication*.
The statement '$P \implies Q$' and the statement 'If P, then Q' are logically same, for
(as we shall see) the truth values of both the statements are always same. Again, 'P'
is called the *antecedent* or the *hypothesis*, and 'Q' is called the *consequent* or the
conclusion. Example 7 is an *implication*.

Equivalence

A statement of the form 'P if and only if Q' (briefly 'P *iff* Q') is called an *equivalence*.
'P implies and implied by Q' (in symbol '$P \iff Q$') is logically same as 'P if and
only if Q'. We also express it by saying that 'P is a necessary and sufficient condition
for Q.' Examples 6 and 7 are equivalences.

1.2   Statement Formula and Truth Functional Rules

A **statement variable** is a variable which can take any value from the class of valid atomic statements (statements without propositional connectives). We use the notations P, Q, R, *etc.* for the statement variables. A **well-formed statement formula** is a finite string of the statement variables, the propositional connectives, and the parenthesis limiting the scopes of connectives. Thus, for example,

$$(P \Longrightarrow Q) \Longleftrightarrow (-P \bigvee Q),$$

$$(P \Longrightarrow Q) \bigvee (Q \Longrightarrow P),$$

and

$$P \bigvee (Q \bigwedge R) \Longleftrightarrow (P \bigvee Q) \bigwedge (P \bigvee R)$$

are well-formed statement formulas.

The rules of dependence of the truth value of a statement formula on the truth values of its statement variables (atomic parts) (which are prompted by our common sense) are called the *truth functional rules*. These rules are illustrated by tables called the *truth tables*.

The truth functional rule for the conjunction '$P \bigwedge Q$'

The statement formula '$P \bigwedge Q$' is true only in case both P as well as Q are true. Thus, the truth functional rule for '$P \bigwedge Q$' is given by the table

P	Q	$P \bigwedge Q$
T	T	T
T	F	F
F	T	F
F	F	F

The truth functional rule for the disjunction '$P \bigvee Q$'

The statement formula '$P \bigvee Q$' is true if at least one of P and Q is true. The table giving the truth functional rule for '$P \bigvee Q$' is

P	Q	$P \bigvee Q$
T	T	T
T	F	T
F	T	T
F	F	F

The truth functional rule for the negation $'(-P)'$

The negation of a true statement is false and that of a false statement is true. Thus,

P	$-P$
T	F
F	T

The truth functional rule for '*If P, then Q*' ('$P \Longrightarrow Q$').

The statement formula '*If P, then Q*' ('$P \Longrightarrow Q$') is false in only one case when P is true but Q is false. Take, for example, the statement 'If a student works hard, then he will pass.' The truth of this statement says that if some student works hard, then he will pass. If there is some student who has not worked hard, then whether he passes, or he fails, the truth of the statement remains unchallenged. Thus, the truth table for '*If P, then Q*' is as follows:

P	Q	*If P, then Q*	$P \Longrightarrow Q$
T	T	T	T
T	F	F	F
F	T	T	T
F	F	T	T

The statement '$P \Longleftrightarrow Q$' is the conjunction of the statement '$P \Longrightarrow Q$' and the statement '$Q \Longrightarrow P$'. Thus, the truth table for the equivalence '$P \Longleftrightarrow Q$' is as follows:

P	Q	$P \Longleftrightarrow Q$	*P iff Q*
T	T	T	T
T	F	F	F
F	T	F	F
F	F	T	T

Example 1.2.1 Truth table for the statement formula '$(P \bigvee Q) \Longrightarrow (Q \bigvee P)$'

P	Q	$P \bigvee Q$	$Q \bigvee P$	$(P \bigvee Q) \Longrightarrow (Q \bigvee P)$
T	T	T	T	T
T	F	T	T	T
F	T	T	T	T
F	F	F	F	T

Example 1.2.2 Truth table for the statement formula '$(P \bigvee Q) \Longrightarrow (P \bigwedge Q)$'

P	Q	$P \bigvee Q$	$P \bigwedge Q$	$(P \bigvee Q) \Longrightarrow (P \bigwedge Q)$
T	T	T	T	T
T	F	T	F	F
F	T	T	F	F
F	F	F	F	T

Example 1.2.3 Truth table for the statement formula '$[(P \vee Q) \wedge -P] \Longrightarrow Q$'

P	Q	$P \vee Q$	$-P$	$(P \vee Q) \wedge -P$	$[(P \vee Q) \wedge -P] \Longrightarrow Q$
T	T	T	F	F	T
T	F	T	F	F	T
F	T	T	T	T	T
F	F	F	T	F	T

Example 1.2.4 Truth table for the statement formula '$(P \Longrightarrow Q) \Longleftrightarrow (-Q \Longrightarrow -P)$'

P	Q	$P \Longrightarrow Q$	$-Q$	$-P$	$-Q \Longrightarrow -P$	$(P \Longrightarrow Q) \Longleftrightarrow (-Q \Longrightarrow -P)$
T	T	T	F	F	T	T
T	F	F	T	F	F	T
F	T	T	F	T	T	T
F	F	T	T	T	T	T

Example 1.2.5 Truth table for the statement formula '$-(P \vee Q) \Longleftrightarrow (-P \wedge -Q)$'

P	Q	$P \vee Q$	$-(P \vee Q)$	$-P$	$-Q$	$-P \wedge -Q$	$-(P \vee Q) \Longleftrightarrow (-P \wedge -Q)$
T	T	T	F	F	F	F	T
T	F	T	F	F	T	F	T
F	T	T	F	T	F	F	T
F	F	F	T	T	T	T	T

Example 1.2.6 Truth table for the statement formula '$P \vee -P$'

P	$-P$	$P \vee -P$
T	F	T
F	T	T

Example 1.2.7 Truth table for the statement formula '$-(P \Longrightarrow Q) \Longleftrightarrow (P \wedge -Q)$'

P	Q	$P \Longrightarrow Q$	$-(P \Longrightarrow Q)$	$-Q$	$P \wedge -Q$	$-(P \Longrightarrow Q) \Longleftrightarrow (P \wedge -Q)$
T	T	T	F	F	F	T
T	F	F	T	T	T	T
F	T	T	F	F	F	T
F	F	T	F	T	F	T

Exercises

Construct the truth tables for the following statement Formulas.

1.2.1 $P \bigwedge -P$.

1.2.2 $(P \bigwedge Q) \Longleftrightarrow (Q \bigwedge P)$.

1.2.3 $(P \bigwedge Q) \Longrightarrow (P \bigvee Q)$.

1.2.4 $P \Longleftrightarrow -(-P)$.

1.2.5 $-(P \bigwedge Q) \Longleftrightarrow (-P \bigvee -Q)$.

1.2.6 $(P \Longrightarrow Q) \Longleftrightarrow (-P \bigvee Q)$.

1.2.7 $Q \Longrightarrow (P \bigvee -Q)$.

1.2.8 $(P \bigwedge -P) \Longrightarrow Q$.

1.2.9 $(P \bigwedge Q) \Longrightarrow Q$.

1.2.10 $P \Longrightarrow (P \bigvee Q)$.

1.2.11 $(P \Longrightarrow Q) \Longrightarrow Q$.

1.2.12 $[-(P \Longleftrightarrow Q)] \Longleftrightarrow [(P \bigwedge -Q) \bigvee (\ P \bigwedge Q)]$.

1.2.13 $(P \bigwedge Q) \Longrightarrow -P$.

1.2.14 $P \Longrightarrow ((-P) \bigwedge Q)$.

1.2.15 $(P \Longrightarrow Q) \bigvee (Q \Longrightarrow P)$.

1.2.16 $(P \Longrightarrow Q) \Longrightarrow (Q \Longrightarrow P)$.

1.2.17 $[P \bigvee (Q \bigwedge -Q)] \Longleftrightarrow P$.

1.2.18 $(P \Longleftrightarrow Q) \bigvee (-P)$.

1.2.19 $P \bigvee (-P \bigwedge Q)$.

1.2.20 $(P \bigvee Q) \bigvee R \Longleftrightarrow P \bigvee (Q \bigvee R)$.

1.2.21 $(P \bigwedge Q) \bigwedge R \Longleftrightarrow P \bigwedge (Q \bigwedge R)$.

1.2.22 $P \bigvee (Q \bigwedge R) \Longleftrightarrow (P \bigvee Q) \bigwedge (P \bigvee R)$.

1.2.23 $P \bigwedge (Q \bigvee R) \Longleftrightarrow (P \bigwedge Q) \bigvee (P \bigwedge R)$.

1.2.24 $[(P \Longleftrightarrow Q) \bigwedge (Q \Longleftrightarrow R)] \Longleftrightarrow [P \Longleftrightarrow R]$.

1.2.25 $[(P \Longrightarrow Q) \bigvee (Q \Longrightarrow R)] \Longleftrightarrow [P \Longrightarrow R]$.

1.3 Quantifiers

Universal Quantifier

Consider the statement 'For every river, there is an origin.' This can be rewritten as 'For every x, 'x is a river' implies 'x has an origin'.' More generally, we have a statement of the form 'For every x, $P(x)$.', where '$P(x)$' is a valid statement involving x. The symbol '$\forall$' is used for 'for every,' and it is called the *universal quantifier*. The example 9 of Sect. 1.1 may be represented by '$\forall x$('x is a river '$\implies$ 'x has an origin')'.

Existential Quantifier

Consider the statement 'There is a man who is immortal.'. More generally, we have statements of the form 'There exists x, $P(x)$.', where '$P(x)$' is a statement involving x. The symbol '$\exists$' stands for 'there exists,' and it is called the *existential quantifier*. The example 10 of Sect. 1.1 may be represented as '$\exists x$, 'x is a man' and 'x is immortal'.'.

Parenthesis '( )' and brackets '[ ]' will be used to limit the scope of propositional connectives and quantifiers to make valid mathematical statements.

Negation of a Statement Formula Involving Quantifiers

Consider the statement '*Every river has an origin.*'. This can be rephrased as '$\forall x$('x is a river' $\implies$ 'x has an origin').'. When can this statement be false? It is false if and only if there is a river which has no origin. Similarly, consider the statement '*Every man is mortal.*'. This can also be rephrased as '$\forall x$('x is a man' $\implies$ 'x is mortal').'. Again this statement can be challenged if and only if there is a man who is immortal. Now, consider the statement '*There is a river which has no origin.*'. To say that this statement is false is to say that '*Every river has an origin.*' This prompts us to have the truth functional rule for the statement formulas involving quantifiers as given by the following table.

$\forall x(P(x) \implies Q(x))$	$\exists x(P(x) \wedge -Q(x))$
T	F
F	T

Thus, '$-[\forall x(P(x) \implies Q(x))] \iff [\exists x(P(x) \wedge -Q(x))]$,' where $P(x)$ and $Q(x)$ are valid statements involving the symbol x, is always a true statement. Also '$-[\exists(P(x) \implies Q(x))] \iff [\forall x((Px) \wedge -Q(x))]$' is always a true statement.

1.4   Tautology and Logical Equivalences

A statement formula is called a **tautology** if its truth value is always T irrespective of the truth values of its atomic statement variables. A statement formula is called a **contradiction** if its truth value is always F irrespective of the truth values of its atomic statement variables. Thus, the negation of a tautology is a contradiction, and the negation of a contradiction is a tautology.

All the examples in Sect. 1.2 except the Example 1.2.2 are tautologies. '$P \bigwedge -P$' is a contradiction.

Example 1.4.1 '$-\forall x(P(x) \implies Q(x)) \iff \exists x(P(x) \bigwedge -Q(x))$' is a tautology.

Example 1.4.2 '$[(P \implies Q) \bigwedge (Q \implies R)] \implies (P \implies R)$' is a tautology (verify by making truth table).

Thus, if the statement formulas '$A \implies B$' and '$B \implies C$' are tautologies, then '$A \implies C$' is also a tautology.

For the given statement formulas 'A' and 'B', we say that A *logically implies B* or *B logically follows from A* if '$A \implies B$' is a tautology. (Here, A and B are not simple statement variables). In fact, if 'P' and 'Q' are statement variables, then '$P \implies Q$' is a tautology if and only if P is same as Q. Further, the statement formula A is said to be *logically equivalent* to B if '$A \iff B$' is a tautology.

In mathematics and logic, we do not distinguish logically equivalent statements. They are taken to be same. If A is logically equivalent to B, we may substitute B for A and A for B in any course of discussion or derivation.

Example 1.4.3 '$P \implies Q$' is logically equivalent to '$-P \bigvee Q$'.

Example 1.4.4 '$-[\forall x(P(x) \implies Q(x))]$' is logically equivalent to '$\exists x(P(x) \bigwedge -Q(x))$' and '$\forall x(P(x) \bigwedge -Q(x))$' is logically equivalent to '$-[\exists x(P(x) \implies Q(x))]$'.

Example 1.4.5 The notation $\lim_{n \to \infty} x_n = x$ stands for the statement
'$\forall \epsilon [\epsilon$ *is a positive real number* $\implies$
$\exists N(N$ *is a natural number* $\implies \forall n(n$ *is greater than N*
$\implies | x_n - x |$ *is less than* $\epsilon))]$'.
If we apply the logical equivalence in Example 1.4.4 repeatedly, then we find that
$-(\lim_{n \to \infty} x_n = x)$ is same as
$\exists \epsilon (\epsilon$ *is a positive real number* $\bigwedge \forall N$
$(N$ *is a natural number* $\bigwedge$
$\exists n(n$ *is greater than N* $\bigwedge | x_n - x |$ *is not less than* $\epsilon)))$.

Exercises

1.4.1 Find out which of the statement formulas in exercises from 1.2.1 to 1.2.25 are tautologies and which of them are contradictions.

1.4.2 Is 'Sun rises from the east' a tautology?

1.4.3 Obtain a logically equivalent statement formula for the negation of '$\forall x[P(x) \implies (R(x) \land T(x))]$'.

1.4.4 Show that the set '$\{\lor, -\}$' of propositional connectives is *Functionally Complete* in the sense that any statement formula is logically equivalent to a statement formula involving only two connectives $\lor$ and $-$. Is the representation thus obtained unique? Support. Similarly, show that the set '$\{\land, -\}$' is also *Functionally Complete*. Thus, the set '$\{\lor, -\}$' of propositional connectives is sufficient to develop the mathematical logic.

1.4.5 Let A be a statement formula which is a tautology. Suppose that '$A \implies B$' is also a tautology. Show that B is also a tautology. Can B be a statement variable? Support.

1.4.6 Let A be a statement formula which is a tautology. Show that '$A \lor B$' and '$B \implies A$' are also tautologies.

1.4.7 Suppose that A and B are tautologies. Show that '$A \land B$' is also a tautology.

1.4.8 Suppose that '$A \implies B$' and '$B \implies C$' are tautologies. Show that '$A \implies C$' is also a tautology.

1.5 Theory of Logical Inference

In any course of mathematical derivations and inferences, we have certain statements termed as axioms, premises, postulates, or hypotheses whose truth values are assumed to be T, and then infer the truth of a statement as a theorem, proposition, corollary, or a lemma. Indeed, a statement is a theorem (proposition, lemma, or a corollary) if and only if the conjunction of premises tautologically imply the statement.

The theory of logical inference is like playing games. Take, for example, a game of chess. The initial position of the chess board corresponds to premises. There are finitely many rules of the game, and the players have to follow these rules while making their moves. These rules of the game correspond to tautological implications. The player 1 initially moves one of his chess pieces as per the rules of the game. The new position of the chess board becomes premises for the player 2. The player 2, in his turn, moves one of his chess pieces as per rules of the game, of course, keeping his eyes on a winning position. Next, the player 1, in his turn, takes this new position as the premises and moves one of his chess pieces as per rules and so on. The player who reaches the winning position (desired theorem) wins the game. However, the game may end in a draw, and the players may reach at a position of the chess board from where no player can ever reach the winning position by moving the chess pieces as per rules. Each position of the chess board where the players reach corresponds to a theorem.

Thus, our main aim is to describe irredundant finite set of rules of inferences which meets the following two criteria.

1. For any set of premises, the rules of logical inferences must allow only those statements which follow tautologically from the conjunction of premises.
2. All statements which follow tautologically from the conjunction of premises can be derived in finitely many steps by applying the rules of inferences.

Indeed, the following **three rules of inferences** are adequate to derive all theorems under given premises.

Rule 1. A premise may be introduced at any point in a derivation.

Rule 2. A statement 'P' may be introduced at any point of derivation if the conjunction of the preceding derivations tautologically implies 'P'.

Rule 3. A statement 'If 'P,' then 'Q' ' may be introduced at any point of derivation provided that 'Q' is derivable from the conjunction of 'P' and some of the premises.

We illustrate these rules of inferences by means of some simple examples.

Example 1.5.1 If 'Shreyansh is a prodigy,' then 'if 'he will become a scientist,' then 'he will win a Nobel prize' '. 'Shreyansh will become a scientist' or 'he will become a cricketer.' If 'Shreyansh is not a prodigy,' then 'he will become a cricketer.' Shreyansh will not become a cricketer. Therefore, 'Shreyansh will win a Nobel prize.'

Here, the statement 'Shreyansh will win a Nobel prize' is to be derived as a theorem. The statements preceding to this statement are premises. We use the rules of inferences to deduce this theorem. We symbolize the statements as follows: Let 'P' stands for the statement 'Shreyansh is prodigy,' 'S' for the statement 'He will become a scientist,' 'N' for the statement 'Shreyansh will win a Nobel prize' and 'C' for the statement 'Shreyansh will become a cricketer'. Thus, '$P \implies (S \implies N)$,' '$S \lor C$,' '$-P \implies C$' and '$-C$' are premises and we have to derive N as a theorem. Now,

1. '$-C$' ....Premise (Rule 1).
2. '$-P \implies C$'........................Premise (Rule 1).
3. 'P'...'$(-P \implies C) \land -C$' tautologically implies 'P' (Rule 2).
4. '$P \implies (S \implies N)$'.......Premise (Rule 1).
5. '$S \implies N$'.........................'$P \land (S \implies N)$' tautologically implies '$S \implies N$' (Rule 2).
6. '$S \lor C$'..................................Premise (Rule 1).
7. 'S'...'$-C \land (S \lor C)$' tautologically implies 'S' (Rule 2).
8. 'N'...'$S \land (S \implies N)$' tautologically implies 'N' (Rule 2).

This establishes 'N' as a theorem. Note that in this derivation we have not used the rule 3.

Example 1.5.2 'If Shreyal is not a genius,' then 'he cannot solve difficult mathematical problems.' 'If he cannot solve difficult mathematical problems,' then 'he will not become a great mathematician.' 'Shreyal will become a business tycoon' or 'he will become a great mathematician.' He will not become a business tycoon. Therefore, 'Shreyal is genius.'

Here, the statement 'Shreyal is genius' is to be derived as a theorem. The statements preceding to this statement are premises. We use the rules of inferences to deduce this theorem. We symbolize the statements as follows: Let 'G' stands for the statement 'Shreyal is genius,' 'P' for the statement 'He can solve difficult mathematical problems,' 'M' for the statement 'He will become a great mathematician,' and 'B' for the statement 'Shreyal will become a business tycoon'. Thus, '$-G \implies -P$', '$-P \implies -M$', '$B \bigvee M$', and '$-B$' are premises, and we have to derive G as a theorem. Now,

1. '$-B$' ....Premise (Rule 1)
2. '$B \bigvee M$'.............Premise (Rule 1)
3. 'M'........................'$-B \bigwedge (B \bigvee M)$' tautologically imply '$M$' (Rule 2)
4. '$-P \implies -M$'.........Premise (Rule 1)
5. 'P' ....'$(-P \implies -M) \bigwedge M$' tautologically imply '$P$' (Rule 2)
6. '$-G \implies -P$'.........Premise (Rule 1)
7. 'G' ....   '$(-G \implies -P) \bigwedge P$' tautologically imply '$G$' (Rule 2)

This establishes G as a theorem. Note that in this derivation also we have not used the rule 3.

The next example uses rule 3 also.

Example 1.5.3 If 'Sachi is honest,' then 'if 'she is brave, then 'she will be intelligent''. 'Sachi is not hard working or she is honest'. 'Sachi is brave'. Therefore, if 'Sachi is hard working,' then 'she will be intelligent.'

Here, the statement 'If 'Sachi is hard working,' then 'she will be intelligent' ' is to be derived as a theorem. The statements preceding to this statement are premises. We use the rules of inferences to deduce this theorem. Symbolize the statements as follows: Let 'H' stands for the statement 'Sachi is honest,' 'B' for the statement 'She is brave,' 'I' for the statement 'She will be a intelligent,' and 'W' for the statement 'Sachi is hard working'. Thus, '$H \implies (B \implies I)$,' '$-W \bigvee H$' and '$B$' are premises, and we have to derive '$W \implies I$' as a theorem. Now,

1. '$H \implies (B \implies I)$'.....Premise (Rule 1).
2. '$-W \bigvee H$'..............................Premise (Rule 1).
3. 'B'...Premise (Rule 1).
4. '$W \implies H$'........................The conjunction 'W' and the premise '$-W \bigvee H$' tautologically imply 'H'(Rule 3).

5. '$W \Longrightarrow I$'........................ The conjunction of 'W,' '$W \Longrightarrow H$,' 'B' and '$H \Longrightarrow$
   $(B \Longrightarrow I)$' tautologically imply 'I' (Rule 2 and Rule 3).

Consistency of Premises

A set of premises is said to be a **consistent** set of premises if the conjunction of
premises has its truth value T for some choice of truth values of each premise.
It is said to be inconsistent, otherwise. Thus, to derive that a set of premises is
inconsistent is to derive that the conjunction of the set of premises tautologically
imply the statement '$P \wedge -P$'. However, in many situations, it is not so easy to
establish the consistency of premises. Easiest way, perhaps, is to have an example
where all the premises happen to be true.

   Often, a lawyer in a court while cross-examining a witness of the other side tries
to establish that the evidences and statements of witness as premises is inconsistent
by producing a paradox out of witness and there by discrediting the witness.

Example 1.5.4 If in the set of premises of Example 1.5.3, we adjoin the statement
'Sachi is hard working and she is not intelligent,' then the set of premises becomes
inconsistent. For, then '$(W \Longrightarrow I) \wedge (W \wedge -I)$' is logically derivable from the
set of premises. Observe that '$(W \Longrightarrow I) \wedge (W \wedge -I)$' is logically equivalent to
'$P \wedge -P$'.

Exercises

1.5.1 If 'the prices of the essential commodities are low,' then 'the government
will become popular.' 'The prices of the essential commodities are low' or 'there
is a shortage of the essential commodities.' If 'there is a shortage of the essential
commodities,' then 'the production of essential commodities is low.' However, 'there
is a huge production of essential commodities.' Using the logical rules of inference,
derive the proposition 'The government will become popular.'

1.5.2 'Sachi is creative' or 'she is intelligent.' If 'Sachi is creative,' then 'she is
imaginative.' 'Sachi is not imaginative' or 'she is not a musician.' In fact, 'Sachi is
a musician.' Derive the statement 'Sachi is intelligent' as a theorem.

1.5.3 Test for consistency the following set of premises.

   If 'Shikhar is good in physics, then 'he is good in mathematics.' If 'he is good
in mathematics, then 'he is good in logic.' 'He is good in logic' or 'he is good in
physics.' He is not good in logic.

Chapter 2
Language of Mathematics 2 (Set Theory)

This chapter contains a brief introduction to set theory which is essential for doing mathematics. There are two main axiomatic systems to introduce sets, viz. Zermelo–Fraenkel axiomatic system and the Gödel–Bernays axiomatic system. Here, in this text, we shall give an account of Zermelo–Fraenkel axiomatic set theory together with the axiom of choice (an axiom which is independent of the Zermelo–Fraenkel axiomatic system). We also discuss some of the important and useful equivalents of the axiom of choice. The ordinal and the cardinal numbers are introduced and discussed in a rigorous way. For the further formal development of the theory, the reader is referred to the '*Set Theory and Continuum hypothesis*' by P.J. Cohen or the '*Axiomatic set theory*' by P. Suppes.

2.1 Set, Zermelo–Fraenkel Axiomatic System

'*Set*', '*belongs to*,' and '*equal to*' are primitive terms of which the reader has intuitive understanding. Their use is governed by some postulates in axiomatic set theory.

To take the help of intuition in ascertaining the use of the primitive terms, we regard a set as a collection of objects. 'A class of students,' 'a flock of sheep,' 'a bunch of flowers,' and 'a packet of biscuits' are all examples of sets of things. The notation '$a \in A$' stands for the statement 'a belongs to A' ('a is an element of A,' or also for 'a is a member of A'). The negation of '$a \in A$' is denoted by '$a \notin A$.' The notation '$A = B$' stands for the statement 'A is equal to B.' The negation of '$A = B$' is denoted by '$A \neq B$.' The following axiom relates '$\in$' and '$=$.'

Axiom 1 (*Axiom of extension*) Let A and B be sets. Then,

'$A = B$' if and only if 'for all x ($x \in A$ if and only if $x \in B$).'

© Springer Nature Singapore Pte Ltd. 2017
R. Lal, *Algebra 1*, Infosys Science Foundation Series in Mathematical Sciences,
DOI 10.1007/978-981-10-4253-9_2

Thus, two sets A and B are equal if they have same members. Two equal sets are treated as same. If $A = B$, then we may substitute A for B and B for A in any course of discussion.

Remark 2.1.1 To be logically sound in the use of primitive terms, axiom of extension is a necessity.

Let A and B be sets. We say that A is a **subset** of B (A *is contained in B* or *B contains A*) if every member of A is a member of B. The statement 'A *is a subset of B*' is the same as the statement '*For all x(if* $x \in A$, *then* $x \in B$).' The notation '$A \subseteq B$' (or also '$B \supseteq A$') stands for the statement 'A is a subset of B.' Thus, '$A = B$' (axiom of extension) if and only if '$A \subseteq B$ and $B \subseteq A$.' The negation of '$A \subseteq B$' is denoted by '$A \nsubseteq B$.' Since the negation of the statement '*For all x(if* $x \in A$, *then* $x \in B$)' is logically same as the statement '*There exists* $x(x \in A$ *and* $x \notin B$),' the notation '$A \nsubseteq B$' stands for the statement '*There exists* $x(x \in A$ *and* $x \notin B$).' Thus, to say that A is not a subset of B is to say that there is an element of A which is not in B.

Every set is a subset of itself, because '*For all x(if* $x \in A$, *then* $x \in A$)' is a tautology (always a true statement). If $A \subseteq B$ and $A \neq B$, then we say that A is a **proper subset** of B. The notation '$A \subset B$' stands for the statement 'A is a proper subset of B.' Thus, A is a proper subset of B if every member of A is a member of B, and there is a member of B which is not a member of A. More precisely, '$A \subset B$' represents the statement '(*For all x(if* $x \in A$, *then* $x \in B$)) and (*there exists* $x(x \in B$ *and* $x \notin A$)).'

Proposition 2.1.2 *If* $A \subseteq B$ *and* $B \subseteq C$, *then* $A \subseteq C$.

Proof Suppose that $A \subseteq B$ and $B \subseteq C$. Let $x \in A$. Since $A \subseteq B$, $x \in B$. Further, since $B \subseteq C$, $x \in C$. Thus, '*for all* $x(if$ $x \in A$, *then* $x \in C$).' This shows that $A \subseteq C$.
♯

Some of the axioms of set theory are designed to produce different sets out of given sets. The first one is to generate subsets of a set.

Consider the set A of all men and the statement 'x *is a teacher*.' Some members of A are teachers, and some of them are not. The condition that 'x *is a teacher*' defines a subset of A, namely the set of all male teachers. To make it more formal, we have:

Axiom 2 (*Axiom of specification*) Let A be a set, and $P(x)$ be a valid statement involving the free symbol x. Then, there is a set B such that

$$\text{'for all } x(x \in B \text{ if and only if } (x \in A \text{ and } P(x)).\text{'}$$

Thus, to every set A, and to every statement $P(x)$, there is a unique set B whose members are exactly those members of A for which $P(x)$ is true.

The set B described above is denoted by $\{x \in A \mid P(x)\}$. Clearly, B is a subset of A.

Proposition 2.1.3 *Let A be a set. Then there is a set B such that $B \notin A$.*

Proof Consider the statement '*x is a set and x* $\notin$ *x*.' By the axiom of specification, there is a unique set $B = \{x \in A$ *such that x is a set and x* $\notin$ *x*$\}$. We show that $B \notin A$. Suppose that $B \in A$. If $B \in B$, then $B \notin B$. Next, if $B \notin B$, then since $B \in A$ (supposition), and B is a set, $B \in B$. Thus, '$B \notin B$ *if and only if* $B \in B$.' This is a contradiction (*P if and only if*—*P* is a contradiction) to the supposition that $B \in A$. Hence, $B \notin A$.   ♮

Corollary 2.1.4 *There is no set containing all sets.*[1]   ♮

Let A and B be sets. Consider the statement '$x \in B$.' The set $\{x \in A \mid x \in B\}$ is denoted by '$A \cap B$,' and it is called the **intersection** of A and B. Thus,

$$x \in A \bigcap B \text{ if and only if } (x \in A \text{ and } x \in B).$$

Since '$[x \in A$ *and* $x \in B]$ *if and only if* $[x \in B$ *and* $x \in A]$' is a tautology, we have the following proposition.

Proposition 2.1.5 $A \bigcap B = B \bigcap A$.   ♮

Proposition 2.1.6 $A \bigcap B \subseteq A$ *and* $A \bigcap B \subseteq B$.

Proof By the definition, $x \in A \bigcap B$ *if and only if* $[x \in A$ *and* $x \in B]$. Further, 'if $[x \in A$ *and* $x \in B]$, *then* $x \in A$' is a tautology. Thus, *if* $x \in A \bigcap B$, *then* $x \in A$. This shows that $A \bigcap B \subseteq A$. Similarly, $A \bigcap B \subseteq B$.   ♮

Proposition 2.1.7 *If* $[C \subseteq A$ *and* $C \subseteq B]$, *then* $[C \subseteq A \bigcap B]$.

Proof Suppose that $C \subseteq A$ and $C \subseteq B$. Let $x \in C$. Since $C \subseteq A$ and $C \subseteq B$, $x \in A$ and $x \in B$. Thus, $x \in A \bigcap B$. Hence, *if* $x \in C$, *then* $x \in A \bigcap B$. This shows that $C \subseteq A \bigcap B$.   ♮

Proposition 2.1.8 $[A \bigcap B = A]$ *if and only if* $[A \subseteq B]$.

Proof Suppose that $A \bigcap B = A$. Since $A \bigcap B \subseteq B$ (Proposition 2.1.6), $A \subseteq B$. Suppose that $A \subseteq B$. Since $A \subseteq A$, $A \subseteq A \bigcap B$ (Proposition 2.1.7). Also, $A \bigcap B \subseteq A$ (Proposition 2.1.6). By the axiom of extension, $A \bigcap B = A$.   ♮

Proposition 2.1.9 $(A \bigcap B) \bigcap C = A \bigcap (B \bigcap C)$.

Proof Let $x \in (A \bigcap B) \bigcap C$. By the definition, $(x \in A$ and $x \in B)$ and $x \in C$. This implies (tautologically) that $x \in A$ and $(x \in B$ and $x \in C)$. It follows that $x \in A \bigcap (B \bigcap C)$. Thus, $(A \bigcap B) \bigcap C \subseteq A \bigcap (B \bigcap C)$. Similarly, $A \bigcap (B \bigcap C) \subseteq (A \bigcap B) \bigcap C$. By the axiom of extension, the result follows.   ♮

---

[1]In pre-axiomatic intuitive development of set theory, people took for granted that there is a set containing all sets. The argument used in the proof of the Proposition 2.1.3 led to a paradox known as '*Russel's paradox*.' In fact, the need for axiomatization of set theory was consequence of such paradoxes.

Let A and B be sets. Consider the statement $x \notin B$. By the axiom of specification, there is a unique set defined by $\{x \in A \mid x \notin B\}$. This set is denoted by $A - B$, and it is called the *complement* of B in A (or A *difference* B). Clearly, $A - B$ is a subset of A.

Proposition 2.1.10 $A - B = A - (A \cap B)$.

Proof Let $x \in A - B$. By the definition, $x \in A$ and $x \notin B$. This implies (tautologically) that $x \in A$ and $(x \in A$ and $x \notin B)$. Thus, $x \in A - (A \cap B)$. This shows that $A - B \subseteq A - (A \cap B)$. Similarly, $A - (A \cap B) \subseteq A - B$. By the axiom of extension, the result follows.   ♯

To have something in our hand, we formally assume the existence of a set as an axiom.

Axiom 3 (*Axiom of existence*) There exists a set.

Let A be a set. Consider $A - A$. If B is any set, then

$$(x \in A \text{ and } x \notin A) \text{ if and only if } (x \in B \text{ and } x \notin B)$$

is a tautology (note that '$(P$ and $-P)$ if and only if $(Q$ and $-Q)$' is a tautology). Thus, $x \in (A - A)$ if and only if $x \in (B - B)$, and so $A - A = B - B$. Therefore, the set $A - A$ is independent of A. This set is called the **empty set**, or the **void set**, or the **null set**, and it is denoted by $\emptyset$. Thus, $\emptyset = \{x \in A \mid x \notin A\}$. Clearly, '$x \in \emptyset$' is a contradiction. Further, the statement '*if* $x \in \emptyset$, *then* Q' is a tautology whatever the statement Q may be.

Let $P(x)$ be any contradiction involving the symbol x. Clearly, then $\emptyset = \{x \in A \mid P(x)\}$. Intuitively, one may think of $\emptyset$ as a set containing no elements.

Proposition 2.1.11 *The empty set* $\emptyset$ *is a subset of every set.*

Proof Let B be a set. We have to show that '*if* $x \in \emptyset$, *then* $x \in B$.' Since $x \in \emptyset$ is a contradiction, '*if* $x \in \emptyset$, *then* $x \in B$' is a tautology. Hence, $\emptyset \subseteq B$.   ♯

Proposition 2.1.12 $A - B = \emptyset$ *if and only if* $A \subseteq B$.

Proof Suppose that $A - B = \emptyset$. Let $x \in A$. Since $A - B = \emptyset$, $x \notin A - B$ (for $x \notin \emptyset$ is a tautology). Further, since $x \in A$ and $x \notin A - B$, $x \in B$. Hence, $A \subseteq B$. Conversely, suppose that $A \subseteq B$. We have to show that $A - B = \emptyset$. Already (Proposition 2.1.11), we have $\emptyset \subseteq A - B$. Let $x \in A - B$. Then, $x \in A$ and $x \notin B$. Since $A \subseteq B$, it follows that $x \in B$ and $x \notin B$. This, in turn, implies that $x \in \emptyset$. Hence, $A - B \subseteq \emptyset$.   ♯

Axiom 4 (*Axiom of replacement*) Let A be a set, and $P(x, y)$ be a statement formula involving x and y such that $\forall x \in A((P(x, y) \text{ and } P(x, z)) \implies y = z)$. Then, there is a set $B = \{y \mid P(x, y) \text{ holds for some } x \in A\}$.

The axiom tells that if A is a set, and there is a correspondence from the members of A to another collection of objects associating each member of A a unique member of the collection, then the image is set. This axiom will be used in our discussions on ordinals.

The following axiom helps us to generate more sets.

Axiom 5 (*Pairing axiom*) Let A and B be sets. Then, there is a set C such that $A \in C$ and $B \in C$.

Consider the statement '$x = A$ or $x = B$.' By the axiom of specification, we have a unique set $\{x \in C \mid x = A \text{ or } x = B\}$. This set is also independent of the set C. It contains A and B as elements and nothing else. We denote this set by $\{A, B\}$. The set $\{A, A\}$ is denoted by $\{A\}$, and it is called a singleton.

We have the empty set $\emptyset$. Consider $\{\emptyset\}$. Since $\emptyset \in \{\emptyset\}$ and $\emptyset \notin \emptyset, \emptyset \neq \{\emptyset\}$. If $\{\emptyset\} = \{\{\emptyset\}\}$, then $\emptyset = \{\emptyset\}$. This is a contradiction. Hence, $\{\emptyset\} \neq \{\{\emptyset\}\}$. Similarly, $\{\{\{\emptyset\}\}\} \neq \{\{\emptyset\}\}$. Axiom of pairing gives us other new sets such as $\{\emptyset, \{\emptyset\}\}, \{\{\emptyset, \{\emptyset\}\} \text{ and,} \{\{\emptyset\}\}\}$. This way we produce several sets.

Axiom 6 (*Union Axiom*) Let A be a set of sets. Then, there is a set U such that '$(X \in A$ and $x \in X)$ implies that $x \in U$.'

By the axiom of specification, we have the unique set given by

$$\{x \in U \mid x \in X \text{ for some } X \in A\}.$$

This set is denoted by $\bigcup_{X \in A} X$, and it is called the **union** of the family A of sets. Thus,

$$x \in \bigcup_{X \in A} X \text{ if and only if } x \in X \text{ for some } X \in A.$$

What is $\bigcup_{X \in \emptyset} X$? If $x \in \bigcup_{X \in \emptyset} X$, then there exists $X \in \emptyset$ such that $x \in X$. But $X \in \emptyset$ is a contradiction. Hence, $\bigcup_{X \in \emptyset} X = \emptyset$. Clearly, $\bigcup_{X \in \{A\}} X = A$.

The set $\bigcup_{X \in \{A, B\}} X$ is denoted by $A \bigcup B$. Thus,

$$x \in A \bigcup B \text{ if and only if } x \in A \text{ or } x \in B.$$

The set $A \bigcup B$ is called the **union** of A and B.

Proposition 2.1.13 $A \subseteq A \bigcup B$.

Proof Suppose that $x \in A$. Then, the statement '$x \in A$ or $x \in B$' is true (*if P, then (P or Q) is a tautology*). Hence, *if $x \in A$, then $x \in A \bigcup B$. Thus, $A \subseteq A \bigcup B$.* ♯

Proposition 2.1.14 $A \bigcup \emptyset = A$.

Proof Since $x \in \emptyset$ is always false, $x \in A$ *if and only if* ($x \in A$ *or $x \in \emptyset$*). Hence, $A \bigcup \emptyset = A$.   ♯

Proposition 2.1.15 $A \bigcup B = B \bigcup A$.

Proof Clearly, '($x \in A$ or $x \in B$) *if and only if* ($x \in B$ or $x \in A$)' is a tautology. Hence, $A \bigcup B = B \bigcup A$.   ♯

Proposition 2.1.16 $A \bigcup A = A$.

Proof Since the statement '($x \in A$ or $x \in A$) *if and only if* $x \in A$' is a tautology, the result follows.   ♯

Proposition 2.1.17 $A \bigcup B = A$ *if and only if* $B \subseteq A$.

Proof Suppose that $A \bigcup B = A$. By the Proposition 2.1.13, $B \subseteq A \bigcup B = A$. Next, suppose that $B \subseteq A$. Then, $A \subseteq A \bigcup B \subseteq A \bigcup A = A$. Hence, $A \bigcup B = A$.   ♯

Proposition 2.1.18 $(A \bigcup B) \bigcup C = A \bigcup (B \bigcup C)$.

Proof Let $x \in (A \bigcup B) \bigcup C$. By the definition, '($x \in A$ or $x \in B$) or $x \in C$.' This implies (tautologically) that '$x \in A$ or ($x \in B$ or $x \in C$).' It follows that '$x \in A \bigcup (B \bigcup C)$.' Thus, '$(A \bigcup B) \bigcup C \subseteq A \bigcup (B \bigcup C)$.' Similarly, '$A \bigcup (B \bigcup C) \subseteq (A \bigcup B) \bigcap C$.' By the axiom of extension, the result follows.   ♯

Proposition 2.1.19 *The union distributes over intersection, and the intersection distributes over union in the following sense:*
1. $A \bigcup (B \bigcap C) = (A \bigcup B) \bigcap (A \bigcup C)$, *and*
2. $(A \bigcap (B \bigcup C) = (A \bigcap B) \bigcup (A \bigcap C)$.

Proof 1. Let $x \in A \bigcup (B \bigcap C)$. By the definition, '$x \in A$ or ($x \in B$ and $x \in C$).' This implies (tautologically) that '($x \in A$ or $x \in B$) and ($x \in A$ or $x \in C$).' In turn, '$x \in (A \bigcup B) \bigcap (A \bigcup C)$.' This shows that '$A \bigcup (B \bigcap C) \subseteq (A \bigcup B) \bigcap (A \bigcup C)$.' Similarly, '$(A \bigcup B) \bigcap (A \bigcup C) \subseteq A \bigcup (B \bigcap C)$.' By the axiom of extension, '$A \bigcup (B \bigcap C) = (A \bigcup B) \bigcap (A \bigcup C)$.'
    Similarly, we can prove 2.   ♯

Theorem 2.1.20 (De Morgan's Law) *Let A, B, and C be sets. Then,*
1. $A - (B \bigcup C) = (A - B) \bigcap (A - C)$.
2. $A - (B \bigcap C) = (A - B) \bigcup (A - C)$.

Proof 1. First observe that the statement '$x \notin (B \bigcup C)$' is logically equivalent to the statement '$x \notin B$ and $x \notin C$.' Let $x \in A - (B \bigcup C)$. Then, by the definition, '$x \in A$ and $x \notin (B \bigcup C)$.' This implies that '$x \in A$ and ($x \notin B$ and $x \notin C$).' In turn, it follows that '($x \in A$ and $x \notin B$) and ($x \in A$ and $x \notin C$).' Thus, '$x \in (A - B) \bigcap (A - C)$.' This shows that '$A - (B \bigcup C) \subseteq (A - B) \bigcap (A - C)$.' Similarly, '$(A - B) \bigcap (A - C) \subseteq A - (B \bigcup C)$.' The result follows by the axiom of extension. The proof of 2 is similar.   ♯

Axiom 7 (*Power Set Axiom*) Given a set A, there is a set Ω such that $B \subseteq A$ implies that $B \in \Omega$.

Consider the statement 'x is a subset of A.' By the axiom of specification, we have a unique set given by

$$\{x \in \Omega \mid x \text{ is a subset of } A\}.$$

This set is independent of the choice of Ω in the power set axiom. We denote this set by $\wp(A)$ and call it the **power set** of A.

Since the empty set $\emptyset$ is a subset of every set, $\wp(A)$ can never be an empty set. What is $\wp(\emptyset)$? Since $\emptyset \subseteq \emptyset$, $\emptyset \in \wp(\emptyset)$. Suppose that $A \in \wp(\emptyset)$. Then, $A \subseteq \emptyset$. But, then *if $x \in A$, then $x \in \emptyset$*. Since $x \in \emptyset$ is a contradiction, $x \in A$ is also a contradiction. Hence, $A = \emptyset$. Thus, $\wp(\emptyset) = \{\emptyset\}$. Further, $A \in \wp(\{\emptyset\})$ *if and only if $A \subseteq \{\emptyset\}$*. This shows that $A = \emptyset$ or $A = \{\emptyset\}$. Thus, $\wp(\{\emptyset\}) = \{\emptyset, \{\emptyset\}\}$. Further, $\wp(\{\emptyset, \{\emptyset\}\}) = \{\emptyset, \{\emptyset\}, \{\{\emptyset\}\}, \{\emptyset, \{\emptyset\}\}\}$, and so on.

The next axiom is the axiom of regularity (also called the axiom of foundation). It is used specially in discussions involving ordinal arithmetic. In axiomatic set theory, the members of sets are also sets. Indeed, any mathematical discussion can be modeled so that all the objects considered are sets of sets. For example, 1 can represented by $\{\emptyset\}$, 2 can be represented by $\{\emptyset, \{\emptyset\}\}$, and so on. The axiom is designed to restrict uncomfortable situations such as $A \in A$, ($A \in B$ and $B \in A$), and ($A \in B$ and $B \in C$ and $C \in A$) in any course of discussion.

Axiom 8 (*Axiom of regularity*) If A is a nonempty set of sets, then '*there exists $X(X \in A$ and $X \bigcap A = \emptyset)$.*'

Thus, given a nonempty set A of sets, there is a set X in A such that no member of X is in A.

Theorem 2.1.21 *Let A be a set of sets. Then, $A \notin A$.*

Proof Let A be a set. $\{A\} \neq \emptyset$. By the axiom of regularity, *there exists $X \in \{A\}$ such that if $x \in X$, then $x \notin \{A\}$*. Now, $X \in \{A\}$ *if and only if $X = A$*. Thus, *if $x \in A$, then $x \notin \{A\}$*. Since $A \in \{A\}, A \notin A$.   ♯

Theorem 2.1.22 *Given sets A and B, $A \notin B$ or $B \notin A$.*

Proof Suppose that $A \in B$ and $B \in A$. Then, $B \in A$, $B \in \{A, B\}$, $A \in B$, and also $A \in \{A, B\}$. Thus, there is no $X \in \{A, B\}$ such that $x \in X$ implies that $x \notin \{A, B\}$. This contradicts the axiom of regularity.   ♯

Let X be a set. The set $X^+ = X \bigcup \{X\}$ is called the **successor** of X.

Proposition 2.1.23 *Let X and Y be sets. Then, $X^+ = Y^+$ if and only if $X = Y$.*

Proof If $X = Y$, then $X^+ = Y^+$. Suppose that $X \neq Y$ and $X^+ = Y^+$. Then, $X \bigcup \{X\} = Y \bigcup \{Y\}$. Since $X \in X \bigcup \{X\}$, $X \in Y \bigcup \{Y\}$, and since $X \neq Y, X \in Y$. Similarly, $Y \in X$. This is a contradiction (Theorem 2.1.22).   ♯

A set S is called a **successor set** if

(i) $\{\emptyset\} \in S$, and
(ii) $X \in S$ implies $X^+ \in S$.

The following axiom asserts that there is an infinite set.

Axiom 9 (*Axiom of infinity*) There exists a successor set.

Proposition 2.1.24 *Let X be a set of successor sets. Then, $\bigcap_{S \in X} S$ is also a successor set.*

Proof Since each S is a successor set, $\{\emptyset\} \in S$, *for all $S \in X$*. Hence, $\{\emptyset\} \in \bigcap_{S \in X} S$. Let $x \in \bigcap_{S \in X} S$. Then, $x \in S$, *for all $S \in X$*. Since each $S \in X$ is a successor set, $x^+ \in S$, *for all $S \in X$*. Hence, $x^+ \in \bigcap_{S \in X} S$. ♯

Corollary 2.1.25 *Let X be a successor set. Then X contains the smallest successor set contained in X.*

Proof The intersection of all successor sets contained in X is the smallest successor set contained in X. ♯

Corollary 2.1.26 *Let X and Y be successor sets. Let A be the smallest successor set contained in X, and B the smallest successor set contained in Y. Then $A = B$.*

Proof $X \cap Y$ is also a successor set. Thus, A and B are both smallest successor sets contained in $X \cap Y$. ♯

Let X be a successor set. The smallest successor set contained in X, which is the smallest successor set contained in any other successor set, is called the set of *natural numbers*. The set of natural numbers is denoted by $\mathbb{N}$. $\{\emptyset\}$ is denoted by 1, and it is called *one*. $\{\emptyset\}^+ = \{\emptyset, \{\emptyset\}\}$ is denoted by 2, and it is called *two*, and so on. The properties of the set $\mathbb{N}$ of natural numbers can be faithfully described in the form of Peano's axioms as given below:

Peano's Axiom

P_1. $1 \in \mathbb{N}$.
P_2. *For all $x \in \mathbb{N}$, $x^+ \in \mathbb{N}$.*
P_3. $x^+ = y^+$ *if and only if $x = y$.*
P_4. *For all $x \in \mathbb{N}$, $1 \neq x^+$.*
P_5. *If M is a set such that $1 \in M$ and $x^+ \in M$ for all $x \in M \cap \mathbb{N}$, then $\mathbb{N} \subseteq M$.*
Further properties of the natural number system $\mathbb{N}$ will be discussed in detail in the next chapter.

Exercises

2.1.1 Show that

(i) $A \cap \emptyset = \emptyset$
(ii) $A \cup \emptyset = A$

(iii) $A - \emptyset = A$

(iv) $\emptyset - A = \emptyset$.

2.1.2 Show that $A - (A - B) = A \cap B$.

2.1.3 Show that $A - (A \cap B) = A - B$.

2.1.4 Show that $A \cup B = A$ *if and only if* $B \subseteq A$.

2.1.5 Show that $(A \cap B) \cup C = (A \cup C) \cap (B \cup C)$.

2.1.6 Show that $(A \cap B) \cup C = A \cap (B \cup C)$ *if and only if* $C \subseteq A$.

2.1.7 Show that $A \subseteq B$ *implies* $C \cup A \subseteq C \cup B$.

2.1.8 Show that $(A - B) - C = (A - C) - B$.

2.1.9 Show that

(i) $A \cap (B \cup A) = A$.

(ii) $A = A \cup (B \cap A)$.

2.1.10 Put $A \oplus B = (A - B) \cup (B - A)$. Show that

(i) $(A \oplus B) \oplus C = A \oplus (B \oplus C)$.

(ii) $A \oplus \emptyset = A = \emptyset \oplus A$.

(iii) $A \oplus B = B \oplus A$.

(iv) $A \oplus B = \emptyset$ *if and only if* $A - B$.

(v) $A \cap (B \oplus C) = (A \cap B) \oplus (A \cap C)$.

(vi) $A \oplus C = B \oplus C$ *if and only if* $A = B$.

2.1.11 $A \subset B$ *if and only if* $\wp(A) \subseteq \wp(B)$.

2.1.12 Show that $\wp(A \cap B) = \wp(A) \cap \wp(B)$.

2.1.13 Show that $\wp(A) \cup \wp(B) \subseteq \wp(A \cup B)$. Show by means of an example that equality need not hold.

2.1.14 Suppose that A contains n elements. Show that $\wp(A)$ contains 2^n elements.

2.1.15 Can $\wp(A)$ be $\emptyset$? Support.

2.1.16 Show that a union of successor sets is a successor set.

2.1.17 Let A be a successor set. Can $\wp(A)$ be a successor set? support.

2.1.18 Let A and B be successor sets. Can $A - B$ be a successor set? Support.

2.1.19 Show that $X^+ \neq X$ for every set X.

2.1.20 $(X^+)^+ \neq X$ for every set X.

2.1.21 Show that the empty set is not successor of any set.

2.2 Cartesian Product and Relations

Let X be a set. Let $a, b \in X$. Then, the set $\{\{a\}, \{a, b\}\}$ is a subset of $\wp(X)$. We denote the set $\{\{a\}, \{a, b\}\}$ by (a, b) and call it an **ordered pair**. Thus, $(a, b) \in \wp(\wp(X))$.

Proposition 2.2.1 $(a, b) = (b, a)$ *if and only if $a = b$.*

Proof Suppose that $(a, b) = (b, a)$. Then, $\{\{a\}, \{a, b\}\} = \{\{b\}, \{b, a\}\}$. Since $\{a, b\} = \{b, a\}, \{a\} = \{b\}$. Hence, $a = b$. Clearly, $a = b$ implies $(a, b) = (a, a) = (b, a)$.   ♯

Observe that $(a, a) = \{\{a\}, \{a, a\}\} = \{\{a\}, \{a\}\} = \{\{a\}\}$.

Let X and Y be sets. Then, the set

$$X \times Y = \{(a, b) \mid a \in X \text{ and } b \in Y\}$$

is called the **cartesian product** of X and Y. Clearly, $X \times Y \subseteq \wp(\wp(X \bigcup Y))$.

Proposition 2.2.2 *Let A, B, and C be sets. Then,*

(i) $(A \bigcup B) \times C = (A \times C) \bigcup (B \times C)$.
(ii) $(A \bigcap B) \times C = (A \times C) \bigcap (B \times C)$.
(iii) $(A - B) \times C = (A \times C) - (B \times C)$.

Proof (i). Let $(x, y) \in (A \bigcup B) \times C$. By the definition, '$x \in A \bigcup B$ *and* $y \in C$.' This implies that '$(x \in A$ *and* $y \in C$) *or* $(x \in B$ *and* $y \in C$).' Thus, '$(x, y) \in (A \times C)$ *or* $(x, y) \in (B \times C)$.' By the definition, $(x, y) \in (A \times C) \bigcup (B \times C)$. It follows that '$(A \bigcup B) \times C \subseteq (A \times C) \bigcup (B \times C)$.' Similarly, it follows that '$(A \times C) \bigcup (B \times C) \subseteq (A \bigcup B) \times C$.' By the axiom of extension, $(A \bigcup B) \times C = (A \times C) \bigcup (B \times C)$.

Similarly, we can prove (ii) and (iii).   ♯

Proposition 2.2.3 $A \times B = \emptyset$ *if and only if* $(A = \emptyset$ *or* $B = \emptyset)$.

Proof Suppose that $A = \emptyset$, and $(x, y) \in A \times B$. Then, $x \in \emptyset$ *and* $y \in B$. Since $x \in \emptyset$ is a contradiction, $(x, y) \in \emptyset \times B$ is also a contradiction. Hence, $\emptyset \times B = \emptyset$. Similarly, $A \times \emptyset = \emptyset$. Now, suppose that $A \neq \emptyset$ and $B \neq \emptyset$. Then, there is an element $x \in A$ and an element $y \in B$. In turn, $(x, y) \in A \times B$. Hence, $A \times B \neq \emptyset$.   ♯

Relations

Consider the relation 'is father of.' Nehru is father of Indira, and Feroze Gandhi is the father of Rajeev Gandhi. This gives us pairs (Nehru, Indira) and (Feroze Gandhi, Rajeev Gandhi). If we look at the set R of all pairs (a, b), where a is father of b, then the set R faithfully describes the relation of 'is father of.' One is genuinely tempted to define a relation as a set of ordered pairs.

Definition 2.2.4 A subset R of $X \times X$ is called a **relation** on X. If $(x, y) \in R$, then we say that x is related to y under the relation R. We also express it by writing xRy.

Example 2.2.5 $\emptyset$ is a relation on X in which no pair of elements in X are related. $X \times X$ is the largest (universal) relation on X in which each pair of elements in X is related.

Example 2.2.6 $\triangle = \{(x, x) \mid x \in X\}$ is a relation on X called the **diagonal** relation on X. This is the most selfish relation on X.

Example 2.2.7 Let $X = \{a, b, c\}$. $R = \{(a, b), (b, a), (a, c)\}$ is a relation on X.

Example 2.2.8 Let X be a set. Then, $R = \{(a, b) \mid a, b \in X \text{ and } a \in b\}$ is a relation on X.

Example 2.2.9 Let X be a set. Then, $R = \{(A, B) \mid A, B \in \wp(X) \text{ and } A \subseteq B\}$ is a relation on $\wp(X)$.

Let R and S be relations on X. Then, $R \bigcup S, R \bigcap S$, and $R - S$ are all subsets of $X \times X$, and hence, they are also relations on X.

Definition 2.2.10 Let R and S be relations on X. The relation

$$RoS = \{(x, z) \in X \times X \mid (x, y) \in S \text{ and } (y, z) \in R \text{ for some } y \in X\}$$

is called the **composition** of R and S.

Proposition 2.2.11 *Let R, S, and T be relations on X. Then,*

$$(RoS)oT = Ro(SoT).$$

Proof Let $(x, y) \in (RoS)oT$. By the definition,

there exists $z \in X$ such that $(x, z) \in T$, and $(z, y) \in RoS$.

Again, by the definition,

there exist z and $u \in X$ such that $(x, z) \in T, (z, u) \in S$, and $(u, y) \in R$.

Thus,

there exists $u \in X$ such that $(x, u) \in SoT$, and $(u, y) \in R$.

Hence, $(x, y) \in Ro(SoT)$. This shows that $(RoS)oT \subseteq Ro(SoT)$. Similarly, $Ro(SoT) \subseteq (RoS)oT$. By the axiom of extension, the result follows.   ♯

Proposition 2.2.12 *$Ro\triangle = R = \triangle oR$.*

Proof Since $(x, x) \in \triangle$ for all $x \in X, (x, y) \in Ro\triangle$ if and only if $(x, y) \in R$. This proves that $Ro\triangle = R$. Similarly, $R = \triangle oR$.   ♯

Proposition 2.2.13 *Let R, S and T be relations on X. Then*

   (i) $Ro(S \bigcup T) = (RoS) \bigcup (RoT)$
  (ii) $Ro(S \bigcap T) \subseteq (RoS) \bigcap (RoT)$
 (iii) $(R \bigcup S)oT = (RoT) \bigcup (SoT)$
 (iv) $(R \bigcap S)oT \subseteq (RoT) \bigcap (SoT)$

Proof (i) Let $(x, y) \in Ro(S \bigcup T)$. By the definition,

$$there\ exists\ z \in X\ such\ that\ (x, z) \in S \bigcup T,\ and\ (z, y) \in R.$$

Thus,

$$there\ exists\ z \in X\ such\ that\ ((x, z) \in S,\ and\ (z, y) \in R)\ or\ ((x, z) \in T,\ and\ (z, y) \in R).$$

In turn, it follows that '$(x, y) \in (RoS)$ or $(x, y) \in (RoT)$.' Hence, $(x, y) \in (RoS) \bigcup (RoT)$. This shows that $Ro(S \bigcup T) \subseteq (RoS) \bigcup (RoT)$. Similarly, $(RoS) \bigcup (RoT) \subseteq Ro(S \bigcup T)$. By the axiom of extension, $Ro(S \bigcup T) = (RoS) \bigcup (RoT)$. Similarly, we can prove the rest. ♯

Example 2.2.14 Let $X = \{a, b, c\}$. Let $R = \{(a, b), (a, c)\}$ *and* $S = \{(b, c), (b, b)\}$. Then $RoS = \emptyset$, and $SoR = \{(a, c), (a, b)\} = R$(verify). Thus, RoS need not be SoR. Observe that $R = SoR = \triangle oR$, and $S \neq \triangle$. If we take $T = \{(a, a), (b, c), (b, b)\}$, then $RoT = \{(a, c), (a, b)\} = R$ and $ToR = R$. But $T \neq \triangle$. Thus, $RoT = R = ToR$ need not imply that $T = \triangle$.

Definition 2.2.15 Let R be a relation on X. Then, the relation

$$R^{-1} = \{(x, y) \in X \times X \mid (y, x) \in R\}$$

is called the **inverse** of R.

Example 2.2.16 Let $R = \{(a, b), (a, c)\}$ be a relation on the set $X = \{a, b, c\}$. Then, $R^{-1} = \{(b, a), (c, a)\}$. Now, $RoR^{-1} = \{(b, b), (c, c)\}$, and $R^{-1}oR = \{(a, a)\}$. Thus, here again, $RoR^{-1} \neq R^{-1}oR$.

Proposition 2.2.17 *Let R and S be relations on X. Then,*
(i) $(R^{-1})^{-1} = R$
(ii) $(RoS)^{-1} = S^{-1}oR^{-1}$.

Proof Clearly, $(x, y) \in R$ *if and only if* $(y, x) \in R^{-1}$. *Also,* $(y, x) \in R^{-1}$ *if and only if* $(x, y) \in (R^{-1})^{-1}$. *Thus,* $R = (R^{-1})^{-1}$. *To prove* (ii), *let* $(x, y) \in (RoS)^{-1}$. *Then,* $(y, x) \in RoS$. *Hence, there exists* $z \in X$ *such that* $(y, z) \in S$ *and* $(z, x) \in R$. *Thus,* $(x, z) \in R^{-1}$, *and* $(z, y) \in S^{-1}$ *for some* $z \in X$. *But, then* $(x, y) \in S^{-1}oR^{-1}$. *This shows that* $(RoS)^{-1} \subseteq S^{-1}oR^{-1}$. *Similarly,* $S^{-1}oR^{-1} \subseteq (RoS)^{-1}$. ♯

Types of Relations

Definition 2.2.18 A relation R on X is said to be
(i) a **reflexive relation** if $(x, x) \in R$ for all $x \in X$, or equivalently if $\triangle \subseteq R$.
(ii) a **symmetric relation** if $(x, y) \in R$ implies that $(y, x) \in R$, or equivalently if $R^{-1} = R$.
(iii) an **antisymmetric relation** if $(x, y) \in R$ and $(y, x) \in R$ implies that $x = y$, or equivalently if $R \cap R^{-1} \subseteq \triangle$.
(iv) a **transitive relation** if when ever $(x, y) \in R$ and $(y, z) \in R$, $(x, z) \in R$, or equivalently if $RoR \subseteq R$.

Example 2.2.19 Let $X = \{a, b, c\}$ and

$$R = \{(a, a), (b, b), (c, c), (a, b), (b, c), (c, b)\}.$$

Then, R is reflexive but none of the rest of the three.

Example 2.2.20 Let $X = \{a, b, c\}$ and $R = \{(a, b), (b, a)\}$. Then, R is symmetric but none of the rest of the three.

Example 2.2.21 Let $X = \{a, b, c\}$ and $R = \{(c, b), (a, c)\}$. Then, R is antisymmetric but none of the rest of the three.

Example 2.2.22 Let $X = \{a, b, c\}$ and

$$R = \{(a, b), (b, a), (a, a), (b, b), (a, c), (b, c)\}.$$

Then, R is transitive but none of the rest of the three.

Example 2.2.23 Let $X = \{a, b, c\}$ and

$$R = \{(a, a), (b, b), (c, c), (a, b), (b, a), (b, c), (c, b)\}.$$

Then, R is reflexive and symmetric but neither antisymmetric nor transitive.

Example 2.2.24 Let $X = \{a, b, c\}$ and

$$R = \{(b, c), (c, b), (b, b), (c, c)\}.$$

Then, R is symmetric and transitive but neither reflexive nor antisymmetric.

Proposition 2.2.25 *Let R be a relation on X which is symmetric and transitive. Suppose that for all $x \in X$, there exists $y \in X$ such that $(x, y) \in R$. Then, R is reflexive.*

Proof Let $x \in X$. Then, $(x, y) \in R$ for some $y \in X$. Since R is symmetric, $(y, x) \in R$. Since R is transitive, $(x, x) \in R$. Thus, R is reflexive. ♯

Example 2.2.26 The relation which is reflexive, symmetric, and antisymmetric is the diagonal relation. Thus, a reflexive, symmetric, and antisymmetric relations are also transitive.

Exercises

2.2.1 Suppose that $A \times C \subseteq B \times C, C \neq \emptyset$. Show that $A \subseteq B$.

2.2.2 Show that $(A \times B = B \times A)$ *if and only if* $(A = \emptyset \text{ or } B = \emptyset \text{ or } A = B)$.

2.2.3 Suppose that A, B, and C are nonempty sets. Is $(A \times B) \times C = A \times (B \times C)$? Support.

2.2.4 Show that $(A \cap B) \times (C \cap D) = (A \times C) \cap (B \times D)$.

2.2.5* Suppose that $A \subseteq A \times A$. Show that $A = \emptyset$.
Hint. Use the axiom of regularity.

2.2.6* Suppose that $A = A \times B$. Show that $A = \emptyset$.

2.2.7 Suppose that A contains n elements and B contains m elements. Show that $A \times B$ contains $n \cdot m$ elements.

2.2.8 Show that the number of relations on a set containing n elements is 2^{n^2}.

2.2.9 Let $X = \{a, b, c\}, R = \{(a, b), (b, c), (c, a)\}$ and $S = \{(a, a), (a, c), (b, b)\}$. Find out (i) $R \cup S$, (ii) $R \cap S$, (iii) RoS, and (iv) R^{-1}.

2.2.10 Show by means of an example that equality in Proposition 2.2.13 (ii) and (iv) need not hold.

2.2.11 Find out the number of reflexive relations on a set containing n elements.
Hint. A reflexive relation on $X \times X$ can be written as $\Delta \cup S$, where $S \subseteq X \times X - \Delta$.

2.2.12 Find out the number of symmetric relations on a set containing n elements.

2.2.13 Find out the number of antisymmetric relations on a set containing n elements.

2.3 Equivalence Relation

The concept of equality in mathematics is best described in terms of equivalence relations.

Definition 2.3.1 A relation R on X which is reflexive, symmetric, and transitive is called an **equivalence relation** on X.

Example 2.3.2 The diagonal relation Δ is the smallest equivalence relation on X. The universal relation $X \times X$ is the largest equivalence relation on X. The relation $R = \{(a, a), (b, b), (c, c), (a, b), (b, a)\}$ is an equivalence relation on $X = \{a, b, c\}$.

Definition 2.3.3 Let R be an equivalence relation on X. Let $x \in X$. The subset

$$R_x = \{y \in X \mid (x, y) \in R\}$$

is called the **equivalence class** of X modulo R determined by the element x.

Thus, for example, the equivalence class Δ_x of X modulo Δ determined by x is the singleton $\{x\}$. For the equivalence relation
$R = \{(a, a), (b, b), (c, c), (a, b), (b, a)\}$ on $X = \{a, b, c\}$, the equivalence classes
are $R_a = \{a, b\} = R_b$ and $R_c = \{c\}$.

Since R is reflexive, $(x, x) \in R$ for all $x \in X$, and hence, $x \in R_x$ for all $x \in X$.

Proposition 2.3.4 *Let R be an equivalence relation on X. Then, the following hold.*

(i) $x \in R_x$ for all $x \in X$.
(ii) $R_x = R_y$ if and only if $(x, y) \in R$.
(iii) $R_x \neq R_y$ if and only if $R_x \cap R_y = \emptyset$.

Proof (i) Since R is reflexive, $(x, x) \in R$ for all $x \in X$, and hence $x \in R_x$ for all $x \in X$.

(ii) Suppose that $R_x = R_y$. Since R is an equivalence relation, $y \in R_y = R_x$. Hence $(x, y) \in R$. Conversely, suppose that $(x, y) \in R$. Since R is symmetric, $(y, x) \in R$. Let $z \in R_x$. Then, $(x, z) \in R$. Since R is transitive, $(y, z) \in R$. Thus, $z \in R_y$. Hence, $R_x \subseteq R_y$. Similarly, $R_y \subseteq R_x$. This shows that $R_x = R_y$.

(iii) Suppose that $R_x \cap R_y \neq \emptyset$. Let $z \in R_x \cap R_y$. Then, $(x, z) \in R$ and $(y, z) \in R$. Since R is symmetric and transitive, $(x, y) \in R$. It follows from (ii) that $R_x = R_y$. Clearly, if $R_x \cap R_y = \emptyset$, then $R_x \neq R_y$, for $x \in R_x$. ♯

Let X be a non emptyset. A set $\wp$ of nonempty subsets of X is called a **partition** of X if the following hold.

(i) Union of members of $\wp$ is X, i.e., $\bigcup_{A \in \wp} A = X$.
(ii) If A and B are distinct members of $\wp$, then $A \cap B = \emptyset$.

Corollary 2.3.5 *Let R be an equivalence relation on X. Then, $\wp_R = \{R_x \mid x \in X\}$ is a partition of X.*

Proof Follows from the above proposition. ♯

The partition $\wp_R$ is the partition determined by the equivalence relation R. The set $\wp_R$ is also denoted by X/R, and it is also called the **quotient set** of X modulo R.

Proposition 2.3.6 *Let $\wp$ be a partition of X. Define a relation $R^\wp$ on X by $R^\wp = \bigcup_{A \in \wp} A \times A$. Then $R^\wp$ is an equivalences relation such that $\wp_{R^\wp} = \wp$.*

Proof Since union of members of $\wp$ is X, given $x \in X$, $x \in A$ for some $A \in \wp$. Hence $(x, x) \in R^\wp$ for all $x \in X$. Thus, $R^\wp$ is reflexive. Suppose that $(x, y) \in R^\wp$. Then, there is an element $A \in \wp$ such that $x, y \in A$, and so $y, x \in A$. Hence, $(y, x) \in R^\wp$. Thus, $R^\wp$ is symmetric. Suppose that $(x, y) \in R^\wp$ and $(y, z) \in R^\wp$. Then, there is an element $A \in \wp$ and an element $B \in \wp$ such that $x, y \in A$ and $y, z \in B$. Since $y \in A \cap B$, $A \cap B \neq \emptyset$. Further, since $\wp$ is a partition, $A = B$. Hence, $x, z \in A \in \wp$. Thus, $(x, z) \in R^\wp$. This shows that $R^\wp$ is transitive.

Next, $R_x^\wp$ is the member A of $\wp$ such that $x \in A$. Hence, $\wp_{R^\wp} = \wp$.   ♯

Proposition 2.3.7 $R^{\wp_R} = R$ *for every equivalence relation R.*

Proof Suppose that $(x, y) \in R$. Then $x, y \in R_x \in \wp_R$. Hence $(x, y) \in R^{\wp_R}$. Suppose that $(x, y) \in R^{\wp_R}$. Then *there exists* $R_z \in \wp_R$ such that $x, y \in R_z$. Hence, there is an element $z \in X$ such that $(x, z) \in R$ and $(y, z) \in R$. Since R is symmetric and transitive, $(x, y) \in R$. This shows that $R = R^{\wp_R}$.   ♯

Remark 2.3.8 It is apparent from the above discussions that every partition can be realized faithfully as an equivalence relation, and every equivalence relation can be realized faithfully as a partition.

Example 2.3.9 Let R be a relation (not necessarily equivalence) on X. Define $R_x = \{y \in X \mid (x, y) \in R\}$. Suppose that $\wp = \{R_x \mid x \in X\}$ is a partition of X. Can we infer that R is an equivalence relation? No. For example, take $X = \{a, b, c\}$, $R = \{(a, b), (b, c), (c, a)\}$. Then, $R_a = \{b\}$, $R_b = \{c\}$, $R_c = \{a\}$. Thus, $\{R_a, R_b, R_c\}$ is a partition of X, whereas R is not an equivalence relation (it is neither reflexive nor symmetric nor transitive).

Example 2.3.10 Let $\wp \subseteq \wp(X)$ (not necessarily a partition). Consider the relation $R^\wp$ on X given by $R^\wp = \{(x, y) \mid$ *such that* $x, y \in A$ *for some* $A \in \wp\}$. Suppose that $R^\wp$ is an equivalence relation. Can we infer that $\wp$ is a partition? Again, no. For example, take $\wp = \{\{a, b\}, \{b, c\}, \{c, a\}\} \subseteq \wp(X)$, where $X = \{a, b, c\}$. Then, $R^\wp = X \times X$ is an equivalence relation.

Exercises

2.3.1 Let R and S be equivalence relations on X. Show that RoS is an equivalence relation if and only if $RoS = SoR$.

2.3.2 Let p_n denote the number of equivalence relations on a set containing n elements. Show that

$$p_{n+1} = \Sigma_{r=0}^{n}(^nC_r)p_r$$

Hint. p_n is the number of partitions of a set containing n elements.

2.3.3 Let $X = \{a, b, c, d\}$ and

$$R = \{(a, a), (b, b), (c, c), (d, d), (a, b), (b, c), (a, c), (b, a), (c, b), (c, a)\}.$$

Show that R is an equivalence relation. Find $\wp_R$. Can we find an other relation S such that $\wp_R = \wp_S$? Support.

2.3.4 Show that the intersection of symmetric relations is symmetric. Deduce that for every relation R on X, there is smallest symmetric relation containing R. This relation is called the symmetric closure of R. Find the symmetric closures of all the relations given above.

2.3.5 Show that the intersection of transitive relations is transitive. Deduce that for every relation R on X, there is smallest transitive relation containing R. This relation is called the transitive closure of R. Find the transitive closures of all the relations given above.

2.3.6 Show that the intersection of equivalence relations is equivalence relation. Deduce that for every relation R on X, there is smallest equivalence relation containing R. This relation is called the equivalence closure of R. Find the equivalence closures of all the relations given above.

2.3.7 Is composite of two symmetric relations always symmetric? If not under what conditions it is symmetric.

2.3.8 Is composite of two transitive relations always transitive? If not under what conditions it is transitive.

2.4 Functions

Let X and Y be sets. A subset f of $X \times Y$ (the Cartesian product) is called a **function** or a **mapping** (or a **map**) from X to Y if the following two conditions hold.

(i) For all $x \in X$, *there exists* $y \in Y$ such that $(x, y) \in f$.
(ii) If $(x, y_1) \in f$ and $(x, y_2) \in f$, *then* $y_1 = y_2$.

X is called the **domain**, and Y is called the **co-domain** of f. If $(x, y) \in f$, we write $y = f(x)$ and call it the **image** of the element $x \in X$ under the map f. Thus, under this notation, $f = \{(x, f(x)) \mid x \in X\}$.

Intuitively, a function f from X to Y is an association or a correspondence which associates to each $x \in X$, a unique $y \in Y$ which we denote by $f(x)$. Thus, to define a map f from X to Y, it is sufficient to give a unique $f(x)$ in Y for all $x \in X$. Any two functions f and g from X to Y are equal if and only if $f(x) = g(x)$ for all $x \in X$.

We also adopt the notation $f : X \longrightarrow Y$ to say that f is a map from X to Y.

Let f be a map from X to Y and g be a map from Y to Z. Then, gof defined by

$$gof = \{(x, z) \mid (x, y) \in f \text{ and } (y, z) \in g \text{ for some } y \in Y\}$$

is also a map from X to Z, and it is called the **composite** of f and g. Thus, the map gof from X to Z is given by $(gof)(x) = g(f(x))$ for all $x \in X$.

The subset Δ of $X \times X$ is also a map from X to X. This map is called the **identity map** on X, and it is denoted by I_X. Thus, $I_X(x) = x$ for all $x \in X$. Clearly, $foI_X = f = I_Y of$ for every map f from X to Y.

Let Y be a subset of X. Then, $i_Y = \{(y, y) \mid y \in Y\}$ is a map from Y to X called the **inclusion** map from Y to X. This map is sometimes denoted by the symbol $Y \hookrightarrow X$.

Let f be a map from X to A, and Y a subset of X. The composition $f \circ i_Y$ is a map from Y to A, and it is called the **restriction** of f to Y. The map $f \circ i_Y$ is also denoted by $f \mid_Y$.

Let X and Y be sets and $y \in Y$. Then, $X \times \{y\}$ is a map f from X to Y such that $f(x) = y$ for all $x \in X$. This map is called a **constant** map.

Let X and Y be sets. Consider the Cartesian product $X \times Y$. The map p_1 from $X \times Y$ to X defined by $p_1((x, y)) = x$ is called the first **projection** and the map p_2 from $X \times Y$ to Y defined by $p_2((x, y)) = y$ is called the second projection map.

Proposition 2.4.1 *Let f be a map from X to Y, g a map from Y to Z, and h a map from Z to U. Then $(h \circ g) \circ f = h \circ (g \circ f)$.*

Proof Clearly, $((h \circ g) \circ f)(x) = (h \circ g)(f(x)) = h(g(f(x))) = h((g \circ f)(x)) = (h \circ (g \circ f))(x)$ *for all $x \in X$. Hence $h \circ (g \circ f) = (h \circ g) \circ f$.*   ♯

Let f be a map from X to Y. Then $f \subseteq X \times Y$. Consider $f^{-1} = \{(y, x) \mid (x, y) \in f\}$. Then $f^{-1} \subseteq Y \times X$ need not be a map from Y to X for two reasons: (i) for $y \in Y$, there may not be any $x \in X$ such that $(x, y) \in f$, and so there may not be any $x \in X$ such that $(y, x) \in f^{-1}$, (ii) $(y, x_1) \in f^{-1}$ and $(y, x_2) \in f^{-1}$ need not imply that $x_1 = x_2$. Thus, f^{-1} will be a map if and only if the following two conditions hold.

(i) *For all $y \in Y$, there is an element $x \in X$ such that $(x, y) \in f$.*
(ii) *If $(x_1, y) \in f$ and $(x_2, y) \in f$, then $x_1 = x_2$.*

A map f from X to Y is called a **surjective** map (*also called an* **onto** *map*) if *for all $y \in Y$, there is an element $x \in X$ such that $(x, y) \in f$*. Thus, f is a surjective map if *for all $y \in Y$, there is an element $x \in X$ such that $f(x) = y$*.

A map f from X to Y is called an **injective** map (*also called a* **one** − **one** *map*) *if $(x_1, y) \in f$, $(x_2, y) \in f$ implies that $x_1 = x_2$*. Thus, f is injective map if whenever $f(x_1) = f(x_2)$, $x_1 = x_2$. In other words, f is injective if whenever $x_1 \neq x_2$, $f(x_1) \neq f(x_2)$.

A map f which is injective as well as surjective is called a **bijective** map (*also called a* **one-one-onto** **map**).

Thus, f^{-1} is a map if and only if f is *bijective*, and then, the map f^{-1} is called the **inverse** of f. The inverse of a bijective map is also bijective.

Example 2.4.2 An injective map need not be surjective. For example, take $X = \{a, b\}$, $Y = \{x, y, z\}$. Define a map f from X to Y by $f(a) = x$ and $f(b) = y$. Then, f is *injective* but it is not *surjective*, for there is no element in X whose image is z.

Example 2.4.3 A *surjective* map need not be *injective*. Take $X = \{a, b, c\}$ and $Y = \{x, y\}$. Define a map f from X to Y by $f(a) = x = f(b)$, $f(c) = y$. Then, f is *surjective*, but it is not *injective*.

Proposition 2.4.4 *Let f be a bijective map from X to Y. Then, f^{-1} is also a bijective map from Y to X. Also (i) $(f^{-1})^{-1} = f$, (ii) $f^{-1}of = I_X$, and $fof^{-1} = I_Y$.*

Proof Let f be a *bijective* map. Then, we have already observed that f^{-1} is a map from Y to X. Suppose that $(y_1, x) \in f^{-1}$ and $(y_2, x) \in f^{-1}$. Then, $(x, y_1) \in f$ and $(x, y_2) \in f$. Since f is a map, $y_1 = y_2$. Thus, f^{-1} is *injective*. Let $x \in X$, then $(x, f(x)) \in f$, and hence, $(f(x), x) \in f^{-1}$. This shows that f^{-1} is *surjective*. We also observe that $(f^{-1}of)(x) = f^{-1}(f(x)) = x$ for all $x \in X$, and $(fof^{-1})(y) = f(f^{-1}(y)) = y$ *for all* $y \in Y$. Thus, $f^{-1}of = I_X$, and $fof^{-1} = I_Y$. The fact that $(f^{-1})^{-1} = f$ follows from the definition of f^{-1}.

Proposition 2.4.5 *(i) The composite of any two injective maps is an injective map, (ii) the composite of any two surjective maps is a surjective, and (iii) the composite of any two bijective maps is a bijective map.*

Proof (i) Let f be an injective map from X to Y and g be an injective map from Y to Z. Suppose $(gof)(x_1) = (gof)(x_2)$. Then, $g(f(x_1)) = g(f(x_2))$. Since g is injective, $f(x_1) = f(x_2)$. Further, since f is injective, $x_1 = x_2$. Hence, gof is *injective*.
(ii) Suppose that f and g are surjective maps. Let $z \in Z$. Since g is surjective, *there exists an element* $y \in Y$ such that $g(y) = z$. Again, since f is surjective, *there exists an element* $x \in X$ such that $f(x) = y$. But, then $(gof)(x) = g(f(x)) = g(y) = z$. Hence, gof is *surjective*.
(iii) Follows from (i) and (ii). ♯

Proposition 2.4.6 *Let f be a map from X to Y and g be a map from Y to Z. Then, the following hold. (i) If gof is surjective, then g is surjective. (ii) If gof is injective, then f is injective.*

Proof (i) Suppose that of gof is surjective. Let $z \in Z$. Since gof is surjective, *there exists* $x \in X$ such that $(gof)(x) = z$, i.e., $g(f(x)) = z$. Hence, g is surjective.
(ii) Suppose that gof is injective and $f(x_1) = f(x_2)$. Then, $g(f(x_1)) = g(f(x_2))$, i.e., $(gof)(x_1) = (gof)(x_2)$. Since gof is injective, $x_1 = x_2$. Hence, f is injective. ♯

Corollary 2.4.7 *If gof is bijective, then g is surjective and f is injective.* ♯

Proposition 2.4.8 *A map f from X to Y is injective if and only if it can be left canceled in the sense that if $fog = foh$, then $g = h$. A map f is surjective if and only if it can be right canceled in the sense that if $gof = hof$, then $g = h$.*

Proof Suppose that f is injective and $fog = foh$. Then, $f(g(z)) = (fog)(z) = (foh)(z) = f(h(z))$ for all $z \in Z$. Since f is injective, $g(z) = h(z)$ for all $z \in Z$. This shows that $g = h$. Now, suppose that f is not injective. Then, *there exist elements* $x_1, x_2 \in X$ such that $x_1 \neq x_2$ and $f(x_1) = f(x_2)$. Take $Z = \{x_1, x_2\}$. Define a map g from Z to X by $g(x_1) = x_1 = g(x_2)$ and a map h from Z to X by $h(x_1) = x_2 = h(x_2)$. Then, $g \neq h$ but $fog = foh$.

Next, suppose that f is surjective and g, h are maps from Y to Z such that $gof = hof$. Then, $g(f(x)) = h(f(x))$ for all $x \in X$. Since f is surjective, $g(y) = h(y)$ for all $y \in Y$. This shows that $g = h$. Now, suppose that f is not surjective. Then, *there exists an element* $y_0 \in Y$ *such that* $y_0 \neq f(x)$ *for all* $x \in X$. Take $Z = \{a, b\}$. Define a map g from Y to Z by $g(y_0) = a$, $g(y) = b$ for all $y \neq y_0$, and a map h from Y to Z by $h(y) = b$ for all $y \in Y$. Clearly, then, $g \neq h$ and $gof = hof$. ♯

Corollary 2.4.9 *A map f from X to Y is bijective if and only if it can be canceled from left as well as from right.* ♯

Proposition 2.4.10 *Let f be a map from X to Y. Then f is bijective if and only if there exists a map g from Y to X such that $gof = I_X$ and $fog = I_Y$. Further, then $g = f^{-1}$.*

Proof If f is bijective, then $f^{-1}of = I_X$ and $fof^{-1} = I_Y$ (Proposition 2.4.4). Let g be a map from Y to X such that $gof = I_X$ and $fog = I_Y$. Since $gof = I_X$ is injective, f is injective. Since $fog = I_Y$ is surjective, f is surjective. Further, then $f^{-1}of = I_X = gof$, and $fof^{-1} = I_Y = fog$. The result follows from the above corollary. ♯

Corollary 2.4.11 *Let f be a bijective map from X to Y, and g be a bijective map from Y to Z. Then $(gof)^{-1} = f^{-1}og^{-1}$.*

Proof Clearly,

$$(f^{-1}og^{-1})o(gof) = (f^{-1}o(g^{-1}og))of = f^{-1}of = I_X.$$

Similarly,

$$(gof)o(f^{-1}og^{-1}) = I_Y.$$

The result follows. ♯

Proposition 2.4.12 *There is no surjective map from any set X to its power set $\wp(X)$.*

Proof Let f be a map from X to $\wp(X)$. Consider the set $A = \{x \in X \mid x \notin f(x)\}$. Then, $A \in \wp(X)$. Suppose that $f(y) = A$ for some $y \in X$. If $y \notin A = f(y)$, then $y \in A$. If $y \in A = f(y)$, then $y \notin f(y) = A$. Hence, the supposition that $f(y) = A$ for some $y \in X$ is false. This shows that f can not be surjective. ♯

Let X and Y be sets. The set of all maps from X to Y is denoted by Y^X. What are $X^\emptyset$ and $\emptyset^X$?

Example 2.4.13 Let X be a set, and 2 denotes the set $\{0,1\}$. Define a map ϕ from $\wp(X)$ to 2^X by $\phi(A)(x) = 0$ if $x \notin A$ and $\phi(A)(x) = 1$ if $x \in A$. Check that the map ϕ is bijective.

Let f be a map from X to Y. Let $A \subseteq X$ and $B \subseteq Y$. The subset $f(A) = \{f(a) \mid a \in A\}$ of Y is called the **image** of A under the map f. The subset $f^{-1}(B) = \{x \in X \mid f(x) \in B\}$ of X is called the **inverse image** of B under f. What are $f^{-1}(Y)$ and $f^{-1}(\emptyset)$? To say that f is surjective is to say that $f(X) = Y$.

Proposition 2.4.14 *Let f be a map from X to Y and $A \subseteq X$. Then $A \subseteq f^{-1}(f(A))$. Also $A = f^{-1}(f(A))$ for all $A \subseteq X$ if and only if f is injective.*

Proof Let $a \in A$. Then, $f(a) \in f(A)$, and hence, by the definition, $a \in f^{-1}(f(A))$. Thus, $A \subseteq f^{-1}(f(A))$. Suppose that f is injective. Let $x \in f^{-1}(f(A))$. Then, $f(x) \in f(A)$ (by def). Hence, *there exists an element $a \in A$* such that $f(x) = f(a)$. Since f is injective, $x = a \in A$. Thus, $f^{-1}(f(A)) \subseteq A$, and therefore, $A = f^{-1}(f(A))$. Suppose that f is not injective. Then, *there exist elements $x_1, x_2 \in X$, $x_1 \neq x_2$* such that $f(x_1) = f(x_2) = y$ (say). Take $A = \{x_1\}$. Then, $f(A) = \{y\}$ and $\{x_1, x_2\} \subseteq f^{-1}(f(A))$. Hence, $A \neq f^{-1}(f(A))$. ♯

Proposition 2.4.15 *Let f be a map from X to Y and $B \subseteq Y$. Then $f(f^{-1}(B)) \subseteq B$. Also $B = f(f^{-1}(B))$ for all $B \subseteq Y$ if and only if f is surjective.*

Proof Let $y \in f(f^{-1}(B))$. Then, $y = f(x)$ for some $x \in f^{-1}(B)$. But then $y = f(x) \in B$. Hence, $f(f^{-1}(B)) \subseteq B$. Suppose that f is surjective and $y \in B$. Then, *there exists an element $x \in X$* such that $f(x) = y$. Clearly, $x \in f^{-1}(B)$, and hence, $y = f(x) \in f(f^{-1}(B))$. Therefore, $B = f(f^{-1}(B))$. Suppose now that f is not surjective. Then, *there exists an element $b \in Y$* such that $b \notin f(X)$. But, then $f^{-1}(\{b\}) = \emptyset$, and hence, $f(f^{-1}(\{b\})) = \emptyset \neq \{b\}$. ♯

Proposition 2.4.16 *Let f be a map from X to Y. Let A_1 and A_2 be subsets of X. Then, the following hold.*

(i) $f(A_1 \bigcup A_2) = f(A_1) \bigcup f(A_2)$.
(ii) $f(A_1 \bigcap A_2) \subseteq f(A_1) \bigcap f(A_2)$.

Further, in (ii), equality holds for every pair of subsets A_1 and A_2 of X if and only if f is injective.

Proof The proof of (i) and (ii) is left as exercises. We prove the last assertion. Suppose now that f is injective. Let $y \in f(A_1) \bigcap f(A_2)$. Then, there is an element $a \in A_1$ and an element $b \in A_2$ such that $y = f(a) = f(b)$. Since f is injective, $a = b \in A_1 \bigcap A_2$, and so $y = f(a) \in f(A_1 \bigcap A_2)$. Thus, $f(A_1) \bigcap f(A_2) \subseteq f(A_1 \bigcap A_2)$. But already (from (ii)) $f(A_1 \bigcap A_2) \subseteq f(A_1) \bigcap f(A_2)$. Thus, equality holds in (ii) if f is injective. Conversely, suppose that f is not injective. Then, we have two distinct elements x_1, x_2 in X such that $f(x_1) = f(x_2) = b$ (say). Take $A_1 = \{x_1\}$, $A_2 = \{x_2\}$. Then, $f(A_1 \bigcap A_2) = f(\emptyset) = \emptyset$, whereas $f(A_1) \bigcap f(A_2) = \{b\} \neq \emptyset$. ♯

Proposition 2.4.17 *Let f be a map from X to Y. Let B_1 and B_2 be subsets of Y. Then, the following hold.*

(i) $f^{-1}(B_1 \bigcap B_2) = f^{-1}(B_1) \bigcap f^{-1}(B_2)$.
(ii) $f^{-1}(B_1 \bigcup B_2) = f^{-1}(B_1) \bigcup f^{-1}(B_2)$.
(iii) $f^{-1}(B_1 - B_2) = f^{-1}(B_1) - f^{-1}(B_2)$.

Proof (i) Let $x \in f^{-1}(B_1 \bigcap B_2)$. By the definition, $f(x) \in B_1 \bigcap B_2$. Thus, $f(x) \in B_1$ and $f(x) \in B_2$. This implies that $x \in f^{-1}(B_1)$ and $x \in f^{-1}(B_2)$. In turn, $x \in f^{-1}(B_1) \bigcap f^{-1}(B_2)$. This shows that $f^{-1}(B_1 \bigcap B_2) \subseteq f^{-1}(B_1) \bigcap f^{-1}(B_2)$. Similarly,

$f^{-1}(B_1) \cap f^{-1}(B_2) \subseteq f^{-1}(B_1 \cap B_2)$. This proves (i). Similarly, we can prove the rest of the two.   ♯

Family of sets.

Let I be a set and X be a set of sets. A surjective map A from I to X is called a **family of sets.** We denote the image $A(\alpha)$ of α by A_α. This family of sets is denoted by $\{A_\alpha \mid \alpha \in I\}$. The set I is called the **indexing set** of the family.

Let $\{A_\alpha \mid \alpha \in I\}$ be a family of sets. Then, the set

$$\bigcup_{\alpha \in I} A_\alpha = \{x \mid x \in A_\alpha \text{ for some } \alpha \in I\}$$

is called the **union** of the family, and

$$\bigcap_{\alpha \in I} A_\alpha = \{x \mid x \in A_\alpha \text{ for all } \alpha \in I\}$$

is called the **intersection** of the family.

Proposition 2.4.18 (De Morgan's Law) *Let X be a set and $\{A_\alpha \mid \alpha \in I\}$ be a family of sets. Then, $\{X - A_\alpha \mid \alpha \in I\}$ is another family of sets and*

(i) $X - (\bigcup_{\alpha \in I} A_\alpha) = \bigcap_{\alpha \in I} (X - A_\alpha)$.
(ii) $X - (\bigcap_{\alpha \in I} A_\alpha) = \bigcup_{\alpha \in I} (X - A_\alpha)$.

The proof of the above proposition is left as an exercise.

Let $\{X_i, \ i \in \{1, 2\}\} = \{X_1, \ X_2\}$ be a family of sets containing only two sets X_1 and X_2. An element $(x_1, \ x_2)$ of the Cartesian product $X_1 \times X_2$ can be faithfully realized as a map x from $\{1, \ 2\}$ to $X_1 \bigcup X_2$ with $x(1) = x_1$ and $x(2) = x_2$. This prompts us to define the Cartesian product of an arbitrary family as follows:

Definition 2.4.19 Let $\{X_\alpha \mid \alpha \in I\}$ be a family of sets. Let $\prod_{\alpha \in I} X_\alpha$ denote the set of all maps x from I to $\bigcup_{\alpha \in I} X_\alpha$ with the property that $x(\alpha) \in X_\alpha$ for all $\alpha \in I$. The set $\prod_{\alpha \in I} X_\alpha$ is called the **Cartesian product** of the family.
Further, for each $\alpha_0 \in I$, the map p_{α_0} from $\prod_{\alpha \in I} X_\alpha$ to X_{α_0} defined by $p_{\alpha_0}(x) = x(\alpha_0)$ is called the α_0^{th} *projection map.*

The Axioms 1–9 constitute the Zermelo–Fraenkel (ZF) axiomatic system for set theory.

Consider the set X of countries in the world. How to select a unique city in each country? More explicitly, how to get a map c from the set X to the set of all cities in the world so that $c(A) \in A$ for all countries A in X. Here, we can give a rule to define the map c by saying that $c(A)$ is the capital of the country A. In general, if $\{X_\alpha \mid \alpha \in I\}$ is a nonempty family of nonempty sets, how to chose a unique member from each class. The following is an other fundamental and important axiom of set theory which ensures the existence of such a map.

Axiom 10 (*Axiom of Choice*) Let $\{X_\alpha \mid \alpha \in I\}$ be a nonempty family of nonempty sets (i.e., I is nonempty, and $X_\alpha \neq \emptyset$ *for all* $\alpha \in I$). Then, $\prod_{\alpha \in I} X_\alpha$ is nonempty set. More explicitly, there exists a map c *from* I *to* $\bigcup_{\alpha \in I} X_\alpha$ (called a choice function) such that $c(\alpha) \in X_\alpha$ *for all* $\alpha \in I$.

Remark 2.4.20 K. Godel in 1932 proved that the axiom of choice is consistent with the *ZF* axiomatic system. More explicitly, the negation of the axiom of choice is not a theorem in the *ZF* axiomatic system. Later, P. Cohn established that the axiom of choice is not a theorem in *ZF* axiomatic system. In turn, the axiom of choice is independent of the *ZF* axiomatic system. It also follows that the ZF axiomatic system is incomplete. The Axioms 1–10 constitute *ZFC* axiomatic system. The axiomatic system ZFC is also incomplete. Consider the following hypothesis: 'If there is an injective map from $\mathbb{N}$ to X, and there is an injective map from X to $2^{\mathbb{N}}$, then there is a bijective from $\mathbb{N}$ to X, or else there is a bijective map from X to $2^{\mathbb{N}}$.' This hypothesis is called the continuum hypothesis (CH). Godel and Cohen proved that the continuum hopothesis is independent of the ZFC axiomatic system. The Whitehead problem in group theory asks: 'Is every abelian group A with $EXT^1(A, \mathbb{Z}) = \{0\}$ a free abelian group?' The Whitehead problem is also an undecidable proposition in ZFC.

Let f be a map from X to Y and g a map from Z to U. Then, the map $f \times g$ from $X \times Z$ to $Y \times U$ defined by $(f \times g)((x, z)) = (f(x), g(z))$ is called the **Cartesian product** of the map f with the map g. Clearly, products of injective maps are injective maps, and those of surjective maps are surjective.

Let f be a map from X to Y and S an equivalence relation on Y. Then, $(f \times f)^{-1}(S)$ is an equivalence relation on X (verify). Let R be an equivalence relation on X. Then, $(f \times f)(R)$ need not be an equivalence relation on Y even if f is surjective (give an example to support this).

The equivalence relation $(f \times f)^{-1}(\Delta)$ on X is called the **kernel** of f, and it is denoted by $\mathbf{ker} f$. It follows from the definitions that f is injective if and only if $ker f = \Delta$ (the diagonal relation on X).

Proposition 2.4.21 *Let f be a surjective map from X to Y. Let R be an equivalence relation on X containing the kernel of f. Then $(f \times f)(R)$ is an equivalence relation on Y such that $(f \times f)^{-1}((f \times f)(R)) = R$.*

Proof Clearly, $(f \times f)(R)$ is symmetric. Since f is surjective, $(f \times f)(R)$ is also reflexive. We prove that it is transitive also. Let $(u, v), (v, w) \in (f \times f)(R)$. Then, *there exist* $(x, y), (z, t) \in R$ such that $(f(x), f(y)) = (u, v)$ and $(f(z), f(t)) = (v, w)$. This shows that $f(y) = f(z) = v$. Hence, $(y, z) \in (f \times f)^{-1}(\Delta) = ker f \subseteq R$. Since R is transitive, $(x, t) \in R$. But, then $(u, w) = (f(x), f(t)) \in (f \times f)(R)$. Thus, $(f \times f)(R)$ is an equivalence relation. Finally, we show that $(f \times f)^{-1}((f \times f)(R)) = R$. Clearly, $R \subseteq (f \times f)^{-1}((f \times f)(R))$. Let $(x, y) \in (f \times f)^{-1}((f \times f)(R))$. Then, $(f(x), f(y)) \in (f \times f)(R)$. Hence, *there exists* $(z, t) \in R$ such that $(f(x), f(y)) = (f(z), f(t))$. But then $f(x) = f(z)$ and $f(y) = f(t)$. This shows that (x, z) and (y, t) belong to $(f \times f)^{-1}(\Delta)$. Since $(f \times f)^{-1}(\Delta)$ is supposed to be contained in R, $(x, z), (y, t)$ and (z, t) are all in R. Since R is an equivalence relation, $(x, y) \in R$. This completes the proof. ♯

Corollary 2.4.22 (Correspondence Theorem) *Let f be a surjective map from X to Y. Let $R(X)$ denote the set of all equivalence relations on X containing $\ker f$ and $R(Y)$ the set of all equivalence relations on Y. Then, f induces a bijective map $\bar{f}$ from $R(X)$ to $R(Y)$ defined by $\bar{f}(R) = (f \times f)(R)$.*

Proof From the above proposition, it follows that $(f \times f)(R) \in R(Y)$ for all $R \in R(X)$. Thus, $\bar{f}$ is a map from $R(X)$ to $R(Y)$. Since f is surjective, $f \times f$ is also surjective. Hence, $(f \times f)((f \times f)^{-1}(S)) = S$ for all $S \in R(Y)$. This shows that $\bar{f}$ is surjective (note that $(f \times f)^{-1}(S) \in R(X)$). Further, suppose that $\bar{f}(R_1) = \bar{f}(R_2)$. Then, $(f \times f)(R_1) = (f \times f)(R_2)$. Since R_1 and R_2 are equivalence relations containing $\ker f$, it follows from the above proposition that $R_1 = (f \times f)^{-1}((f \times f)(R_1)) = (f \times f)^{-1}((f \times f)(R_2)) = R_2$. This proves that $\bar{f}$ is injective.   ♯

Let X be a set and R be an equivalence relation on X. Consider the quotient set $X/R = \{R_x \mid x \in X\}$. The map ν from X to X/R defined by $\nu(x) = R_x$ is called the **quotient map**. Clearly, ν is surjective and $(\nu \times \nu)^{-1}(\Delta) = \{(x, y) \mid R_x = \nu(x) = \nu(y) = R_y\} = R$. Thus, every equivalence relation is kernel of a map. We shall show that if f is a surjective map from X to Y, then Y can be realized as a quotient set through a bijective map.

Theorem 2.4.23 *Let f be a surjective map from X to Y. Let R be an equivalence relation on X containing $\ker f$. Let $S = (f \times f)(R)$. Then, there is a bijective map $\bar{f}$ from X/R to Y/S such that the diagram*

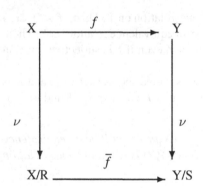

is commutative.

Proof Suppose that $R_{x_1} = R_{x_2}$. Then, $(x_1, x_2) \in R$, and so $(f(x_1), f(x_2)) \in (f \times f)(R) = S$. Hence, $S_{f(x_1)} = S_{f(x_2)}$. This shows that we have a map $\bar{f}$ from X/R to Y/S defined by $\bar{f}(R_x) = S_{f(x)}$. Further, since f is surjective, every member of Y/R is of the form $S_{f(x)} = \bar{f}(R_x)$. This shows that $\bar{f}$ is surjective. Suppose that $\bar{f}(R_{x_1}) = \bar{f}(R_{x_2})$. Then, $S_{f(x_1)} = S_{f(x_2)}$. This means that $(f(x_1), f(x_2)) \in S = (f \times f)(R)$. In turn, $(x_1, x_2) \in (f \times f)^{-1}((f \times f)(R))$. From the Proposition 2.4.21, it follows that $(x_1, x_2) \in R$. This means that $R_{x_1} = R_{x_2}$, and so $\bar{f}$ is also injective. The commutativity of the diagram is evident.   ♯

Corollary 2.4.24 (Fundamental Theorem of Maps) *Let f be a surjective map from X to Y, and $R = (f \times f)^{-1}(\Delta) = \ker f$. Then there is a bijective map ϕ from X/R to Y such that $\phi o \nu = f$.*

Proof Clearly, $(f \times f)((f \times f)^{-1}(\Delta)) = \Delta$. Take $S = \Delta$ in the above theorem. One also observes that the quotient map ν from Y to Y/Δ is bijective map given by $\nu(y) = \{y\}$. Take $\phi = \nu^{-1}o\bar{f}$. The result follows from the above theorem. ♯

Exercises

2.4.1 Let X be a finite set containing n elements and Y be a set containing m elements. Suppose that $n \leq m$. Find the number of injective maps from X to Y. What happens if $m < n$?

2.4.2 Find the number of surjective maps from a set containing n elements to a set containing m elements.

2.4.3 Let X be a set. Show that there is no injective map from $P(X)$ to X.

2.4.4 Let X^Y denote the set of all maps Y to X. Suppose that $X \neq \emptyset$. Show that there is a surjective map from Y to X^Y if and only if X is a singleton set.

2.4.5 Let X, Y, and Z be sets. Show that there is a bijective map from $X^{Y \times Z}$ to $(X^Y)^Z$.

2.4.6 Let R and S be two equivalence relations on a set X such that $R \subseteq S$. Show that there is a bijective map $\phi : X/S \longrightarrow (X/R)/(\nu \times \nu)(S)$ such that the diagram formed by quotient maps is commutative.

2.4.7 Let $\{X_\alpha \mid \alpha \in I\}$ be a family of nonempty sets. Show that each projection map is a surjective map.
Hint. Use the axiom of choice.

2.4.8 Let $f : X \longrightarrow Y$ be a surjective map. Show that there is an injective map t from Y to X such that $fot = I_Y$.
Hint. Use the axiom of choice.

2.4.9 Let $f : X \longrightarrow Y$ be an injective map. Show that there is a surjective map $s : Y \longrightarrow X$ such that $sof = I_X$.

2.4.10 Let X be a nonempty set. Show that the following conditions on X are equivalent:

  (i) Every injective map from X to X is surjective.
 (ii) Every surjective map from X to X is injective.
(iii) Every injective map from X to X is bijective.

Hint. Use the Exercises 2.4.8 and 2.4.9.
A set satisfying the condition in Exercise 2.4.10 is called a **finite set**. A set which is not finite is called an **infinite** set.

2.4.11 Show that every subset of a finite set is finite, and every set containing an infinite set is infinite.

2.4.12 Show that the union of two finite sets is finite.

2.4.13 Show that every successor set is infinite (This justifies the name 'Axiom of Infinity' for the existence of successor set).
Hint. If X is a successor set, then the map $x \longleftarrow x^+$ is an injective map from $X \bigcup \{\emptyset\}$ to itself which is not surjective.

2.4.14 Show that $f(A) - f(B) \subseteq f(A - B)$. Show further that the equality holds provided that f is injective.

2.5 Partial Order

Let X be a set. A relation R on X is called a **partial order** if it is reflexive, anti-symmetric, and transitive. Usually, a partial order is denoted by '$\leq$.' A pair $(X, \leq)$, where $\leq$ is a partial order on X, is called a **partially ordered set**.

Example 2.5.1 Let Y be a set and $X = \wp(Y)$. Then, the relation $\{(A, B) \mid A \subseteq B\}$ is a partial order, and it is called the inclusion relation on X. We denote this relation also by $\subseteq$. Thus, $(X, \subseteq)$ is a partial ordered set. Note that the inverse of a partial order is also a partial order. Thus, $\supseteq$ is also a partial order on X.

Example 2.5.2 Let $X = \{a, b, c, d\}$. Then,

$$R = \{(a, a), (b, b), (c, c), (d, d), (a, b), (c, d)\}$$

is a partial order on X.

Let $(X, \leq)$ be a partially ordered set and Y be a subset of X. Then, the induced relation on Y is also a partial order on Y which is denoted by $\leq_Y$.

A partial order $\leq$ on X is called a **total order** if given x, y in X, $x \leq y$ or $y \leq x$. Example 2.5.2 is not a total order. Example 2.5.1 is a total order if and only if Y is singleton (prove it).

Example 2.5.3 Let $X = \{a, b, c, d\}$ and

$$R = \{(a, a), (b, b), (c, c), (d, d), (a, b), (b, c), (a, c), (c, d), (a, d), (b, d)\}.$$

Then, R is a total order on X

Let $(X, \leq)$ be a partially ordered set. A subset Y of X is called a **chain** in X if the induced partial order on Y is a total order on Y.

Example 2.5.4 Let $Y = \{a, b, c\}$ and $X = \wp(Y)$. Then, the inclusion relation is a partial order on X. The subset $Z = \{\emptyset, \{a\}, \{a, b\}, \{a, b, c\}\}$ is a chain in X.

Let $(X, \leq)$ be a partially ordered set and $A \subseteq X$. An element $x \in X$ is called an **upper bound** (**lower bound**) of A if $a \leq x(x \leq a)$ for all a in A.

Remark 2.5.5 A subset of a partially ordered set need not have any upper bound (lower bound). It may have several upper bounds (lower bounds). Give examples to support it.

Let $(X \leq)$ be a partially ordered set. An element $a \in X$ is called a **maximal** (**minimal**) **element** if $a \leq x(x \leq a)$ *implies that* $x = a$.

In Example 2.5.2, b and d are maximal elements, whereas a and c are minimal elements of X. Thus, there may be so many maximal or minimal elements of a partially ordered set. There may not be any maximal or minimal elements (give examples to support it).

Example 2.5.6 Let $X = \wp(Y) - \{Y, \emptyset\}$, where $Y = \{a, b, c\}$. Then, X is a partially ordered set with respect to inclusion. Clearly, $\{a, b\}, \{b, c\}, \{a, c\}$ are maximal elements and $\{a\}, \{b\}, \{c\}$ are minimal elements.

Example 2.5.7 Let $X = \{a, b, c\}$ and $\triangle = \{(a, a), (b, b), (c, c)\}$. Then, $\triangle$ is a partial order on X such that each element is maximal and also each element is minimal.

Example 2.5.8 Let Y be an infinite set and X be the set of all finite subsets of Y. Then, X is a partially ordered set with respect to inclusion relation which has no maximal element. If we take the set Z of infinite subsets of Y, then it has no minimal elements.

Let $(X \leq)$ be a partially ordered set. An element $a \in X$ is called the **largest** (**least**) element of X if $x \leq a(a \leq x)$ *for all* $x \in X$. If x_1 and x_2 are largest (least) elements of X, $x_1 \leq x_2$ and $x_2 \leq x_1$. By the antisymmetry of $\leq$, $x_1 = x_2$. Thus, there is a unique largest (least) element in a partially ordered set provided it exists.

It may be observed that a largest (least) element is also a maximal(minimal) but a maximal (minimal) element need not be the largest (least). In Example 2.5.2, b and d (a and c) are maximal (minimal) but none of them are largest (least). It may also be noticed that largest (least) need not exist(see Example 2.5.6).

Let $(X \leq)$ be a partially ordered set and $A \subseteq X$. Let $U(A)(L(A))$ denote the set of all upper (lower) bounds of A (note that $U(A)(L(A))$ may be empty sets also). Then, $\leq$ induces a partial order on $U(A)(L(A))$. Note that all elements of A are lower (upper) bounds of $U(A)(L(A))$. Thus, $A \subseteq L(U(A))(A \subseteq U(L(A)))$. The least (largest) element of $U(A)(L(A))$ (if exists) is called the **least upper bound(greatest lower bound)** of A. The least upper bound (greatest lower bound) of A is denoted by **l.u.b(A)(g.l.b(A))** or **supA(infA)**. If A has the largest (least) element, then that is the l.u.b(g.l.b) of A.

Remark 2.5.9 Least upper bound (greatest lower bound) need not exist even if A has upper (lower) bounds: Let $Y = \{a, b, c, d\}$ and $X = \wp(Y) - \{\{a, b\}\}$. Then, $\subseteq$ defines a partial order on X. Take $A = \{\{a\}, \{b\}\}$. Then, $U(A) = \{\{a, b, c\}, \{a, b, d\}, Y\}$. Clearly, $U(A)$ has no least element. Thus, A has no l.u.b.

Theorem 2.5.10 *Let* $(X \leq)$ *be a partially ordered set. Then, the following conditions are equivalent.*
(1) Every nonempty subset of X which has an upper bound has least upper bound in X.
(2) Every nonempty subset of X which has a lower bound has greatest lower bound.

Proof Assume 1. Let A be a nonempty subset of X which has a lower bound. Then, $L(A) \neq \emptyset$. Clearly, $\emptyset \neq A \subseteq U(L(A))$. Hence, $L(A)$ has an upper bound. By 1, $L(A)$ has the least upper bound a (say). Since a is the least element of $U(L(A))$ (by the definition of l.u.b) and $A \subseteq U(L(A))$, $a \leq x$ for all $x \in A$. Thus, $a \in L(A)$. Further, if $y \in L(A)$, then $y \leq x$ for all $x \in U(L(A))$. In particular, $y \leq a$. Thus, a is the largest element of $L(A)$. This shows that a is the g.l.bA.
The proof of 2 $\implies$ 1 is similar.   ♯

A partial order $\leq$ on X is called a **complete order** if it satisfies any one (and hence both) of the equivalent conditions of the above theorem.

Let $(X, \leq)$ be a partially ordered set. A subset Y of X is called an **initial segment** of X if $y \in Y$ and $x \leq y$ implies that $x \in Y$. Thus, X itself is an initial segment of $(X, \leq)$. For each $x \in X$, the subset $\sigma_x = \{y \in Y \mid y \leq x\}$ is an initial segment of X associated with the element $x \in X$. The map σ from X to $\wp(X)$ defined by $\sigma(x) = \sigma_x$ is an injective map from X to $\wp(X)$ which is order preserving in the sense that '$x \leq y \iff \sigma_x \subseteq \sigma_y$.' Again, for each $x \in X$, the subset $\eta_x = \{y \in X \mid y < x\}$ is also an initial segment. This initial segment is called the **strict initial segment** associated with x.

A partial order $\leq$ on X is called a **well-order** if every nonempty subset of X has the least element. A pair $(X, \leq)$, where $\leq$ is a well-order, is called a **well-ordered set**. Every well-order is a total order: Let $\leq$ be a well-order on X. Let $x, y \in X$. Then, $\{x, y\}$ is a nonempty subset of X. Since $\leq$ is a well-order, $\{x, y\}$ has a least element. If x is the least element, then $x \leq y$; if y is the least element, then $y \leq x$. This proves that every well-order is a total order. Indeed, a well-order is a complete order. For, suppose that $\leq$ is a well-order on X. Let A be a nonempty subset of X which has an upper bound. Then, the set $U(A)$ of upper bounds of A is a nonempty subset of X. Since $\leq$ is a well-order on X, $U(X)$ has the least element a (say). Evidently, a is the least upper bound of A. A complete order need not be a well-order. For example, the inclusion relation on the power set $\wp(X)$ of $X = \{a, b, c\}$ is a complete order, but it is not a well-order.

Proposition 2.5.11 *Let* $(X, \leq)$ *be a well-ordered set. Then, a proper subset Y of X is an initial segment if and only if it is strict initial segment η_x for some $x \in X$. It need not be σ_x for any $x \in X$.*

Proof Let Y be a proper subset of X. Then, $X - Y \neq \emptyset$. Since $(X, \leq_X)$ is a well-ordered set, $X - Y$ has least element x (say). Clearly, $\eta_x \subseteq Y$. Since Y is an initial segment, $x \nleq_X y$ for any $y \in Y$. This shows that $Y \subseteq \eta_x$. Thus, $Y = \eta_x$. Note that the successor $\mathbb{N}^+$ of $\mathbb{N}$ is a well-ordered set with usual inclusion ordering, and $\mathbb{N}$ is a proper subset of $\mathbb{N}^+$ which is an initial segment, but it is not σ_x for any $x \in \mathbb{N}^+$.   ♯

Finally, we state and prove the two important equivalents of axiom of choice which are commonly used in mathematics.

Zorn's Lemma: Let $(X, \leq)$ be a nonempty partially ordered set in which every chain has an upper bound. Then, $(X, \leq)$ has a maximal element.

Well-ordering principle: On every set, there is a well-order.

Theorem 2.5.12 *The following are equivalent:*

(1) Axiom of choice.
(2) Zorn's lemma.
(3) Well-ordering principle.

Proof The following is the scheme of the proof. We shall prove that $2 \implies 3$, $3 \implies 1$, and then $1 \implies 2$.

$(2 \implies 3)$. Assume 2. Let X be a set. We have to show the existence of a well-order on X. If $X = \emptyset$, then there is nothing to do. Assume that X is a nonempty set. Consider the set Σ given by

$$\Sigma = \{(Y, \leq_Y) \mid \leq \text{ is a well-order on } Y, \text{ where } Y \subseteq X\}.$$

If $x \in X$, then there is the unique partial order $\leq_{\{x\}}$ on $\{x\}$ which is a well-order. Thus, $(\{x\}, \leq_x) \in \Sigma$. Hence, Σ is nonempty set. We say that $(Y, \leq_Y) \leq (Z, \leq_Z)$ if $(Y, \leq_Y) = (Z, \leq_Z)$ or else $Y \subset Z$, $\leq_Z / Y = \leq_Y$, and $Y = \eta_z$ for some $z \in Z$. Clearly, $(\Sigma, \leq)$ is a nonempty partially ordered set. Let $\Omega = \{(Y_\alpha, \leq_{Y_\alpha}) \mid \alpha \in \Lambda\}$ be a chain in $(\Sigma, <)$. Take $Y_0 = \bigcup_{\alpha \in \Lambda} Y_\alpha$. Then, there is a unique order $\leq_{Y_0}$ on Y_0 whose restriction to each Y_α is $\leq_{Y_\alpha}$. If A is a nonempty subset of Y_0, then $A \cap Y_{\alpha_0} \neq \emptyset$ for some $\alpha_0 \in \Lambda$. If $(Y_\alpha, \leq_{Y_\alpha}) \leq (Y_{\alpha_0}, \leq_{Y_{\alpha_0}})$ for all $\alpha \in \Lambda$, then $Y_0 = Y_{\alpha_0}$, and so A has the least element. If not, then there is an element $\alpha \in \Lambda$ such that $(Y_{\alpha_0}, \leq_{Y_{\alpha_0}}) < (Y_\alpha, \leq_{Y_\alpha})$. Hence, there is an element $x \in Y_0$ such that $Y_{\alpha_0} = \eta_x$.

Let a be the least element of $A \cap Y_{\alpha_0}$. Let b be any element of A. Then, b is not strictly less that a, for then b will be a member of $A \cap Y_{\alpha_0}$. Hence, $a \leq_{Y_0} b$. Thus, a is the least element of A. It follows that $(Y_0, \leq_{Y_0})$ is a well-ordered set, and it is an upper bound of Ω. This shows that every chain in $(\Sigma, \leq)$ has an upper bound. By the Zorn's lemma, there is a maximal element $(M, \leq_M)$ of $(\Sigma, \leq)$. We show that $M = X$. Suppose not. Then, there is an element $x_0 \in X - M$. Consider the set $L = M \bigcup \{x_0\}$. Extend the well-order $\leq_M$ on M to the well-order $\leq_L$ on L by defining $x \leq_L x_0$ for all $x \in M$. Clearly, $(L, \leq_L) \in \Sigma$, and it is larger than $(M, \leq_M)$. This is a contradiction to the maximality of $(M, \leq_M)$. Thus, $M = X$, and $\leq_X$ is a well-order on X. This completes the proof of $2 \implies 3$.

$(3 \implies 1)$. Assume 3. Let $\{X_\alpha \mid \alpha \in \Lambda\}$ be a nonempty family of nonempty sets. By the well-ordering principle, there is a well-order $\leq_\alpha$ on X_α for each α. For each $\alpha \in \Lambda$, let $c(\alpha)$ denote the least element of X_α. This gives us a map c from Λ to $\bigcup_{\alpha \in \Lambda} X_\alpha$ such that $c(\alpha) \in X_\alpha$. This completes the proof of $3 \implies 1$.

$(1 \implies 2)$. Assume 1. Let $(X, \leq)$ be a nonempty partially ordered set in which every chain has an upper bound. We need to show the existence of a maximal element.

Recall the map σ from X to $\wp(X)$ given by $\sigma(x) = \sigma_x$, where $\sigma_x = \{y \in X \mid y \leq x\}$ is the initial segment associated with x. Clearly, σ is an injective map which is order preserving in the sense that '$x \leq y$ *if and only if* $\sigma(x) \subseteq \sigma(y)$.' Consider $Y = \sigma(X)$. Then, $(X, \leq)$ is order isomorphic to $(Y, \subseteq)$. It is sufficient, therefore, to show that $(Y, \subseteq)$ has a maximal element. Let $C(X)$ denote the set of all chains in X. Then, $(C(X), \subseteq)$ is also a partially ordered set. Further, since every chain in $(X, \leq)$ has an upper bound, every member of $C(X)$ is contained in a member of Y. Thus, Y is co-final in $(C(X), \subseteq)$. It follows that the maximal members of $(Y, \subseteq)$ are same as those of $(C(X), \subseteq)$. It is sufficient, therefore, to show that $(C(X), \subseteq)$ has a maximal element.

Now, $C(X)$ satisfies the following two properties:

(i) If $A \in C(X)$, then all subsets of A also belong to $C(X)$. In particular, $\emptyset \in C(X)$.
(ii) If Γ is a chain in $(C(X), \subseteq)$, then $\bigcup_{A \in \Gamma} A \in C(X)$.

By the axiom of choice, we have a map c from $\wp(X) - \{\emptyset\}$ to X such that $c(A) \in A$ for all $A \in \wp(X) - \{\emptyset\}$. For each $A \in C(X)$, consider the set $\tilde{A} = \{x \in X \mid A \bigcup\{x\} \in C(X)\}$. To say that A is maximal in $(C(X), \subseteq)$ is to say that $\tilde{A} = A$. Define a map χ from $C(X)$ to X by $\chi(A) = A$ if $\tilde{A} - A = \emptyset$, and $\chi(A) = A \bigcup\{c(\tilde{A} - A)\}$ if $\tilde{A} - A \neq \emptyset$. We need to show that there is an element $A \in C(X)$ such that $\chi(A) = A$.

Let us call a subset Σ of $C(X)$ to be a tower in $C(X)$ if the following 3 conditions hold.

(i) $\emptyset \in \Sigma$.
(ii) $\chi(A) \in \Sigma$ for all $A \in \Sigma$.
(iii) If Γ is a chain in $(\Sigma, \subseteq)$, then $\bigcup_{A \in \Gamma} A \in \Sigma$.

Clearly, $C(X)$ is a tower, and the intersection of a family of towers is a tower. Let Σ_0 denote the smallest tower in $C(X)$. Indeed, it is the intersection of all towers in $C(X)$. It is sufficient to show that Σ_0 is a chain in $(C(X), \subseteq)$. For, then $B = \bigcup_{A \in \Sigma_0} A \in \Sigma_0$, and so $\chi(B) \in \Sigma_0$. Since $\chi(B) \subseteq B$, it follows that $\chi(B) = B$.

Now, we show that Σ_0 is a chain in $C(X)$. More explicitly, we need to show that for any pair $A, B \in \Sigma_0, A \subseteq B$, or $B \subseteq A$. Let

$$\Gamma = \{A \in \Sigma_0 \mid \text{ for all } B \in \Sigma_0, \ A \subseteq B \text{ or } B \subseteq A\}.$$

Clearly, $\emptyset \in \Gamma$. Let $A \in \Gamma$. Consider

$$\Gamma_A = \{B \in \Sigma_0 \mid B \subseteq A \text{ or } \chi(A) \subseteq B\}.$$

We show that Γ_A is tower. Clearly, $\emptyset \in \Gamma_A$. Let $B \in \Gamma_A$. Then, $B \subseteq A$ or $\chi(A) \subseteq B$. Suppose that $B \subseteq A$. If $B = A$, then $\chi(A) = \chi(B)$, and so in this case, $\chi(B) \in \Gamma_A$. Suppose that $B \subset A$. Then, $\chi(B) \subseteq A$. For, if not, then, since $A \in \Gamma, A \subset \chi(B)$. This is not true, for $\chi(B)$ contains at the most one more element than B. Thus, in this case also, $\chi(B) \in \Gamma_A$. Finally, if $\chi(A) \subseteq B$, then $\chi(A) \subseteq \chi(B)$. In this case also, $\chi(B) \in \Gamma_A$.

Let $\{B_\alpha \mid \alpha \in \Lambda\}$ be a chain in Γ_A. Then, from the definition of Γ_A, either each B_α is contained in A or $\chi(A)$ is contained in some B_α. This shows that $\bigcup_{\alpha \in \Lambda} B_\alpha \subseteq A$ or $\chi(A) \subseteq \bigcup_{\alpha \in \Lambda} B_\alpha$. Hence, $\bigcup_{\alpha \in \Lambda} B_\alpha \in \Gamma_A$.

This completes the proof of the fact that Γ_A is a tower contained in Σ_0. Since Σ_0 is the smallest tower, it follows that $\Gamma_A = \Sigma_0$.

Finally, we prove that $\Gamma = \Sigma_0$. Indeed, again, we prove that Γ is a tower. Clearly, $\emptyset \in \Gamma$. Let $A \in \Gamma$. Consider $\chi(A)$. Let $B \in \Sigma_0$. Then, from what we have proved above, $\Gamma_A = \Sigma_0$, and so $B \in \Gamma_A$. Hence, $B \subseteq A \subseteq \chi(A)$, or $\chi(A) \subseteq B$. This shows that $\chi(A) \in \Gamma$. Let $\{A_\alpha \mid \alpha \in \Lambda\}$ be a chain in Γ. Let $B \in \Sigma_0$. Then, either each A_α is contained in B, or $B \subseteq A_\alpha$ for some α. This means that $\bigcup_{\alpha \in \Lambda} A_\alpha \subseteq B$, or $B \subseteq \bigcup_{\alpha \in \Lambda} A_\alpha$. This means that $\bigcup_{\alpha \in \Lambda} A_\alpha \in \Gamma$. Hence, Γ is a tower. In turn, $\Gamma = \Sigma_0$. Hence, Σ_0 is a chain.   ♯

Exercises

2.5.1 Let $(X, \leq)$ be a partially ordered set. Let $A \subseteq B$. Show that $U(A) \supseteq U(B)$ and $L(A) \supseteq L(B)$.

2.5.2 Show that $U(A) = U(L(U(A)))$ and $L(A) = L(U(L(A)))$.

2.5.3 Show that g.l.b need not exist.

2.5.4 Let $A \subseteq B$. Show that

(i) g.l.b$B \leq$ g.l.bA
(ii) l.u.b$A <$ l.u.bB.

2.5.5 Show by means of an example that l.u.bA need not belong to A.

2.5.6 Show that $(P(X), \subseteq)$ is order complete.

2.5.7 Give an example of a partially ordered set which is not complete.

2.5.8 A partially ordered set $(L, \leq)$ is called a **lattice** if any pair of points a, b has the least upper bound denoted by $a \bigvee b$ as well as the greatest lower bound $a \bigwedge b$. Show that $(P(X), \subseteq)$ is a lattice.

2.5.9 Let $(X, \leq_X)$ and $(Y, \leq_Y)$ be well-ordered sets. Show that $X \times Y$ with dictionary order is a well-ordered set.

2.5.10 Let f be a surjective map from X to Y. Use axiom of choice to show the existence of an injective map g from Y to X such that fog is the identity map on Y.

2.6 Ordinal Numbers

Definition 2.6.1 A well-ordered set $(\alpha, \leq)$ is called an **ordinal number** if for each $x \in \alpha$, the strict initial segment $\eta_x = \{a \in \alpha \mid a < x\}$ is same as x.

There is a unique well-order $\phi = \phi \times \phi$ on the set ϕ. Clearly, the statement '$\forall x \in \phi, \eta_x = \{a \in \alpha \mid a < x\} = x$' is vacuously satisfied. Thus, the $\emptyset$ together with this ordering is an ordinal. This ordinal number is denoted by 0. However, if $(\alpha, \leq)$ is an ordinal number, where $\alpha \neq \emptyset$, then $\emptyset \in \alpha$, and indeed, ϕ is the least element of α. For, by the definition of an ordinal number, if x is the least element of α, then $\emptyset = \{a \in \alpha \mid a < x\} = \eta_x = x$.

Example 2.6.2 Consider the set $1 = \{\emptyset\}$. There is only one well-order $\leq$ on 1. The strict initial segment $\eta_\emptyset = \{x \mid x < \emptyset\} = \emptyset$. It follows that $(1, \leq)$ is an ordinal number. The set $2 = \{\phi, \{\phi\}\}$ with the inclusion ordering is clearly an ordinal number. Indeed, all natural numbers are ordinals.

Example 2.6.3 Consider the set $\mathbb{N}$ of natural numbers together with inclusion ordering. Let A be a nonempty subset of $\mathbb{N}$. If $\emptyset \in A$, then ϕ is the least element of A. Let $x \in A$. Then, since $x \in \mathbb{N}$ is an ordinal, $\eta_x = x$ is a well-ordered set. Clearly, the least element of $A \cap \eta_x$ is the least element of A. It follows that $\mathbb{N}$ together with the usual inclusion ordering is a well-ordered set. By the definition, $\eta_x = x$. Thus, $\mathbb{N}$ with usual ordering is an ordinal. This ordinal will be denoted by ω.

Example 2.6.4 Consider the set the successor $\mathbb{N}^+$ of the set $\mathbb{N}$ of natural numbers. We extend the well-ordering of $\mathbb{N}$ to the ordering $\leq_{\mathbb{N}^+}$ on $\mathbb{N}^+$ by defining $n \leq_{\mathbb{N}^+} \mathbb{N}$ for all $n \in \mathbb{N}$. Clearly, $(\mathbb{N}^+, \leq_{\mathbb{N}^+})$ is a well-order. Further, it is also an ordinal number, for the strict initial segment $\eta_{\mathbb{N}} = \{n \in \mathbb{N}^+ \mid n <_{\mathbb{N}^+} \mathbb{N}\} = \mathbb{N}$. This ordinal is the continuation of ω, and it is denoted by $\omega + 1$. Similarly, we have the ordinal number $\omega + 2$, and so on. The axiom of replacement ensures the existence of the set $\{\omega + n \mid n \in \omega\}$ of ordinal numbers such that $\omega + n^+ = (\omega + n)^+$. This is a well-ordered set of ordinal numbers with ω the least ordinal number. Indeed, $\omega + n^+$ is the continuation of $\omega + n$. There is a unique well-order $\leq_{\omega(2)}$ on the union $\omega 2 = \bigcup_{n \in \omega} (\omega + n)$ subject to the condition that their restriction to each $\omega + n$ is the order $\leq_{\omega + n}$ of the ordinal number $\omega + n$. This process continues to generate different ordinal numbers.

Definition 2.6.5 Two partially ordered sets $(X, \leq_X)$, and $(Y, \leq_Y)$ are said to be **order isomorphic** (also called similar) if there is a bijective map f from X to Y such that $a \leq_X b$ implies that $f(a) \leq_Y f(b)$.

Proposition 2.6.6 *Let f be an order isomorphism from a partially ordered set $(X, \leq_X)$ to a partially ordered set $(Y, \leq_Y)$. Then*

   (i) *$a <_X b$ implies that $f(a) <_Y f(b)$,*
  (ii) *f^{-1} is an order isomorphism from $(Y, \leq_Y)$ to $(X, \leq_X)$, and*
 (iii) *the relation of being 'order isomorphic to' is an equivalence relation on any set of partially ordered sets.*

Proof (i) Suppose that $a <_X b$. Then, by the definition, $f(a) \leq_Y f(b)$. Suppose that $f(a) = f(b)$. Since f is bijective, $a = b$. This is a contradiction to the supposition that $a <_X b$.

(ii) Suppose that f is an order isomorphism. Suppose that $c \leq_Y d$, where $c, d \in Y$. Suppose that $a = f^{-1}(c)$, and $b = f^{-1}(d)$. Then, $f(a) = c$ and $f(b) = d$. If $b <_X a$, then from (i), $d = f(b) <_Y f(a) = c$. This is a contradiction. Hence, $f^{-1}(c) \leq_X f^{-1}(d)$.

(iii) Since I_X is an order isomorphism from $(X, \leq_X)$ to itself, the relation is reflexive. If f is an order isomorphism from $(X, \leq_X)$ to $(Y, \leq_Y)$, then, from (ii), it follows that f^{-1} is an order isomorphism from $(Y, \leq_Y)$ to $(X, \leq_X)$. This shows that the relation is symmetric. Since the composition of two-order isomorphism is an order isomorphism, it follows that the relation is transitive.

Proposition 2.6.7 *A well-ordered set $(X, \leq_X)$ may be order isomorphic to a proper subset Y with induced well-ordering. If f is an injective order preserving map from a well-ordered subset $(X, \leq_X)$ to itself, then $a \leq f(a)$ for all $a \in X$.*

Proof $\mathbb{N}$ is a well-ordered set with usual ordering, and the successor map s from $\mathbb{N}$ to its proper subset $\mathbb{N} - \{\emptyset\}$ is an order isomorphism. Let f be an injective order preserving map from a well-ordered subset $(X, \leq_X)$ to itself. Let $A = \{x \in X \mid f(x) <_X x\}$. Suppose that $A \neq \emptyset$. Since $(X, \leq_X)$ is a well-ordered set, A has the least element b (say). Then, $f(b) <_X b$. From the above proposition, $f(f(b)) <_X f(b)$. This means that $f(b) \in A$. This is a contradiction. Hence, $A = \emptyset$, and so $a \leq_X f(a)$ for all $a \in X$.   ♯

Corollary 2.6.8 *Let $(X, \leq_X)$ and $(Y, \leq_Y)$ be two well-ordered sets which are order isomorphic. Then, there is a unique order isomorphism from X to Y.*

Proof Let f and g be two-order isomorphisms from X to Y. Then, $g^{-1}of$ is an order isomorphism from X to itself. From the previous proposition, $a \leq_X g^{-1}(f(a))$ for all $a \in X$. This means that $g(a) \leq_X f(a)$ for all $a \in X$. Similarly, considering the order isomorphism $f^{-1}og$, we conclude that $f(a) \leq_X g(a)$ for all $a \in X$. This shows that $f = g$.   ♯

Corollary 2.6.9 *A well-ordered set can not be order isomorphic to any of its strict initial segment.*

Proof Let $(X, \leq_X)$ be a well-ordered set. Let $x \in X$. Consider the strict initial segment η_x. Let f be a map from X to η_x. Then, $f(x) <_X x$. From the Proposition 2.6.7, it follows that f can not be an order isomorphism.   ♯

Corollary 2.6.10 *The only order isomorphism from a well-ordered set $(X, \leq_X)$ to itself is the identity map.*   ♯

Corollary 2.6.11 *Let $(X, \leq_X)$ and $(Y, \leq_Y)$ be well-ordered sets. Then, one and only one of the following hold:*

(i) $(X, \leq_X)$ is order isomorphic to a strict initial segment of $(Y, \leq_Y)$.
(ii) $(Y, \leq_Y)$ is order isomorphic to a strict initial segment of $(X, \leq_X)$.
(iii) $(X, \leq_X)$ is order isomorphic to $(Y, \leq_Y)$.

Proof From the Corollary 2.6.9, it follows that at the most one of the above condition can hold. We need to prove that at least one of the three conditions hold. Let $(X, \leq_X)$ and $(Y, \leq_Y)$ be well-ordered sets. Further, from the Corollary 2.6.9, it again follows that in any well-ordered set the strict initial segment associated with a is order isomorphic to an strict initial segment associated with b if and only if $a = b$. Let

$$\Sigma = \{x \in X \mid \eta_x \text{ is order isomorphic to } \eta_y \text{ for some, } y \in Y\}.$$

Let $a \in \Sigma$, and $x <_X a, x \in X$. Then, there is a unique element $b \in Y$ such that η_a is order isomorphic to η_b. Let f be the unique order isomorphism from η_a to η_b. Then, $f(x) \in \eta_b$, and the restriction of f to η_x is an order isomorphism from η_x to $\eta_{f(x)}$. Hence, $x \in \Sigma$. This ensures that Σ is an initial (not necessarily proper) segment of X. We have the map χ from Σ to Y given by the condition that η_x is order isomorphic to $\eta_{\chi(x)}$. Clearly, χ is an injective order preserving map. Observe that the image $\chi(\Sigma)$ is also an initial segment of Y. If $\Sigma = X$, then X will either be order isomorphic to Y or it is order isomorphic to a proper initial segment of Y. Suppose that $\Sigma \neq X$. Then, Σ is a proper segment, and hence, there is an element $x \in X - \Sigma$ such that $\eta_x = \Sigma$. Suppose that $\chi(\Sigma) \neq Y$. Then $\chi(\Sigma)$ is a proper segment of Y. Hence, there exists an element $y \in Y$ such that $\chi(\Sigma) = \eta_y$. But then η_x is order isomorphic to η_y, where $x \notin \Sigma$. This is a contradiction to the choice of Σ. Hence, $\chi(\Sigma) = Y$. This means that Y is order isomorphic to the initial segment Σ of X. ♯

Proposition 2.6.12 *Let $(\alpha, \leq_\alpha)$ and $(\beta, \leq_\beta)$ be ordinal which are order isomorphic as well-ordered set. Then $(\alpha, \leq_\alpha) = (\beta, \leq_\beta)$.*

Proof Let f be an order isomorphism from α to β. We need to show that $f(x) = x$ for all $x \in \alpha$. Consider $\gamma = \{x \in \alpha \mid f(x) = x\}$. Suppose that $\gamma \neq \alpha$. Then, $\alpha - \gamma \neq \emptyset$. Since $(\alpha, \leq_\alpha)$ is a well-ordered set, $\alpha - \gamma \neq \emptyset$ has the least element a (say). Then, $f(x) = x$ for all $x \in \eta_a$. Since f is an order isomorphism, and α and β are ordinals, $a = \eta_a = f(\eta_a) = \eta_{f(a)} = f(a)$. This is a contradiction. Hence, $\gamma = \alpha$. This shows that $(\alpha, \leq_\alpha) = (\beta, \leq_\beta)$. ♯

Corollary 2.6.13 *Every set of ordinal numbers is a total order.*

Proof Follows from Corollary 2.6.11 and the above proposition. ♯

Corollary 2.6.14 *Every set of ordinal numbers is well-ordered.*

Proof Let Ω be a set of ordinal numbers. Let Σ be a nonempty subset of Ω. Let $\alpha \in \Sigma$. If $\alpha \leq \beta$ for all $\beta \in \Sigma$, then α is the least element of Σ, and there is nothing to do. Suppose that there is a $\beta \in \Sigma$ such that $\beta < \alpha$. From the definition of ordinal number, $\beta \in \alpha$. This means that $\alpha \cap \Sigma$ is non empty subset of α. Since α is a well-ordered set, it has the least element γ (say). We show that γ is the least element of Σ. Let $\delta \in \Sigma$. If $\alpha \leq \delta$, then $\gamma \leq \delta$. If not, then $\delta < \alpha$, and so $\delta \in \alpha \cap \Sigma$. Since γ is the least element of $\alpha \cap \Sigma$, $\gamma \leq \delta$. This shows that γ is the least element of Σ. ♯

The ordinals are of two types: Consider the ordinal ω. For all $n < \omega$, there is an ordinal number m such that $n < m < \omega$. In other words, there is no immediate predecessor of ω. Such ordinals are called the **limit ordinals**. All the natural numbers have immediate predecessors. These are not limit ordinals. The ordinal $\omega 2$ is also a limit ordinal.

Corollary 2.6.15 *Let Ω be a set of ordinal numbers. Then, Ω is a order complete with respect to the ordering of ordinal numbers.*

Proof The result follows from the fact that every well-ordered set is order complete. ♯

Corollary 2.6.16 *Let Ω be a set of ordinal numbers. Then, there is an ordinal number $\alpha \notin \Omega$. In other words, there is no set containing all ordinal numbers.*

Proof Let Ω be a set of ordinal numbers. Let $\beta = \bigcup_{\alpha \in \Omega} \alpha$. Then, there is a unique order $\leq_\beta$ on β whose restriction to each $\alpha \in \Omega$ is the order $\leq_\alpha$ on α. Consider $(\beta, \leq_\beta)$. Let $a \in \beta$. Then, $a \in \alpha$ for some $\alpha \in \Omega$. Hence, the strict initial segment η_a is a itself. This shows that $(\beta, \leq_\beta)$ is an ordinal number which is an upper bound (indeed, l.u.b of Ω) of Ω. β may be a member of Ω in case it is a limit ordinal. However, the successor β^+ of β is an ordinal number which does not belong to Ω. ♯

Proposition 2.6.17 *Let $(X, \leq_X)$ be a well-ordered set. Then, there is a unique ordinal $(\alpha, \leq_\alpha)$ which is order isomorphic to $(X, \leq_X)$.*

Proof The uniqueness part is evident from the Proposition 2.6.12. We show the existence of an ordinal which is order isomorphic to $(X, \leq_X)$. Let a be an element of X such that for each $x \in \eta_a$, there is, of course, unique ordinal α_x which is order isomorphic to η_x. Clearly, the least element of X is such an element. It is also clear that η_a is order isomorphic to the ordinal β, where β is the l.u.b of the set $\{\alpha_x \mid x \in \eta_a\}$ of ordinals. This shows that if each strict initial segment of η_a is order isomorphic to an ordinal number, then η_a is also order isomorphic to an ordinal number.
    Let

$$\Sigma = \{a \in X \mid \forall x \in \eta_a, \eta_x \text{ is order isomorphic to an ordinal number } \alpha_x\}.$$

Clearly, the least element of X belongs to Σ. We first show that $\Sigma = X$. Suppose not. Then, $X - \Sigma$ is a nonempty subset of X. Since $(X, \leq_X)$ is a well-ordered set, it has the least element a (say). Then, for all $x <_X a, x \in \Sigma$. This means that for all $y \in \eta_x, \eta_y$ is order isomorphic to an ordinal α_y. From what we have already proved, it follows that η_x is also order isomorphic to an ordinal number α_x. We arrive at a contradiction that $a \in \Sigma$. Thus, $\Sigma = X$. Repeating again the previous arguments, we see that X is order isomorphic to an ordinal. ♯

Remark 2.6.18 The above proposition may prompt us to introduce an ordinal number as an equivalence class of well-ordered sets. But the equivalence classes are not sets. As such, one needs to select unique members from each equivalence classes. Indeed, this is what we have done in our approach.

Arithmetic of Ordinal Numbers

Let $(A, \leq_A)$ and $(B, \leq_B)$ be two well-ordered sets. We have an order $\leq_{AB}$ on $A \times \{0\} \bigcup B \times \{1\}$ defined as follows: (i) $(a, 0) \leq_{AB} (a', 0)$ if and only if $a \leq_A a'$, (ii) $(b, 1) \leq_{AB} (b', 1)$ if and only if $b \leq_B b'$, and (iii) $(a, 0) \leq_{AB} (b, 1)$ for all $a \in A$ and $b \in B$. Clearly, $\leq_{AB}$ is a well-order. If $(A, \leq_A)$ is order isomorphic to $(C, \leq_C)$ and $(B, \leq_B)$ is order isomorphic to $(D, \leq_D)$, then it is clear that $(A \times \{0\} \bigcup B \times \{1\}, \leq_{AB})$ is order isomorphic $(C \times \{0\} \bigcup D \times \{1\}, \leq_{CD})$. Thus, we can define, unambiguously, the sum $\alpha + \beta$ of two ordinal numbers as follows: Suppose that α is order isomorphic to $(A, \leq_A)$ and β is order isomorphic to $(B, \leq_B)$. Define $\alpha + \beta$ to be the unique ordinal number which is order isomorphic to $(A \times \{0\} \bigcup B \times \{1\}, \leq_{AB})$. The following properties addition of ordinal numbers can be easily observed:

(i) $\alpha + 0 = \alpha = 0 + \alpha$, and
(ii) $(\alpha + \beta) + \gamma = \alpha + (\beta + \gamma)$
    for all ordinals α, β, and γ.

The addition of ordinal number is not commutative. Indeed, $1 + \omega = \omega$ is limit ordinal, where as $\omega + 1 = \omega^+$ is not limit ordinal.

Next, suppose that $(A, \leq_A)$ and $(B, \leq_B)$ are two well-ordered sets. We have the lexicographic ordering $\leq_{A \times B}$ on $A \times B$ defined as follows: $(a, b) \leq_{A \times B} (c, d)$ if $b <_B d$ or $b = d$ and $a \leq_A c$. It can be checked that this order is a well-order. Further, if $(A, \leq_A)$ is order isomorphic to $(C, \leq_C)$ and $(B, \leq_B)$ is order isomorphic to $(D, \leq_D)$, then $(A \times B, \leq_{A \times B})$ is order isomorphic to $(C \times D, \leq_{C \times D})$. This prompts us to define the multiplication $\cdot$ on ordinals as follows: Suppose that the ordinal α is order isomorphic to the well-ordered set $(A, \leq_A)$ and the ordinal β is order isomorphic to $(B, \leq_B)$. Define $\alpha \cdot \beta$ to be the unique ordinal which is order isomorphic to the well-ordered set $(A \times B, \leq_{A \times B})$. The following properties of $\cdot$ can be easily observed.

(i) $\alpha \cdot 0 = 0 = 0 \cdot \alpha$,
(ii) $\alpha \cdot 1 = \alpha = 1 \cdot \alpha$,
(iii) $(\alpha \cdot \beta) \cdot \gamma = \alpha \cdot (\beta \cdot \gamma)$, and
(iv) $\alpha \cdot (\beta + \gamma) = \alpha \cdot \beta + \alpha \cdot \gamma$
    for all ordinal numbers α, β, and γ. Note that the left distributivity of $\cdot$ over $+$ need not hold.

For further arithmetical properties of ordinals, refer to Naive set theory by Halmos, or to the set theory by Vipul Kakkar.

2.7 Cardinal Numbers

The abstraction of the counting process of finite sets leads to the concept of the ordinal numbers. On a finite set X, any two well-order structure is order isomorphic. Indeed, two finite well-ordered sets $(X, \leq_X)$ and $(Y, \leq_Y)$ define the same ordinal numbers if and only if their sizes are same in the sense that there is a bijective map from X to Y. However, this is not the situation in infinite case. Indeed, on the same infinite set X, we can have different well-order structures which define different ordinal numbers. For example, the usual well-order $\leq$ on the set $\mathbb{N}$ of natural numbers determines the ordinal number ω. We have another order $\leq'$ on $\mathbb{N}$ defined as follows: If $m \neq 1 \neq n$, then $m \leq' n$ if and only if $m \leq n$. Also, $m \leq' 1$ for all $m \in \mathbb{N}$. Clearly, $(\mathbb{N}, \leq')$ is a well-ordered set, and the initial segment η_1 in $(\mathbb{N}, \leq')$ is $\mathbb{N} - \{1\}$. Note that $(\mathbb{N}, \leq)$ and $(\mathbb{N}, \leq')$ are not order isomorphic. $(\mathbb{N}, \leq')$ is order isomorphic to the ordinal $\omega + 1$. Similarly, we have another well-order on $\leq''$ on $\mathbb{N}$ defined as follows: If $m, n \in \mathbb{N} - \{1, 2\}$, $m \leq'' n$ if and only if $m \leq n$. Also, $m \leq'' 1 \leq'' 2$ for all $m, n \in \mathbb{N} - \{1, 2\}$. Then, $(\mathbb{N}, \leq'')$ is order isomorphic to $\omega + 2$, and so on. Thus, there are infinitely many nonorder isomorphic well-order structures on $\mathbb{N}$ corresponding to different ordinals. Thus, ordinals are good for counting, but it does not distinguish the size of the infinite sets. This prompts us to look for an other concept, the concept of cardinals which measures the size of sets.

Definition 2.7.1 Let X and Y be sets. We say that X dominates Y if there is an injective map from Y to X. X and Y are said to be **equipotent** or **equinumerous** if there is a bijective map from X to Y. We use the notation $X \approx Y$ to say that X is equipotent to Y.

Theorem 2.7.2 (Schröder-Bernstein Theorem) *If X dominates Y and Y also dominates X, then X is equipotent to Y. More explicitly, if there is an injective map from X to Y, and also there is an injective map from Y to X, then there is a bijective map from X to Y.*

Proof Let f be an injective map from X to Y and g be an injective map from Y to X. We have to show that X and Y are equipotent. Put $X - g(Y) = Z$. Then, $X = g(Y) \bigcup Z$, where $g(Y)$ and Z are disjoint. Since Y and $g(Y)$ are equipotent, it is sufficient to show that $g(Y)$ and X are equipotent. Let $u \in (gof)^r(Z) \bigcap (gof)^s(Z)$, where $r < s$. Then, there exist $x, y \in Z$ such that $u = (gof)^r(x) = (gof)^s(y)$. Since gof is injective, $x = (gof)^{s-r}(y)$. This means that $x \in Z \bigcap g(Y)$. This is impossible. It follows that $(gof)^r(Z) \bigcap (gof)^s(Z) = \emptyset$ for all $r \neq s$. Put $U = \bigcup_{r \in \mathbb{N}}(gof)^r(Z)$, and $V = \bigcup_{r \in \mathbb{N} - \{1\}}(gof)^r(Z)$. Then, $U = (gof)(Z) \bigcup V$. From what we have observed, it follows that $(gof)(Z)$ and V are disjoint. Also, U and V are equipotent. Indeed, gof is a bijective map from U to V. Also Z and $(gof)(Z)$ are equipotent. Since $U = (gof)(Z) \bigcup V, Z \bigcup U$ is equipotent to U(note that Z and U are disjoint). Now, put $g(Y) - U = W$. Then, $g(Y) = U \bigcup W$. Hence, $X = U \bigcup W \bigcup Z$. Since $U \bigcup Z$ is equipotent to U, $U \bigcup W$ is equipotent to X. This shows that $g(Y)$ is equipotent to X.   ♯

Definition 2.7.3 An ordinal number α is said to be a **cardinal number** if whenever an ordinal number β is equipotent to α, $\alpha \leq \beta$.

Thus, all natural numbers are cardinal numbers. ω is a cardinal number, whereas $\omega + 1$ is not a cardinal number. The ordinal number ω considered as a cardinal number is denoted by $\aleph_0$. Indeed, an infinite cardinal number is a limit ordinal. It follows from the properties of ordinal numbers that a set of cardinal numbers is totally ordered.

Let X be a set. From the well-ordering principle, there is a well-order on X. The set of all ordinal numbers which are order isomorphic to the different well-order structures on X has the least element. This least element is clearly a cardinal number, and it is called the cardinal number of X. The cardinal number of X is denoted by $\mid X \mid$. Evidently, $\mid X \mid = \mid Y \mid$ if and only if X is equipotent to Y. Further, if $a = \mid X \mid$ and $b = \mid Y \mid$ are two cardinal numbers, then $a \leq b$ if and only if there is an injective map from X to Y.

Definition 2.7.4 A set X is said to be a **countable set** if $\mid X \mid$ is a natural number, or it is $\aleph_0$. It is said to be an infinite countable set if $\mid X \mid = \aleph_0$. Thus, X is countably infinite if and only if there is a bijective map from X to $\mathbb{N}$. A set X is said to be **uncountable** if it is not countable.

Since there is no surjective map from $\mathbb{N}$ to the power set $\wp(\mathbb{N})$, $\wp(\mathbb{N})$ is uncountable. Observe that $\wp(\mathbb{N})$ and $2^{\mathbb{N}}$ are equipotent, and so $\mid \wp(\mathbb{N}) \mid = \mid 2^{\mathbb{N}} \mid$. The cardinal number $\mid 2^{\mathbb{N}} \mid$ is denoted by $\aleph_1$. The cardinal number $\mid 2^{2^{\mathbb{N}}} \mid$ is denoted by $\aleph_2$, and so on. If $\aleph$ is an cardinal number, then the cardinal number $\mid 2^{\aleph} \mid$ is denoted by $2^{\aleph}$. If A is equipotent to B, and C is equipotent to D, then A^C is equipotent to B^D. Thus, we can, unambiguously, define the power a^b as follows: Suppose that $a = \mid A \mid$ and $b = \mid B \mid$. Define $a^b = \mid A^B \mid$. In turn, for each cardinal number $\aleph$, we have the cardinal number $2^{\aleph}$, and we have a chain of infinite cardinal numbers $\aleph_0, \aleph_1, \aleph_2, \ldots, \aleph_\alpha, \aleph_{\alpha+1}, \ldots$, where $\aleph_{\alpha+1} = 2^{\aleph_\alpha}$ of infinite cardinal numbers, where α runs over a chain of ordinal numbers.

Continuum hypothesis. The continuum hypothesis (CH) asserts that there is no cardinal number in between $\aleph_0 = \mid \mathbb{N} \mid$ and $\aleph_1 = \mid 2^{\mathbb{N}} \mid$. More precisely, it asserts that if there is a set A such that there is an injective map from $\mathbb{N}$ to A, and there is an injective map from A to $2^{\mathbb{N}}$, then A is equipotent to $\mathbb{N}$ or it is equipotent to $2^{\mathbb{N}}$.

$K.G\ddot{o}del$ in 1939 proved that if the ZF axiomatic system is consistent, then adjunction of CH in ZF does not lead to any contradiction. In other words, CH is consistent with the ZF axiomatic system. Further, in 1963, P. Cohen proved that ZF axiomatic system does not lead to a proof of CH. Consequently, CH is independent of the ZF axiomatic system.

Generalized continuum hypothesis. The generalized continuum hypothesis (GCH) asserts that for each ordinal α, there is no cardinal number between $\aleph_\alpha$ and $\aleph_{\alpha+1} = 2^{\aleph_\alpha}$. The topologist Sierpinski proved that GCH implies axiom of choice. $K.G\ddot{o}del$ also showed that GCH is consistent with the ZF axiomatic system.

Arithmetic of Cardinal Numbers

Let (A, C) and (B, D) be pairs of equipotent sets. Suppose that $A \cap B = \emptyset = C \cap D$. It is evident that $A \bigcup C$ is equipotent to $B \bigcup D$. Thus, we have the addition $+$ on a suitable set Ω of cardinal numbers defined by $a + b = \mid (A \times \{0\}) \bigcup (B \times \{1\}) \mid$, where $a = \mid A \mid$ and $b = \mid B \mid$. The following properties of $+$ can be verified easily.

(i) $a + 0 = a = 0 + a$,
(ii) $(a + b) + c = a + (b + c)$,
(iii) $a + b = b + a$, and
(iii) $a \leq b$ and $c \leq d$ implies that $a + c \leq b + d$
for all $a, b, c, d \in \Omega$.

We can also define sum of an arbitrary family of cardinal numbers as follows: Let $\{a_\alpha = |A_\alpha| \mid \alpha \in \Lambda\}$ be a family of cardinal numbers. We define

$$\Sigma_{\alpha \in \Lambda} a_\alpha = \left| \bigcup_{\alpha \in \Lambda} A_\alpha \times \{\alpha\} \right|.$$

Recall that a set X is said to be a finite set if every injective map from X to itself is a surjective map. It is easily observed that a set X is finite if and only if $|X|$ is a natural number.

Proposition 2.7.5 *A set X is infinite if and only if there is an injective map from the set $\mathbb{N}$ of natural numbers to X.*

To prove this result, we need the following recursion theorem whose proof can be found in the next chapter.

Recursion Theorem. Let X be a set and $a \in X$. Let f be a map from X to X. Then, there is a unique map g from $\mathbb{N}$ to X such that $g(1) = a$ and $g(n^+) = f(g(n))$ for all $n \in \mathbb{N}$.

Proof of the proposition 2.7.5: Let X be an infinite set. Let f be an injective map from X to X which is not surjective. Let $a \in X$ which is not in the image of f. By the recursion theorem, there is a unique map g from $\mathbb{N}$ to X with $g(1) = a$ and is such that $g(n^+) = f(g(a))$. Let

$$M = \{m \in \mathbb{N} \mid g(m) = g(n) \text{ implies that } m = n\}.$$

Since a is not in the image of f, $1 \in M$. Suppose that $m \in M$. Then, $g(m) = g(n)$ implies that $m = n$. Suppose that $g(m^+) = g(n)$. Then, $n \neq 1$. Hence, there is an element $r \in \mathbb{N}$ such that $n = r^+$. By the definition of g, $f(g(m)) = f(g(r))$. Since f is injective, $g(m) = g(r)$. This means that $m = r$, and so $m^+ = n$. It follows that $m^+ \in M$. By P_5, $M = \mathbb{N}$. This shows that g is injective. ♯

Proposition 2.7.6 $\aleph_0 + \aleph_0 = \aleph_0$.

Proof It is sufficient to give a bijective map from $\mathbb{N} \times \{0\} \bigcup \mathbb{N} \times \{1\}$ to $\mathbb{N}$. Let X denote the set $\{2n \mid n \in \mathbb{N}\}$ of even natural numbers and Y denote the set $\{2n + 1 \mid n \in \mathbb{N}\}$ of odd natural numbers. Then, X and Y are disjoint. Further, $n \rightsquigarrow 2n$ is a bijective map from $\mathbb{N}$ to X, and $n \rightsquigarrow 2n + 1$ is a bijective map from $\mathbb{N}$ to Y. This shows that $\mathbb{N} \times \{0\} \bigcup \mathbb{N} \times \{1\}$ is equipotent to $\mathbb{N}$. The result follows. ♯

Corollary 2.7.7 *For every natural number n, $\aleph_0 + n = \aleph_0$.*

Proof $(\mathbb{N} \times \{0\} \bigcup n \times \{1\}) \subset (\mathbb{N} \times \{0\} \bigcup \mathbb{N} \times \{1\})$. From the above proposition, it follows that $(\mathbb{N} \times \{0\} \bigcup n \times \{1\})$ is equipotent to a subset of $\mathbb{N}$. Also, the map $n \rightsquigarrow (n, 0)$ is an injective map from $\mathbb{N}$ to $(\mathbb{N} \times \{0\} \bigcup n \times \{1\})$. By the Schröder–Berstein theorem, $\mathbb{N}$ is equipotent to $(\mathbb{N} \times \{0\} \bigcup n \times \{1\})$. The result follows.   ♯

Proposition 2.7.8 *If a is an infinite cardinal, then $a + a = a$.*

Proof Let us suppose that $a = |A|$, where A is an infinite set. We need to show that $A \times \{0\} \bigcup A \times \{1\}$ is equipotent to A. Let

$$\Sigma = \{(X,f) \mid X \subseteq A, \ f \ is \ a \ bijective \ map \ from \ X \ to \ X \times \{0\} \bigcup X \times \{1\}\}.$$

Since A is infinite, by the Proposition 2.7.5, there is a subset X of A which is equipotent to $\mathbb{N}$. From the Proposition 2.7.6, X and $X \times \{0\} \bigcup X \times \{1\}$ are equipotent. Hence, $\Sigma \neq \emptyset$. Define a partial order $\leq$ on Σ by putting $(X,f) \leq (Y,g)$ if $X \subseteq Y$ and $g/X = f$. Clearly, $(\Sigma, \leq)$ is a nonempty partially ordered set. Let $\{(X_\alpha, f_\alpha) \mid \alpha \in \Lambda\}$ be a chain in $(\Sigma, \leq)$. Let $X = \bigcup_{\alpha \in \Lambda} X_\alpha$ and f be the map whose restriction to each X_α is f_α. It is an easy observation that f is a bijective map from X to $X \times \{0\} \bigcup X \times \{1\}$. This shows that (X,f) is an upper bound of the chain. By the Zorn's lemma, $(\Sigma, \leq)$ has a maximal element (X_0, f_0) (say). Now, we show that $A - X_0$ is a finite set. Suppose not. Again, by the Proposition 2.7.5, there is a subset Z of $A - X_0$ which is equipotent to $\mathbb{N}$. But, then there is a bijective map h from Z to $Z \times \{0\} \bigcup Z \times \{1\}$. Take $U = X_0 \bigcup Z$, and the map ϕ from U to $U \times \{0\} \bigcup U \times \{1\}$ whose restriction to X_0 is f_0, and whose restriction to Z is h. Clearly, $(U, \phi) \in \Sigma$. This is a contradiction to the maximality of (X_0, f_0). Thus, $A - X_0$ is finite. From the Corollary 2.7.7,

$$a = |A| = |X_0| = |X_0 \times \{0\} \bigcup X_0 \times \{1\}| = |X_0| + |X_0| = a + a.$$

♯

Now, we define the product $\cdot$ of two cardinal numbers as follows: First observe that $|A| = |C|$ and $|B| = |D|$ imply that $|A \times B| = |C \times D|$. Thus, we can, unambiguously, define the product $a \cdot b$ of two cardinal numbers $a = |A|$ and $b = |B|$ by $a \cdot b = |A \times B|$. The following properties of the multiplication $\cdot$ can be easily observed:

(i) $a \cdot 0 = 0 = 0 \cdot a$,
(ii) $(a \cdot b) \cdot c = a \cdot (b \cdot c)$,
(iii) $a \cdot b = b \cdot a$, and
(iv) $a \cdot (b + c) = a \cdot b + a \cdot c$
    for all ordinal numbers a, b, and c.

The proof of the following proposition uses Zorn's lemma, and it is similar to the proof of the Proposition 2.7.8.

Proposition 2.7.9 *If a is an infinite cardinal number, then* $a \cdot a = a$. ♯

As a corollary, we obtain the following:

Proposition 2.7.10 *If* $a \leq b$, *then* $a \cdot b = b$. ♯

Gödel–Bernays Axiomatic system

In 1920, John von Neumann attempted an other axiomatic system for set theory. His axiomatic system significantly differed from the ZF axiomatic system. Indeed, for him, the primitive term (concept) was that of a correspondence (a map) instead of a set. Later, Gödel and Bernays modified it to make it more appealing and near to *ZF* system. For them, the primitive term is class instead of set. A member of a class in this axiomatic system is a set. Most of the axioms of the Gödel–Bernays system is same as those of *ZF* axiomatic system with set replaced by class except the axiom of replacement. Further, in this axiomatic system, a set may be a class, but then it does contain all sets or all ordinal numbers. Sets are those classes which are adequate to develop mathematics. The Gödel–Bernays axiomatic system is most suitable for the categorical discussions.

Chapter 3
Number System

3.1 Natural Numbers

What is one? One pen, one man. These all reflect the idea of being single which is the common property of all singletons. One may be tempted to represent 'one' by all singletons. But, if there is a set X containing all singletons, then $\{X\} \in X$ and also $X \in \{X\}$. This is a contradiction to the axiom of regularity in set theory. Thus, instead of looking at all singletons, we choose a canonical representative $\{\emptyset\}$ of the class of all singletons to define **one**. Similarly, the representative $\{\emptyset, \{\emptyset\}\} = \{\emptyset\}^+$ of all doubletons is chosen to represent **two** and so on. It has been seen in the previous chapter that successor sets contain all these and the axiom of infinity ensures the existence of a successor set. We have also seen in the previous chapter that there is a unique successor set contained in all successor sets.

Recall that the set $\mathbb{N}$ of natural numbers is the smallest successor set, and in turn, it satisfies the following properties termed as Peano's axioms.

P_1. $1 \in \mathbb{N}$.
P_2. *For all* $x \in \mathbb{N}$, $x^+ \in \mathbb{N}$.
P_3. $x^+ = y^+$ *if and only if* $x = y$.
P_4. *For all* $x \in \mathbb{N}$, $1 \neq x^+$.
P_5. *If* M *is a set such that* $1 \in M$ *and* $x^+ \in M$ *for all* $x \in M \bigcap \mathbb{N}$, *then* $\mathbb{N} \subseteq M$.

The properties P_1, P_2 *and* P_3 follow from the fact that $\mathbb{N}$ is a successor set. To prove P_4, suppose that $1 = x^+$ for some $x \in \mathbb{N}$. Clearly, $x \neq 1$, for $1^+ \neq 1$ (Exercise 2.1.19). Thus, $1 \in \mathbb{N} - \{x\}$. Suppose that $y \in \mathbb{N} - \{x\}$. If $y^+ = x$ then $(y^+)^+ = x^+ = 1 = \{\emptyset\} = \emptyset^+$. But, then $y^+ = \emptyset$. This means that $y \in \emptyset$ (a contradiction). Hence, $y^+ \in N - \{x\}$. This shows that $N - \{x\}$ is a successor set, a contradiction to the fact that $\mathbb{N}$ is the smallest successor set. This proves P_4. Under the hypothesis of P_5, $M \bigcap \mathbb{N}$ becomes a successor set. Since $\mathbb{N}$ is the smallest successor set, $M \subseteq \mathbb{N}$.

© Springer Nature Singapore Pte Ltd. 2017
R. Lal, *Algebra 1*, Infosys Science Foundation Series in Mathematical Sciences,
DOI 10.1007/978-981-10-4253-9_3

The property P_5 is called the **principle of induction**. A proof using P_5 is called a **proof by induction**.

Now, instead of looking $\mathbb{N}$ as the smallest successor set, we shall be using only the properties $P_1 - P_5$ which is sufficient for the further course of developments. We shall call x^+ the successor of x.

Proposition 3.1.1 $x^+ \neq x$ for all $x \in \mathbb{N}$.

Proof Let $M = \{x \in \mathbb{N} \mid x^+ \neq x\}$. Since $1^+ \neq 1 (P_4)$, $1 \in M$. Let $y \in M$. Then $y^+ \neq y$. Now $(y^+)^+ \neq y^+$ for, otherwise by P_3, $y^+ = y$. Thus, $y^+ \in M$. By $P_5, \mathbb{N} \subseteq M \subseteq \mathbb{N}$. Hence $M = \mathbb{N}$.   ♮

Proposition 3.1.2 *1 is the only element in $\mathbb{N}$ which is not successor of any element in $\mathbb{N}$.*

Proof From P_4, 1 is not successor of any element in $\mathbb{N}$. If $a \neq 1$ is not successor of any element in $\mathbb{N}$, then by P_5, $\mathbb{N} \subseteq \mathbb{N} - \{a\}$. This is a contradiction.   ♮

Recall that a set X is called **finite** if every injective map from X to X is surjective (equivalently every surjective map from X to X is injective (Exercise 2.4.10)). A set which is not finite is called an **infinite set**.

Proposition 3.1.3 $\mathbb{N}$ *is an infinite set.*

Proof The map $s : \mathbb{N} \longrightarrow \mathbb{N}$ defined by $s(x) = x^+$ is an injective map (P_3). It is not surjective for 1 is not successor of any element in $\mathbb{N}$.   ♮

The following theorem gives sound and rigorous footing for the definitions by induction.

Theorem 3.1.4 (Recursion Theorem) *Let X be a set and $a \in X$. Let f be a map from X to X. Then, there is a unique map g from $\mathbb{N}$ to X such that $g(1) = a$ and $g(n^+) = f(g(n))$ for all $n \in \mathbb{N}$.*

Proof We first show the uniqueness.[1] Let g and h be maps from $\mathbb{N}$ to X such that $g(1) = a = h(1)$ and $g(n^+) = f(g(n))$, $h(n^+) = f(h(n))$ for all $n \in \mathbb{N}$. Let $M = \{n \in \mathbb{N} \mid g(n) = h(n)\}$. Since $g(1) = h(1) = a$, $1 \in M$. Suppose that $n \in M$. Then, $g(n^+) = f(g(n)) = f(h(n)) = h(n^+)$. Hence, $n^+ \in M$. By P_5, $M = \mathbb{N}$ and so $g(n) = h(n)$ for all $n \in \mathbb{N}$. This shows that $g = h$.

Now, we show the existence. Let

$$A = \{h \subseteq \mathbb{N} \times X \mid (1, a) \in h \text{ and whenever } (n, x) \in h, (n^+, f(x)) \in h\}.$$

Clearly, $\mathbb{N} \times X \in A$ and so $A \neq \emptyset$. Let $g = \bigcap_{h \in A} h$. Then, g is also a member of A. Hence, it is sufficient to show that g is a map. Let $M = \{n \in \mathbb{N} \mid \text{there is a unique } x \in X \text{ such that } (n, x) \in g\}$. Now $(1, a) \in g$. Suppose that $(1, b) \in g$, where $a \neq b$.

---

[1] The reader may skip the proof.

Clearly, $g - \{(1, b)\} \in A$. This shows that $1 \in M$. Suppose that $n \in M$. Then, there is a unique $x \in X$ such that $(n, x) \in g$. Since $g \in A$, $(n^+, f(x)) \in g$. Suppose that $(n^+, b) \in g$, where $b \neq f(x)$. Then, $g - \{(n^+, b)\} \in A$ (verify). This shows that $n^+ \in M$. By P_5, $\mathbb{N} = M$. Hence, g is a map with the required property. ♯

Any use of recursion theorem is called a **definition by induction**.

Definition 3.1.5 Let X be a set. A map $o : X \times X \longrightarrow X$ is called a **binary operation** in X. The image of (a, b) under o is denoted by aob.

We define binary operations $+$ and $\cdot$ in $\mathbb{N}$ as follows:

Let s be the successor map from $\mathbb{N}$ to $\mathbb{N}$ given by $s(n) = n^+$. Let $m \in \mathbb{N}$. Then, by the recursion theorem, there is a unique map f_m from $\mathbb{N}$ to $\mathbb{N}$ such that $f_m(1) = m^+$ and $f_m(n^+) = s(f_m(n)) = f_m(n)^+$. Now, define a binary operation $+$ in $\mathbb{N}$ by

$$m + n = f_m(n).$$

Then, evidently

(i) $m + 1 = f_m(1) = m^+$.
(ii) $m + n^+ = f_m(n^+) = (f_m(n))^+ = (m + n)^+$.

Next, consider the map s^m from $\mathbb{N}$ to $\mathbb{N}$ defined by $s^m(n) = n + m$. Again, by the recursion theorem, there is a unique map f^m from $\mathbb{N}$ to $\mathbb{N}$ such that $f^m(1) = m$ and $f^m(n^+) = s^m(f^m(n)) = f^m(n) + m$. Define a binary operation $\cdot$ in $\mathbb{N}$ by

$$m \ n = f^m(n).$$

Evidently,

(i) $m \cdot 1 = m$.
(ii) $m \cdot n^+ = f^m(n^+) = s^m(f^m(n)) = m \cdot n + m$.

Theorem 3.1.6 *The triple* $(\mathbb{N}, +, \cdot)$ *has the following properties.*

(i) $+$ *is associative in the sense that* $(n+m)+r = n+(m+r)$ *for all* $n, m, r \in \mathbb{N}$.

(ii) $+$ *is commutative in the sense that* $m + n = n + m$ *for all* $m, n \in \mathbb{N}$.

(iii) $1 \cdot n = n = n \cdot 1$ *for all* $n \in \mathbb{N}$.

(iv) $\cdot$ *distributes over* $+$ *from left as well as right in the sense that*

$$m \cdot (n + r) = m \cdot n + m \cdot r,$$

and

$$(m + n) \cdot r = m \cdot r + n \cdot r$$

for all m, n, r ∈ ℕ.

(v) *· is associative in the sense that* $(m \cdot n) \cdot r = m \cdot (n \cdot r)$ *for all* $m, n, r \in \mathbb{N}$.

(vi) *· is commutative in the sense that* $m \cdot n = n \cdot m$ *for all* $m, n \in \mathbb{N}$.

(vii) $n + m \neq n$ *for all* $m, n \in \mathbb{N}$.

(viii) *Cancellation law holds for* + *in the sense that* $m + n = m + r$ *implies that* $n = r$ *and* $m + n = r + n$ *implies that* $m = r$.

(ix) *Cancellation law holds for* · *in the sense that* $m \cdot n = m \cdot r$ *implies that* $n = r$ *and* $n \cdot m = r \cdot m$ *implies that* $n = r$ *for all* $m, n, r \in \mathbb{N}$.

Proof We prove (ii), (iv), (vii), and (ix). The rest can be proved similarly and is left as an exercise.

(ii) Let $M = \{n \in \mathbb{N} \mid 1 + n = n + 1\}$. Clearly, $1 \in M$. Suppose that $n \in M$. Then, $1 + n^+ = (1 + n)^+ \, (by \, def) = (n + 1)^+ \, (for \, n \in M) = (n^+)^+ \, (by \, def) = n^+ + 1$. Thus, $n^+ \in M$. By P_5, $M = N$, and so $n + 1 = 1 + n$ *for all* $n \in \mathbb{N}$. Let $M' = \{n \in \mathbb{N} \mid n + m = m + n$ *for all* $m \in \mathbb{N}\}$. We have already proved that $1 \in M'$. Let $n \in M'$. Then, $m + n^+ = (m + n)^+ \, (by \, def) = (n + m)^+ \, (for \, n \in M') = n + m^+ \, (by \, def) = n + (m + 1) = n + (1 + m) \, (for \, 1 \in M') = (n + 1) + m \, (by \, (i)) = n^+ + m$. Thus, $n^+ \in M'$ and hence $M' = \mathbb{N}$.

(iv) Take $M = \{r \in \mathbb{N} \mid m \cdot (n + r) = m \cdot n + m \cdot r$ *for all* $m, n \in \mathbb{N}\}$. Since $m \cdot (n + 1) = m \cdot n^+ = m \cdot n + m \, (by \, def) = m \cdot n + m \cdot 1$, it follows that $1 \in M$. Suppose that $r \in M$. Then, $m \cdot (n + r^+) = m \cdot (n + r)^+ \, (by \, def) = m \cdot (n + r) + m \, (by \, def) = (m \cdot n + m \cdot r) + m \, (for \, r \in M) = m \cdot n + (m \cdot r + m) \, (by \, (i)) = m \cdot n + m \cdot r^+ \, (by \, def)$. This shows that $r^+ \in M$. By $P_5, M = \mathbb{N}$.

(vii) Let $M = \{n \in \mathbb{N} \mid n + m \neq n$ *for all* $m \in \mathbb{N}\}$. $1 \in M$, for $1 + m = m + 1 = m^+ \neq 1 \, (by \, P_4)$. Suppose that $n \in M$. Then, $n^+ + m = n^+$ *implies that* $(m + n)^+ = n^+$. By P_3, $m + n = n$, a contradiction to the supposition that $n \in M$. This shows that $n^+ \in M$. By $P_5, M = \mathbb{N}$.

(ix) Let $M = \{m \in \mathbb{N} \mid n \cdot m = n \cdot r$ *implies that* $m = r\}$. Suppose that $n \cdot 1 = n \cdot r$. If $r \neq 1$, then, by Proposition 3.1.2, *there is a* $q \in \mathbb{N}$ such that $r = q^+$. But, then $n = n \cdot q^+ = n \cdot q + n$. This is a contradiction to (vii). This shows that $1 \in M$. Suppose that $m \in M$ and $n \cdot m^+ = n \cdot r$. If $r = 1$, then from the earlier argument, $m^+ = 1$, a contradiction. Suppose that $r \neq 1$. Then, again, $r = q^+$ for some $q \in \mathbb{N}$. Now, $n \cdot m^+ = n \cdot q^+$ implies that $n \cdot m + n = n \cdot q + n$. By (viii) $n \cdot m = n \cdot q$. Since $m \in M, m = q$ and so $m^+ = q^+ = r$. Thus, $m^+ \in M$. By $P_5, M = \mathbb{N}$.   ♯

3.2   Ordering in $\mathbb{N}$

We define relations '$<$' (called 'less than') and '$\leq$' (called 'less than or equal to') on $\mathbb{N}$ as follows:
$$< = \{(a, b) \in \mathbb{N} \times \mathbb{N} \mid a + c = b \text{ for some } c \in \mathbb{N}\},$$
and
$$\leq\; = < \bigcup \Delta = \{(a, b) \in \mathbb{N} \times \mathbb{N} \mid a = b \text{ or } a + c = b \text{ for some } c \in \mathbb{N}\}.$$
We write '$a < b$' to say that $(a, b) \in <$ and '$a \leq b$' to say that $(a, b) \in \leq$. Thus,

$$a < b \text{ if and only if } a + c = b \text{ for some } c \in \mathbb{N},$$

and

$$a \leq b \iff a = b \text{ or } a + c = b \text{ for some } c \in \mathbb{N}.$$

Proposition 3.2.1   (i) $a \not< a$ for all $a \in \mathbb{N}$.

(ii) $a < b$ implies that $b \not< a$ for all $a, b \in \mathbb{N}$.

(iii) $[a < b \text{ and } b < c]$ implies that $a < c$ for all $a, b, c \in \mathbb{N}$.

Proof   (i) By Theorem 3.1.6 (vii), it follows that there is no $c \in \mathbb{N}$ such that $a + c = a$. Hence, by the definition, $a \not< a$.

(ii) Suppose that $a < b$ and $b < a$. Then, by the definition of $<$, *there exist* $c, d \in \mathbb{N}$ such that $b = a + c$ and $a = b + d$. But, then $b = b + (c + d)$. This contradicts Theorem 3.1.6 (vii).

(iii) Suppose that $a < b$ and $b < c$. Then, *there exist elements* $u, v \in \mathbb{N}$ such that $b = a + u$ and $c = b + v$. But, then $c = a + (u + v)$ for some $u, v \in \mathbb{N}$. By the definition of $<$, $a < c$.   ♯

Remark 3.2.2 The above proposition implies that the relation '$<$' is nonreflexive, nonsymmetric, antisymmetric, and transitive.

Corollary 3.2.3 *The relation '$\leq$' is a partial order in* $\mathbb{N}$.

Proof Follows from Proposition 3.2.1.   ♯

Theorem 3.2.4 (Law of Trichotomy) *Given* $a, b \in \mathbb{N}$, *one and only one of the following holds.*

(i) $a = b$
(ii) $a < b$
(iii) $b < a$

Proof It follows from the Proposition 3.2.1 (i), (ii) that at most one of the above conditions will be satisfied for any pair $a, b \in \mathbb{N}$. We have to show that at least one of the above three conditions hold for any pair $a, b \in \mathbb{N}$. Let $M = \{a \in \mathbb{N} \mid$ *for any* $b \in \mathbb{N}$, $a = b$ *or* $a < b$ *or* $b < a\}$. Let $b \in \mathbb{N}$. If $b \neq 1$, then by Proposition 3.1.2, $b = c^+ = 1 + c$ for some $c \in \mathbb{N}$. By the definition, $1 < b$. This shows that $1 \in M$. Let $a \in M$. Given any $c \in \mathbb{N}$, $a = c$ or $a < c$ or $c < a$. Let $b \in \mathbb{N}$. If $b = 1$ then as before $a = b$ or $b < a$. Suppose that $b \neq 1$. Then again by Proposition 3.1.2, $b = c^+$ for some $c \in \mathbb{N}$. Since $a \in M$, $a = c$ or $a < c$ or $c < a$. If $a = c$, *then* $a^+ = c^+ = b$. If $a < c$, *then* $a^+ < c^+ = b$, and if $c < a$, *then* $b = c^+ < a^+$. This shows that $a^+ \in M$. By P_5, $M = \mathbb{N}$, and the proof is complete.   ♯

Corollary 3.2.5 '$\leq$' *is a total order in* $\mathbb{N}$.

Proof Follows from the law of trichotomy.   ♯

Theorem 3.2.6 (Well-ordering Property of $\mathbb{N}$) $(\mathbb{N}, \leq)$ *is a well-ordered set.*

Proof We know (Corollary 3.2.5) that $(\mathbb{N}, \leq)$ is a totally ordered set. Thus, it is sufficient to show that every nonempty subset of $\mathbb{N}$ has the least element. Let S be a nonempty subset of $\mathbb{N}$. Let $M = \{a \in \mathbb{N} \mid a \leq x \text{ for all } x \in S\}$. Since $S \neq \emptyset$, *there is an element* $a \in S$. But, then $a^+ \notin M$, for $a^+ \not\leq a$. Thus, $M \neq \mathbb{N}$. Since $1 \leq x$ for all $x \in \mathbb{N}$, $1 \in M$. Now, *there is an element* $b \in M$ such that $b^+ \notin M$, for otherwise, by P_5, $M = \mathbb{N}$. Clearly, $b \leq x$ for all $x \in S$. The proof will be complete if we show that $b \in S$. Suppose that $b \notin S$. Then by the law of trichotomy, $b < x$ for all $x \in S$, and so $b^+ \leq x$ for all $x \in S$. This contradicts the fact that $b^+ \notin M$.   ♯

Corollary 3.2.7 (Second Principle Of Induction) *Let M be a set of natural numbers such that*

(i) $1 \in M$, *and*
(ii) $\{x \in \mathbb{N} \mid x < n\} \subseteq M$ *implies that* $n \in M$.

Then, $M = \mathbb{N}$.

Proof Suppose that $M \neq \mathbb{N}$. Then, $S = \mathbb{N} - M \neq \emptyset$. By the well-ordering property of $\mathbb{N}$, S has the least element n (say). Since $1 \notin S$, $n \neq 1$. Since $n \in S$, $n \notin M$. Let $x \in \mathbb{N}$ and $x < n$. Then $x \notin S = \mathbb{N} - M$, for n is the least element of S. Thus, $x < n$ implies that $x \in M$. This means that $\{x \in \mathbb{N} \mid x < n\} \subseteq M$ whereas $n \notin M$. This is a contradiction to the hypothesis.   ♯

Exercises

3.2.1 Show that $(x^+)^+ \neq x$ *for all* $x \in \mathbb{N}$.

3.2.2 Prove the remaining part of the Theorem 3.1.6.

3.2.3 Let $n, m \in \mathbb{N}$ such that $n \cdot m = 1$. Show that $n = 1$ and $m = 1$.

3.2.4 Show that $m \leq m \cdot n$ *for all* $m, n \in \mathbb{N}$.

3.2.5 Let f be an injective map from A to B. Suppose that A is infinite. Show that B is also infinite.

3.2.6 Show that a subset of a finite set is finite.

3.2.7 Show that every successor set is infinite.

3.2.8 Let f be a surjective map from A to B. Suppose that B is infinite. Then show that A is also infinite.

3.2.9 Let f be an injective map from X to X. Suppose that $a \notin f(X)$. Let g be a map from $\mathbb{N}$ to X such that $g(1) = a$ and $g(n^+) = f(g(n))$. Show that g is injective.

3.2.10 Let X be an infinite set. Show that there is an injective map from $\mathbb{N}$ to X.

3.2.11 Let $n \in \mathbb{N}$. Show that $A_n = \{r \in \mathbb{N} \mid r \leq n\}$ is finite.
Hint. Use induction on n. Observe that there is a bijection from A_n to $A_{n+1} - \{r\}$, $1 \leq r \leq n+1$.

3.2.12*. Let X be an infinite set. Show that there is a bijection from $X \times X$ to X.

3.2.13 Suppose that there is an injective map from $\bigwedge$ to X. Let $\{f_\alpha : A_\alpha \longrightarrow X \mid \alpha \in \bigwedge\}$ be a family of injective maps such that for all $\alpha, \beta \in \bigwedge$, $f_\alpha/(A_\alpha \cap A_\beta) = f_\beta/(A_\alpha \cap A_\beta)$. Show that there is an injective map f from $\bigcup_{\alpha \in \bigwedge} A_\alpha$ to X such that $f/A_\alpha = f_\alpha$ for all α.

3.2.14*. Give a bijective map from $\mathbb{N} \times \mathbb{N}$ to $\mathbb{N}$.

3.2.15 Call a set A to be **countable** if there is a bijective map from $\mathbb{N}$ to A. Show that if A and B are countable then $A \times B$ is also countable.

3.2.16 Show that $P(\mathbb{N})$ is not countable.

3.2.17 Show that finite union of countable sets is countable.

3.2.18*. Show that countable unions of countable sets are countable.

3.3 Integers

We have observed in Sect. 3.1 that an equation $a + x = b$ need not have any solution in $\mathbb{N}$ for $a, b \in \mathbb{N}$. However, if it has a solution, then there is a unique solution. Our aim is to enlarge the system $(\mathbb{N}, +, \cdot)$ so that equations $a + x = b$ in the enlarged system has always a unique solution. The solution of $a + x = b$ will of course depend on the pair (a, b). It is natural temptation to consider the set $X = \mathbb{N} \times \mathbb{N}$ for the purpose. The solutions of $a + x = b$ and $c + x = d$ should be same if and only if $a + d = b + c$. Thus, we should identify the pairs (a, b) and (c, d) whenever $a + d = b + c$. We therefore define a relation $\sim$ on X by

$$(a, b) \sim (c, d) \text{ if and only if } a + d = b + c.$$

Using the property of $+$ in $\mathbb{N}$, we can verify that the relation $\sim$ is an equivalence relation on X. The equivalence class determined by (a, b) is denoted by $\overline{(a, b)}$. Thus,

$$\overline{(a, b)} = \{(c, d) \in \mathbb{N} \times \mathbb{N} \mid (a, b) \sim (c, d)\} = \{(c, d) \in \mathbb{N} \times \mathbb{N} \mid a + d = b + c\}$$

and

$$\overline{(a, b)} = \overline{(c, d)} \text{ if and only if } a + d = b + c.$$

Definition. The equivalence class $\overline{(a, b)}$ is called an **integer**, and the quotient set $X/\sim = \{\overline{(a, b)} \mid (a, b) \subset \mathbb{N} \times \mathbb{N}\}$ is called the **set of integers** and is denoted by $\mathbb{Z}$.

Binary operations in $\mathbb{Z}$.

Proposition 3.3.1 *Suppose that* $\overline{(a, b)} = \overline{(u, v)}$ *and* $\overline{(c, d)} = \overline{(w, x)}$. *Then*

$$\overline{(a + c, b + d)} = \overline{(u + w, v + x)},$$

and

$$\overline{(ac + bd, ad + bc)} = \overline{(uw + vx, ux + vw)}.$$

Proof Under the hypothesis of the proposition, $a + v = b + u$ and $c + x = d + w$. But, then

$$a + c + v + x = b + d + u + w$$

Thus,

$$\overline{(a + c, b + d)} = \overline{(u + w, v + x)}.$$

Similarly,

$$\overline{(ac + bd, ad + bc)} = \overline{(uw + vx, ux + vw)}. \qquad \sharp$$

The above proposition allows to define binary operations $\oplus$ and $\star$ in $\mathbb{Z}$ as follows:

$$\overline{(a, b)} \oplus \overline{(c, d)} = \overline{(a + c, b + d)}$$

and

$$\overline{(a, b)} \star \overline{(c, d)} = \overline{(ac + bd, ad + bc)}$$

Theorem 3.3.2 *The triple $(\mathbb{Z}, \oplus, \star)$ has the following properties:*

(i) $\oplus$ is associative in the sense that

$$(x \oplus y) \oplus z = x \oplus (y \oplus z)$$

for all $x, y, z \in \mathbb{Z}$.

(ii) $\oplus$ is commutative in the sense that

$$x \oplus y = y \oplus x$$

for all $x, y \in \mathbb{Z}$.

(iii) There is a unique element in $0 \in \mathbb{Z}$ such that

$$0 \oplus x = x = x \oplus 0$$

for all $x \in \mathbb{Z}$.

(iv) For all $x \in \mathbb{Z}$, there is a unique element $(-x) \in \mathbb{Z}$ such that

$$x \oplus (-x) = 0 = (-x) \oplus x$$

(v) $\star$ is associative in the sense that

$$(x \star y) \star z = x \star (y \star z)$$

for all $x, y, z \in \mathbb{Z}$.

(vi) ⋆ *is commutative in the sense that*

$$x \star y \; = \; y \star x$$

for all $x, y \in \mathbb{Z}$.

(vii) *There is unique element* $1 \in \mathbb{Z}$ *such that*

$$1 \star x \; = \; x \; = \; x \star 1$$

for all $x \in \mathbb{Z}$.

(viii) ⋆ *distributes over* ⊕ *in the sense that*

$$x \star (y \oplus z) \; = \; (x \star y) \oplus (x \star z)$$

and

$$(x \oplus y) \star z \; = \; (x \star z) \oplus (y \star z)$$

for all $x, y, z \in \mathbb{Z}$.

Proof We prove some of them and leave the rest as exercises.

(iii). Since $a + 1 = 1 + a$, $\overline{(1, 1)} = \overline{(a, a)}$ for all $a \in \mathbb{N}$. Take $0 = \overline{(1, 1)}$. Let $x = \overline{(a, b)} \in \mathbb{Z}$. Then

$$x \oplus 0 \; = \; \overline{(a, b)} \oplus \overline{(1, 1)} \; = \; \overline{(a + 1, b + 1)} \; = \; \overline{(a, b)} \; = \; x,$$

for $a + 1 + b = b + 1 + a$.

Similarly, $0 \oplus x = x$ for all $x \in \mathbb{Z}$.

Suppose that there is an element $\bar{0} \in \mathbb{Z}$ such that

$$\bar{0} \oplus x \; = \; x \; = \; x \oplus \bar{0}$$

for all $x \in \mathbb{Z}$. Then

$$\bar{0} \; = \; \bar{0} \oplus 0 \; = \; 0$$

This proves (iii).

(iv) Let $x = \overline{(a, b)} \in \mathbb{Z}$. Take $-x = \overline{(b, a)}$. Then

$$x \oplus (-x) \; = \; \overline{(a, b)} \oplus \overline{(b, a)} \; = \; \overline{(a + b, b + a)} \; = \; \overline{(1, 1)} \; = \; 0.$$

Similarly, $(-x) \oplus x \; = \; 0$.

Suppose next that there is an element $y \in \mathbb{Z}$ such that $y \oplus x = 0$. Then

$$y = y \oplus 0 = y \oplus (x \oplus (-x)) = (y \oplus x) \oplus (-x) = 0 \oplus (-x) = (-x).$$

This proves (iv).

(vii) Note that $\overline{(1+1, 1)} = \overline{(a+1, a)}$ *for all* $a \in \mathbb{N}$. We denote $\overline{(1+1, 1)}$ also by 1. Then

$$1 \star \overline{(a, b)} = \overline{(a \cdot (1+1) + b \cdot 1, a \cdot 1 + b \cdot (1+1))} = \overline{(a, b)}$$

Further, if $\bar{1}$ also satisfies the same property, then

$$\bar{1} = \bar{1} \star 1 = 1$$

This proves (vii). ♯

The unique element 0 is called the **zero** of $\mathbb{Z}$ or the **additive identity** of $(\mathbb{Z}, \oplus)$. The unique element 1 is called **one** or the **multiplicative identity** of $\mathbb{Z}$. The element $-x$ is called the **additive inverse** of x or the **negative** of x. The element $x \oplus (-y)$ will be denoted by $x - y$.

Corollary 3.3.3 *Cancellation law holds for* $\oplus$ *in the sense that*

$$x \oplus y = x \oplus z \implies that \ y = z,$$

and

$$y \oplus x = z \oplus x \implies that \ y = z$$

Proof Suppose that $x \oplus y = x \oplus z$. Then

$$y = 0 \oplus y = ((-x) \oplus x) \oplus y = (-x) \oplus (x \oplus y) = (-x) \oplus (x \oplus z) =$$
$$((-x) \oplus x) \oplus z = 0 \oplus z = z$$

Similarly, the second part follows. ♯

Corollary 3.3.4 *The equation* $a \oplus X = b$, *where X is unknown has a unique solution in* $\mathbb{Z}$ *for all* $a, b \in \mathbb{Z}$.

Proof Check that $-a \oplus b$ is a solution of the equation. The fact that the solution is unique follows from the cancellation law for $\oplus$. ♯

Corollary 3.3.5 *(i)* $-(-x) = x$

(ii) $-(x \oplus y) = -x - y$

(iii) $x \star 0 = 0 = 0 \star x$ *and*

(iv) $x \star (-y) = -(x \star y) = (-x) \star y$

for all $x, y \in \mathbb{Z}$.

Proof (i) Since $x \oplus (-x) = 0 = -x \oplus x$, $-(-x) = x$ for all $x \in \mathbb{Z}$.

(ii) Using the associativity and commutativity of $\oplus$ in $\mathbb{Z}$,

$$(x \oplus y) \oplus (-x - y) = (x \oplus -x) \oplus (y \oplus (-y)) = 0 \oplus 0 = 0$$

Hence $-(x \oplus y) = -x - y$.

(iii) $(0 \star x) \oplus 0 = 0 \star x = (0 \oplus 0) \star x = (0 \star x) \oplus (0 \star x)$.
By the cancellation law for $\oplus$, $0 = 0 \star x$. Similarly, $x \star 0 = 0$ for all $x \in \mathbb{Z}$.
Similarly, we can prove (iv).   ♯

Corollary 3.3.6 *Let* $x, y \in \mathbb{Z}$. *Then*

$$x \star y = 0 \text{ implies that } x = 0 \text{ or } y = 0.$$

Proof Suppose that $x = \overline{(a, b)} \neq 0$. Then $a \neq b$. By the law of trichotomy in $\mathbb{N}$, $a < b$ or $b < a$. Suppose that $a < b$. Then, *there is an element* $u \in \mathbb{N}$ such that $a + u = b$. Suppose that $\overline{(a, b)} \star \overline{(c, d)} = 0$. Then, $\overline{(ac + bd, ad + bc)} = 0$. This means that $ac + bd = ad + bc$. Substituting $b = a + u$ and using cancellation law in $\mathbb{N}$, we find that $c = d$, and so $\overline{(c, d)} = 0$. Similarly, if $b < a$, we can show that $\overline{(c, d)} = 0$.   ♯

Corollary 3.3.7 *The restricted cancellation law holds for* $\star$ *in* $\mathbb{Z}$ *in the following sense.*

$$\text{If } x \neq 0 \text{ and } x \star y = x \star z, \text{ then } y = z,$$

and

$$\text{if } x \neq 0 \text{ and } y \star x = z \star x, \text{ then } y = z.$$

Proof Suppose that $x \neq 0$ and $x \star y = x \star z$. Then

$$x \star y - x \star z = 0 = x \star (y - z)$$

From the previous corollary, it follows that $y - z = 0$. Hence $y = z$. Similarly, the second part follows.   ♯

Embedding of $\mathbb{N}$ in $\mathbb{Z}$
Define a map f from $\mathbb{N}$ to $\mathbb{Z}$ by $f(n) = \overline{(n + 1, 1)}$. Then, f is injective (verify). Also it is easy to check that

$$f(n + m) = f(n) \oplus f(m)$$

and

$$f(n \cdot m) = f(n) \star f(m)$$

Thus, f is an operation preserving injective map. Such a map is called an **embedding**. Also observe that $f(1) = 1$.

Let $\overline{(a, b)} \in \mathbb{Z}$, $a, b \in \mathbb{N}$. By the law of trichotomy in $\mathbb{N}$, one and only one of the following holds:

(i) $a = b$
(ii) $a < b$
(iii) $b < a$

If $a = b$, then $\overline{(a, b)} = 0 \in \mathbb{Z}$. Suppose that $a < b$. Then, *there is an element* $c \in \mathbb{N}$ such that $a + c = b$. In this case

$$\overline{(a, b)} = \overline{(a, a + c)} = \overline{(1, 1 + c)} = -\overline{(c + 1, 1)} = -f(c),$$

where f is the embedding of $\mathbb{N}$ in to $\mathbb{Z}$ defined above. Thus, in the case (ii) $-\overline{(a, b)} \in f(\mathbb{N})$. Similarly, in case (iii), we find that $\overline{(a, b)} \in f(\mathbb{N})$. It follows that for any member $x \in \mathbb{Z}$, one and only one of the following holds:

(i) $x = 0$
(ii) $x \in f(\mathbb{N})$.
(iii) $-x \in f(\mathbb{N})$.

Since f is an injective map which preserves operations, there is no loss in identifying $f(n)$ by n for all n in $\mathbb{N}$. As such,

$$\mathbb{Z} = \mathbb{N} \bigcup \{0\} \bigcup -\mathbb{N},$$

where $-\mathbb{N} = \{x \in \mathbb{Z} \mid -x \in \mathbb{N}\}$. Also, given any $x \in \mathbb{Z}$, one and only one of the following holds:

(i) $x = 0$
(ii) $x \in \mathbb{N}$
(iii) $-x \in \mathbb{N}$

There is no loss in denoting the operations $\oplus$ and $\star$ by the original $+$ and $\cdot$ respectively.

Order in $\mathbb{Z}$

We define a relation '$<$' called **less than** and a relation '$\leq$' called **less than or equal to** in $\mathbb{Z}$ as follows:

$$< = \{(a, b) \in \mathbb{Z} \times \mathbb{Z} \mid b = a + c \text{ for some } c \in \mathbb{N}\}$$

and

$$\leq\, =\, <\bigcup \triangle,$$

where $\triangle$ is the diagonal relation on $\mathbb{Z}$.

Thus, $a < b$ if *there is an element $c \in \mathbb{N}$ such that $b = a + c$. $a \leq b$ if $a < b$* or $a = b$. In other words, $a < b$ if $-a + b \in \mathbb{N}$, and $a \leq b$ if $-a + b \in \mathbb{N} \bigcup \{0\}$.

Theorem 3.3.8 (Law of Trichotomy in $\mathbb{Z}$) *Let $a, b \in \mathbb{Z}$. Then, one and only one of the following holds:*

(i) $a = b$
(ii) $a < b$
(iii) $b < a$

Proof Consider the element $-a + b$ of $\mathbb{Z}$. From the fact already seen, one and only one of the following holds:

(i) $-a + b = 0$
(ii) $-a + b \in \mathbb{N}$
(iii) $-(-a + b) = -b + a$ *belongs to* $\mathbb{N}$. Evidently, the result follows.   ♯

Remark 3.3.9   (i) The embedding f of $\mathbb{N}$ in to $\mathbb{Z}$ is also order preserving.
   (ii) It follows from the above theorem that $\leq$ is a total order, where as $<$ is anti-symmetric and transitive.
   (iii) $\mathbb{N} = \{x \in \mathbb{Z} \mid 0 < x\}$. Thus, $\mathbb{N}$ is also termed as the set of positive integers.

Absolute Value
The map $|\ |$ from $\mathbb{Z}$ to $\mathbb{N} \bigcup \{0\}$ defined by

$$|a| = \begin{cases} 0 & \text{if } a = o \\ a & \text{if } a \in \mathbb{N} \\ -a & \text{if } -a \in \mathbb{N} \end{cases}$$

is called the **absolute value** on $\mathbb{Z}$. We denote $|\ | \ (a)$ by $|a|$ and call it the **absolute value of a**.

Theorem 3.3.10 *Let $a, b \in \mathbb{Z}$. Then, the following hold.*

(i) $|a| = 0$ *if and only if $a = 0$.*
(ii) $||a|| = |a|$.
(iii) $|a| = |-a|$.
(iv) $a \leq |a|$.
(v) $|ab| = |a||b|$.
(vi) $|a + b| \leq |a| + |b|$.
(vii) $||a| - |b|| \leq |a - b|$.
(viii) $|a| \leq |ab|$ *for all $b \neq 0$.*
(ix) $|a - b| \leq max(a, b)$.

Proof (i), (ii), and (iii) follow from the definition itself.

(iv). If $a \in \mathbb{N} \bigcup \{0\}$, then $a = |a|$. Suppose that $-a \in \mathbb{N}$. Then $|a| = -a$, and since $-a = a + (-a + -a)$, it follows that $a < -a = |a|$. This proves the (iv).

(v). If $a = 0$ or $b = 0$, then both sides of (v) are 0. If $a, b \in \mathbb{N}$, then both sides are ab. If $-a, -b \in \mathbb{N}$, then again both sides are ab. Finally, if $a \in \mathbb{N}$ and $-b \in \mathbb{N}$, then both sides are $-ab$. This proves (v).

(vi). If $a, b \in \mathbb{N} \bigcup \{0\}$, then both sides of (vi) are $a + b$. If $-a, -b \in \mathbb{N}$, then both sides are $-a - b$. Suppose that $a \in \mathbb{N}$ and $-b \in \mathbb{N}$. If $a = -b$, then the left hand side is 0, the right hand side is in $\mathbb{N}$, and so the inequality holds. Suppose that $a \neq -b$. Then $a + b \in \mathbb{N}$ or $-a - b \in \mathbb{N}$. In the first case,

$$|a| + |b| = a - b = a + b - b - b = |a+b| + (-b - b),$$

and hence the inequality holds. Similarly, in the second case also the inequality holds.

(vii). $|a| = |a - b + b| \leq |a - b| + |b|$ (by (vi)).
Hence

$$|a| - |b| \leq |a - b|$$

Similarly, $|b| - |a| \leq |a - b|$. This proves (vii).

(viii) If $a = 0$, then $|a| = 0 |ab|$. Suppose $a \neq 0$, $b \neq 0$. Then $|a|, |b| \in \mathbb{N}$, and also $|ab| = |a||b| \in \mathbb{N}$. The result follows from Exercise 3.2.4.

(ix) Let $a, b \in \mathbb{Z}$. Suppose that $a < b$. Then $|a - b| = b - a < b$, for $b = b - a + a$. ♯

Corollary 3.3.11 *Let $a, b \in \mathbb{Z}$ such that $ab = 1$ or $(ab = -1)$. Then $a = \pm 1$, $b = \pm 1$ $(a = \pm 1, b = \mp 1)$.*

Proof Suppose that $ab = 1$. Then $|a||b| = 1$. From the Exercise 3.2.3, it follows that $|a| = 1 = |b|$. This shows that $a = \pm 1, b = \pm 1$. ♯

Theorem 3.3.12 (Division Algorithm) *Let $a, b \in \mathbb{Z}$ and $b \neq 0$. Then, there exists a unique pair $(q, r) \in \mathbb{Z} \times \mathbb{Z}$ such that*

$$a = bq + r,$$

where $0 \leq r < |b|$.

Proof We first prove the existence of the pair (q, r) with the required property. If $a = bq + r$, where $0 \leq r < |b|$, then $a = (-b)(-q) + r$, where r again has the property $0 \leq r < |b| = |-b|$. Hence, we can assume that $b \in \mathbb{N}$. Further, if $a = 0$, then $0 = b \cdot 0 + 0$, and there is nothing to prove. Suppose that $a \neq 0$. If $a = bq + r$ and $r = 0$, then $-a = b(-q) + 0$. If $0 < r < b$, then $-a = b(-q - 1) + b - r$, where $0 < b - r < b$. Therefore, without any loss, we can assume that $a \in \mathbb{N}$. The proof is by induction on a. If $b = 1$ then $a = 1 \cdot a + 0$, and there is nothing to do. If $1 < b$ and $a = 1$, then $a = b \cdot 0 + a$. Thus, the result is true for $a = 1$. Assume that the result is true for a, and $a = bq + r$, where

$0 \leq r < b$. Then $a + 1 = bq + r + 1$, where $r + 1 \leq b$. If $r + 1 < b$, then there is nothing to do. If $r + 1 = b$, then $a + 1 = b(q + 1) + 0$. Thus, we have proved the existence of a pair (q, r) with the required property.

Now, we prove the uniqueness of the pair (q, r). Suppose that

$$a = bq_1 + r_1 = bq_2 + r_2,$$

where $0 \leq r_1 < |b|$ and $0 \leq r_2 < |b|$ Then

$$b(q_1 - q_2) = r_2 - r_1$$

If $r_1 \neq r_2$, then

$$|b| \leq |b| |q_1 - q_2| < max(r_2, r_1) < |b|$$

This is a contradiction. Hence $r_1 = r_2$. Since $b \neq 0$, $q_1 = q_2$.   ♯

Alternate proof for the existence of a pair (q, r):
Consider the set $X = \{a - bq \mid bq \leq a\} \subseteq \mathbb{N} \bigcup \{0\}$. It is easy to show that $X \neq \emptyset$ (If $b \leq a$, *take* $q = 1$, *and if* $b > a$, *then take* $q = -1$). If $0 \in X$, then $a = bq$ for some q and there is nothing to do. If $0 \notin X$, then X is a nonempty subset of $\mathbb{N}$. By the well-ordering property of $\mathbb{N}$, X has the least element r (say). Then $a = bq + r$, where $r < |b|$, for otherwise $r - |b| = a - bq - |b| \in X$ is a contradiction to the choice of r.   ♯

Remark 3.3.13 The first proof for the existence of the pair (q, r) is algorithmic whereas the alternate proof is the existential proof. The integer q is called the **quotient** and r is called the **remainder** obtained when a is divided by b.

Let $a \in \mathbb{Z}$, $a \neq 0$. Let $b \in \mathbb{Z}$. We say that **a divides b** if there exists a $c \in \mathbb{Z}$ such that $b = ac$. We use the notation a/b to say that a divides b. Since $0 = a \cdot 0$, a divides 0. Again, since $a = 1 \cdot a = -1 \cdot (-a)$, $1/a$ and $-1/a$ for all $a \in \mathbb{Z}$. If b/a *for all* $a \in \mathbb{Z}$, then, in particular, $b/1$. Hence, *there is an element* $c \in \mathbb{Z}$ such that $1 = bc$. It follows from the Corollary 3.3.11 that $b = \pm 1$. This shows that 1 and -1 are the only integers which divide each integer. The elements 1 and -1 are called the **units** of $\mathbb{Z}$. Thus, units divide each integer.

The relation 'divides' is reflexive and transitive, but it is neither symmetric nor antisymmetric (verify).

Suppose that a/b and b/a. Then, *there exist* $c, d \in \mathbb{Z}$ such that $b = ac$ and $a = bd$. But, then

$$b \cdot 1 = b = b \cdot d \cdot c$$

By the restricted cancellation law $cd = 1$. Hence c, d are units. Conversely, if a and b differ by a unit, then a/b and b/a. Thus, a/b and b/a if and only if a and b differ by a unit.

We say that a, $b \in \mathbb{Z}$ are **associate** to each other if a/b and b/a or equivalently $a = ub$, where $u = \pm 1$ is a unit.

We have the relation '$\sim$' called "**is associate to**" on $\mathbb{Z}^* = \mathbb{Z} - \{0\}$ defined by

$$\sim = \{(a, b) \in \mathbb{Z}^* \times \mathbb{Z}^* \mid a/b \text{ and } b/a\}$$

Thus, $a \sim b$ *if and only if* a and b differ by a unit.
The following proposition is immediate.

Proposition 3.3.14 *The relation '$\sim$' is an equivalence relation on $\mathbb{Z}^*$.*   ♯

Given $a \in \mathbb{Z}^*, 1, -1, a, -a$ are divisors of a. They are called the **improper divisors** of a. Other divisors of a are called the **proper divisors** of a.

An integer $p \notin \{0, 1, -1\}$ is called an **irreducible integer** if it has no proper divisors.

Example 3.3.15 $2 = 1 + 1$ is an irreducible integer.

Proof Suppose that $2 = a \cdot b$, a, $b \in \mathbb{Z}$. We have to show that a or b is a unit. Clearly, a and b are nonzero, and so $\mid a \mid$ and $\mid b \mid$ are in $\mathbb{N}$. Further, $2 = \mid a \mid \cdot \mid b \mid$. Suppose that $\mid a \mid \neq 1$ and $\mid b \mid \neq 1$. Then, since 1 is the only element in $\mathbb{N}$ which is not successor of any element in $\mathbb{N}$, $\mid a \mid = n + 1$ and $\mid b \mid = m + 1$ for some n, $m \in \mathbb{N}$. But, then

$$1 + 1 = 2 = (n + 1) \cdot (m + 1) = m + n + n \cdot m + 1$$

By the cancellation law $1 = n + m + n \cdot m$. This means that $n < 1$ which is a contradiction to the fact that 1 is the least element of $\mathbb{N}$.   ♯

3.4   Greatest Common Divisor, Least Common Multiple

Let $a, b \in \mathbb{Z}^*$. An element $d \in \mathbb{Z}$ is called a **greatest common divisor (g.c.d)** or **greatest common factor (g.c.f)** of a and b if

  (i) d/a, d/b, and
 (ii) (d'/a and d'/b) implies that d'/d.

An element m is called a **Least common multiple (l.c.m)** if

  (i) a/m, b/m, and
 (ii) (a/m' and b/m') implies that m/m'.

Proposition 3.4.1 *If d_1 and d_2 are the greatest common divisors of a and b, then $d_1 \sim d_2$. If m_1 and m_2 are least common multiples of a and b, then $m_1 \sim m_2$*

Proof Since d_1 and d_2 are both assumed to be the greatest common divisors, d_1/d_2 and d_2/d_1. Hence $d_1 \sim d_2$. Similarly, any two least common multiples are associates to each other.   ♯

Remark 3.4.2 It follows from the above proposition that a positive (an integer which is in $\mathbb{N}$ is also called a positive integer and if its negative is in N, then it is also called a negative integer) greatest common divisor (least common multiple) is unique and some times this is called the greatest common divisor (least common multiple).

A g.c.d of a and b is usually denoted by (a, b) and a l.c.m is usually denoted by $[a, b]$.

Theorem 3.4.3 *Let m be a least common multiple and d be a greatest common divisor of integers a and b. Then $a \cdot b \sim m \cdot d$.*

Proof Since $a/a \cdot b$ and $b/a \cdot b$, $m/a \cdot b$. Suppose that $a \cdot b = m \cdot u$. Since m is l.c.m of a and b, a/m. Suppose that $m = a \cdot v$ for some $v \in \mathbb{Z}$. Then

$$a \cdot b = a \cdot v \cdot u$$

By the restricted cancellation law, $b = v \cdot u$. This shows that u/b. Similarly, u/a. Hence u/d, and so $a \cdot b/m \cdot d$. Next, since d is a g.c.d of a and b, $a = d \cdot s$ and $b = d \cdot t$ for some integers s and t. But, then $a \cdot b = d \cdot s \cdot t \cdot d$. Now, $a/d \cdot s \cdot t$ and $b/d \cdot s \cdot t$ and hence $m/d \cdot s \cdot t$. Thus, $m \cdot d/a \cdot b$. Hence $a \cdot b \sim m \cdot d$.   ♯

Proposition 3.4.4 *Let a, $b \in \mathbb{Z}^*$. Suppose that $a = b \cdot q + r$, where $q, r \in \mathbb{Z}$. Then d is a g.c.d. of a and b if and only if it is a g.c.d. of b and r.*

Proof Follows from the fact that the set of common divisors of a and b is same as the set of common divisors of b and r.   ♯

Theorem 3.4.5 (Euclidean Algorithm) *Let $a, b \in \mathbb{Z}^*$. Then a greatest common divisor of a and b exists. If d is a greatest common divisor of a and b, then there exist $u, v \in \mathbb{Z}$ such that*

$$d = u \cdot a + v \cdot b$$

Proof We prove it by induction on $\min(\mid a \mid, \mid b \mid)$. If $\min(\mid a \mid, \mid b \mid) = 1$, then $\mid a \mid = \mid b \mid = 1$. In this case, 1 and -1 are the greatest common divisors of $a = \pm 1$ and $b = \pm 1$, and there is nothing to do. Assume that the result is true for all those pairs c, d for which $\min(\mid c \mid, \mid d \mid) < \min(\mid a \mid, \mid b \mid)$. Then, we have to prove the result for a, b. If $\mid a \mid = \mid b \mid$, then $a = \pm b$. In this case, a and $-a$ are the greatest common divisors, and there is nothing to do ($a = 1 \cdot a + 0 \cdot b$ and $-a = -1 \cdot a + 0 \cdot b$). Suppose that $\mid b \mid = min(\mid a \mid, \mid b \mid) < \mid a \mid$. If b/a, then b and $-b$ are the greatest common divisors of a and b, and there is nothing to do. If b does not divide a, then by the division algorithm *there exist $q, r \in \mathbb{Z}$ such that*

$$a = b \cdot q + r,$$

where $0 < r < |b|$. By the induction hypothesis, b and r have a g.c.d, and hence by Proposition 3.4.4, a and b also have a g.c.d. Further, the set of greatest common divisors of a and b are same as those of b and r. Let d be a greatest common divisor of a and b. Then, d is also a greatest common divisor of b and r. By the induction hypothesis, *there exist* $u, v \in \mathbb{Z}$ such that

$$d = u \cdot b + v \cdot r = u \cdot b + v \cdot (a - b \cdot q) = v \cdot a + (u - v \cdot q) \cdot b$$

This completes the proof.   ♮

Remark 3.4.6 The representation of a greatest common divisor d of a and b as $d = u \cdot a + v \cdot b$ is not unique for, $u \cdot a + v \cdot b = (u + x \cdot b) \cdot a + (v - x \cdot a) \cdot b$ for all integers x.

Corollary 3.4.7 *Suppose that a/c and b/c. Then $a \cdot b/c \cdot d$, where d is g.c.d of a and b.*

Proof By the Euclidean algorithm *there exist* $u, v \in \mathbb{Z}$ such that $d = u \cdot a + v \cdot b$. Hence $c \cdot d = u \cdot a \cdot c + v \cdot b \cdot c$. Since b/c and a/c, it follows that $a \cdot b/c \cdot d$.   ♮

Corollary 3.4.8 *Let $a, b \in \mathbb{Z}^*$. Then l.c.m of a and b exists. Suppose that $a = d \cdot u$ and $b = d \cdot v$, where d is g.c.d of a and b. Then greatest common divisor (u, v) of u and v is a unit and $d \cdot u \cdot v$ is a least common multiple of a and b.*

Proof If d' is a g.c.d of u and v, then $d \cdot d'/a$ and $d \cdot d'/b$. But, then, $d \cdot d'/d$, and so d' is a unit. Now, we show that $m = d \cdot u \cdot v$ is a least common multiple of a and b. Clearly, $a = d \cdot u$ and $b = d \cdot v$ divide m. Suppose that a/m' and b/m'. Then $m' = a \cdot k$ and $m' = b \cdot l$ for some $k, l \in \mathbb{Z}$. Thus, $m' = d \cdot u \cdot k = d \cdot v \cdot l$. By the restricted cancellation law $u \cdot k = v \cdot l$. Now u and v both divide $u \cdot k = v \cdot l$. Since $(u, v) \sim 1$, it follows from the above corollary that $u \cdot v$ divides $u \cdot k$. Hence, $m = duv$ divides $d \cdot u \cdot k = m'$. This proves that m is l.c.m of a and b.   ♮

Illustration of the Euclidean algorithm
Now, we describe and illustrate the Euclidean algorithm to find the positive greatest common divisor d of a pair of integers a, b and also a pair of integers u, v such that

$$d = u \cdot a + v \cdot b$$

Let a and b be nonzero integers. By the division algorithm, we can find integers q and r such that
$$a = b \cdot q + r,$$

where $0 \leq r < |b|$. If $r = 0$, then $(a, b) \sim b = 0 \cdot a + 1 \cdot b$. Suppose that $r \neq 0$. Then $(a, b) \sim (b, r)$. By the division algorithm we can find q_1 and r_1 such that

$$b = q_1 \cdot r + r_1.$$

where $0 \le r_1 < |r|$. If $r_1 = 0$, then

$$(a, b) \sim (b, r) \sim r = 1 \cdot a + (-q) \cdot b.$$

Suppose that $r_1 \neq 0$. Then

$$(a, b) \sim (b, r) \sim (r, r_1)$$

By the division algorithm, again we can find q_2 and r_2 such that

$$r = q_2 \cdot r_1 + r_2,$$

where $0 \le r_2 < r_1 < r < |b|$. If $r_2 = 0$, then,

$$(a, b) \sim (b, r) \sim (r, r_1) \sim r_1 = (-q_1) \cdot r + b = (-q_1) \cdot a + (q \cdot q_1 + 1) \cdot b.$$

If $r_2 \neq 0$, proceed further. This process stops after finitely many steps giving us a greatest common divisor d and integers u and v such that

$$d = u \cdot a + v \cdot b$$

As an example, we find g.c.d of 578 and 250 and also find integers u and v such that $578u + 250v = (578, 250)$. Now,

$$578 = 250 \cdot 2 + 78$$
$$250 = 78 \cdot 3 + 16$$
$$78 = 16 \cdot 4 + 14$$
$$16 = 14 \cdot 1 + 2$$
$$14 = 2 \cdot 7 + 0$$

Thus, $(578, 250) \sim 2$. Also,

$$
\begin{aligned}
2 &= 16 - 1 \cdot 14 \\
&= 16 - 1 \cdot (78 - 4 \cdot 16) \\
&= -1 \cdot 78 + 5 \cdot 16 \\
&= -1 \cdot 78 + 5 \cdot (250 - 3 \cdot 78) \\
&= 5 \cdot 250 - 16 \cdot 78 \\
&= 5 \cdot 250 - 16 \cdot (578 - 2 \cdot 250) \\
&= -16 \cdot 578 + 37 \cdot 250.
\end{aligned}
$$

From now onward, the notation of multiplication $\cdot$ will be omitted unless there is any confusion. Thus, $a \cdot b$ will usually be written as ab.

Proposition 3.4.9 *Let a, b and c be nonzero integers. Then the following hold.*

  (i) $(a, ab) \sim a$.
  (ii) $(a, b) \sim (b, a)$.

(iii) $((a, b), c) \sim (a, (b, c))$.

(iv) $(ca, cb) \sim c(a, b)$.

Proof (i), (ii), and (iii) are simple and left as exercises. For the (iv), let $d = (a, b)$. Since d/a and d/b, cd/ca and cd/cb. Hence $cd/(ca, cb)$. Further, since c/ca and c/cb, $c/(ca, cb)$. Suppose that $(ca, cb) \sim cu$. Then cu/ca and cu/cb. But, then u/a and u/b, and so u/d. This shows that $cu \sim (ca, cb)$ divides cd. ♮

Corollary 3.4.10 *If* $(a, c) \sim 1$ *and* $(b, c) \sim 1$, *then* $(ab, c) \sim 1$.

Proof By the Proposition 3.4.9, we have

$$(ab, c) \sim (ab, (bc, c)) \sim ((ab, bc), c) \sim ((a, c)b, c) \sim (b, c) \sim 1$$

♮

Remark 3.4.11 If p is an irreducible integer, then $(p, a) \sim 1$ or $(p, a) \sim p$ according as p does not divides a or p divides a.

A pair of integers a and b are said to be **co-prime** if $(a, b) \sim 1$.

Theorem 3.4.12 *Let* $p \neq \pm 1$ *be an integer. Then the following two conditions are equivalent.*

(i) p *is an irreducible integer.*

(ii) p/ab *implies that* p/a *or* p/b.

Proof Let p be an irreducible integer. Suppose that p does not divide a and it also does not divide b. Then $(p, a) \sim 1$ and $(p, b) \sim 1$. By Corollary 3.4.10, $(p, ab) \sim ((p, pa), ab) \sim (p, (pa, ab)) \sim (p, (p, b)a) \sim (p, a) \sim 1$. Hence, p does not divide ab. Thus, (i) implies (ii).

Assume (ii). Suppose that $p = ab$. Then p/ab. But, then p/a or p/b. If $a = pc$, then $p = pcb$. By the restricted cancellation law $cb = 1$. But, then $b = \pm 1$ and $c = \pm 1$. Thus, $a = \pm p$. This shows that p has no proper divisors, and so it is irreducible. ♮

An irreducible integer is also called a **prime** integer.

Corollary 3.4.13 *Let* p *be a prime integer such that* $p/a_1 a_2 \cdots a_r$. *Then* p/a_i *for some i.*

Proof Use induction on r. ♮

Theorem 3.4.14 (Fundamental Theorem of Arithmetic) *Every nonzero nonunit integer a can be written as a finite product of irreducible integers. Further, the representation of a as product of irreducible integers is unique in the sense that if*

$$a = p_1 p_2 \cdots p_r = p_1' p_2' \cdots p_s',$$

where p_i and p'_j are irreducible integers, then

   (i) $r = s$,
      and
   *(ii) there is a bijective correspondence f from 1, 2, ..., r to itself such that $p_i \sim p'_{f(i)}$
       for all i. In other words, $p_i \sim p'_i$ after some rearrangement.*

Proof We first prove that every nonzero nonunit integer can be written as product
of irreducible integers. The proof is by induction on $| a |$. If $| a | = 1$, then a is a
unit, and there is nothing to do. Assume that the result is true for all those integers
whose absolute values are less than n. Let a be an integer such that $| a | = n$. If a is
irreducible, there is nothing to do. If not, then $a = bc$, where b and c are nonunits.
Clearly, $| b | < | a |$ and $| c | < | a |$. By the induction hypothesis, b and c are
products of irreducible elements. Hence, a is also product of irreducible elements.

Uniqueness. Suppose that

$$p = p_1 p_2 \cdots p_r = p'_1 p'_2 \cdots p'_s,$$

where p_i and p'_j are irreducible integers. Then $p_1 / p'_1 p'_2 \cdots p'_s$. It follows from the
above corollary that p_1 / p'_j for some j. Rearranging the factors, we may assume that
p_1 / p'_1. Since p'_1 is irreducible and p_1 is a nonunit, $p_1 \sim p'_1$. Hence

$$p_1 p_2 \cdots p_r = p_1 u p'_2 \cdots p'_s,$$

where u is a unit. By the restricted cancellation law,

$$p_2 p_3 \cdots p_r = p''_2 p'_3 \cdots p'_s.$$

From the previous argument, we may assume that $p_2 \sim p''_2 \sim p'_2$. Again canceling
p_2 and p''_2, we find that $p_3 \ldots p_r \sim p'_3 \ldots p'_s$. Proceeding this way, we find that
$p_1, p_2, \ldots, p_r$ and $p'_1, p'_2, \ldots, p'_s$ both will exhaust simultaneously, for otherwise
we shall arrive at a product of irreducible integers equal to ± 1. This, however, is
impossible. Hence $r = s$, and after some rearrangement $p_i \sim p'_i$.   ♯

Proposition 3.4.15 *The set of positive primes is infinite.*

Proof Let $p_1, p_2, \ldots, p_n$ be first n primes in ascending order. Then, $a = p_1 p_2 \cdots \cdot$
$p_n + 1$ is a nonunit, and so from the fundamental theorem of arithmetic, there is a
prime p which divides a. Clearly, $p \neq p_i$ for all i. Thus, the set of positive primes
contains more than n elements for every $n \in \mathbb{N}$. This shows that the set of positive
primes is infinite.   ♯

Positive multiplicative integral powers of an integer
We define a^n by induction for all $a \in \mathbb{Z}$ and $n \in \mathbb{N} \bigcup \{0\}$.
Define $a^0 = 1$, *and* $a^1 = a$. Assuming that a^n has already been defined, define
$a^{n+1} = a^n \cdot a$. It is easy to prove (by induction) the following law of exponents:

(i) $a^{n+m} = a^n \cdot a^m$

(ii) $(a^n)^m = a^{n \cdot m}$

Exercises

3.4.1 Show that $\overline{(5, 6)} = \overline{(7, 8)}$.

3.4.2 Show that the subtraction is not associative.

3.4.3 Show that in $\mathbb{Z}$, $x \cdot y = 0$ *implies that* $x = 0$ *or* $y = 0$.

3.4.4 Show that $-(x - y) = (y - x)$.

3.4.5 Show that $(-x) \cdot (-y) = x \cdot y$.

3.4.6 Show that $(\mathbb{Z}, \leq)$ is a totally ordered set in which every nonempty subset bounded from above (below) has largest (least) element. Deduce that it is an order complete set. Show also that it is not a well-ordered set.

3.4.7 Show that $a < b$ *if and only if* $a + c < b + c$ *for all* $c \in \mathbb{Z}$.

3.4.8 Show that $a \leq b$ *if and only if* $-b \leq -a$.

3.4.9 Show that $\mid a - b \mid \leq max(\mid a \mid, \mid b \mid)$, whenever $a, b \in \mathbb{N}$.

3.4.10 Show that divisibility is a reflexive as well as a transitive relation on $\mathbb{Z}$.

3.4.11 Show that $3 = 1 + 1 + 1$ is a prime integer.

3.4.12 Show that if we divide an integer by 2, then the remainder is either 0 or it is 1. Deduce that an integer is of the form $2n$ or of the form $2m + 1$ but not both. An integer is called **even** if it is of the form $2n$ and it is called **odd** if it is of the form $2n + 1$.

3.4.13 Show that $\mathbb{Z}$ is countable in the sense that there is a bijective map from $\mathbb{Z}$ to $\mathbb{N}$. **Hint.** Show that the map f from $\mathbb{N}$ to $\mathbb{Z}$ given by $f(2n) = n$, and $f(2n+1) = -n$ is bijective.

3.4.14 Suppose that m/a, n/a and $(m, n) \sim 1$. Show that mn/a.

3.4.15 Let a and b be nonzero nonunit integers. Show that there exist distinct primes $p_1, p_2, \ldots, p_n$ and nonnegative integers $\alpha_1, \alpha_2, \ldots, \alpha_n, \beta_1, \beta_2, \ldots, \beta_n$ such that

$$a = p_1^{\alpha_1} p_2^{\alpha_2} \cdots p_n^{\alpha_n}$$

and

$$b = p_1^{\beta_1} p_2^{\beta_2} \cdots p_n^{\beta_n}$$

Further, show that

$$(a, b) \sim p_1^{\gamma_1} p_2^{\gamma_2} \cdots p_n^{\gamma_n}$$

and

$$[a, b] \sim p_1^{\delta_1} p_2^{\delta_2} \cdots p_n^{\delta_n},$$

where $\gamma_i = min(\alpha_i, \beta_i)$ and $\delta_i = max(\alpha_i, \beta_i)$.

3.4.16 Let $a_1, a_2, \ldots, a_n$ be integers such that

$$a_1^2 + a_2^2 + \cdots + a_n^2 = 0.$$

Show that each $a_i = 0$.

3.4.17 Show that $| a |^2 = a^2$

3.4.18 Show that $(a, b) \sim 1$ *implies that* $(a^n, b) \sim 1$.

3.4.19 Suppose that $(a, b) \sim 1$. Show that a/bc implies that a/c.

3.4.20 Show that the set of all odd integers is in bijective correspondence with the set $\mathbb{Z}$ of all integers.

3.4.21 Find g.c.d of 238 and 55, and also integers u and v such that $238u + 55v = (238,55)$. Express it in two different ways as $238u + 55v$.

3.4.22 Let a, b, c be nonzero nonunit integers. Show that

$$(a, [b, c]) = [(a, b), (a, c)]$$

3.4.23 Let $n \in \mathbb{N}$. Show that the equation

$$X^2 - Y^2 = n$$

has a solution in $\mathbb{N}$ if and only if whenever n is divisible by 2, it is divisible by 4. When can the solution be unique?

3.4.24 Find all possible integral solutions of the equation

$$p^2 X^n + a_1 X^{n-1} + \cdots + a_{n-1} X + p^2 = 0,$$

where $a_1, \ldots, a_{n-1}$ are integers, and p is a positive prime. Determine the conditions on the coefficients for a possible solution.

3.4.25 Let n be a nonzero nonunit positive integer which is not a prime(such an integer is called a **composite** integer). Show that there is a prime p such that $p^2 < n$ and p/n.

3.4.26 Suppose that $(m, n) \sim 1$ but $(m+n, m-n) \not\sim 1$. Show that $(m+n, m-n) \sim 2$.

3.4.27 Suppose that m and n are co-prime and mn is a square. Show that m and n are both squares.

3.4.28 Establish the following identities using induction on n.

(i) $x^n - y^n = (x - y)(x^{n-1} + x^{n-2}y + \cdots + y^{n-1})$, where $n \in \mathbb{N}$
(ii) If n is odd, then, $x^n + y^n = (x + y)(x^{n-1} - x^{n-2}y + x^{n-3} - \cdots + y^{n-1})$.

3.4.29 Use the above identities to show that if $a^n - 1$ is prime, then $a = 2$ and n is a prime. The primes of the forms $2^p - 1$ are called the **Mersenne primes**. It is not known, if there are infinitely many Mersenne primes.

3.4.30 Suppose that $a^n + 1$ is an odd prime. Show that a is even and n is a power of 2. The prime numbers of the forms $2^{2^n} + 1$ are called the **Fermat primes**. It is also not known if there are infinitely many Fermat primes.

3.4.31 Suppose that n is odd. Show that $8/n^2 - 1$. Suppose further that n is co-prime to 3. Show that $6/n^2 - 1$.

3.4.32 Show that $30/n^5 - n$ for all $n \in \mathbb{N}$, and for all odd number n, $120/n^5 - n$.

3.4.33 Let a, b, c be set of pairwise co-prime integers such that $a^2 + b^2 = c^2$. Suppose that a is even. Show that there is a pair of co-prime integers u, v such that $a = 2uv, b = v^2 - u^2$, and $c = v^2 + u^2$.

3.5 Linear Congruence, Residue Classes

Let m be a fixed positive integer. Let $a, b \in \mathbb{Z}$. We say that a is **congruent to b modulo m**, if m divides $a - b$. We use the notation

$$a \equiv b(mod\ m)$$

to say that a is congruent to b modulo m. Consider the relation R on $\mathbb{Z}$ given by

$$R = \{(a, b) \in \mathbb{Z} \times \mathbb{Z} \mid a \equiv b(mod\ m)\}.$$

Thus, $(a, b) \in R$ *if and only if* $m/a - b$. It is easy to verify that R is an equivalence relation on $\mathbb{Z}$. The equivalence class R_a determined by a is denoted by $\bar{a}$. Thus,

$$\bar{a} = \{b \in \mathbb{Z} \mid m/a - b\}$$

From the properties of equivalence classes, we have the following:

(i) $a \in \bar{a}$ for all $a \in \mathbb{Z}$.
(ii) $\bar{a} = \bar{b}$ if and only if $m/a - b$.
(iii) $\bar{a} \neq \bar{b}$ if and only if $\bar{a} \cap \bar{b} = \emptyset$.

To say that $m/a - b$ is to say that the remainder obtained when a is divided by m is the same as the remainder obtained when b is divided by m. More explicitly, $\bar{a} = \bar{r}$, where r is the remainder obtained when we divide a by m. Further, if $0 \leq r_1, r_2 < m$, then $\bar{r_1} = \bar{r_2}$ if and only if $r_1 = r_2$. Let $\mathbb{Z}_m$ denote the quotient set $\mathbb{Z}/R$. Then, it follows that $\mathbb{Z}_m$ contains m elements. Indeed,

$$\mathbb{Z}_m = \{\bar{a} \mid a \in \mathbb{Z}\} = \{\bar{0}, \bar{1}, \bar{2}, \cdots, \overline{m-1}\}$$

The set $\mathbb{Z}_m$ is called the set of **residue classes** modulo m.

Proposition 3.5.1 *There are binary operations $\oplus$ and $\star$ on $\mathbb{Z}_m$ given by*

$$\bar{a} \oplus \bar{b} = \overline{a+b}$$

and

$$\bar{a} \star \bar{b} = \overline{a \cdot b},$$

where $a, b \in \mathbb{Z}$.

Proof Suppose that $\bar{a} = \bar{a'}$ and $\bar{b} = \bar{b'}$. Then $m/a - a'$ and $m/b - b'$. But, then $m/(a+b) - (a'+b')$. Hence

$$\bar{a} \oplus \bar{b} = \overline{a+b} = \overline{a'+b'} = \bar{a'} \oplus \bar{b'}$$

This shows that $\oplus$ is indeed a binary operation. Further, $ab - a'b' = ab - a'b + a'b - a'b' = (a-a')b + a'(b-b')$ is divisible by m. This shows that $\bar{a} \star \bar{b} = \bar{a'} \star \bar{b'}$.   ♯

The proof of the following proposition is straightforward and is left as an exercise.

Theorem 3.5.2 *The triple $(\mathbb{Z}_m, \oplus, \star)$ satisfies the following properties.*

(i) $(\bar{a} \oplus \bar{b}) \oplus \bar{c} = \bar{a} \oplus (\bar{b} \oplus \bar{c})$ *for all $a, b, c \in \mathbb{Z}$. Thus, $\oplus$ is associative.*
(ii) $\bar{a} \oplus \bar{b} = \bar{b} \oplus \bar{a}$ *for all $a, b \in \mathbb{Z}$. Thus, $\oplus$ is commutative.*
(iii) $\bar{0} \oplus \bar{a} = \bar{a} = \bar{a} \oplus \bar{0}$ *for all a in $\mathbb{Z}$. Thus, $\bar{0}$ is the identity for $\oplus$.*
(iv) $\bar{a} \oplus (\overline{-a}) = \bar{0} = (\overline{-a}) \oplus \bar{a}$ *for all $a \in \mathbb{Z}$. Thus, every element $\bar{a} \in \mathbb{Z}_m$ has inverse $\overline{-a}$ in $\mathbb{Z}_m$ with respect to $\oplus$.*
(v) $\bar{a} \oplus \bar{b} = \bar{a} \oplus \bar{c}$ *implies that $\bar{b} = \bar{c}$. Thus, the cancellation law holds for $\oplus$ in $\mathbb{Z}_m$.*
(vi) $(\bar{a} \star \bar{b}) \star \bar{c} = \bar{a} \star (\bar{b} \star \bar{c})$ *for all $a, b, c \in \mathbb{Z}$. Thus, $\star$ is associative operation.*

(vii) $\bar{a} \star \bar{b} = \bar{b} \star \bar{a}$ *for all* $a, b \in \mathbb{Z}$. *Thus, $\star$ is commutative.*
(viii) $\bar{a} \star (\bar{b} \oplus \bar{c}) = (\bar{a} \star \bar{b}) \oplus (\bar{a} \star \bar{c})$ *for all* $a, b, c \in \mathbb{Z}$. *Thus, $\star$ distributes over $\oplus$*
  from left.
  (ix) $(\bar{a} \oplus \bar{b}) \star \bar{c} = (\bar{a} \star \bar{c}) \oplus (\bar{b} \star \bar{c})$ *for all* $a, b, c \in \mathbb{Z}$. *Thus, $\star$ distributes over $\oplus$*
  from right also.
   (x) $\bar{1} \star \bar{a} = \bar{a} = \bar{a} \star \bar{1}$ *for all* $a \in \mathbb{Z}$. *Thus, $\bar{1}$ is the identity for $\star$.*   ♯

An equation

$$aX \equiv b(mod\ m),$$

where $a, b \in \mathbb{Z}$ and X is unknown is called a **linear congruence**. This equation can be read in $\mathbb{Z}_m$ as

$$\bar{a} \star \bar{X} = \bar{b}$$

An integer c is said to be a solution of the above congruence if

$$ac \equiv b(mod\ m),$$

or equivalently $m/ac - b$.

Remark 3.5.3 A linear congruence $aX \equiv b(mod\ m)$ need not have any solution. For example $2x \equiv 1(mod\ 4)$ has no solution in $\mathbb{Z}$ *(for $2c - 1$ is never divisible by 4)*.

The following is a necessary and sufficient condition for a linear congruence to have a solution.

Theorem 3.5.4 *A linear congruence $aX \equiv b(mod\ m)$ has a solution in $\mathbb{Z}$ if and only if (a, m) divides b. (Equivalently an equation $\bar{a}X = \bar{b}$ in $\mathbb{Z}_m$ has a solution in $\mathbb{Z}_m$ if and only if (a, m) divides b.). Further, if c is a solution of $aX \equiv b(mod\ m)$, then (c, m) divides b.*

Proof Suppose that $aX \equiv b(mod\ m)$ has a solution $c \in \mathbb{Z}$. Then $ac \equiv b(mod\ m)$, and so m divides $ac - b$. Suppose that $ac - b = qm$, or equivalently $ac - qm = b$ for some q. Then, any common divisor of a and m divides b, in particular (a, m) divides b.

Conversely, suppose that (a, m) divides b. Let $d = (a, m)$. Then, there is a $q \in \mathbb{Z}$ such that $b = qd$. By the Euclidean algorithm, *there exist $u, v \in \mathbb{Z}$ such that*

$$ua + vm = d$$

Thus,

$$b = qd = qua + qvm$$

But, then $a(qu) \equiv b(mod\ m)$. Hence, qu is a solution of the given congruence.

Further, let c be a solution of $aX \equiv b(mod\ m)$, then a is a solution of the congruence $cX \equiv b(mod\ m)$. From what we have just proved, it follows that $(c, m)/b$. ♯

Corollary 3.5.5 $aX \equiv 1(mod\ m)$ *has a solution if and only if a and m are co-prime (i.e.$(a, m) \sim 1$). Further, if c is a solution, then c and m are also co-prime. In particular, for $\bar{a} \in Z_m$, there is a $\bar{b} \in Z_m$ such that $\bar{a} \star \bar{b} = \bar{1}$ if and only if $(a, m) \sim 1$.*

Proof $(a, m)/1$ *if and only if* $(a, m) \sim 1$. The result follows from the above theorem. ♯

Algorithm to find solutions of a linear congruence
Two solutions X_1 and X_2 are said to be congruent if $X_1 \equiv X_2(mod\ m)$. The solutions X_1 and X_2 are said to be incongruent, otherwise. The purpose is to give an algorithm to find all incongruent solutions of the equation $aX \equiv b(mod\ m)$. If (a, m) does not divide b, then there is no solution of the congruence.

   Suppose that $(a, m)/b$. Let d be the positive greatest common divisor of a and m. Then d/a, d/m and also d/b. Suppose that $a = du$, $b = dv$ and $m = dw$. Since d is a g.c.d of a and m, $(u, w) \sim 1$. Further, then m divides $ac - m$ if and only if w divides $uc - v$. Thus, c is a solution of $aX \equiv b(mod\ m)$ if and only if it is a solution $uX \equiv v(mod\ w)$. More explicitly, the solutions of $aX \equiv b(mod\ m)$ are same as those of $uX \equiv v(mod\ w)$.

Theorem 3.5.6 *If $(u, w) \sim 1$, then the linear congruence*

$$uX \equiv v(mod\ w)$$

has a unique incongruent solution. In other words, there is a unique c, $0 < c < w$ such that $uc \equiv v(mod\ w)$.

Proof By the Euclidean algorithm, we can find integers y and z such that $uy + wz = 1$. But, then $uyv + wzv = v$. Hence $u(yv) \equiv v(mod\ w)$. Let c be the remainder obtained(division algorithm) when we divide yv by w. Then $0 < c < w$ and $uc \equiv v(mod\ w)$. Let c_1, c_2, $0 < c_1 < w$ and $0 < c_2 < w$ be such that $uc_1 \equiv v(mod\ w)$ and $uc_2 \equiv v(mod\ w)$. Then $uc_1 \equiv uc_2(mod\ w)$, and so $w/uc_1 - uc_2$. Since $(u, w) \sim 1$, $w/c_1 - c_2$. This shows that $c_1 = c_2$. ♯

Theorem 3.5.7 *Let d be a positive g.c.d of a and m which divides b. Let $a = du$, $b = dv$, and $m = dw$. Let c be the unique solution of the congruence $uX \equiv v(mod\ w)$, where $0 < c < w$. Then c, $c + w, c + 2w, \cdots, c + (d - 1)w$ are precisely the least positive incongruent solutions of $aX \equiv b(mod\ m)$. In particular, there are d incongruent solutions of $aX \equiv b(mod\ m)$. Any solution of the given congruence is of the form $c + iw + qm$ for some i, q, where $0 < i < w$ and $q \in \mathbb{Z}$.*

Proof Clearly, c, $c + w$, $c + 2w, \cdots$, $c + (d - 1)w$ are solutions of $uX \equiv v(mod\ w)$, and so also of $aX \equiv b(mod\ m)$ which are pairwise incongruent modulo

m. It is sufficient to show that any solution of $aX \equiv b(mod\ m)$ is congruent to one of the above solutions. Let l be a solution of $aX \equiv b(mod\ m)$. Then, it is also a solution of $uX \equiv v(mod\ w)$. From the previous theorem, $l \equiv c(mod\ w)$. Suppose that $l - c = kw$. By the division algorithm, $k = qd + r$, where $0 \leq r \leq d - 1$. Thus, $l - c = rw + qm$. This shows that $l \equiv (c + rw)(mod\ m)$.   ♯

We illustrate the above algorithm by finding the least positive incongruent solutions of the congruence

$$51x \equiv 12(mod\ 87)$$

The positive greatest common divisor(use Euclidean algorithm) of 51 and 87 is 3 which divides 12. Thus, the solutions of $51X \equiv 12(mod\ 87)$ are same as those of $17X \equiv 4(mod\ 29)$. Further if c is the least positive incongruent solution of $17X \equiv 4(mod\ 29)$, then c, $c + 29$, $c + 58$ is the complete list of least positive incongruent solutions of $51X \equiv 12(mod\ 87)$. Since $(17, 29) \sim 1$, using Euclidean algorithm we find integers 12 and -7 such that

$$1 = 12 \times 17 - 7 \times 29$$

Hence $4 = 48 \times 17 - 28 \times 29$. This shows that 48 is a solution of $17X \equiv 4(mod\ 29)$. The least positive solution is obtained by dividing 48 by 29 and taking the remainder. Thus, 19 is the least positive solution of $17X \equiv 4(mod\ 29)$. Therefore, The complete list of least positive incongruent solutions of $51X \equiv 12(mod\ 87)$ is $\{19, 48, 67\}$.

Linear Diophantine Equations
An equation

$$a_1X_1 + a_2X_2 + \cdots + a_rX_r = b,$$

where $a_1, a_2, \cdots, a_r, b$ are integers and $X_1, X_2, \cdots, X_r$ are unknowns, is called a **linear diophantine equation** in r variables. Solutions of this equation are to be determined (if they exist) from the set of integers. For simplicity, we consider linear Diophantine equations

$$aX + bY = c$$

in two variables only. We are interested in solving this equation in $\mathbb{Z}$. Geometrically, this means finding out lattice points (points in $\mathbb{Z} \times \mathbb{Z}$) on the straight line $aX + bY = c$.

Solutions of the equation.

$$aX + bY = c \tag{3.5.1}$$

If the above Eq. (3.5.1) has a solution, then g.c.d of a and b divides c. Let d be the positive greatest common divisor of a and b. Suppose that $a = du$, $b = dv$, $c = dw$. Then the solutions of (3.5.1) are same as those of

$$uX + vY = w \tag{3.5.2}$$

Clearly, $(u, v) \sim 1$. By the Euclidean algorithm, we can find integers m, n such that

$$um + vn = 1$$

But, then

$$umw + vnw = w$$

Thus, $x_0 = mw$ and $y_0 = nw$ give a solution of (3.5.2), and so of (3.5.1). Further, since x_0, y_0 is a solution of (3.5.2), for any integer t, $x_0 + tv$, $y_0 - tu$ also constitutes a solution of (3.5.2) (check it). Conversely, if x', y' also constitutes a solution of (3.5.1) and so of (3.5.2), then

$$ux_0 + vy_0 = w = ux' + vy',$$

and so

$$u(x' - x_0) + v(y' - y_0) = 0$$

Since $(u, v) \sim 1$, $u/y' - y_0$ and $v/x' - x_0$. Suppose that $x' - x_0 = tv$. Then $utv + v(y' - y_0) = 0$, and so $y' - y_0 = -tu$. Thus, $x' = x_0 + tv$ and $y' = y_0 - tu$ for some integer t. It follows that the set

$$\{(x_0 + t\frac{b}{d}, y_0 - t\frac{a}{d}) \mid t \in Z\}$$

is precisely the set of all solutions of the equation (3.5.1).

To illustrate the above algorithm, we find all solutions of the equation

$$51X + 87Y = 12$$

The positive greatest common divisor of 51 and 87 is 3 which divides 12. Thus, it has a solution and its solutions are same as those of

$$17X + 29Y = 4$$

By Euclidean algorithm, we find that

$$17 \times 48 + (-28) \times 29 = 4$$

This gives us a solution $(48, -28)$ of the given equation. Thus, the set $\{(48 + 29t, -28 - 17t) \mid t \in Z\}$ is precisely the set of all solutions of the above equation.

Exercises

3.5.1 Let p be a prime which does not divide a. Show that there is an integer b such that $ab \equiv 1 (mod\ p)$.

3.5.2 Suppose that $(a, m) \sim 1$. Let $b \in \mathbb{Z}$. Show that there is an integer c such that $ac \equiv b (mod\ m)$.

3.5.3 Show that the congruence $25X \equiv 6 (mod\ 30)$ has no solution in $\mathbb{Z}$.

3.5.4 Find the set of least positive incongruent solutions of $22X \equiv 4 (mod\ 18)$.

3.5.5 Find the set of least positive incongruent solutions of $12X \equiv 15 (mod\ 9)$.

3.5.6 Find out solutions of the congruence $3X + 20 \equiv 17 (mod\ 15)$.

3.5.7 Find out the solutions of $X^2 \equiv 1 (mod\ p)$, where p is a prime number.

3.5.8 Let $m \in \mathbb{N}$. Let $\{a_1, a_2, \cdots, a_m\}$ be a set of integers such that every integer is congruent to some a_i modulo m. Let $l \in \mathbb{Z}$ be such that $(l, m) \sim 1$. Show that every integer is congruent to some la_i.

3.5.9 Consider the linear congruence

$$aX + bY \equiv c (mod\ m)$$

Show that it has a solution if and only if (a, b, m) divides c. Call two solutions x_1, y_1 and x_2, y_2 congruent if $x_2 \equiv x_1 (mod\ m)$ and $y_2 \equiv y_1 (mod\ m)$. Find an algorithm to find the least positive incongruent solutions of the above linear congruence. Show that there are dm least positive incongruent solutions of the above congruence, where d is the positive g.c.d of a, b, m.

3.5.10 Generalize the above exercise to the congruence in n variables. Show that there are $m^{n-1}d$ incongruent solutions, where d is positive g.c.d of $a_1, a_2, \cdots, a_n$ which divides b.

3.5.11 Determine least positive incongruent solutions of $3X + 7Y \equiv 8 (mod\ 15)$.

3.5.12 Find a general solution of the linear Diophantine equation $641X + 372Y = 1254$.

3.5.13* Let $a, b \in \mathbb{N}$ and $(a, b) \sim 1$. Let r be the smallest nonnegative integer such that $aX + bY = n$ is solvable in $\mathbb{N}$ for all $n \geq r$. Show that $r = (a - 1)(b - 1)$. Show further that for half of the nonnegative integers $< r$, the equation will have a solution in $\mathbb{N}$ and for other half not. (This is a particular case of the Frobenius problem in two variables.)
Hint. Look at the lines parallel to $aX + bY = 0$ and see when it always passes through a lattice point in the first quadrant.

3.6   Rational Numbers

This is a brief section in which we introduce the rational number system $\mathbb{Q}$ and their properties through a set of graded exercises.

The equation $aX = b$, where $a, b \in \mathbb{Z}$ need not have any solution in $\mathbb{Z}$. For example, $2X = 1$ has no solution in $\mathbb{Z}$ (prove it). The purpose of this section is to enlarge the system of integers to a bigger system $\mathbb{Q}$ of rational numbers in which we can solve these equations whenever $a \neq 0$, and without loosing any other algebraic properties of $\mathbb{Z}$.

The verification of most of the claims in this section is simple, straightforward (the proofs are similar to those in the introduction of integers from natural numbers) and will be left as exercises.

The solution of the equation $aX = b$, where $a \neq 0$ depends on the pair (b, a). This prompts us to consider the set $X = \mathbb{Z} \times \mathbb{Z}^*$. Since the solution of $aX = b$ and that of $caX = cb, c \neq 0$ are same, we define relation to identify pairs (b, a) and (bc, ca) for all $c \in \mathbb{Z}^*$. More precisely, define a relation $\sim$ on X by

$$\sim = \{((a, b), (c, d)) \in X \times X \mid ad = bc\}.$$

3.6.1. Show that $\sim$ is an equivalence relation on X.

The equivalence class determined by (a, b) will be denoted by $\dfrac{a}{b}$. Thus,

$$\frac{a}{b} = \{(c, d) \in X \mid ad = bc\}$$

a is called the numerator and b is called the denominator. Clearly,

$$\frac{a}{b} = \frac{c}{d} \text{ if and only if } ad = bc$$

Let $\mathbb{Q}$ denote the quotient set $X/\sim$ and call it the set of rational numbers.

3.6.2. Show that we have two binary operations on $\mathbb{Q}$ denoted by $+$ and $\cdot$ which are given by

$$\frac{a}{b} + \frac{c}{d} = \frac{ad + bc}{bd},$$

and

$$\frac{a}{b} \cdot \frac{c}{d} = \frac{ac}{bd}$$

3.6.3. Show that for every nonzero member x of $\mathbb{Q}$, there is a unique pair a, b, $a \in \mathbb{Z}, b \in \mathbb{N}$ and $(a, b) \sim 1$ such that $x = \dfrac{a}{b}$.

3.6.4. Show that the map f from $\mathbb{Z}$ to $\mathbb{Q}$ defined by $f(a) = \dfrac{a}{1}$ is embedding in the sense that it is injective and respects addition and multiplication.

We shall identify a and $f(a) = \dfrac{a}{1}$.

3.6.5. Show that the order $\leq$ in $\mathbb{Z}$ can also be extended to that in $\mathbb{Q}$ by defining $\dfrac{a}{b} \leq \dfrac{c}{d}$ $(b, d \in \mathbb{N})$ if $ad \leq bc$. Check that it is a total order. Show also that the law of trichotomy holds here also.

The proof of the following theorem is left as an exercise.

Theorem 3.6.1 $(\mathbb{Q}, +, \cdot, \leq)$ *is an ordered field in the sense that it satisfies the following conditions:*

(i) $+$ *is associative.*

(ii) $+$ *is commutative.*

(iii) $x + 0 = x = 0 + x$ *for all $x \in \mathbb{Q}$.*

(iv) *For all $x = \dfrac{a}{b} \in \mathbb{Q}$, there exists $-x = \dfrac{-a}{b} \in \mathbb{Q}$ such that $x + -x = 0 = -x + x$.*

(v) $\cdot$ *is associative.*

(vi) $\cdot$ *is commutative.*

(vii) $x \cdot 1 = x = 1 \cdot x$ *for all $x \in \mathbb{Q}$*

(viii) *For all $x = \dfrac{a}{b} \in \mathbb{Q}$, $a \neq 0$, there exists $x^{-1} = \dfrac{b}{a} \neq 0 \in \mathbb{Q}$ (called the inverse of x) such that $x \cdot x^{-1} = 1 = x^{-1} \cdot x$.*

(ix) $\cdot$ *distributes over $+$ from left as well as right.*

(x) *Let $P = \{x \in \mathbb{Q} \mid x > 0\}$. Given any $x \in \mathbb{Q}$, show that one and only one of the following holds:*

(a) $x \in P$.

(b) $x = 0$.

(c) $-x \in P$.

(xi) $x, y \in P$ *implies that $x + y$ and $x \cdot y \in P$.*

3.6.6. Show that if $x_1^2 + x_2^2 + \cdots + x_n^2 = 0$, then $x_i = 0$ for all i.

3.6.7. Let a, b, c and d be members of $\mathbb{Q}$ such that $a < b$ and $c < d$. Show that there is a bijective map from the set $\{x \in \mathbb{Q} \mid a < x < b\}$ to the set $\{x \in \mathbb{Q} \mid c < x < d\}$. Indeed, there is a bijective map from these sets to $\mathbb{Q}$ also.

3.6.8. Show that $\mathbb{Q}$ is countable in the sense that there is a bijective map from $\mathbb{N}$ to $\mathbb{Q}$.

Hint. It suffices to show (from the above exercise) the countability of $\{x \in \mathbb{Q} \mid 0 < x < 1\}$. Observe that any such rational number is expressible as $\frac{m}{n}$ where $0 < m < n$. Apply double induction first on m and then on n.

3.6.9. Let p be a prime number in $\mathbb{N}$. Then, $X^2 = p$ has no solution in $\mathbb{Q}$. For, suppose that $(\frac{m}{n})^2 = p$, where $(m, n) \sim 1$. Then $m^2 = n^2 p$. But then p/m and so $p^2/n^2 p$. This means that p/n, a contradiction.

3.6.10. Let $m \in \mathbb{Z}$. Show that $X^2 = m$ has a solution in $\mathbb{Q}$ if and only if it has a solution in $\mathbb{N} \bigcup \{0\}$.

Hint. Use the above exercise.

3.6.11. Show that every equation $aX = b$, where $a \in \mathbb{Q}^*$ and $b \in \mathbb{Q}$, has a solution in $\mathbb{Q}$.

3.6.12. Show that an equation $X^n + a_1 X^{n-1} + a_2 X^{n-2} + \cdots + a_n$, where $a_i \in \mathbb{Z}$, has a rational solution if and only if it has an integral solution.

3.6.13. Show that $\mathbb{Q}$ is not order complete with respect to the order $\leq$.

Hint. Show that $\{x \in \mathbb{Q} \mid 0 < x \text{ and } x^2 < 2\}$ has an upper bound but it has no least element.

3.6.14. Define absolute value of a rational number as we defined for $\mathbb{Z}$. Also prove all the properties which were proved to be true over $\mathbb{Z}$.

3.7 Real Numbers

In this section, we enlarge the system $\mathbb{Q}$ to a bigger system $\mathbb{R}$ of real numbers which has all the property of $\mathbb{Q}$ as ordered field with an extra important property of being order complete. As a consequence, it will also have solutions of the equations $X^2 = a$ for all $a > 0$. Here also we shall develop the system with the help of graded exercises. The reader may take the help of a book on real analysis, for example, 'Principles of Real Analysis' by W. Rudin, in case of any difficulty in proving the facts.

A map f from $\mathbb{N}$ to a set X is called a **sequence** in X. A sequence f is also denoted by $\{f(n)\}$.

A sequence f in $\mathbb{Q}$ is called a **Cauchy sequence** in $\mathbb{Q}$ if

$$\text{for all } \epsilon \in \mathbb{Q}, \ \epsilon > 0, \ \text{there exists } n_0 \in \mathbb{N} \text{ such that}$$
$$n, m \geq n_0 \text{ implies that } |f(n) - f(m)| < \epsilon.$$

Let us denote the set of all Cauchy sequences in $\mathbb{Q}$ by Γ.

3.7.1. Let f and g be members of Γ. Show that $f + g$ and $f \cdot g$ defined by $(f + g)(n) = f(n) + g(n)$ and $(f \cdot g)(n) = f(n) \cdot g(n)$ are also members of Γ.

3.7.2. Show that the system $(\Gamma + \cdot)$ is a commutative ring in the sense that it satisfies the following properties.

(i) $+$ *and* $\cdot$ are associative as well as commutative.
(ii) $+$ has the identity 0 (called zero) given by $0(n) = 0 \in \mathbb{Q}$ *for all* $n \in \mathbb{N}$. Thus, $f + 0 = f = 0 + f$ for all $f \in \Gamma$.
(iii) $\cdot$ has the identity 1 given by $1(n) = 1 \in \mathbb{Q}$ *for all* $n \in \mathbb{N}$. Thus, $f \cdot 1 = f = 1 \cdot f$ for all $f \in \Gamma$.
(iv) $\cdot$ distributes over $+$ from left as well as from right. Thus, $f \cdot (g+h) = f \cdot g + f \cdot h$ for all $f, g, h \in \Gamma$.

Call a sequence $f \in \Gamma$ to be a **null sequence** if *for all* $\epsilon > 0$, $\epsilon \in \mathbb{Q}$, *there exists* $n_0 \in \mathbb{N}$ such that $n \geq n_0$ *implies that* $|f(n)| < \epsilon$. Define a relation $\sim$ on Γ as follows:

$$f \sim g \text{ if and only if } f - g \text{ is a null sequence.}$$

3.7.3. Show that $\sim$ is an equivalence relation.
Let $\mathbb{R}$ denote the quotient set $\Gamma/\sim$. Let us denote the equivalence class determined by f by $\bar{f}$.
3.7.4. Suppose that $f \sim f'$ and $g \sim g'$. Show that $f + g \sim f' + g'$ and $f \cdot g \sim f' \cdot g'$. Deduce the existence of binary operations $+$ and $\cdot$ on $\mathbb{R}$ given by

$$\bar{f} + \bar{g} = \overline{(f+g)}$$

and

$$\bar{f} \cdot \bar{g} = \overline{fg}$$

3.7.5. For each $r \in \mathbb{Q}$, let f_r denote the constant sequence given by $f_r(n) = r$ *for all* $n \in \mathbb{N}$. Show that the map ϕ from $\mathbb{Q}$ to $\mathbb{R}$ defined by $\phi(r) = \bar{f_r}$ is an embedding (injective map which preserves operations).
We further identify r in $\mathbb{Q}$ by $\phi(r)$ in $\mathbb{R}$.
3.7.6. Show that the order $\leq$ in $\mathbb{Q}$ can also be extended to that in $\mathbb{R}$ by defining $\bar{f} \leq \bar{g}$ *if and only if there exists* $n_0 \in \mathbb{N}$ such that $n \geq n_0$ *implies that* $f(n) \leq g(n)$.
3.7.7. Show that $(\mathbb{R}, +, \cdot, P)$, where $P = \{x \in R \mid x > 0\}$, is an ordered field in the sense that it satisfies all the properties of $\mathbb{Q}$ listed in the Theorem 3.6.1.
The following facts can be proved with a little more effort (The reader may consult a book on elementary real analysis, for example, "Principles of Mathematical Analysis" by W. Rudin, in case of any difficulty.):
3.7.8. Show that $\mathbb{R}$ is order complete with respect to the order $\leq$.
3.7.9. The concept of absolute value can be introduced on $\mathbb{R}$ in the similar manner as it was done for $\mathbb{Q}$. Show that it also obeys the same laws.
3.7.10. Define the concept of Cauchy sequence in $\mathbb{R}$ as we did it in $\mathbb{Q}$. If we repeat again the process by taking $\mathbb{R}$ at the place of $\mathbb{Q}$, then we will not be getting any thing bigger that $\mathbb{R}$. More precisely, show that if f is a Cauchy sequence in $\mathbb{R}$, then *there exists* $r \in \mathbb{R}$ such that $f \sim f_r$, where f_r is given by $f_r(n) = r$, *for all* $n \in \mathbb{N}$.

3.7.11. A sequence f in $\mathbb{R}$ is said to converge to an element $a \in \mathbb{R}$ if given any $\epsilon > 0$, $\epsilon \in \mathbb{R}$, there is a natural number $n_0 \in \mathbb{N}$ such that $n \geq n_0$ *implies that* $|f(n) - a| < \epsilon$. a is called the limit of the sequence f, and we express it by writing $lim_{n \longrightarrow \infty} f(n) = a$. Show that $\mathbb{R}$ is complete in the sense that every Cauchy sequence in $\mathbb{R}$ converges to a unique point in $\mathbb{R}$.

We state the following facts. The proofs of these facts can be found in the book "Principles of Mathematical Analysis" by W. Rudin.

1. Let $n \in \mathbb{Z}$ and $a \in \mathbb{R}$, $a > 0$. Then, the equation $X^n = a$ has a solution in $\mathbb{R}$. If n is an odd integer, then the equation $X^n = a$ will have a solution for all $a \in \mathbb{R}$. Thus, $a^{\frac{1}{n}}$ can be defined for all $n \in \mathbb{N}$ and $a \in \mathbb{R}$, $a \geq 0$. In turn, a^r can be defined for all $r \in \mathbb{Q}$ and $a \in \mathbb{R}$, $a \geq 0$.

2. **Archemidean property of** $\mathbb{R}$. Given any real number a, there is a natural number n such that $n > a$.

3. $\mathbb{Q}$ is dense in $\mathbb{R}$ in the sense that every real number is limit of a sequence in $\mathbb{Q}$. This is equivalent to say that between any two distinct real numbers there is a member of $\mathbb{Q}$.

4. The real numbers which are not rational numbers are called irrational numbers. Thus, the solutions of $X^2 = 2$ which are denoted by $\pm\sqrt{2}$ are all irrational numbers. The set $\mathbb{R} - Q$ of irrational numbers is also dense in the same sense.

5. Let $a \in \mathbb{R}$, $a > 0$ and $b \in \mathbb{R}$. Suppose f is a sequence in $\mathbb{Q}$ which converges to b. Then, we have a sequence a^f defined by $a^f(n) = a^{f(n)}$. It can be checked that this is also a Cauchy sequence. The limit of this sequence is defined to be a^b. The law of indices is true here also.

The map $x \rightsquigarrow a^x$, $a > 0$ is a bijective map from the set $\mathbb{R}$ to the set $\mathbb{R}^+$ of positive real numbers. The inverse of this map is denoted by log_a which is a bijective map from $\mathbb{R}^+$ to $\mathbb{R}$. This map is called logarithm to the base a.

6. We define $n!$ for all $n \in \mathbb{N} \bigcup \{0\}$ inductively as follows. Define $0! = 1$ and also $1! = 1$. Assuming that $n!$ has already been defined, define $n + 1! = (n + 1) \cdot n!$. Thus, $n! = 1 \cdot 2 \cdots n$. The sequence $\{1 + \frac{1}{1!} + \frac{1}{2!} + \cdots + \frac{1}{n!}\}$ can be seen to be a Cauchy sequence. In turn, it will converge to a unique real number denoted by e. This number e is called the exponential. For any real number x, the sequence $\{1 + x + \frac{x^2}{2!} + \cdots + \frac{x^n}{n!}\}$ is also a Cauchy sequence, and its limit is in fact e^x.

7. The sequence $\{x - \frac{x^3}{3!} + \cdots + (-1)^{2n-1}\frac{x^{2n-1}}{2n-1!}\}$ is also a Cauchy sequence, and its limit is denoted by $sin\ x$. This defines a function *sine* from the set of reals to itself. Similarly, the sequence $\{1 - \frac{x^2}{2!} + \cdots + (-1)^n\frac{x^{2n}}{2n!}\}$ is also a Cauchy sequence, and its limit is denoted by $cos\ x$. This defines a function *cosine* from $\mathbb{R}$ to $\mathbb{R}$. These functions obey all trigonometrical identities with which the reader is familiar.

8. The sequence $\{1 + \frac{1}{2^2} + \cdots + \frac{1}{n^2}\}$ is also a Cauchy sequence. The number π is defined by the requirement that $\frac{\pi^2}{6}$ is the limit of this sequence.

9. $\mathbb{R}$ is uncountable.

10. The equation $X^2 + 1 = 0$ has no solution in $\mathbb{R}$. In fact, $X^2 + 1 = 0$ has no solution in any ordered field.

3.8 Complex Numbers

The equation $X^2 + 1 = 0$ has no solution in $\mathbb{R}$. We wish to enlarge the system $\mathbb{R}$ to a system to include the solution of the above equation. The best that we can expect is to retain the field properties of $\mathbb{R}$. Obviously, the order property cannot be retained. Let us denote the enlarged system (if possible) by $\mathbb{C}$ and denote the solution of $X^2 + 1 = 0$ by i. Then $i^2 = -1$. We denote the extended operations by same notations $+$ and $\cdot$. The set $\{a + bi \mid a, b \in \mathbb{R}\}$ is contained in $\mathbb{C}$. If $a + bi = c + di$, then $a - c = (d - b)\iota$. Hence if $d \neq b$, then $i = (a - c)(d - b)^{-1}$ belongs to $\mathbb{R}$ which is impossible. Thus, $b = d$ and also $a = c$. Clearly,

$$(a + bi) + (c + di) = (a + c) + (b + d)i,$$

and

$$(a + bi) \cdot (c + di) = (ac + bdi^2 + (ad + bc)i = (ac - bd) + (ad + bc)i$$

Thus, the set $\{a + bi \mid a, b \in R\}$ is closed under the binary operations. Further, it can be checked that $(a + bi) \cdot \left(\frac{a}{a^2 + b^2} - i\frac{b}{a^2 + b^2}\right) = 1$ (observe that we are identifying $a \in \mathbb{R}$ by $a + i0$ in $\mathbb{C}$). Thus, we have the required enlarged system $\mathbb{C} = \{a + bi \mid a, b \in \mathbb{R}\}$. The following fact is known as the **fundamental theorem of algebra**. The proof of this fact will be given in the Chap. 9 of algebra 2.

Theorem 3.8.1 *Any equation*

$$a_0 + a_1 X + a_2 X^2 + \cdots + a_n X^n = 0,$$

where all $a_i \in \mathbb{C}$, has a solution (in fact all solutions) in $\mathbb{C}$. ♯

Given any complex number $z = a + bi$, a is called the real part and b is called the imaginary part of z. The complex number $a - bi$ is called the conjugate of $a + bi$, and it is denoted by $\overline{a + bi}$. The real number $a^2 + b^2$ is nonnegative, and its nonnegative real square root is called the **modulus** of $a + bi$, and it is denoted by $\mid a + bi \mid$. The modulus on $\mathbb{C}$ satisfies the same properties as the absolute value satisfies on $\mathbb{R}$.

Cauchy sequences can be defined in $\mathbb{C}$ also as it was defined in $\mathbb{R}$. $\mathbb{C}$ is also complete in the sense that every Cauchy sequence in $\mathbb{C}$ converges in $\mathbb{C}$. e^z, $\cos z$, $\sin z$ can be defined for all complex numbers z as it was defined for real numbers. It follows that $e^{ix} = \cos x + i\sin x$ for all real numbers x. The set $S^1 = \{z \in C \mid\mid z \mid = 1\}$ is called the **circle or torus of dimension 1**. Thus, every member of S^1 is of the form $\cos x + i\sin x$, and so it is of the form e^{ix}, where $x \in \mathbb{R}$. For details, refer to any book on Complex Analysis, for example, "Complex Analysis" by Ahlfors.

Chapter 4
Group Theory

One of the most fundamental concepts in mathematics today is that of a group. Germs of group were present, even in ancient times, in the study of congruences of geometric figures and also in the study of motions in space. It started taking shape in the beginning of the nineteenth century. One of the most challenging problems at that time was the problem of solvability of general polynomial equations of degree $n, n \geq 5$ by the field and radical operations (addition, subtraction, multiplication, and division by nonzero elements and taking mth roots for different m). Paulo Ruffini (1736–1813) and Niels Henrik Abel (1802–1829), using the structure of a set of permutations on the set of roots of the polynomials, proved that a general nth degree equation $n \geq 5$ is not solvable by the field and the radical operations. Evariste Galois (1811–1832) discovered that the key factor behind the algebraic solvability of a polynomial equation is an structure (called the Galois group of the polynomial equation) and proved that a polynomial equation is solvable by the field and the radical operations if and only if the corresponding structure (namely the Galois group) possesses a property (called the solvability).

In the second half of the nineteenth century, the notion of the congruences of geometric objects was further generalized. The development during this period was influenced by the works of Sophus Lie (1842–1899), Felix Klein (1849–1925), Henri Poincare (1854–1912), and Max Dehn (1878–1952) on geometry and topology. The importance of the study of permutation groups, continuous groups, groups of homeomorphisms, and fundamental groups was realized, and this lead to the formulation of an abstract group. The notion of an abstract group is present in the works of Arthur Caley (1821–1895) and von Dyck (1856–1934).

Theory of groups developed slowly but steadily in the first half of the twentieth century with some very significant contributions by G. Frobenius (1849–1917), William Burnside (1852–1957), Isai Schur (1875–1936), O. Schreier (1901–1929), P. Hall (1904–1982), and others. Theory of finite groups picked up momentum with the works of R. Brauer (1901–1977) and his students in 1955. Theory of groups, now, has tremendous applications and interest in itself.

© Springer Nature Singapore Pte Ltd. 2017
R. Lal, *Algebra 1*, Infosys Science Foundation Series in Mathematical Sciences,
DOI 10.1007/978-981-10-4253-9_4

4.1 Definition and Examples

Let G be a set. Recall that a map $o : G \times G \longrightarrow G$ is called a **binary operation** on G. The image of (a, b) under this binary operation is denoted by aob.

Definition 4.1.1 A **groupoid** is a pair (G, o), where G is a set and o is a binary operation on G.

The pairs $(\mathbb{N}, +)$, $(\mathbb{N}, \cdot)$, $(\mathbb{Z}, +)$, $(\mathbb{Z}, -)$, $(\mathbb{Z}, \cdot)$, $(\mathbb{Z}_m, \oplus)$, $(\mathbb{Z}_m, \star)$, $(\mathbb{Q}, +)$, $(\mathbb{Q}, \cdot)$, $(\mathbb{Q}, -)$, $(\mathbb{R}, +)$, $(\mathbb{R}, \cdot)$, $(\mathbb{R}, -)$, $(\mathbb{C}, +)$, $(\mathbb{C}, \cdot)$, and $(\mathbb{C}, -)$ are all examples of groupoid.

The first and the second projections from $G \times G$ to G are distinct binary operations on G provided that G is different from singleton.

Union, intersection, relative compliment, and symmetric difference are all binary operations on the power set of a set Y.

If G is a finite set, then a binary operation on G can be defined by a table called **multiplication table**. Thus, the following table defines a binary operation on $G = \{a, b, c, d\}$.

o	a	b	c	d
a	b	a	d	c
b	a	a	c	d
c	b	b	c	a
d	c	d	b	c

The binary operation o is evident from the table. For example, $boc = c$ and $cob = b$.

Recall that a binary operation o on G is said to be an **associative** operation if

$$(aob)oc = ao(boc) \text{ for all } a, b, c \in G,$$

and it is said to be a **commutative** operation if

$$aob = boa \text{ for all } a, b \in G.$$

The usual addition '+' and multiplication '·' on $\mathbb{N}, \mathbb{Z}, \mathbb{Q}, \mathbb{R}$ and $\mathbb{C}$ are all associative as well as commutative operations. However, the subtraction $-$ is neither associative nor commutative. The projection maps on $G \times G$ are associative but not commutative unless G is singleton. The operation o on $\mathbb{N}$ defined by $aob = a^2 + b^2$ is commutative but not associative. The binary operations $\oplus$ and $\star$ on $\mathbb{Z}_m$ (see Proposition 3.5.2) are associative as well as commutative. The $\bigcup, \bigcap, \triangle$ are all associative as well as commutative on the power set of a set.

Definition 4.1.2 A pair (G, o), where G is a set and o is an associative binary operation on G, is called a **semigroup**. If the binary operation o is also commutative, then we say that it is a **commutative semigroup**.

The pairs $(\mathbb{N}, +)$, $(\mathbb{N}, \cdot)$, $(\mathbb{Z}, +)$, $(\mathbb{Z}, \cdot)$, $(\mathbb{Q}, +)$, $(\mathbb{Q}, \cdot)$ $(\mathbb{R}, +)$, $(\mathbb{R}, \cdot)$, $(\mathbb{C}, +)$, $(\mathbb{C}, \cdot)$, $(\mathbb{Z}_m, \oplus)$, and $(\mathbb{Z}_m, \star)$ are all commutative semigroups. The pair (G, p_1), where p_1 is the first projection, is a noncommutative semigroup. The pair $(\mathbb{Z}, -)$ is not a semigroup.

Let X be a set having more than 2 elements and $G = R(X)$ be the set of all relations on X. Then, the pair (G, o), where o is the composition of relations on X, is a noncommutative semigroup. Similarly, if we denote by $F(X)$ the set of all maps from X to X, then $(F(X), o)$ is a noncommutative semigroup.

Definition 4.1.3 Let (G, o) be a groupoid. An element $e \in G$ is called a **left(right)** **identity** of (G, o) if $eoa = a(aoe = a)$ *for all* $a \in G$. If e is left as well as right identity, then we say that it is **both-sided identity**.

0 is both-sided identity of $(\mathbb{Z}, +)$, *and* 1 is both-sided identity of $(\mathbb{Z}, \cdot)$. $\overline{0}$ is both-sided identity of $(\mathbb{Z}_m, \oplus)$, and $\overline{1}$ is both-sided identity of $(\mathbb{Z}_m, \star)$. The empty set $\emptyset$ is both-sided identity of $(P(Y), \bigcup)$ and also of $(P(Y), \triangle)$. Y is both-sided identity of $(P(Y), \bigcap)$. The semigroup $(\mathbb{N}, +)$ has no identity. 0 is the right identity of the groupoid $(\mathbb{Z}, -)$ which has no left identity. Every element of G is a right identity of (G, p_1), where p_1 is the first projection on $G \times G$. Similarly, in (G, p_2), every element is left identity but no element is right identity. However, we have the following:

Proposition 4.1.4 *Let (G, o) be a groupoid. Let e_1 be a left identity and e_2 a right identity of (G, o). Then $e_1 = e_2$ and it is both-sided identity.*

Proof Since e_2 is a right identity and o_1 is a left identity, $e_1 - e_1oe_2 - e_1$.   ‖

In case a groupoid (G, o) has both-sided identity, it is unique, and we call it **the identity**.

Definition 4.1.5 A semigroup with identity is also called a **monoid**.

Definition 4.1.6 Let (G, o) be a groupoid. Let e be a left(right) identity of (G, o). Let $a \in G$. An element $a' \in G$ is called a **left(right) inverse** of a with respect to e if $a'oa = e(aoa' = e)$. If e is the identity of (G, o), then a' is called **both-sided** **inverse** of a if

$$aoa' = e = a'oa$$

Example 4.1.7 If (G, o) is a groupoid and e is a left(right) identity of (G, o), then e is left(right) inverse of e with respect to e. In $(\mathbb{N}, \cdot)$, 1 is the identity and 1 is the only element which has inverse (both-sided). In $(\mathbb{Z}, \cdot)$, 1 is the identity, and the inverse of 1 is 1 and that of -1 is -1. No other element has inverse (verify). In $(\mathbb{Z}, +)$, 0 is the identity element, and $-a$ is the both-sided inverse of a. In $(\mathbb{Z}_m, \oplus)$, $\overline{0}$ is the identity and $\overline{-a}$ is the inverse (both-sided) of $\overline{a}$. In $(\mathbb{Z}_m, \star)$, $\overline{1}$ is the identity and $\overline{a}$ has a inverse $\overline{b}$ if and only if $(a, m) \sim 1$ (Corollary 3.5.5). In $(P(Y), \triangle)$, $\emptyset$ is the identity and inverse of every element $A \in P(Y)$ is A itself (verify). In $(\mathbb{Z}, -)$, 0 is a right identity and every element of $\mathbb{Z}$ is right inverse of itself with respect to 0.

Left inverse and right inverse of an element in a groupoid with identity may be different. For example, consider the groupoid (G, o), where $G = \{e, a, b, c\}$ and o is defined by the following multiplication table.

o	e	a	b	c
e	e	a	b	c
a	a	c	e	b
b	b	b	a	b
c	c	e	c	b

Here, e is the identity, $aob = e = coa$. Thus, b is a right inverse, and c is a left inverse of a. However, the following proposition says that this can not happen in a semigroup.

Proposition 4.1.8 *Let (G, o) be a semigroup with the identity e. Let $a \in G$. Let a' be a left inverse and a'' a right inverse of a. The $a' = a''$.*

Proof Under the hypothesis of the proposition, $a' = a'oe = a'o(aoa'') = (a'oa)oa''$ $= eoa'' = a''$.   ♯

The above proposition says that both-sided inverse of an element a in a semigroup with identity, if exists, is unique. The inverse of an element a in a semigroup (G, o) with identity is usually denoted by a^{-1}. Thus,

$$a^{-1}oa = e = aoa^{-1}$$

If we denote the binary operation additively by +, which we usually do when the binary operation is commutative, then the identity is denoted by 0 and the inverse of an element a is denoted by $-a$. Thus,

$$-a + a = 0 = a + -a$$

Definition 4.1.9 Let (G, o) be a semigroup. Let $a, b \in G$. An equation $aoX = b$ $(Xoa = b)$ is said to be **solvable** if *there exists $c \in G$* (called a **solution** of the equation) such that $aoc = b(coa = b)$.

Theorem 4.1.10 *Let (G, o) be a semigroup, where G is a nonempty set. The following conditions on (G, o) are equivalent.*

1. *Equations $aoX = b$ and $Xoa = b$ are solvable for all $a, b \in G$.*
2. *(G, o) has a left identity e such that every element of G has a left inverse with respect to e, i.e.*

    (i) there exist $e \in G$ such that

$$eoa = a \text{ for all } a \in G,$$

and

(ii) *for all a ∈ G, there exists a' ∈ G such that*

$$a'oa = e.$$

3. *(G, o) has a right identity e such that every element of G has a right inverse with respect to e, i.e.*

(i) *there exists e ∈ G such that*

$$aoe = a \text{ for all } a \in G,$$

and

(ii) *for all a ∈ G, there exists a' ∈ G such that*

$$aoa' = e.$$

4. *(G, o) has the identity and every element of G has the inverse, i.e.*

(i) *there exists e ∈ G such that*

$$eoa = a = aoe \text{ for all } a \in G,$$

and

(ii) *for all a ∈ G, there exists a^{-1} ∈ G such that*

$$a^{-1}oa = e = aoa^{-1}.$$

5. *Equations aoX = b and Xoa = b have unique solutions in G for all a, b ∈ G.*

Proof (1 $\Longrightarrow$ 2). Assume 1. Since $G \neq \emptyset$, *there is an element a ∈ G.* By (1), the equation $Xoa = a$ is solvable. Let e be a solution of $Xoa = a$. Then,

$$eoa = a \tag{4.1.1}$$

We show that e is a left identity. Let $b \in G$. By (1), the equation $aoX = b$ has a solution c (say). Then,

$$aoc = b \tag{4.1.2}$$

Now, using 4.1.1 and 4.1.2, and the associativity of o,
$eob = eo(aoc) = (eoa)oc = aoc = b$

This shows that e is a left identity. Let $a \in G$. A solution of the equation $Xoa = e$ gives an element a' such that

$$a'oa = e$$

This proves that $1 \implies 2$. Similarly, $1 \implies 3$.

($2 \implies 4$). Assume 2. Let e be a left identity and a' a left inverse of a with respect to e. We show that e is also a right identity and a' is also a right inverse of a. We first show that

$$aoa' = e$$

Since every element of G has a left inverse with respect to e, for a' *there* $\exists a'' \in G$ such that

$$a''oa' = e \tag{4.1.3}$$

Now, using the Eq. 4.1.3 and the associativity of the binary operation,
$aoa' = eo(aoa') = (a''oa')o(aoa') = a''o(a'o(aoa')) = a''o((a'oa)oa') = a''o(eoa') = a''oa' = e$.
Thus,

$$aoa' = e \tag{4.1.4}$$

In turn, using 4.1.4,
$aoe = ao(a'oa) = (aoa')oa = eoa = a$
Thus, e is the identity and a' is the inverse of a, which we denote by a^{-1}. This proves $2 \implies 4$.

The proof of $3 \implies 4$ is similar to that of $2 \implies 4$.

($4 \implies 5$). Assume 4. Then, $a^{-1}ob$ is a solution of the equation $aoX = b$, and boa^{-1} is a solution of $Xoa = b$. Further, if c and d are solutions of $aoX = b$, then $aoc = b = aod$. But, then $c = (a^{-1}oa)oc = a^{-1}o(aoc) = a^{-1}o(aod) = (a^{-1}oa)od = eod = d$. Thus, $aoX = b$ has a unique solution. Similarly, $Xoa = b$ has a unique solution.

$5 \implies 1$ is evident.   ♮

Definition 4.1.11 A semigroup (G, o) satisfying any one (and hence all) of the above 5 equivalent conditions in the theorem is called a **group**. A group (G, o) is said to be an **abelian group** (after the name of Abel) or a **commutative group** if the operation o is commutative.

In an abelian group, a binary operation is usually denoted by $+$ called the addition, the identity is denoted by 0 called 0, and the inverse of an element a is denoted by $-a$ called the negative of a.

Example 4.1.12 $(\mathbb{Z}, +)$, $(\mathbb{Q}, +)$, $(\mathbb{R}, +)$, $(\mathbb{C}, +)$ are all infinite abelian groups. 0 is the identity and the inverse of a is $-a$.

Example 4.1.13 The usual multiplication · in $\mathbb{Q}$, $\mathbb{R}$, and $\mathbb{C}$ induces multiplications in $\mathbb{Q}^* = \mathbb{Q} - \{0\}$, $\mathbb{R}^* = \mathbb{R} - \{0\}$, $\mathbb{C}^* = \mathbb{C} - \{0\}$ with respect to which all of these are abelian groups. 1 is the identity, and the inverse of an element a is $a^{-1} = \frac{1}{a}$.

Definition 4.1.14 If a group (G, o) is finite, then the number of elements in G is called the **order** of the group G and is denoted by $\mid G \mid$ or $o(G)$.

Example 4.1.15 The multiplication · in Z induces multiplication in $\{1, -1\}$ with respect to which it is a finite abelian group of order 2.

Example 4.1.16 Let m be a positive integer. It follows from Theorem 3.5.2 (i)–(iv) (Chap. 3) that $(\mathbb{Z}_m, \oplus)$ is a finite abelian group of order m. $\overline{0}$ is the identity, and the inverse of $\overline{a}$ is $\overline{-a}$.

Example 4.1.17 Let m be a positive integer. Let

$$U_m = \{\overline{a} \in \mathbb{Z}_m \mid (a, m) \sim 1\}$$

Note that if $(a, m) \sim 1$ and $\overline{a} = \overline{b}$ in $\mathbb{Z}_m$, then $(b, m) \sim 1$. Let $\overline{a} \in U_m$ and $\overline{b} \in U_m$. Then, $(a, m) \sim 1$ and $(b, m) \sim 1$. By Corollary 3.4.10, $(ab, m) \sim 1$. Hence, $\overline{a} \star \overline{b} = \overline{ab}$ belongs to U_m. Thus, $\star$ induces a binary operation on U_m which we again denote by $\star$. It again follows from Theorem 3.5.2 ((vi), (vii), and (x)) (Chap. 3) that $(U_m, \star)$ is a commutative semigroup with identity. Further, it follows from Corollary 3.5.5 that if $(a, m) \sim 1$, then there exists an integer b such that $(b, m) \sim 1$ and $ab \equiv 1 (mod\ m)$. In other words, there is a $\overline{b} \in U_m$ such that $\overline{a} \star \overline{b} = \overline{1} = \overline{b} \star \overline{a}$. Thus, every element of $(U_m, \star)$ has the inverse, and so it is a finite abelian group. The group $(U_m, \star)$ is called the **group of prime residue classes modulo m**.

Definition 4.1.18 The function ϕ from $\mathbb{N}$ to $\mathbb{N}$ given by $\phi(1) = 1$, and for $n > 1$, $\phi(n) =$ the number of positive integers less than n and co-prime to n is called the **Euler's phi function or Euler's totient function**.

Thus, the order of the group U_m of prime residue classes modulo m is $\phi(m)$.

Formula for ϕ (m)

Let p be a prime number and $r \geq 1$. The set of positive integers less than or equal to p^r and not co-prime to p^r is $\{p, 2p, 3p, \ldots, p^{r-1} \cdot p\}$. Thus, the number of positive integers less than p^r and co-prime to p^r is $p^r - p^{r-1}$. This means that $\phi(p^r) = p^r - p^{r-1}$. In particular, $\phi(p) = p - 1$ for all prime p. It will be shown later in Chap. 7 that the function ϕ is multiplicative in the sense that if $(m, n) = 1$, then $\phi(mn) = \phi(m)\phi(n)$. Thus, if

$$m = p_1^{r_1} p_2^{r_2} \cdots p_k^{r_k},$$

where $p_1, p_2, \ldots, p_k$ are distinct primes, then

$$\phi(m) = (p^{r_1} - p^{r_1-1})(p^{r_2} - p^{r_2-1}) \ldots (p^{r_k} - p^{r_k-1}).$$

For example, $\phi(100) = \phi(5^2) \cdot \phi(2^2) = (5^2 - 5) \cdot (2^2 - 2) = 40$.

Example 4.1.19 Let $S^1 = \{z \in \mathbb{C} \mid |z| = 1\}$ denote the unit circle in the complex plane. The multiplication of complex numbers induces multiplication in S^1 with respect to which it is a group, called the **circle group** or the **torus group** of dimension 1. This is an infinite abelian group.

Example 4.1.20 Let P denote the set of roots of unity. Thus, $P \{z \in C \mid z^n = 1 \text{ for some } n \in \mathbb{Z}\}$. The multiplication of complex numbers induces a multiplication in P with respect to which it is a group (verify). This is also an infinite abelian group.

Example 4.1.21 Let $n \in \mathbb{N}$. Let P_n denote the set of nth roots of unity. Thus, $P_n = \{e^{\frac{2\pi r i}{n}} \mid 0 \leq r < n\}$. The multiplication of complex numbers again induces a multiplication in P_n with respect to which it is an abelian group of order n (verify).

Example 4.1.22 (Klein's four group). Let $V_4 = \{e, a, b, c\}$. Define a binary operation o on V_4 by the following table:

o	e	a	b	c
e	e	a	b	c
a	a	e	c	b
b	b	c	e	a
c	c	b	a	e

It can be checked that o is associative. Clearly, e is the identity, and every element is its own inverse. Thus, (V_4, o) is a group called the **Klein's four group** (after the name of the great geometer Felix Klein). This group is a finite abelian group containing four elements in which every element is its own inverse.

So far, we had examples of abelian groups only. Following few examples are those of nonabelian groups.

Example 4.1.23 Let $Q_8 = \{1, -1, i, j, k, -i, -j, -k\}$. Define a multiplication $\cdot$ in Q_8 as follows: 1 acts as the identity of the operation, multiplication by -1 changes sign, $i^2 = j^2 = k^2 = (-i)^2 = (-j)^2 = (-k)^2 = -1$, and the multiplication between $i, j, k, -i, -j, -k$ is obtained by treating them as unit vectors along the three axises and taking vector product between them. Thus, the multiplication table for binary operation $\cdot$ is given by

·	1	−1	i	j	k	−i	−j	−k
1	1	−1	i	j	k	−i	−j	−k
−1	−1	1	−i	−j	−k	i	j	k
i	i	−i	−1	k	−j	1	−k	j
j	j	−j	−k	−1	i	k	1	−i
k	k	−k	j	−i	−1	−j	i	1
−i	−i	i	1	−k	j	−1	k	−j
−j	−j	j	k	1	−i	−k	−1	i
−k	−k	k	−j	i	1	j	−i	−1

$(Q_8, \cdot)$ is a group of order 8 (verify) called the **Quaternion group** of degree 8 or the **Hamiltonian group** (after the name of Hamilton). This group is nonabelian $(i \cdot j \neq j \cdot i)$ group of order 8.

Example 4.1.24 **Transformation group or symmetric group.** Let X be a nonempty set. A bijective map from X to X is called a **permutation** on X. Let $Sym(X)$ denote the set of all permutations on X. Then, $Sym(X)$ is a group with respect to composition of maps. This group is called the **transformation group on X** or the **symmetric group on X** or the **permutation group on X**. If $X = \{1, 2, \ldots, n\}$, then $Sym(X)$ is denoted by S_n and called the **symmetric group of degree n** or the **permutation group of degree n**. The number of bijective maps from X to X in this case is $n!$ (prove it by induction). If X contains just one element, then $Sym(X)$ also contains one element, and hence, it is abelian. If X contains two elements, then $Sym(X)$ contains $2! = 2$ elements of which one is the identity element, and hence, in this case also, $Sym(X)$ is abelian. Suppose that X contains at least three elements. Let a, b, c be distinct elements of X. Define a map f from X to X as follows: $f(a) = b$, $f(b) = c$, $f(c) = a$, and $f(x) = x$ whenever $x \notin \{a, b, c\}$. Clearly, $f \in Sym(X)$. Let g be another map from X to X given by $g(a) = b$, $g(b) = a$, and $g(x) = x$ whenever $x \notin \{a, b\}$. Then, $(g o f)(a) = a$ and $(f o g)(a) = c$. Hence, $g o f \neq f o g$. Thus, in this case, $Sym(X)$ is nonabelian.

Remark 4.1.25 There are two main resources of groups, viz. the symmetric groups and the matrix groups. The symmetric group will be studied in detail in Sects. 6.1 and 6.2 of Chap. 6. The matrix groups will be introduced and studied in the Sects. 6.3 and 6.4 of the Chap. 6.

Example 4.1.26 (**Group of isometries** or **group of motions**). Consider the Euclidean space $\mathbb{R}^3$. The distance $d(x, y)$ between two points $x = (x_1, x_2, x_3)$ and $y = (y_1, y_2, y_3)$ of $\mathbb{R}^3$ is given by

$$d(x, y) = +\sqrt{(x_1 - y_1)^2 + (x_2 - y_2)^2 + (x_3 - y_3)^2}$$

A bijective map[1] f from $\mathbb{R}^3$ to $\mathbb{R}^3$ is called an **isometry** also called a **rigid motion** if $d(f(x), f(y)) = d(x, y)$ *for all* $x, y \in \mathbb{R}^3$. Let $Iso(\mathbb{R}^3)$ denote the set of all isometries of $\mathbb{R}^3$. Then, $Iso(\mathbb{R}^3)$ is a group with respect to composition of maps, and it is called the **isometry group of** $\mathbb{R}^3$ or the **group of rigid motions**.

Example 4.1.27 **Group of symmetries of geometric objects**. Let $X \subseteq \mathbb{R}^3$ (e.g., X may be a regular tetrahedron or X may be an ellipsoid, sphere, or any object). Let

$$Iso_X(\mathbb{R}^3) = \{f \in Iso(\mathbb{R}^3) \mid f(X) = X\} \subseteq Iso(\mathbb{R}^3)$$

Then, $Iso_X(\mathbb{R}^3)$ is a group with respect to composition of maps. This group is called the **group of symmetries of X**. This group measures the symmetry of X. More symmetrical is the geometrical object X, larger is the group of isometries of X.

Example 4.1.28 **Group of symmetries of plane geometric figures**. Consider the group $Iso(\mathbb{R}^2)$ of isometries of $\mathbb{R}^2$. Thus, $Iso(\mathbb{R}^2)$ is the set of all bijective (the condition of being bijective is redundant) distance preserving maps from $\mathbb{R}^2$ to $\mathbb{R}^2$, and it is a group under composition of maps. There are three types of fundamental isometries of $\mathbb{R}^2$: (i) rotations about different points, (ii) reflections about different lines in $\mathbb{R}^2$, and (iii) translations (write down transformations representing these isometries). It is a fact (prove it or see algebra 2) that every isometry of $\mathbb{R}^2$ is obtained by composing these fundamental isometries. In fact, an isometry of $\mathbb{R}^2$ is either a rotation or a reflection or a translation or a composition of a reflection and a translation.

Let X be a bounded subset of $\mathbb{R}^2$ (e.g., a circle, an ellipse, a triangle, or a polygon). Then, translation or composition of translations and reflections can not leave X invariant. Thus, only rotations and reflections may belong to the group $Iso_X(\mathbb{R}^2)$ of isometries of X. Let us describe the group $Iso_X(\mathbb{R}^2)$, where X is a regular polygon of n sides, $n \geq 3$. Since X is bounded, only rotations and reflections can belong to $Iso_X(\mathbb{R}^2)$. Suppose that n is even. The rotations through angles $\frac{2\pi r}{n}$, $0 \leq r < n$ about the center of the polygon are the only rotations which keep X invariant. The reflections which keep X invariant are reflections in lines joining opposite vertices of the regular polygon and also reflections in the line joining middle points of the opposite sides. These are also n in number. Thus, $Iso_X(\mathbb{R}^2)$ contains exactly $2n$ elements.

Next, suppose that n is odd. Again, rotations through angles $\frac{2\pi r}{n}$, $0 \leq r < n$ about the center are the only rotations which keep X invariant. The reflections in lines joining vertices with middle points of their opposite edges are the only reflections which keep X invariant. Thus, again $Iso_X(R^2)$ contains $2n$ elements. This group (in both cases) is called the **dihedral group** and is denoted by D_n. The structure of this group will be studied later.

---

[1]The condition of f being bijective is redundant. In fact, if a map f from $\mathbb{R}^3$ to $\mathbb{R}^3$ satisfies the condition $d(f(x), f(y)) = d(x, y)$ *for all* $x, y \in \mathbb{R}^3$, then as a consequence f is bijective.

Exercises

4.1.1 Let G be a set containing n elements. Find the number of binary operations on G. How many of them are commutative?

4.1.2 Define a binary operation o on $\mathbb{Z}$ by $xoy = |x - y|$. Show that o is commutative but not associative.

4.1.3 Define a binary operation o on $\mathbb{Z}$ by $xoy = x + y - x \cdot y$, where $+$ and $\cdot$ are usual addition and multiplication in $\mathbb{Z}$. Show that $(\mathbb{Z}, o)$ is a semigroup. Is it a group?

4.1.4 Let Y be a set. Let G be the set of all maps from Y to Y. Then, (G, o) is a semigroup with identity. Let f be a surjective map on Y. Let $g \in G$. Show that the equation $foX = g$ is solvable, where X is unknown in the equation. If $f \in G$ is injective, then show that $Xof = g$ is solvable. Is (G, o) a group?

4.1.5 Let (G, o) be a group. Define another binary operation o' on G by

$$xo'y = y^{-1}oxoy^2$$

Show that (G, o') is a groupoid in which equations $Xo'x = y$ and $xo'X = y$ have unique solutions for all $x, y \in G$. Show that if o' is commutative, then $o' = o$. Show that (G, o') may be a group even if (G, o) is nonabelian. Find a necessary and sufficient condition so that it becomes a group.

4.1.6 Describe the group of symmetry of an isosceles triangle.

4.1.7 Describe the group of symmetry of a square.

4.1.8 Describe the group of symmetry of a rectangle which is not a square and observe that square is more symmetrical.

4.1.9 Describe the group $Iso_X(\mathbb{R}^3)$, where X is a regular tetrahedron with origin as centroid. How many elements in this group are there?

4.1.10 Show that $Iso_{S^1}(\mathbb{R}^2) = \{f_\theta : \mathbb{R}^2 \longrightarrow \mathbb{R}^2 \mid \theta \in \mathbb{R}\} \cup \{\rho_\theta : \mathbb{R}^2 \longrightarrow \mathbb{R}^2 \mid \theta \in \mathbb{R}\}$, where f_θ is the map defined by

$$f_\theta((x, y)) = (x \cos\theta + y \sin\theta \quad -x \sin\theta + y \cos\theta)$$

and ρ_θ is the map defined by

$$\rho_\theta((x, y)) = (x\cos\theta + y\sin\theta \quad -x\sin\theta - y\cos\theta)$$

4.1.11 Describe the group of symmetries of the ellipse

$$\frac{x^2}{4} + \frac{y^2}{9} = 1$$

4.1.12 Describe the group of symmetries of $\mathbb{Z}$.

4.1.13 Describe the group of symmetries of the unit sphere in $\mathbb{R}^3$ with center as origin.

4.1.14 Let (G, o) be a group and X a nonempty set. Let G^X denote the set of all maps from X to G. Define a binary operation $\star$ on G^X by $(f \star g)(x) = f(x)og(x)$ for all $x \in X$. Show that $(G^X, \star)$ is a group. Show that it is commutative if and only is G is commutative.

4.1.15 Let (G_1, o_1) and (G_2, o_2) be groups. Define a binary operation $\star$ on $G_1 \times G_2$ by $(a, b) \star (c, d) = (ao_1c, bo_2d)$. Show that $(G_1 \times G_2, \star)$ is a group called the **external direct product** of G_1 and G_2.

4.1.16 Let $G = \mathbb{Q} - \{1\}$. Define a binary operation o on G by $aob = a+b-ab$, where $+$ and $\cdot$ are usual addition and multiplications in $\mathbb{Q}$. Show that (G, o) is a group. What is the identity and what is the inverse of an element a in G?

4.1.17 Let (G, o) be an abelian group and $c \in G$. Define a binary operation $\star$ on G by $a \star b = (aob)oc^{-1}$. Show that $(G, \star)$ is a group. What is its identity and what is the inverse of an element $a \in G$? What happens if we drop the condition of the group being abelian?

4.1.18 Let (G, o) be a group and f a bijective map from X to G. Define a binary operation $\star$ on X by the requirement that $f(x \star y) = f(x)of(y)$. Show that $(X, \star)$ is a group.

4.1.19 Let X be a nonempty finite set. Show that we can always define a binary operation on X so that it becomes a group.

4.1.20 Show that $(P(X), \triangle)$ is an abelian group in which every element is its own inverse.

4.1.21* Let X be an infinite set and Y the set of finite subsets of X. Show that there is a bijective map from X to Y. Show that $(Y, \triangle)$ is an abelian group. Use it to show that every nonempty set can be given an abelian group structure. Deduce that there is no set containing all abelian groups.

4.1.22 Let p be a positive prime. Let G_p denote the set of all complex numbers which are p power roots of unity. Show that the multiplication of complex numbers induces a multiplication in G_p with respect to which it is an infinite abelian group.

4.1.23 Examine whether the set $\{\frac{1+2m}{1+2n} \mid m, n \in \mathbb{Z}\}$ of rational numbers form a group with respect to the multiplication of rational numbers.

4.1.24 Show that the set $\{\frac{m}{n} \mid m, n \in \mathbb{Z}, n \neq 0 \text{ and } (n, p) \sim 1\}$, where p is a given prime, is a group with respect to the usual addition of rational numbers.

4.1.25 Let (G, o) be a group. Define a binary operation $\star$ on G by

$$x \star y = y^{-2} \circ x \circ y^3$$

Show that $(G, \star)$ is a group if and only if every element of the form $y^{-2} \circ (x \circ y)^2 \circ x^{-2}$ commutes with each element of G.

4.1.26 Let

$$S^3 = \{a_0 + a_1 i + a_2 j + a_3 k \mid a_0, a_1, a_2, a_3 \in \mathbb{R} \text{ and } a_0^2 + a_1^2 + a_2^2 + a_3^2 = 1\}$$

This set is called the set of unit Quaternions (observe that this set can also be identified with the unit sphere in the Minkowski space R^4). Define a multiplication $\star$ in S^3 as follows:

$$(a_0 + a_1 i + a_2 j + a_3 k) \star (b_0 + b_1 i + b_2 j + b_3 k) = c_0 + c_1 i + c_2 j + c_3 k$$

where $c_0 = a_0 b_0 - a_1 b_1 - a_2 b_2 - a_3 b_3$, $c_1 = a_0 b_1 + a_1 b_0 + a_2 b_3 - a_3 b_2$, $c_2 = a_0 b_2 + a_2 b_0 - a_1 b_3 + a_3 b_1$ and $c_3 = a_0 b_3 + a_3 b_0 + a_1 b_2 - a_2 b_1$. Show that $(S^3, \star)$ is a nonabelian group (compare this group with circle group).

4.1.27 Find the inverse of $\overline{7}$ in U_{19}.

4.1.28 Find the inverse of $\overline{10}$ in U_{21}.

4.1.29 Find the inverse of $\overline{250}$ in U_{641}.

4.1.30 Find the order of U_{640}.

4.2   Properties of Groups

As already observed, identity of a group is unique. The inverse of an element a in a group (G, o) is unique and is denoted by a^{-1}. If the binary operation is written (which we usually do when the operation is commutative) additively, then the identity is denoted by 0 (called zero) and the inverse of an element a is denoted by $-a$ (called the negative of a).

Proposition 4.2.1 *In a group* (G, o), $(a^{-1})^{-1} = a$ *and* $(aob)^{-1} = b^{-1}oa^{-1}$ *for all* $a, b \in G$.

Proof Since the inverse of an element in a group is unique, it is sufficient to observe

$$a^{-1}oa = e = aoa^{-1}$$

and

$$(b^{-1}oa^{-1})o(aob) = e = (aob)o(b^{-1}oa^{-1})$$

The second observation follows from the associativity of the binary operation.   ♯

Proposition 4.2.2 *Cancelation law holds in a group* (G, o) *in the sense that*

$$(aob = aoc) \Longrightarrow b = c$$

and

$$(boa = coa) \Longrightarrow b = c$$

Proof Suppose that $aob = aoc$. Then,

$$b = eob = (a^{-1}oa)ob = a^{-1}o(aob) = a^{-1}o(aoc) = (a^{-1}oa)oc = eoc = c$$

Second part follows similarly.   ♯

Proposition 4.2.3 *Let* (G, o) *be a finite semigroup in which cancelation law holds. Suppose that* $G \neq \emptyset$. *Then,* (G, o) *is a group.*

Proof It is sufficient to show that equations $aoX = b$ and $Xoa = b$ have solutions in G for all $a, b \in G$. Let $a \in G$. Define a map L_a from G to G (called the left multiplication by a) by $L_a(g) = aog$. Suppose that $L_a(g_1) = L_a(g_2)$. Then, $aog_1 = aog_2$. Since (G, o) satisfies cancelation law, $g_1 = g_2$. Hence, L_a is injective. Since G is finite, L_a is also surjective. Thus, given an element b in G, there is an element c in G such that $L_a(c) = b$. This means that $aoc = b$. Similarly, considering the right multiplication R_a by a, we can show that equations $Xoa = b$ is solvable for all $a, b \in G$. ♯

Remark 4.2.4 $(\mathbb{N}, +)$ is an infinite semigroup in which cancelation law holds, but $(\mathbb{N}, +)$ is not a group. Thus, the finiteness condition in the above proposition is essential.

Remark 4.2.5 The associativity of a binary operation o in a groupoid implies that product of a finite sequence $a_1, a_2, \ldots, a_n$ taken in same order is independent of the manner in which we put parenthesis. If in addition it is commutative, then the product is independent of the order also. A precise proof of the above assertion follows by induction on n.

Integral Powers of Elements of a Group

Let (G, o) be a group and $a \in G$. We first define nonnegative integral powers of a. This we do by induction. Define $a^0 = e$, the identity of the group. Assuming that a^n has already been defined, define $a^{n+1} = a^n o a$. If n is negative, then a^{-n} has already been defined, and then, define $a^n = (a^{-n})^{-1}$. Thus, all integral powers of a have been defined. Clearly, for $n > 0$,

$$a^n = \underbrace{aoao\cdots oa}_{n}$$

and

$$a^{-n} = \underbrace{a^{-1}oa^{-1}o\cdots oa^{-1}}_{n}$$

Proposition 4.2.6 (Law of Exponents) *Let (G, o) be a group and $a \in G$. Then,*

(i) $a^{n+m} = a^n o a^m$
(ii) $(a^n)^m = a^{n \cdot m}$ *for all $n, m \in \mathbb{Z}$.*

Proof The proof for $n \geq 0$, $m \geq 0$ follows by induction on m and is left as an exercise. If $n \leq 0$, $m \leq 0$, then from the previous case, $a^{-m-n} = a^{-m}oa^{-n}$, or $a^{-(n+m)} = a^{-m}oa^{-n}$. Taking the inverses,

$$a^{n+m} = (a^{-n})^{-1}o(a^{-m})^{-1} = a^n o a^m$$

Next, suppose that $n \geq 0$ and $m \leq 0$. Then, $n+m \geq 0$ or $n+m \leq 0$. Suppose first that $n + m \geq 0$. Then, from what we have just proved, it follows that $a^{n+m}oa^{-m} = a^n$, and hence, $a^{n+m} = a^n o a^m$. The case $n + m \leq 0$ follows similarly. This completes the proof of (i). Similarly, by induction on m, we can prove the (ii).   ♯

Remark 4.2.7 Let (G, o) be a group and $a \in G$. Then,

(i) $a^n o a^m = a^{n+m} = a^{m+n} = a^m o a^n$
(ii) $(a^n)^m = a^{n \cdot m} = a^{m \cdot n} = (a^m)^n$

Thus, integral powers of an element commute with each other.

Illustrations

2.1. Let (G, o) be a group and $a \in G$. Suppose that $a^m = e = a^n$, $m \neq 0 \neq n$. Then $a^d = e$, where d is g.c.d of m and n. In particular, if m and n are co-prime, then $a = e$.

Proof By the Euclidean algorithm, *there exist* $u, v \in \mathbb{Z}$ such that $d = um + vn$. Hence, using the law of exponents, we get that

$$a^d = a^{um+vn} = a^{um} o a^{vn} = (a^m)^u o (a^n)^v = e^u o e^v = e. \qquad ♯$$

2.2. Let (G, o) be a group in which every element is its own inverse or equivalently $a^2 = e$ for all $a \in G$. Then, (G, o) is abelian.

Proof $aob = (aob)^{-1}$ (by the hypothesis) $= b^{-1} o a^{-1}$ (by Proposition 4.2.1) $= boa$ (by the hypothesis) for all $a, b \in G$. This proves that (G, o) is abelian.   ♯

2.3. Let (G, o) be a group such that $(aob)^2 = a^2 o b^2$ for all $a, b \in G$. Then, (G, o) is abelian.

Proof $aoaobob = a^2 o b^2 = (aob)^2 = aoboaob$ $\forall a, b \in G$. By the cancelation law, $aob = boa$ $\forall a, b \in G$.   ♯

Remark 4.2.8 1. A group in which $a^3 = e$ for all members a of the group may not be abelian.

2. A group G in which $(aob)^3 = a^3 o b^3$ *for all* $a, b \in G$ need not be abelian. In fact, for each $n \geq 3$, there is a nonabelian group in which $a^n = e$ (and so $(aob)^n = a^n o b^n$ for all a, b) for all a in the group. Example to support this will be given later.

2.4. Let (G, o) be a group and n an integer such that $(aob)^m = a^m o b^m$ for all $m \in \{n, n + 1, n + 2\}$, and all $a, b \in G$. Then, (G, o) is an abelian group.

Proof $aobo(aob)^n = (aob)^{n+1} = a^{n+1} o b^{n+1}$. Since $(aob)^n = a^n o b^n$, we have $aoboa^n o b^n = a^{n+1} o b^{n+1}$. By the cancelation law, we get

$$boa^n = a^n ob \qquad (4.2.1)$$

for all $a, b \in G$. Using the same argument by putting $n + 1$ at the place of n, we get

$$boa^{n+1} = a^{n+1}ob \tag{4.2.2}$$

for all $a, b \in G$. Further, using the above two equations, we get

$$a^{n+1}ob = boa^{n+1} = boa^n oa = a^n oboa$$

for all $a, b \in G$. Canceling a^n from left, we get that $aob = boa$ for all $a, b \in G$. ♯

2.5. Let (G, o) be a finite group containing even number of elements. Then, there is a an element $a \in G, a \neq e$ such that $a^{-1} = a$ (i.e., $a^2 = e$). Further, odd number of such elements exists.

Proof Define a relation $\approx$ on G by

$$a \approx b \text{ if and only if } a = b \text{ or } a = b^{-1}$$

It is easily seen that $\approx$ is an equivalence relation. The equivalence class determined by a is $\{a, a^{-1}\}$. It is singleton, if $a = a^{-1}$, and otherwise, it is doubleton. Let S be the union of all those equivalence classes which are doubletons. Then, S is disjoint union of doubletons, and hence, it contains even number of elements $2m$ (say). Clearly, $e \notin S$. Thus, $S \cup \{e\}$ contains $2m + 1$ elements. Since G contains even number of elements, $G - (S \cup \{e\}) = \{a \in G \mid a \neq e \text{ and } a^{-1} = a\}$ is nonempty and contains odd number of elements.   ǁ

2.6. Converse of 2.5 is also true: Let (G, o) be a finite group in which there is an element $a \neq e$ such that $a = a^{-1}$ $(a^2 = e)$. Then, G contains even number of elements.

Proof Let $a \in G, a \neq e$ and $a^2 = e$. Define a relation $\approx$ on G by

$$x \approx y \text{ if and only if } x = y \text{ or } x = ay$$

It is easy to see that this is an equivalence relation and the equivalence class determined by x is $\{x, ax\}$. Clearly, $x \neq ax$, and so each equivalence class contains 2 elements. Since G is union of disjoint equivalence classes, it contains $2n$ elements for some n.   ♯

2.7. Let (G, o) be a group and $\mid G \mid \leq 4$. Then, G is abelian.

Proof If $\mid G \mid = 1$, then $G = \{e\}$, and nothing to do. If $\mid G \mid = 2$, then it contains two elements of which one is identity, and again, there is nothing to do. Suppose that $\mid G \mid = 3$. Then, $G = \{e, a, b\}$, where e is the identity element and a and b are distinct nonidentity elements of the group. We need to show that $aob = boa$. Now, $aob \neq a$, for otherwise $b = e$. Similarly, $aob \neq b$, for otherwise $a = e$. Thus, $aob = e$. Similarly, $boa = e$. Hence, in this case, also G is abelian.

Next, suppose that $| \, G \, | = 4$. Let $G = \{e, a, b, c\}$, where e is the identity of G and a, b, c are distinct non identity elements of G. Since G is of even order, by illustration 2.5, it contains odd number of nonidentity elements which are their own inverses. There are two cases:

(i) $a^{-1} = a$, $b^{-1} = b$ and $c^{-1} = c$
(ii) Only one of a, b, c is its own inverse.

In case (i), every element of G is its own inverse and so, by illustration 2.2, G is abelian (in fact, in this case, it is left as an exercise to prove that the group is the Klein's four group). Consider the case (ii). Without any loss we can assume that $a^{-1} = a$ and $b^{-1} \neq b$, $c^{-1} \neq c$. Now, $ab \neq e$, for otherwise $b = a^{-1} = a$. Next, $ab \neq a$, for otherwise $b = e$. Also, $ab \neq b$, for otherwise $a = e$. Thus, $ab = c$. Consider b^2. Clearly, $b^2 \neq e$, for otherwise $b^{-1} = b$. $b^2 \neq b$, for otherwise $b = e$. Also $b^2 = c$ implies that $b^2 = ab$. This, in turn, implies that $a = b$. Thus, $b^2 = a$ and $b^3 = ab = c$. This shows that $G = \{e = b^0, b, b^2, b^3\}$, and so it is abelian (see Remark 2.3).   ♯

An Application to Number Theory

Theorem 4.2.9 (Wilson Theorem). *Let p be a positive prime. Then*

$$(p - 1)! + 1 \equiv 0(mod\ p)$$

Equivalently $(p - 1)! - (p - 1)$ is divisible by p (if we divide $(p - 1)!$ by p the remainder is $p - 1$).

Proof Consider the group $(U_p, \star)$ of prime residue classes modulo p. This is an abelian group. Let $\bar{a} \in U_p$, $1 \leq a \leq p - 1$ be its own inverse. Then, $\overline{a^2} = \bar{a}^2 = \bar{1}$. This is equivalent to say that p divides $a^2 - 1 = (a - 1)(a + 1)$. Since p is prime, p divides $a - 1$ or p divides $a + 1$. Since $1 \leq a \leq p - 1$, $a = 1$ or $a = p - 1$. Thus, $\bar{1}$ and $\overline{p - 1}$ are the only elements in U_p which are their own inverses. If we take the product of all elements in U_p (observe that the order will not matter for the group is commutative), the elements which are not their own inverses will cancel with their inverses, and we get,

$$\bar{1} \cdot \bar{2} \cdots \overline{p - 1} = \bar{1} \cdot \overline{p - 1} = \overline{p - 1}.$$

Thus, $\overline{(p - 1)!} = \overline{p - 1} = \overline{-1}$. Equivalently, $\overline{(p - 1)! + 1} = \bar{0}$ in U_p. This means that

$$(p - 1)! + 1 \equiv 0(mod\ p).$$ ♯

Converse of the Wilson, the theorem is also true:

Theorem 4.2.10 *Let $n > 1$. Suppose that $(n - 1)! + 1 \equiv 0(mod\ n)$. Then n is prime.*

Proof Suppose that n is not prime. Then, $n = r \cdot s$, where $1 < r < n$. Clearly, r divides $(n-1)!$. Since n is supposed to divide $(n-1)! + 1$, r also divides $(n-1)! + 1$. This is impossible.   ♯

As an application, we find that 1 is the remainder obtained when $28! \times 30 + 2$ is divided by 29. For, in $\mathbb{Z}_{29}$, $\overline{28! \times 30 + 2} = \overline{28!} \star \overline{30} \oplus \overline{2} = \overline{-1} \star \overline{1} \oplus \overline{2} = \overline{1}$ (by Wilson theorem, $\overline{28!} = \overline{-1}$).

Similarly, when we divide 27! by 29, the remainder is 1. For, $\overline{-1} = \overline{28!} = \overline{27!} \star \overline{28} = \overline{-(27!)}$ (observe that $\overline{28} = \overline{-1}$).

Exercises

4.2.1 Show that a group (G, o) is abelian if and only if $(aob)^{-1} = a^{-1}ob^{-1}$ *for all* $a, b \in G$.

4.2.2 Give an example of a semigroup in which left cancelation law holds but right cancelation does not hold.

4.2.3 In the group $(\mathbb{Z}, +)$, show that a^n is $n \cdot a$ for all $n \in \mathbb{Z}$ and $a \in \mathbb{Z}$, where $\cdot$ denotes the usual product of integers. In particular, $1^n = n$ *for all* $n \in \mathbb{Z}$ (observe that the power is taken with respect to +).

4.2.4 Let (G, o) be a group. Suppose that $a^m = a^n = a^r = e$. Show that $a^d = e$, where d is g.c.d of $m, n,$ and r.

4.2.5 Let (G, o) be a group and $a, b \in G$. Show that $(boaob^{-1})^n = boa^nob^{-1}$ *for all* $n \in \mathbb{Z}$.

4.2.6 Let (G, o) be a group and $a, b \in G$. Suppose that $boaob^{-1} = a^r$. Show that

(i) $b^m oaob^{-m} = a^{(r^m)}$
(ii) $b^m oa^t ob^{-m} = a^{r^m \cdot t}$

for all integers $m, r,$ and t.

4.2.7 Let (G, o) be a group. Let $a, b \in G$ such that $a^5 = e = b^3$. Suppose that $boaob^{-1} = a^2$. Show that $a = e$.

4.2.8 Let (G, o) be a group. Find solutions of equations

(i) $aoXob^2 = b$
(ii) $aoXoa^2 = b$
(iii) $aoXoaoX = boX$

4.2.9 Show that in a group the equation $X^2 = e$ has even number of solutions. Show that there may be more than two solutions.

4.2.10 Do we always have solutions of equation $XoaoX = b$ in a group (G, o)? Support.

4.2.11 Show that every group of order 5 is abelian.

4.2.12 Let X be a finite set. Let G be a nonempty subset of $Sym(X)$ such that $f o g \in G$ for all $f, g \in G$. Show that G is a group with respect to composition of maps.

4.2.13 Show that if p is prime, then $(p - 2)! - 1$ is divisible by p.

4.2.14 Find the remainder when $18! \times 17! + 3$ is divided by 19.

4.2.15 Find the remainder when $100!$ is divided by 101.

4.2.16 Let (G, o) be a finite group containing n elements. Let $a \in G$. Show that there exists $m, 1 \leq m \leq n$ such that $a^m = e$.

4.2.17 Let p be a prime number. Let a be a nonidentity element of a group such that $a^p = e$. Let $i \in \mathbb{N}, 1 \leq i < p$. Show that $a^i \neq e$. Deduce that in this case if G is finite, then the number of elements in G is a multiple of p.
Hint. Define a relation $\approx$ on G by $x \approx y \iff x = a^i y$ for some $i, 0 \leq i < p$. Observe that it is an equivalence relation. Look at the equivalence classes.

4.2.18 Let (G, o) be a finite group of odd order such that 3 does not divide the order of the group. Suppose that $(aob)^3 = a^3 o b^3$ for all $a, b \in G$. Show that G is abelian. Hint. Show that the maps $a \rightsquigarrow a^2$ and $a \rightsquigarrow a^3$ are injective, and so surjective. Further, use cancelation law to show that $a^2 o b^3 = b^3 o a^2$ for all $a, b \in G$.

4.2.19 Show by means of an example that the conclusion of Exercise 4.2.17 need not hold if p is not assumed to be prime.

4.2.20 Suppose that m is the smallest positive integer such that $a^m = e$ (such a m if exists is called the order of a). Show that the conclusion of Exercise 4.2.17 holds good if p is replaced by m (observe by means of an example that such a number m need not be prime).

4.2.21 Let (G, o) be a finite group of odd order. Suppose further that $(aob)^5 = a^5 o b^5$ and $(aob)^3 = b^3 o a^3$ for all $a, b \in G$. Show that (G, o) is abelian.

4.2.22 Let (G, o) be a finite abelian group. Let a be the product of all elements of G. Show that $a^2 = e$.

4.3 Homomorphisms and Isomorphisms

Whenever we have some mathematical object, the first and the foremost thing to study that object is to make it explicit as to when we are going to identify such mathematical objects. For example, to study triangles, the place where the triangle is situated in the space is immaterial. More precisely, two triangles $\triangle_1$ and $\triangle_2$ are taken to be same if there is a rigid motion σ which takes $\triangle_1$ to $\triangle_2$. We identify congruent triangles. If we are interested only in the shape but not the size of the triangles, then we identify similar triangles. In groups, we identify two groups (G_1, o_1) and (G_2, o_2) if they are isomorphic in the sense that there is a bijective map f from G_1 to G_2 which preserves binary operations (i.e., $f(a o_1 b) = f(a) o_2 f(b) \, \forall a, b \in G_1$). More generally, we have the following:

Definition 4.3.1 Let (G_1, o_1) and (G_2, o_2) be groups. A map f from G_1 to G_2 is called a **homomorphism** if

$$f(a o_1 b) = f(a) o_2 f(b)$$

for all $a, b \in G_1$.

An injective homomorphism is called a **monomorphism**. A surjective homomorphism is called an **epimorphism**. A bijective homomorphism is called an **isomorphism**. A homomorphism from a group (G, o) to itself is called an **endomorphism**. An isomorphism from (G, o) to itself is called an **automorphism** of the group.

Remark 4.3.2 Every branch of mathematics, apart from having its applications in other branches of knowledge, has some of its own guiding problems. The researches in that branch are centered around these problems. Of course, one never dreams of solving the problem completely. However, in attempts to solve the problem partially, one develops literature and tools in the subject, and this is how the subject develops. The main problem in group theory is to classify groups up to isomorphism. For example, one may ask: 'How many nonisomorphic groups of order n are there?' and 'what are they?' Theory of finite groups is centered around this problem.

Proposition 4.3.3 *Let (G_1, o_1) and (G_2, o_2) be groups and f a homomorphism from G_1 to G_2. Let e_1 be the identity of (G_1, o_1) and e_2 the identity of (G_2, o_2). Then,*

(i) $f(e_1) = e_2$,
(ii) $f(a^{-1}) = (f(a))^{-1}$ *for all $a \in G_1$,*
(iii) $f(a^n) = (f(a))^n$ *for all $a \in G_1$ and $n \in \mathbb{Z}$, and*
(iv) $f(a_1^{n_1} o_1 a_2^{n_2} o_1 \ldots o_1 a_r^{n_r}) = f(a_1)^{n_1} o_2 f(a_2)^{n_2} o_2 \ldots o_2 f(a_r)^{n_r}$
    for all $a_1, a_2, \ldots, a_r \in G_1$ and $n_1, n_2, \ldots, n_r \in \mathbb{Z}$.

Proof Since e_2, e_1 are identities in the corresponding groups and f is a homomorphism,

$$e_2 o_2 f(e_1) = f(e_1) = f(e_1 o_1 e_1) = f(e_1) o_2 f(e_1).$$

By the cancelation law in a group, $e_2 = f(e_1)$. This proves (i). Further,

$$f(a^{-1})o_2 f(a) = f(a^{-1}o_1 a) = f(e_1) = e_2 = f(a)^{-1}o_2 f(a).$$

By the cancelation law, $f(a^{-1}) = f(a)^{-1}$. This proves (ii). The proof of (iii) for $n \geq 0$ follows by induction and the fact that f is a homomorphism. Suppose that $n = -m$, where $m \geq 0$. Then $f(a^n) = f(a^{-m}) = f((a^m)^{-1}) = (f(a^m)^{-1})$ (by (ii)) $= (f(a)^m)^{-1} = f(a)^{-m} = f(a)^n$. This proves (iii). The proof of (iv) follows from (iii) and the induction on r.   ♯

Proposition 4.3.4 *Composite of any two homomorphisms is a homomorphism.*

Proof Let f be a homomorphism from a group (G_1, o_1) to a group (G_2, o_2) and g a homomorphism from (G_2, o_2) to (G_3, o_3). Then,

$$(gof)(ao_1 b) = g(f(ao_1 b)) = g(f(a)o_2 f(b)) = g(f(a))o_3 g(f(b)) =$$
$$(gof)(a)o_3(gof)(b).$$

Hence, gof is a homomorphism.   ♯

Since the composite of injective maps is injective, the composite of surjective maps is surjective, and the composite of bijective maps is bijective, it follows that the composite of monomorphisms is a monomorphism, the composite of epimorphisms is an epimorphism, and the composite of isomorphisms is an isomorphism.

Proposition 4.3.5 *If f is an isomorphism from a group (G_1, o_1) to a group (G_2, o_2), then f^{-1} is also an isomorphism from (G_2, o_2) to (G_1, o_1).*

Proof Since f is bijective, f^{-1} is also bijective. It is sufficient, therefore, to show that f^{-1} is also a homomorphism. Since f is a homomorphism,

$$f(f^{-1}(c)o_1 f^{-1}(d)) = f(f^{-1}(c))o_2 f(f^{-1}(d)) = co_2 d$$

for all $c, d \in G_2$. Further, since f is bijective

$$f^{-1}(co_2 d) = f^{-1}(c)o_1 f^{-1}(d)$$

for all $c, d \in G_2$. This shows that f^{-1} is an isomorphism.   ♯

It follows from the above propositions that the relation 'is isomorphic to' is an equivalence relation on any set of groups. The notation $G_1 \approx G_2$ will stand to say that G_1 is isomorphic to G_2.

Remark 4.3.6 The counterpart of homomorphisms in the category of metric spaces (topological spaces) are continuous maps. The reader may note the difference between an algebraic category and a topological category by observing that the inverse of a bijective continuous map need not be a continuous map.

Example 4.3.7 Let (G_1, o_1) and (G_2, o_2) be groups. The constant map from G_1 to G_2 which maps each element of G_1 to the identity e_2 of G_2 is a homomorphism. This homomorphism is called the **trivial homomorphism or zero homomorphism**.

Example 4.3.8 Let (G, o) be a group. The identity map I_G on G is clearly an isomorphism. Thus, it is also an automorphism of (G, o).

Example 4.3.9 Let $m \in \mathbb{Z}$. The map f_m from $\mathbb{Z}$ to $\mathbb{Z}$ given by $f_m(a) = ma$ is a homomorphism from the additive group $(\mathbb{Z}, +)$ to itself(verify). We show that every homomorphism from $(\mathbb{Z}, +)$ to itself is f_m for some m in $\mathbb{Z}$. Let f be a homomorphism from $(\mathbb{Z}, +)$ to itself. Suppose that $f(1) = m$. Then, as observed earlier, n is the nth additive power of 1 in the group $(\mathbb{Z}, +)$. From Proposition 4.3.3, it follows that $f(n)$ is the nth additive power of $f(1) = m$. Thus, $f(n) = nm$, and so $f = f_m$. Clearly, f_m is the zero homomorphism if $m = 0$, and it is injective if $m \neq 0$. This also shows that a homomorphism from $(\mathbb{Z}, +)$ to itself is either zero homomorphism or a monomorphism. Further, the map $m \rightsquigarrow f_m$ is a bijective map from $\mathbb{Z}$ to the set $End(\mathbb{Z}, +)$ of all endomorphisms of the additive group $(\mathbb{Z}, +)$ of integers. Note that f_m is an automorphism if and only if $m = \pm 1$. Also, observe that $End(\mathbb{Z}, +)$ is a semigroup with respect to composition of maps and $m \rightsquigarrow f_m$ is an isomorphism from the semigroup $(\mathbb{Z}, \cdot)$ to $End(Z, +)$.

Example 4.3.10 Let $(\mathbb{R}, +)$ denote the additive group of real numbers and $(\mathbb{R}^+, \cdot)$ the multiplicative group of positive real numbers. Let a be a positive real number. Define a map $f_a : \mathbb{R} \longrightarrow \mathbb{R}^+$ by $f_a(x) = a^x$. Then, it follows from the law of exponents that f_a is a homomorphism. Also, f_a is bijective, and the inverse of f_a is the map log_a which maps y to $log_a y$. Thus, f_a is an isomorphism, and so the groups $(\mathbb{R}, +)$ and $(\mathbb{R}^+, \cdot)$ are isomorphic.

Example 4.3.11 The map $f : \mathbb{R} \longrightarrow S^1$ defined by $f(x) = e^{\iota x}$ is a homomorphism(law of exponents). Further, if $z = a + b\iota \in S^1$, then $a^2 + b^2 = 1$. Hence, there is an angle $\theta \in \mathbb{R}$ such that $a = cos\theta, b = sin\theta$. But, then $z = cos\theta + \iota sin\theta = e^{\iota \theta} = f(\theta)$. This shows that f is a surjective homomorphism.

Example 4.3.12 Let (G, o) be a group and $a \in G$. Define a map f_a from $\mathbb{Z}$ to G by $f(n) = a^n$. Then, from the law of exponents, it follows that f_a is a homomorphism with $f_a(1) = a$. This homomorphism, therefore, is completely determined by its image on 1. Thus, $a \rightsquigarrow f_a$ defines a bijective map from G to the set $Hom(\mathbb{Z}, G)$ of all homomorphisms from the additive group of integers to the group G.

Illustrations

3.1. Any homomorphism from the additive group $(\mathbb{Q}, +)$ of rational numbers to itself is multiplication by a rational number. In particular, it is a zero homomorphism or it is an isomorphism.

Proof Let f be a homomorphism from $(\mathbb{Q}, +)$ to itself. Let $f(1) = r \in \mathbb{Q}$. Then, as in Example 4.3.9, $f(m) = rm$ for all $m \in \mathbb{Z}$. Let $m \neq 0$ and $f(\frac{1}{m}) = t$. Then, if $m > 0$,

$$f(1) = \underbrace{f(\frac{1}{m}) + f(\frac{1}{m}) + \cdots + f(\frac{1}{m})}_{m}$$

and if $m < 0$,

$$f(1) = \underbrace{f(-\frac{1}{m}) + f(-\frac{1}{m}) + \cdots + f(-\frac{1}{m})}_{-m}$$

This means that $r = f(1) = mt$. In turn, $t = f(\frac{1}{m}) = r \cdot \frac{1}{m}$. Hence, $f(\frac{n}{m}) = n \cdot f(\frac{1}{m}) = n \cdot \frac{r}{m} = r \cdot \frac{n}{m}$. Thus, $f(s) = r \cdot s$ for all $s \in \mathbb{Q}$. If $r = 0$, then f is a zero homomorphism. If $r \neq 0$, then it is bijective, and so it is an isomorphism. ♯

3.2. Let (G, o) be a finite group. Then the only homomorphism from (G, o) to $(\mathbb{Z}, +)$ (or to $(\mathbb{Q}, +)$ or to $(\mathbb{R}, +)$) is the zero homomorphism. In particular, there is no nontrivial homomorphism from the additive group $\mathbb{Z}_m$ of residue classes modulo m to any of the groups $(\mathbb{Z}, +)$, $(\mathbb{Q}, +)$ *or* $(\mathbb{R}, +)$.

Proof Let f be a homomorphism from G to $\mathbb{Z}$. Suppose that $| G | = n$. Let $a \in G$. Then *there exists* m, $1 \le m \le n$ such that $a^m = e$ (see ex 4.2.16). Since f is a homomorphism, $0 = f(e) = f(a^m) = m \cdot f(a)$. Since $m \neq 0$, $f(a) = 0$. This shows that f is a zero homomorphism. ♯

3.3. There is no nontrivial homomorphism from

(i) $(\mathbb{Q}, +)$ to $(\mathbb{Z}, +)$
(ii) $(\mathbb{Q}, +)$ to $(\mathbb{Z}_m, \oplus)$
(iii) $(\mathbb{Q}, +)$ to $(\mathbb{Q}^*, \cdot)$.

In particular, these pair of groups are not isomorphic.

Proof Any homomorphism from $(\mathbb{Q}, +)$ to $(\mathbb{Z}, +)$ can viewed as a homomorphism from $(\mathbb{Q}, +)$ to $(\mathbb{Q}, +)$ such that the image is contained in $\mathbb{Z}$. From the illustration 3.1, it follows that any such homomorphism is multiplication by a rational number. The result (i) follows if we note that there is no nonzero rational number such that all rational multiples of that rational numbers are integers. For (ii), let f be a homomorphism from $(\mathbb{Q}, +)$ to $(\mathbb{Z}_m, \oplus)$. Let $r \in \mathbb{Q}$. Then $f(r) = m \cdot f(\frac{r}{m}) = m\bar{u} = \overline{mu} = \bar{0}$. This proves that f is the trivial homomorphism. Further, to prove (iii) let f be a homomorphism from $(\mathbb{Q}, +)$ to $(\mathbb{Q}^*, \cdot)$. Let $r \in \mathbb{Q}$. Then, since f is a homomorphism, $f(r) = f(\frac{r}{m})^m$. Thus, $f(\frac{r}{m}) = f(r)^{\frac{1}{m}} \in \mathbb{Q}^*$ *for all* $m \in \mathbb{Z}$. Since there is no rational number except 1 all of whose roots are rational, it follows that $f(r) = 1$ for all $r \in \mathbb{Q}$. ♯

3.4. There is no homomorphism f from $(\mathbb{Z}, +)$ to $(\mathbb{Q}^*, \cdot)$ such that $f(2) = \frac{1}{3}$

Proof If $f(1) = r$, then $f(2) = r^2 = \frac{1}{3}$. Since there is no rational number r such that $r^2 = \frac{1}{3}$, it follows that there is no such homomorphism. ♯

Recall that an endomorphism of a group (G, o) is a homomorphism from (G, o) to itself. $End(G, o)$ denotes the set of all endomorphisms of the group (G, o). Since composition of homomorphisms is homomorphisms, $End(G, o)$ is a semigroup with identity with respect to the composition of maps. Further, an automorphism of the group (G, o) is an isomorphism from (G, o) to itself. $Aut(G, o)$ denotes the set of all automorphisms of (G, o) which is a group (composition of automorphisms is automorphisms, inverse of an automorphism is an automorphism, and identity map on G is also an automorphism) with respect to composition of maps. We compute $End(G, o)$ and $Aut(G, o)$ for some groups.

3.5. (i) $End(\mathbb{Z}, +) \approx (\mathbb{Z}, \cdot)$ as a semigroup.

(ii) $Aut(\mathbb{Z}, +) \approx (\{1, -1\}, \cdot)$ (note that $\{1, -1\}$ is a group with respect to multiplication).

(iii) $End(\mathbb{Q}, +) \approx (\mathbb{Q}, \cdot)$ as a semigroup.

(iv) $Aut(\mathbb{Q}, +) \approx (\mathbb{Q}^\star, \cdot)$.

(v) $End(\mathbb{Z}_m, \oplus) \approx (\mathbb{Z}_m, \star)$ as a semigroup.

(vi) $Aut(\mathbb{Z}_m, \oplus) \approx (U_m, \star)$.

Proof (i) Define a map $\eta : End(\mathbb{Z}, +) \longrightarrow \mathbb{Z}$ by $\eta(f) = f(1)$. It follows from Example 4.3.9 that η is bijective and $\eta(gof) = gof(1) = g(f(1)) = g(1) \cdot f(1) = \eta(g) \cdot \eta(f)$. Thus, η is an isomorphism.

(ii) Let $f \in Aut(\mathbb{Z}, +)$. Then, again by Example 4.3.9, $f(a) = a \cdot f(1)$. Since f is bijective, $f(1) = \pm 1$. Thus, η defined in (i) when restricted to $Aut(\mathbb{Z}, +)$ defines an isomorphism from $Aut(\mathbb{Z}, +)$ to the group $(\{1, -1\}, \cdot)$.

The proof of (iii) is similar to that of (i), and the proof of (iv) is similar to that of (ii) provided we note that multiplication by a rational number r from $\mathbb{Q}$ to $\mathbb{Q}$ is bijective if and only if $r \neq 0$.

(v) Define a map η from $End(\mathbb{Z}_m, \oplus)$ to $\mathbb{Z}_m$ by $\eta(f) = f(\overline{1})$. Suppose that $\eta(f) = \eta(g)$. Then, $f(\overline{1}) = g(\overline{1})$. But, then

$$f(\overline{r}) = f(\underbrace{\overline{1} \oplus \overline{1} \oplus \cdots \oplus \overline{1}}_{r}) = \underbrace{f(\overline{1}) \oplus f(\overline{1}) \oplus \cdots \oplus f(\overline{1}))}_{r} =$$

$$\underbrace{g(\overline{1}) \oplus g(\overline{1}) \oplus \cdots \oplus g(\overline{1}))}_{r} = g(\underbrace{\overline{1} \oplus \overline{1} \oplus \cdots \oplus \overline{1}}_{r}) = g(\overline{r})$$

Thus, $f = g$. This shows that η is injective. Next, let $\overline{r} \in \mathbb{Z}_m$. Define a map f from $\mathbb{Z}_m$ to $\mathbb{Z}_m$ by $f(\overline{a}) = \overline{r} \star \overline{a}$. Then, $f \in End(\mathbb{Z}_m, \oplus)$ (verify) and $\eta(f) = \overline{r}$. Hence, η is also surjective. Further,

$$\eta(gof) = gof(\overline{1}) = g(f(\overline{1})) = g(\overline{1}) \star f(\overline{1}) = \eta(g) \star \eta(f)$$

This proves that η is an isomorphism.

(vi) We first show that an endomorphism $f \in End(\mathbb{Z}_m, \oplus)$ defined by $f(\bar{a}) = \bar{r} \star \bar{a}$ is an automorphism if and only if $\bar{r} \in U_m$. Suppose that $\bar{r} \in U_m$. Then, there exists a $\bar{s} \in U_m$ such that $\bar{r} \star \bar{s} = \bar{1}$. Define a map g from $\mathbb{Z}_m$ to itself by $g(\bar{a}) = \bar{s} \star \bar{a}$. Then,

$$gof(\bar{a}) = g(f(\bar{a})) = g(\bar{r} \star \bar{a}) = \bar{s} \star \bar{r} \star \bar{a} = \bar{1} \star \bar{a} = \bar{a}$$

Thus, gof is the identity map. Similarly, it follows that fog is also the identity map. Hence, f is an automorphism. It follows that the map η defined in (v) induces an isomorphism from $Aut(\mathbb{Z}_m, \oplus)$ to $(U_m, \star)$.   ♯

3.6. Let (G, o) be a group and $a \in G$. Define a map f_a from G to G by $f_a(x) = aoxoa^{-1}$. Then, f_a is an automorphism of (G, o) (verify) called an **inner automorphism** of (G, o) determined by a. It is easy to check that $f_{aob} = f_a of_b$ for all $a, b \in G$. This shows that the map f from the group G to the group $Aut(G, o)$ defined by $f(a) = f_a$ is a homomorphism.

3.7. Let (G_1, o_1) and (G_2, o_2) be groups. Let $Hom(G_1, G_2)$ denote the set of all homomorphisms from G_1 to G_2. Let f and g be members of $Hom(G_1, G_2)$. Define a map $f \star g$ from G_1 to G_2 by

$$(f \star g)(x) = f(x)o_2g(x), \ x \in G_1.$$

It is easy to check that $f \star g \in Hom(G_1, G_2)$ provided that each element of $f(G_1)$ commutes with each element of $g(G_1)$. Assume that (G_2, o_2) is an abelian group. Then, $\star$ defines a binary operation on $Hom(G_1, G_2)$, and $Hom(G_1, G_2)$ is an abelian group with respect to $\star$. The constant map $\bar{e}$ from G_1 to G_2 defined by $\bar{e}(x) = e_2$ is the identity of the group. The inverse f^{-1} of f is given by $f^{-1}(x) = f(x)^{-1}$. The commutativity of $\star$ follows from that of o_2. We have the following:

 (i) For all abelian group (G, o), $Hom(\mathbb{Z}, G) \approx G$.
 (ii) $Hom(\mathbb{Z}_m, \mathbb{Z}_n) \approx \mathbb{Z}_d$, where d is the greatest common divisor of m and n.

Proof (i) The map η from $Hom(Z, G)$ to G defined by $\eta(f) = f(1)$ is easily
        seen to be an isomorphism.   ♯
 (ii) We define a map η from $Hom(\mathbb{Z}_m, \mathbb{Z}_n)$ to $\mathbb{Z}_d$ as follows: Let $f \in Hom(\mathbb{Z}_m, \mathbb{Z}_n)$.
        Suppose that $f(\bar{1}) = \bar{r} \in \mathbb{Z}_n$. Then,

$$\bar{0} = f(\bar{0}) = f(\bar{m}) = m\bar{r}$$

This implies that n divides $m \cdot r$. Let $n = du$ and $m = dv$. Then, $(u, v) = 1$ and du divides dvr. Hence, u divides vr, Since $(u, v) = 1$, u divides r. Thus, $\frac{r}{u} \in \mathbb{Z}$. Now, define $\eta(f) = \frac{\bar{r}}{u} = \frac{dr}{n}$ in $\mathbb{Z}_d$. If $\bar{r} = \bar{s}$ in $\mathbb{Z}_n$, then n divides $r - s$, and so d divides $\frac{dr}{n} - \frac{ds}{n}$. Thus, η is a map from $Hom(\mathbb{Z}_m, \mathbb{Z}_n)$ to $\mathbb{Z}_d$. Next, let $f, g \in Hom(\mathbb{Z}_m, \mathbb{Z}n)$. Suppose that $f(\bar{1}) = \bar{r}$ and $g(\bar{1}) = \bar{s}$. Then, $(f + g)(\bar{1}) = \bar{r} \oplus \bar{s} = \overline{r + s}$. Hence,

$$\eta(f+g) = \frac{\overline{d(r+s)}}{n} = \frac{\overline{dr}}{n} \oplus \frac{\overline{ds}}{n} = \eta(f) \oplus \eta(g)$$

This shows that η is a homomorphism. Next, suppose that

$$\eta(f) = \frac{\overline{dr}}{n} = \eta(g) = \frac{\overline{ds}}{n},$$

where $f(\overline{1}) = \overline{r}$ and $g(\overline{1}) = \overline{s}$. Then, d divides $\frac{dr}{n} - \frac{ds}{n}$. But, then $\frac{r-s}{n}$ is an integer. This means that $\overline{r} = \overline{s}$ in $\mathbb{Z}_n$. In turn,

$$f(\overline{t}) = t\overline{r} = t\overline{s} = g(\overline{t})$$

for all $\overline{t} \in \mathbb{Z}_m$. This shows that $f = g$, and so η is injective. Finally, let $\overline{a} \in \mathbb{Z}_d$. Define a relation f from $\mathbb{Z}_m$ to $\mathbb{Z}_n$ by $f(\overline{t}) = \frac{\overline{tna}}{d}$. It is easily seen that f is a map which is also a homomorphism. Further, since $\frac{n}{d}$ and d are co-prime, $f(\overline{1}) = \frac{\overline{na}}{d} = \overline{a}$. Thus, η is also surjective. ♯

Exercises

4.3.1 Let (G, o) be a group. Show that the map $a \rightsquigarrow a^{-1}$ from G to itself is a homomorphism if and only if the group is abelian.

4.3.2 Show that the map $a \rightsquigarrow a^2$ from a group to itself is a homomorphism if and only if the group is abelian.

4.3.3 Show that the conjugation map $z \rightsquigarrow \overline{z}$ from $(\mathbb{C}, +)$ to itself is an automorphism.

4.3.4 Show that $(U_5, \star) \approx (\mathbb{Z}_4, \oplus)$.

4.3.5 Show that $(\mathbb{R}^*, \cdot)$ is not isomorphic to the group $(\mathbb{R}, +)$.

4.3.6 Find all homomorphisms from the Quaternion group Q_8 to the Klein's four group V_4 and also all homomorphisms from V_4 to Q_8. Do we have a monomorphism from V_4 to Q_8?

4.3.7 Find out all homomorphisms from $(\mathbb{Z}, +)$ to V_4.

4.3.8 Find the number of homomorphisms from $(\mathbb{Z}_{16}, \oplus)$ to $(\mathbb{Z}_{12}, \oplus)$. What is the group $Hom(\mathbb{Z}_{16}, \mathbb{Z}_{12})$?

4.3.9 Let (G, o) be a group and $Hom(Z, G)$ denote the set of all homomorphisms from the additive group of integers to the group (G, o). Show that the map η from $Hom(Z, G)$ to G defined by $\eta(f) = f(1)$ is a bijective map.

4.3.10 Let (G, o) be a group and $X = \{a \in G \mid a^m = e\}$. Define a map η from $Hom(\mathbb{Z}_m, G)$ to G by $\eta(f) = f(\overline{1})$. Show that η is injective map and its image is X.

4.3.11 Show, by means of an example, that if (G_2, o_2) in illustration 3.7 is not abelian, then $f \star g$ need not belong to $Hom(G_1, G_2)$ even if f and g are homomorphisms.

4.3.12 Show, by means of an example, that even if the group (G_2, o_2) in the illustration 3.7 is nonabelian, the product $f \star g$ of a pair of homomorphisms f and g may be a homomorphism.

4.3.13 A pair f, g of homomorphisms from a group (G_1, o_1) to a group (G_2, o_2) is said to be **summable** if $f \star g$ is also a homomorphism. Show that a pair f, g of homomorphisms from a group (G_1, o_1) to a group (G_2, o_2) is **summable** if and only if each element of $f(G_1)$ commutes with each element of $f(G_2)$.

4.3.14 Show that any two endomorphisms of Q_8 which are not automorphisms are summable. Indeed, image of any endomorphism which is not an automorphism is trivial or else $\{1, -1\}$.

4.3.15 Characterize groups in which all pairs of endomorphisms are summable.

4.3.16 Show that there is no nontrivial homomorphism from the circle group S^1 (or the multiplicative group P of roots of unity) to the additive group $\mathbb{Z}$ of integers.

4.3.17 Show that every continuous homomorphism from the additive group $\mathbb{R}$ of real numbers to itself is multiplication by a real number. Deduce that the group $Hom_c(\mathbb{R}, \mathbb{R})$ of all continuous homomorphisms from $\mathbb{R}$ to $\mathbb{R}$ is isomorphic to the additive group $\mathbb{R}$ of real numbers.

4.3.18 Show that the set of all continuous automorphisms of the additive group $\mathbb{R}$ of real numbers is a group with composition of maps which is isomorphic to the multiplicative group $\mathbb{R}^\star$ of nonzero real numbers.

4.3.19 Find all continuous homomorphisms from the additive group $\mathbb{R}$ of real numbers to the multiplicative group $\mathbb{R}^+$ of positive real numbers. Show that the group $Hom_c(\mathbb{R}, \mathbb{R}^+)$ (with pointwise multiplication) of all continuous homomorphisms from the additive group $\mathbb{R}$ of real numbers to the multiplicative group $\mathbb{R}^+$ of positive real numbers is isomorphic to the multiplicative group of positive real numbers.

4.3.20 Show that there is no homomorphism f from $(\mathbb{Z}, +)$ to $(\mathbb{Q}^\star, \cdot)$ such that $f(5) = 2^6$.

4.3.21 Find all continuous homomorphisms from S^1 to itself.

4.3.22 Show that the group $Hom(Q_8, V_4)$ is isomorphic to the external direct product $V_4 \times V_4$.

4.3.23 Find out all members of $Aut Q_8$ and also $Aut V_4$. Are they isomorphic? Support.

4.3.24 Show that $G_1 \approx G_2$ implies that $Aut G_1 \approx Aut G_2$. Show further by means of an example that $Aut G_1 \approx Aut G_2$ does not imply that $G_1 \approx G_2$.

4.3.25 Let f be a homomorphism from $(\mathbb{Q}^+, \cdot)$ to $(\mathbb{Q}, +)$. Suppose that $f(3) = -1$. Find $f(27)$.

4.3.26 Show that $(\mathbb{C}^*, \cdot)$ and $(\mathbb{R}^*, \cdot)$ are not isomorphic.

4.3.27 Let X denote a square in $\mathbb{R}^2$. Show that $Iso_X(\mathbb{R}^2)$ is not isomorphic to Q_8, whereas they are nonabelian groups of same order.

4.3.28 Let (G, o) be an abelian group and $c \in G$. Consider the group $(G, \star)$, where $a \star b = aoboc^{-1}$. Define a map f from G to G by $f(a) = aoc$. Show that f is an isomorphism.

4.3.29 Show that $\mathbb{Z}_4$ is not isomorphic to V_4. Show also that $Hom(\mathbb{Z}_4, V_4) \approx V_4$.

4.3.30 Show that a homomorphism f from a group (G_1, o_1) to (G_2, o_2) is injective if and only if $f^{-1}(\{e_2\}) = \{e_1\}$.

4.3.31 Let (G, o) be a group such that $a \rightsquigarrow a^n$, $a \rightsquigarrow a^{n+1}$, and $a \rightsquigarrow a^{n+2}$ are homomorphisms, $n \geq 1$. Show that (G, o) is abelian.

4.3.32 Let (A_1, o_1) and (A_2, o_2) be abelian groups and f a homomorphism from A_1 to A_2. Let (G, o) be a group. Define a map $f_\star$ from $Hom(G, A_1)$ to $Hom(G, A_2)$ by $f^\star(g) = fog$. Show that $f_\star$ is a homomorphism which is injective provided that f is injective. Suppose further that (G, o) is also abelian. Define a map $f^\star$ from $Hom(A_2, G)$ to $Hom(A_1, G)$ by $f^\star(g) = gof$. Show that $f^\star$ is a homomorphism which is surjective provided that f is surjective.

4.3.33 Show that $(\mathbb{R}, +)$ and $(S^1, \cdot)$ are not isomorphic.

4.3.34 Let G be a finite group and f an automorphism of G such that $f(x) = x$ *implies that* $x = e$. Show that the map g from G to G defined by $g(x) = x^{-1} f(x)$ is bijective. Suppose that $fof = I_G$. Show that G is abelian and $f(x) = x^{-1}$ *for all* $x \in G$.

4.3.35 Let (G, o) be a group containing more than two elements. Show that there is a nonidentity automorphism of (G, o).
Hint. If there is an element $a \in G$ which does not commute with all elements of G, then the inner automorphism determined by a is nonidentity automorphism. If G is abelian and $a^{-1} \neq a$ for some a in G, then $a \rightsquigarrow a^{-1}$ is a nonidentity automorphism. If $a = a^{-1}$ for all $a \in G$, then take any two distinct nonidentity elements $a, b \in G$ and define a map f from G to G by $f(a) = b$, $f(b) = a$, and $f(x) = x$ for all x different from a, b. Check that f is a nonidentity automorphism.

4.3.36 Determine the groups $Hom((\mathbb{Q}, +), (\mathbb{R}, +))$, $Hom((\mathbb{Q}, +), (\mathbb{C}, +))$, $Hom_c((\mathbb{R}, +), (\mathbb{R}, +))$, $Hom_c((\mathbb{R}, +), (S^1, \cdot))$, and $Hom_c((S^1, \cdot), (S^1, \cdot))$.

4.3.37 Let f be a nontrivial homomorphism from $(\mathbb{Q}, +)$ to $(\mathbb{R}, +)$. Show that $f(\mathbb{Q})$ is dense in $\mathbb{R}$.

4.3.38 Let f be a homomorphism from $(\mathbb{R}, +)$ to $(\mathbb{R}, +)$ $((S^1, \cdot))$ which is continuous at a point. Show that f is continuous (indeed differentiable) at all points.

4.3.39 Let f be an automorphism of $(\mathbb{R}, +)$ which is also semigroup homomorphism from the semigroup $(\mathbb{R}, \cdot)$ to itself. Show that f is the identity map.

4.4 Generation of Groups

To study a mathematical structure, it is always good to have sufficiently many examples of that structure. This is because examples give deep insight to the structure. Thus, one tries to develop different construction processes by which one can construct different examples from known examples. The first step in this direction would be to see as to how to construct groups out of subsets of a group G.

Let (G, o) be a group and H a nonempty subset of G. To make H a group, we need a binary operation on H. For that purpose, we may look at the binary operation o of G. If $aob \in H$ for all $a, b \in H$, then the binary operation o of G induces a binary operation o' on H defined by $ao'b = aob$ for all $a, b \in H$. Even now, H together with the induced binary operation o' need not be a group. For example, $(\mathbb{Z}, +)$ is a group, where $+$ induces binary operation on $\mathbb{N}$ but $\mathbb{N}$ together with the induced addition is not a group. This motivates to have the following definition.

Definition 4.4.1 Let (G, o) be a group. A nonempty subset H of G is called a **subgroup** of (G, o) if the binary operation o of G induces a binary operation on H with respect to which H is a group.

Thus, if H is a subgroup, then $aob \in H$ for all $a, b \in H$.

Proposition 4.4.2 *Let (G, o) be a group and H a subgroup of G. Denote the induced binary operation on H by o'. Let e be the identity of the group (G, o) and e' the identity of the group (H, o'). Let $a \in H$. Let a^{-1} denote the inverse of a considered as an element of the group (G, o) and a_H^{-1} the inverse of a considered as an element of the group (H, o'). Then,*

(i) $e = e'$
   and
(ii) $a^{-1} = a_H^{-1}$

Proof Since o' is the induced operation on H, $aob = ao'b$ for all $a, b \in H$. Further, since e is the identity of G and e' is the identity of H, we have

$$e'oe' = e'o'e' = e' = eoe'.$$

By cancellation law in (G, o), $e = e'$. Further,

$$a_H^{-1} oa = a_H^{-1} o'a = e' = e = a^{-1}oa.$$

By cancelation law, $a_H^{-1} = a^{-1}$.   ♯

The following corollary is immediate.

Corollary 4.4.3 *Let (G, o) be a group. Then a subset H of G is a subgroup if and only if*

(i) $aob \in H \ \forall a, b \in H$.
(ii) $e \in H$.
(iii) $a^{-1} \in H$ for all $a \in H$.   ♯

Proposition 4.4.4 *Let (G, o) be a group. A nonempty subset H of G is a subgroup if and only if $aob^{-1} \in H (a^{-1}ob \in H)$ for all $a, b \in H$.*

Proof If H is a subgroup and $a, b \in H$, then from the above corollary $a, b^{-1} \in H$, and so $aob^{-1} \in H (a^{-1}ob \in H)$. Conversely, suppose that $H \neq \emptyset$ and aob^{-1} for all $a, b \in H$. Since $H \neq \emptyset$, there is an element $a \in H$. Hence, $e = aoa^{-1} \in H$. Further, since $e \in H$, if $a \in H$, then $a^{-1} = eoa^{-1} \in H$. Finally, if $a, b \in H$, then as already proved, $b^{-1} \in H$. Hence, $aob = ao(b^{-1})^{-1} \in H$. From the above corollary, it follows that H is a subgroup of G.   ♯

The proof of the following proposition is by the induction and is left as an exercise.

Proposition 4.4.5 *Let H be a subgroup of (G, o). Then,*

(i) $a^n \in H$ for all $a \in H$ and $n \in \mathbb{Z}$.
(ii) $a_1^{n_1} oa_2^{n_2} o \cdots oa_r^{n_r} \in H$ whenever $a_1, a_2, \ldots, a_r \in H$ and $n_1, n_2, \ldots, n_r \in \mathbb{Z}$. ♯

Proposition 4.4.6 *Let (G, o) be a group. Then a finite nonempty subset H of G is a subgroup if and only if $aob \in H$ for all $a, b \in H$.*

Proof The condition is necessary because of Corollary 4.4.2 (i) Conversely, suppose that H is a nonempty finite subset of G such that $aob \in H$ for all $a, b \in H$. Then, the binary operation o induces a binary operation o' on H. Clearly, (H, o') is a nonempty finite semigroup in which cancelation law holds. Hence, (H, o') is a group at its own right. This means that H is a subgroup of G.   ♯

Example 4.4.7 $\{e\}$, where e is the identity of the group (G, o), is a subgroup of (G, o) called the **trivial subgroup**. G is also a subgroup of G called the **improper subgroup** of (G, o). Other subgroups of G are called **proper subgroups** of (G, o).

Example 4.4.8 $\mathbb{Z}$ is a subgroup of $(\mathbb{Q}, +)$, $\mathbb{Q}$ is a subgroup of $(\mathbb{R}, +)$, and $\mathbb{R}$ is a subgroup of $(\mathbb{C}, +)$.

Example 4.4.9 S^1 is a subgroup of $(\mathbb{C}^*, \cdot)$, and the set P of roots of unity is a subgroup of $(S^1, \cdot)$.

Example 4.4.10 $\{e, a\}, \{e, b\}, \{e, c\}$ are all proper subgroups of V_4 (verify). Thus, there are five subgroups of V_4: one trivial, one improper, and three proper subgroups of V_4.

Example 4.4.11 Nontrivial proper subgroups of the Quaternion group Q_8 can be enumerated as $\{1, -1\}, \{1, i, -1, -i\}, \{1, j, -1, -j\}$, and $\{1, k, -1, -k\}$. Thus, there are 6 subgroups of Q_8.

Example 4.4.12 In this example, we find all subgroups of $(\mathbb{Z}, +)$. Let $m \in \mathbb{N} \bigcup \{0\}$. Then, $m\mathbb{Z} = \{mr \mid r \in \mathbb{Z}\}$ is a subgroup of $\mathbb{Z}$ (verify). They are all distinct, for if $m\mathbb{Z} = n\mathbb{Z}$, then $m = nr$ and $n = ms$ for some $r, s \in \mathbb{N} \bigcup \{0\}$. But then $m = msr$. Clearly, $m = 0$ *if and only if* $n = 0$. If $m \neq 0$, then, $rs = 1$. Since $r, s \in \mathbb{N}, r = 1 = s$. Thus, $m\mathbb{Z} = n\mathbb{Z}$ if and only if $m = n$. Lastly, we show that every subgroup of $(\mathbb{Z}, +)$ is of the form $m\mathbb{Z}$ for some $m \in \mathbb{N} \bigcup \{0\}$. Let H be a subgroup of $(\mathbb{Z}, +)$. If $H = \{0\}$, then $H = 0 \cdot \mathbb{Z}$. Suppose that $H \neq \{0\}$. Then, *there exists* $r \in H - \{0\}$. Since H is a subgroup of $(\mathbb{Z}, +)$, the inverse $-r$ of r belongs to H. Hence, there is an element $r \in \mathbb{N} \bigcap H$, and so $\mathbb{N} \bigcap H \neq \emptyset$. By the well-ordering principle in $\mathbb{N}$, $\mathbb{N} \bigcap H$ has the least element m (say). We show that $H = m\mathbb{Z}$. Since $m \in H$ and H is a subgroup of $(\mathbb{Z}, +)$, $mr \in H$ *for all* $r \in \mathbb{Z}$. Thus, $m\mathbb{Z} \subseteq H$. Let $h \in H$. Since $m \neq 0$, by the division algorithm, *there exist* $q, r \in \mathbb{Z}$ such that $h = mq + r$, where $r = 0$ or else $0 < r < m$. Since $h \in H$ and also $mq \in H$, and H is a subgroup, $r = h - mq \in H$. Hence, $r = 0$, for otherwise $r \in \mathbb{N} \bigcap H$ and $r < m$, a contradiction to the choice of m. Thus, $h = mq \in m\mathbb{Z}$. To summarize, the set $S(\mathbb{Z}, +)$ of all subgroups of $(\mathbb{Z}, +)$ is $\{m\mathbb{Z} \mid m \in \mathbb{N} \bigcup \{0\}\}$ and there is a bijection from the set $S(\mathbb{Z}, +)$ to $\mathbb{N} \bigcup \{0\}$ given by $m\mathbb{Z} \rightsquigarrow m$.

Example 4.4.13 In this example, we find all subgroups of $(\mathbb{Z}_m, \oplus)$, where $m > 0$. $\{\overline{0}\}$ is the trivial subgroup. Let $H \neq \{\overline{0}\}$ be a subgroup of $(\mathbb{Z}_m, \oplus)$. Then, *there exists* $r, 0 < r < m$ such that $\overline{r} \in H$. Thus, the set $\{r \in \mathbb{N} \mid \overline{r} \in H\}$ is nonempty set. By well-ordering property in $\mathbb{N}$, this set has the least element t (say). We first show that t divides m. By the division algorithm, *there exist* $q, r \in \mathbb{Z}$ such that $m = tq + r$, where $r = 0$ or else $0 < r < t$. But, then $\overline{0} = \overline{m} = q\overline{t} + \overline{r}$. Thus, $\overline{r} = -q\overline{t} \in H$. Hence, $r = 0$, for otherwise $\overline{r} \in H$ and $0 < r < t$. Thus, t divides m. Let $m = qt$. Then, we show that

$$H = \{\overline{0}, \overline{t}, 2\overline{t}, \ldots, (q - 1)\overline{t}\}.$$

Since $\overline{t} \in H$ and H is a subgroup,

$$\{\overline{0}, \overline{t}, 2\overline{t}, \ldots, (q - 1)\overline{t}\} \subseteq H$$

Let $\bar{a} \in H$, $0 \leq a \leq m - 1$. By the division algorithm *there exist* $l, r \in \mathbb{Z}$ such that

$$a = lt + r,$$

where $r = 0$ or else $0 < r < t$. Since $\bar{a}$ and $\bar{t}$ belong to H and H is a subgroup, $\bar{r} = \bar{a} - l\bar{t} \in H$. Hence, $r = 0$, for otherwise $\bar{r} \in H$ and $0 < r < t$, a contradiction to the choice of t. Hence, $\bar{a} = \overline{lt} = l\bar{t}$, where $l \in \{0, 1, 2, \ldots, q-1\}$. This shows that

$$H = \{\bar{0}, \bar{t}, 2\bar{t}, \ldots, (q-1)\bar{t}\}.$$

Thus, every subgroup H of $(\mathbb{Z}_m, \oplus)$ is determined by a unique divisor t of m in the sense described above (note that $\{\bar{0}\}$ is determined by m). Next, we observe that distinct divisors determine distinct subgroups, for the subgroups determined by distinct divisors have distinct orders (the order of the subgroup determined by the divisor t is q, where $m = qt$). To summarize, every subgroup of $(\mathbb{Z}_m, \oplus)$ determines and is uniquely determined by a positive divisor t in the sense described above. In particular, there are as many subgroups of $(\mathbb{Z}_m, \oplus)$ as many divisors of m.

Remark 4.4.14 We have a function $\tau : \mathbb{N} \longrightarrow \mathbb{N}$ (called the **divisor function**) defined by $\tau(n) = $ the number of divisors of n. Clearly, the number $\tau(n)$ of divisors of $n = p_1^{\alpha_1} p_2^{\alpha_2} \ldots p_r^{\alpha_r}$, where $p_1, p_2, \ldots, p_r$ are distinct primes, is given by

$$\tau(n) = (\alpha_1 + 1)(\alpha_2 + 1) \ldots (\alpha_r + 1).$$

Thus, $\tau(n)$ is the number of subgroups of $(\mathbb{Z}_n, \oplus)$.

Definition 4.4.15 Let (G, o) be a group. Then, $Z(G) = \{a \in G \mid aox = xoa \text{ for all } x \in G\}$ is a subgroup of (G, o) (verify) called the **center** of (G, o). Let $a \in G$. Then, $C_G(a) = \{x \in G \mid aox = xoa\}$ is a subgroup of (G, o) (verify) called the **centralizer** of a in (G, o). Again, given a subset S of G, consider $N_G(S) = \{x \in G \mid xoS = Sox\}$, where $xoS = \{xoy \mid y \in G\}$. Then, $N_G(S)$ is a subgroup of (G, o) (verify) called the **normalizer** of S in (G, o).

The center $Z(Q_8)$ of the Q_8 is the subgroup $\{1, -1\}$. The centralizer $C_{Q_8}(i)$ of i in Q_8 is the subgroup $\{1, i, -1, -i\}$ of Q_8. The normalizer of $\{1, i, -1, -i\}$ in Q_8 is Q_8 itself.

Example 4.4.16 In this example, we describe all subgroups of $(\mathbb{R}, +)$. Let H be a nontrivial subgroup of $(\mathbb{R}, +)$. If $a \in H$, then $-a \in H$. Thus, $H \cap \mathbb{R}^+ \neq \emptyset$. There are two cases:

Case (i) $H \cap \mathbb{R}^+$ has the least element h (say).

Case (ii) $H \cap \mathbb{R}^+$ has no least element (i.e., 0 is the limit point of H).

Consider the case (i). We show that $H = \{nh \mid n \in \mathbb{Z}\}$. Since $h \in H$, $\{nh \mid n \in \mathbb{Z}\} \subseteq H$. Let $a \in H$ and $a > 0$. Since $h > 0$, by the Archimedean property of $\mathbb{R}$, *there exists* $n \in \mathbb{N}$ such that $\frac{a}{h} < n$. Thus, the set $\{n \in \mathbb{N} \mid \frac{a}{h} < n\} \neq \emptyset$. By the well-ordering property of $\mathbb{N}$, it has least element m (say). Then, $0 \leq m-1 \leq \frac{a}{h} < m$.

This means that $(m-1)h \leq a < mh$. We show that $a = (m-1)h$. Suppose not. Then $(m-1)h < a < mh$. Since $a \in H$ and $(m-1)h \in H$, it follows that $a - (m-1)h \in H$. Clearly, $0 < a - (m-1)h < mh - (m-1)h = h$. This is a contradiction to the supposition that h is the least element of $H \cap \mathbb{R}^+$. Thus, $a \in \{nh \mid n \in \mathbb{Z}\}$. If $a < 0$, then from what we have proved above, it follows that $-a = nh$ for some $n \in \mathbb{Z}$, and so $a = (-n)h$. This shows that $H = \{nh \mid n \in \mathbb{Z}\}$.

Next, consider the case(ii). In this case, we show that between any two distinct real numbers there is a member of H (such sets are called dense sets). If $a < 0 < b$, then there is nothing to do ($0 \in H$). Suppose that $0 \leq a < b$. Take $\epsilon = \frac{b-a}{2}$. Since $H \cap \mathbb{R}^+$ has no least element, there is $h \in H$, $h \neq 0$ and $h < \epsilon$. By the Archimedean property of $\mathbb{R}$ and well-ordering property of $\mathbb{N}$, we have the smallest $n \in \mathbb{N}$ such that $a < nh$. Suppose that $nh \geq b$. Then, $(n-1)h > a$, a contradiction to the choice of n. Thus, $a < nh < b$. Since H is a subgroup, $nh \in H$. Next, suppose that $a < b \leq 0$. Then, $0 \leq -b < -a$. Hence, from the previous case, there $\exists k \in H$ such that $-b < k < -a$. But, then $a < -k < b$. Since H is a subgroup, $-k \in H$. To summarize, a subgroup H of $(\mathbb{R}, +)$ either consists of integral multiples of a fixed real number, or it is dense in the sense that between any two real numbers there is a member of H. In particular, a proper closed subgroup of $(\mathbb{R}, +)$ consists of all integral multiples of a fixed real number.

Operations on Subgroups

Proposition 4.4.17 *Intersection of a family of subgroups is a subgroup.*

Proof Let $\{H_\alpha \mid \alpha \in I\}$ be a family of subgroups of a group (G, o). Since H_α is a subgroup *for all* $\alpha \in I$, $e \in H_\alpha$ for all $\alpha \in I$. Hence, $e \in \bigcap_{\alpha \in I} H_\alpha$. Thus, $\bigcap_{\alpha \in I} H_\alpha \neq \emptyset$. Let $a, b \in \bigcap_{\alpha \in I} H_\alpha$. Then, $a, b \in H_\alpha$ for all $\alpha \in I$. Since each H_α is a subgroup, $aob^{-1} \in H_\alpha$ for all $\alpha \in I$. Hence, $aob^{-1} \in \bigcap_{\alpha \in I} H_\alpha$. This shows that $\bigcap_{\alpha \in I} H_\alpha$ is a subgroup.   ♯

Remark 4.4.18 Union of subgroups need not be a subgroup. For example, $\{e, a\}$ and $\{e, b\}$ are subgroups of the Klein's four group, but their union $\{e, a, b\}$ is not a subgroup(it does not contain $ab = c$).

Proposition 4.4.19 *Let H_1 and H_2 be subgroups of a group (G, o). Then $H_1 \bigcup H_2$ is a subgroup if and only if $H_1 \subseteq H_2$ or $H_2 \subseteq H_1$.*

Proof If $H_1 \subseteq H_2$, then $H_1 \bigcup H_2 = H_2$ is a subgroup. If $H_2 \subseteq H_1$, $H_1 \bigcup H_2 = H_1$ is a subgroup. Conversely, suppose that $H_1 \bigcup H_2$ is a subgroup and $H_1 \nsubseteq H_2$. Let $h \in H_1 - H_2$. Let k be any element of H_2. Then, $h, k \in H_1 \bigcup H_2$. Since $H_1 \bigcup H_2$ is a subgroup, $hok \in H_1 \bigcup H_2$. Now, $hok \notin H_2$, for otherwise $h = (hok)ok^{-1}$ would become a member of H_2, a contradiction to the choice of h. Since $h \in H_1$ and H_1 is a subgroup, $k = h^{-1}ohok$ belongs to H_1. This shows that $H_2 \subseteq H_1$.   ♯

Definition 4.4.20 A family $\{H_\alpha \mid \alpha \in I\}$ of subgroups of a group (G, o) is called a **chain** of subgroups if given $\alpha, \beta \in I$, $H_\alpha \subseteq H_\beta$ or $H_\beta \subseteq H_\alpha$.

Proposition 4.4.21 *Union of a chain of subgroups is a subgroup.*

Proof Let $\{H_\alpha \mid \alpha \in I\}$ be a chain of subgroups of a group (G, o). Clearly, $e \in \bigcup_{\alpha \in I} H_\alpha$. Let $a, b \in \bigcup_{\alpha \in I} H_\alpha$. Then, $a \in H_\alpha$ and $b \in H_\beta$ for some $\alpha, \beta \in I$. Since the given family is a chain, $H_\alpha \subseteq H_\beta$ or $H_\beta \subseteq H_\alpha$. Thus, $a, b \in H_\alpha$ or $a, b \in H_\beta$. Since H_α and H_β are subgroups, aob^{-1} belongs to H_α or to H_β. This shows that ab^{-1} belongs to the union $\bigcup_{\alpha \in I} H_\alpha$. ♯

Definition 4.4.22 Let (G, o) be a group and A, B be subsets of G. Then, the set $AoB = \{aob \mid a \in A \text{ and } b \in B\}$ is called the product of A and B.

Thus, the product of subsets of G defines a binary operation on the power set $\wp(G)$ of G. Is it a group? Is it a semigroup?

Proposition 4.4.23 *Let H and K be subgroups of a group (G, o). Then, HoK is a subgroup if and only if $HoK = KoH$.*

Proof Suppose that $HoK = KoH$. Clearly, $e = eoe$ belongs to HoK. Thus, $HoK \neq \emptyset$. Let $aob, cod \in HoK$, where $a, c \in H$ and $b, d \in K$. Then, $(aob)o(cod)^{-1} = aobod^{-1}oc^{-1} = aokoc^{-1}$, where $k = bod^{-1} \in K$. Now, $koc^{-1} \in KoH$. Since $KoH = HoK$, $koc^{-1} \in HoK$. Suppose that $koc^{-1} = uov$, where $u \in H$ and $v \in K$. Then, $(aob)o(cod)^{-1} = aouov \in HoK$. This shows that HoK is a subgroup.

Conversely, suppose that HoK is a subgroup. Let $hok \in HoK$, where $h \in H$ and $k \in K$. Then, $k^{-1}oh^{-1} = (hok)^{-1} \in HoK$. Suppose that $k^{-1}oh^{-1} = uov$, where $u \in H$ and $v \in K$. Then, $hok = v^{-1}ou^{-1} \in KoH$. This shows that $HoK \subseteq KoH$. Further, let $koh \in KoH$, where $k \in K$ and $h \in H$. Then $koh = (h^{-1}ok^{-1})^{-1} \subset HoK$, for $h^{-1}ok^{-1} \in HoK$ and HoK is a subgroup. This shows that $KoH \subseteq HoK$. It follows that $HoK = KoH$. ♯

Remark 4.4.24 In general, HoK need not be a subgroup. Consider the symmetric group S_3 of degree 3 of all bijective maps from $\{1, 2, 3\}$ to itself (see Example 4.1.24). Let $H = \{I, f\}$, where I is the identity map on $\{1, 2, 3\}$ and f is the map given by $f(1) = 2$, $f(2) = 1$, and $f(3) = 3$. Clearly, $fof = I$ and hence H is a subgroup of S_3. Let $K = \{I, g\}$, where g is given by $g(1) = 1$, $g(2) = 3$, and $g(3) = 2$. Then as above, K is also a subgroup. Now, $HoK = \{I, f, g, fog\}$ is not a subgroup for $gof \notin HoK$ (verify).

Homomorphisms and Subgroups

Proposition 4.4.25 *Let f be a homomorphism from a group (G_1, o_1) to (G_2, o_2). Let H_1 be a subgroup of (G_1, o_1) and H_2 a subgroup of (G_2, o_2). Then $f(H_1)$ is a subgroup of (G_1, o_1) and $f^{-1}(H_2)$ is a subgroup of (G_1, o_1).*

Proof Since $e_1 \in H_1$ (H_1 is a subgroup), $e_2 = f(e_1) \in f(H_1)$. Thus, $f(H_1) \neq \emptyset$. Let $f(a), f(b) \in f(H_1)$, where $a, b \in H_1$. Since f is a homomorphism, $f(a)o_2 f(b)^{-1} = f(ao_1 b^{-1})$. Further, since H_1 is a subgroup, $ao_1 b^{-1} \in H_1$ and so $f(a)o_2 f(b)^{-1} = f(ao_1 b^{-1}) \in f(H_1)$. This shows that $f(H_1)$ is a subgroup.

Again, since $f(e_1) = e_2$ (for, f is a homomorphism) and $e_2 \in H_2$ (for, H_2 is a subgroup), $e_1 \in f^{-1}(H_2)$. Thus, $f^{-1}(H_2) \neq \emptyset$. Let $a, b \in f^{-1}(H_2)$. Then,

$f(a), f(b) \in H_2$. Since H_2 is a subgroup, $f(ao_1b^{-1}) = f(a)o_2f(b)^{-1} \in H_2$. In turn, $ao_1b^{-1} \in f^{-1}(H_2)$. This shows that $f^{-1}(H_2)$ is a subgroup.   ♯

Corollary 4.4.26 *Let f be a homomorphism from a group (G_1, o_1) to a group (G_2, o_2). Then, $f^{-1}(\{e_2\})$ is a subgroup of (G_1, o_1).*

Proof Since $\{e_2\}$ is a subgroup of (G_2, o_2), the result follows from the above proposition.   ♯

Definition 4.4.27 Let f be a homomorphism from a group (G_1, o_1) to a group (G_2, o_2). Then, $f^{-1}(\{e_2\}) = \{a \in G_1 \mid f(a) = e_2\}$ is called the **Kernel** of f and is denoted by *ker f*.

Thus, *ker f* is a subgroup of (G_1, o_1). If f is a homomorphism, then $f(a) = f(b)$ if and only if $a^{-1}ob \in ker\ f$. Evidently, f is injective if and only if $ker\ f = \{e_1\}$.

Proposition 4.4.28 *Let H be a subgroup of a group (G_1, o_1) which contains the kernel of a homomorphism f from (G_1, o_1) to a group (G_2, o_2). Then*

$$f^{-1}(f(H)) = H.$$

Proof Clearly, $H \subseteq f^{-1}(f(H))$. Let $a \in f^{-1}(f(H))$. Then, $f(a) \in f(H)$, and so $f(a) = f(h)$ for some $h \in H$. But, then $ao_1h^{-1} \in ker\ f \subseteq H$. Thus, $a = ao_1h^{-1}o_1h$ belongs to H. This shows that $f^{-1}(f(H)) \subseteq H$.   ♯

Theorem 4.4.29 (Correspondence Theorem) *Let f be a surjective homomorphism from a group (G_1, o_1) to a group (G_2, o_2). Let $S(G_1)$ denote the set of all subgroups of (G_1, o_1) which contain kernel of f and $S(G_2)$ denote the set of all subgroups of (G_2, o_2). Then f induces bijective map ϕ from $S(G_1)$ to $S(G_2)$ defined by $\phi(H) = f(H)$ (the image of H under f).*

Proof Suppose that $\phi(H_1) = \phi(H_2)$. Then, $f(H_1) = f(H_2)$. From the above proposition, $H_1 = f^{-1}(f(H_1)) = f^{-1}(f(H_2)) = H_2$. Thus, ϕ is injective. Next, let $K \in S(G_2)$. Then, $f^{-1}(K)$ is a subgroup containing $f^{-1}(\{e_2\}) = ker\ f$. Hence, $f^{-1}(K) \in S(G_1)$. Since f is surjective, $\phi(f^{-1}(K)) = f(f^{-1}(K)) = K$. Thus, ϕ is surjective also.   ♯

Subgroup Generated by a Subset S

We have observed that a subset S of a group G need not be a subgroup of G. A natural question is: 'How far is the subset S from being a subgroup?' In other words, how can we make S a subgroup with minimum effort? Taking out elements from S will not help in general (for if $e \notin S$, taking out elements from S will never make it a subgroup). We can always make S a subgroup by putting some members of G in to S (if worst comes, by putting all members of $G - S$ in S). Again, as desired, we should add minimum number of elements of G to make it a subgroup. In other words, we are interested in the smallest subgroup of (G, o) containing S. The existence of such a subgroup follows from the following theorem.

Theorem 4.4.30 *Let (G, o) be a group and S a subset of G. Then the smallest subgroup of G containing S exists and is unique.*

Proof The family of all subgroups containing S is nonempty. The intersection of this family of subgroups is the smallest subgroup of (G, o) containing S.   ♯

Definition 4.4.31 Let S be a subset of a group G. Then, the smallest subgroup of G containing S is called the **subgroup generated by S**. This subgroup is denoted by $< S >$.

Remark 4.4.32 A subset H of a group (G, o) is the subgroup generated by S if and only if

(i) H is a subgroup.
(ii) $S \subseteq H$.
(iii) If K is a subgroup of (G, o) containing S, then $H \subseteq K$

Proposition 4.4.33 *Let (G, o) be a group and S a nonempty subset of G. Then the subgroup generated by S is the set of all finite products of integral powers of elements of S. More explicitly,*

$$< S > = \{a_1^{n_1} o a_2^{n_2} o \cdots o a_r^{n_r} \mid a_i \in S \text{ and } n_i \in \mathbb{Z}\}$$

Proof Let X denote the set of all finite products of integral powers of S. Since every element of S is an integral power of the same element, $S \subseteq X$. Next, if H is a subgroup of G containing S, then by Proposition 4.4.5, $a_1^{n_1} o a_2^{n_2} o \ldots o a_r^{n_r}$ belong to H for all $a_1, a_2, \ldots, a_r$ in S and integers $n_1, n_2, \ldots, n_r$. Further, since products of integral powers of elements of S are again a product of integral powers of elements of S, and also the inverse of a product of integral powers of elements of S is again a product of integral powers of elements of S, it follows that X is a subgroup. Thus, X is the smallest subgroup of G containing S, and so $< S > = X$.   ♯

Following is an immediate corollary of the above proposition.

Corollary 4.4.34 *Let (G, o) be a group and $a \in G$. Then, $< \{a\} > = \{a^n \mid n \in \mathbb{Z}\}$.*   ♯

The subgroup generated by a singleton $\{a\}$ is conveniently denoted by $< a >$ instead of $< \{a\} >$.

Example 4.4.35 $< \emptyset > = \{e\}$, for $\{e\}$ is the smallest subgroup of G which contains the $\emptyset$.

Example 4.4.36 The subgroup generated by $\{a\}$ in V_4 is $\{e, a\}$, for $\{e, a\}$ is the smallest subgroup which contains $\{a\}$. The subgroup generated by $\{a, b\}$ is the group V_4 itself, for any subgroup of V_4 containing a, b will contain $c = a o b$ and e also.

Example 4.4.37 The subgroup of the Quaternion group Q_8 generated by i is $\{1, -1, i, -i\}$, for it is a subgroup and any subgroup which contains i contains $1, -1 = i^2, -i = i^3$ and i also. The subgroup $< \{i, j\} >$ is Q_8 itself, for if a subgroup of Q_8 contains $\{i, j\}$, then it also contains $1, -1, i, -i, j, -j, k = i \cdot j$ and $-k = j \cdot i$.

Example 4.4.38 The subgroup of $(\mathbb{Z}, +)$ generated by $\{m\}$ is $m\mathbb{Z}$ (the set of all integral multiples of m), for $m\mathbb{Z}$ is the smallest subgroup of $(\mathbb{Z}, +)$ containing $\{m\}$. In particular, $\mathbb{Z} = < 1 > = < -1 >$.

Example 4.4.39 In this example, we determine subgroup generated by an arbitrary subset S of the additive group $\mathbb{Z}$ of integers. If $S = \emptyset$, then $< S > = \{0\}$. Since $< S > = < S - \{0\} >$, we may assume that $0 \notin S$. Let d be the positive greatest common divisor of elements of S (g.c.d of an infinite set can be defined in the same manner as it is done for finite set of integers). Then, we show that $< S > = d\mathbb{Z}$. Clearly, $S \subseteq d\mathbb{Z}$ as all members of S are multiples of d. Also, $d\mathbb{Z}$ is a subgroup of $(\mathbb{Z}, +)$. Next, let H be a subgroup of $(\mathbb{Z}, +)$ containing S. We can find a finite subset $\{m_1, m_2, \ldots, m_r\}$ of S such that g.c.d of S is the same as that of $m_1, m_2, \ldots, m_r$. By the Euclidean algorithm, there exist integers $u_1, u_2, \ldots, u_r$ such that

$$d = u_1 m_1 + u_2 m_2 + \cdots + u_r m_r$$

Let $dt \in d\mathbb{Z}, t \in \mathbb{Z}$. Then,

$$dt = t u_1 m_1 + t u_2 m_2 + \cdots + t u_r m_r$$

is a member of H, for H is a subgroup and $m_1, m_2, \ldots, m_r \in H$

Example 4.4.40 In this example, we determine the subgroup generated by an arbitrary subset of the group $(\mathbb{Z}_m, \oplus)$. Let S be a subset of $\mathbb{Z}_m$. If $S = \emptyset$ or $S = \{\bar{0}\}$, then $< S > = \{\bar{0}\}$. Let
$$S = \{\bar{r_1}, \bar{r_2}, \ldots, \bar{r_t}\},$$

where we can assume that $\bar{r_i} \neq \bar{0}$ for all $i = 1, 2, \ldots, t$. This means that m does not divide r_i for all i. Let d be g.c.d of $m, r_1, r_2, \ldots, r_t$. We show that

$$< S > = \{\bar{0}, \bar{d}, 2\bar{d}, \ldots, (q-1)\bar{d}\},$$

where $m = qd$. Since d divides r_i for all i, it is clear that

$$\{\bar{r_1}, \bar{r_2}, \ldots, \bar{r_t}\} \subseteq \{\bar{0}, \bar{d}, 2\bar{d}, \ldots, (q-1)\bar{d}\}$$

Further, it is easy to observe that $\{\bar{0}, 2\bar{d}, \ldots, (q-1)\bar{d}\}$ is a subgroup of $\mathbb{Z}_m$. Next, let H be a subgroup of $(\mathbb{Z}_m, \oplus)$ containing $\{\bar{r_1}, \bar{r_2}, \ldots, \bar{r_t}\}$. By the Euclidean algorithm, there exist integers $u, u_1, u_2, \ldots, u_t$ such that

$$d = um + u_1 r_1 + \cdots + u_t r_t$$

But, then

$$\overline{d} = u_1\overline{r_1} + u_2\overline{r_2} + \cdots + u_t\overline{r_t}$$

belongs to H, and since H is a subgroup, $\{\overline{0}, \overline{d}, 2\overline{d}, \ldots, (q-1)\overline{d}\} \subseteq H$. This shows that

$$<S> = \{\overline{0}, \overline{d}, 2\overline{d}, \ldots, (q-1)\overline{d}\}.$$

Example 4.4.41 As a particular case of the above example, we find that the subgroup of $\mathbb{Z}_m$ generated by $\{\overline{r}\}$ is $\{\overline{0}, \overline{d}, 2\overline{d}, \ldots, (q-1)\overline{d}\}$, where d is g.c.d of m and r and $qd = m$. In particular, singleton $\{\overline{r}\}$ generates $\mathbb{Z}_m$ if and only if m and r are co-prime. Thus, there are as many singleton generators of $\mathbb{Z}_m$ as many positive integers less than m and co-prime to m. It follows that there are $\phi(m)$ singleton generators of $\mathbb{Z}_m$.

Definition 4.4.42 Let (G, o) be a group. A subset S of G is said to be a **set of generators** of G if $<S> = G$. A group (G, o) is said to be **finitely generated** if it has a finite set of generators.

Example 4.4.43 $\{1\}$ and $\{-1\}$ are the only singleton generators of $(\mathbb{Z}, +)$ (prove it). It follows from examples above that the set $\{m_1, m_2, \ldots, m_r\}$ is a set of generators of $\mathbb{Z}$ if and only if the positive g.c.d of the above set is 1. It also follows that the set $\{\overline{r_1}, \overline{r_2}, \ldots, \overline{r_t}\}$ is a set of generators of the group $(\mathbb{Z}_m, \oplus)$ if and only if the positive g.c.d of $m, r_1, r_2, \ldots, r_t$ is 1. There are $\phi(m)$ singleton generators of $(\mathbb{Z}_m, \oplus)$.

Example 4.4.44 The group $(\mathbb{Q}, +)$ is not finitely generated: Let S be a finite subset of $\mathbb{Q}$. Let P be the set of primes appearing in the denominators of the members of S. Then, P is finite and so there are primes outside P. It is clear that $<S>$ contains no rational number in reduced form whose denominator is a prime not in P. Hence, $<S> \neq Q$.

Proposition 4.4.45 *Let f be a surjective homomorphism from a group (G_1, o_1) to a group (G_2, o_2). Let S be a set of generators of (G_1, o_1). Then, $f(S)$ is a set of generators of (G_2, o_2).*

Proof Let K be a subgroup of G_2 containing $f(S)$. Then, $f^{-1}(K)$ is a subgroup of G_1 containing S. Since S generates G_1, it follows that $f^{-1}(K) = G_1$. Since f is surjective, $K = f(f^{-1}(K)) = f(G_1) = G_2$. This shows that $f(S)$ generates G_2.   ♯

Corollary 4.4.46 *Any homomorphic image of a finitely generated group is finitely generated.*   ♯

Exercises

4.4.1 Determine all subgroups of $(\mathbb{Z}_8, \oplus)$.

4.4.2 Determine all subgroups of $(U_{12}, \star)$.

4.4.3 Let H be a subgroup of (G, o). Show that $g o H o g^{-1}$ is a subgroup of G for all $g \in G$.

4.4.4 Find the normalizers of all subgroups of the Quaternion group. Find also the centralizers of all elements in the symmetric group S_3.

4.4.5 Let (G, o) be a group. Show that the set $I(G)$ of all inner automorphisms of G is a subgroup of $Aut(G)$.

4.4.6 Find the center of the symmetric group S_3 and also of the Dihedral group D_4.

4.4.7 Show that a group cannot be written as union of two proper subgroups.

4.4.8 Find the number of subgroups of $(\mathbb{Z}_{200}, \oplus)$.

4.4.9 Find the center of the Quaternion group Q_8.

4.4.10 Let (G_1, o_1) and (G_2, o_2) be groups. Show that a map f from G_1 to G_2 is a homomorphism if and only if f considered as a subset of the direct product $G_1 \times G_2$ is a subgroup.

4.4.11 Find the kernel of the homomorphism f from the additive group $\mathbb{R}$ of real numbers to the circle group S^1 defined by $f(x) = e^{ix}$.

4.4.12 Find the kernel of the homomorphism ν from $(\mathbb{Z}, +)$ to $(\mathbb{Z}_m, \oplus)$ defined by $\nu(a) = \bar{a}$.

4.4.13 Show that a homomorphism f is injective if and only if $ker\ f = \{e\}$

4.4.14 Describe all subgroups of $(\mathbb{R}^+, \cdot)$ and also its closed subgroups.

4.4.15 Describe all subgroups of $(P, \cdot)$ and also of $(S^1, \cdot)$.

4.4.16 Describe all subgroups and also the closed subgroups of the circle group S^1.

4.4.17 Describe all subgroups and also all the closed subgroups of $(\mathbb{C}, +)$.

4.4.18 Let f be a surjective homomorphism from (G_1, o_1) to (G_2, o_2). Suppose that G_2 is finitely generated and also the kernel of f is finitely generated. Show that G_1 is also finitely generated.

4.4.19 Describe all subgroups of $\mathbb{Z} \times \mathbb{Z}$. Generalize it for the products of finitely many copies of the additive group $\mathbb{Z}$ of integers.

4.4.20 (a) Use the above result to show that subgroup of a finitely generated abelian group is finitely generated. (observe that every finitely generated abelian group is homomorphic image of products of finitely many copies of Z). It may be mentioned that the above result is not true for nonabelian groups. In fact, every countably generated group is subgroup of a group which is generated by two elements.
(b) Show that every finitely generated group is finite or at most countable. Deduce that the additive group $\mathbb{R}$ of reals and the circle group S^1 are not finitely generated. Observe that the additive group $\mathbb{Q}$ of rational numbers is countable, but it is not finitely generated.

4.4.21 Let S be a subset of a group (G, o). Let

$$C_G(S) = \{g \in G \mid gox = xog \ \forall x \in G\}$$

Show that $C_G(S)$ is a subgroup of G. This subgroup is called the **centralizer** of S in G. Observe that $C_G(S) \subseteq N_G(S)$. Show by means of an example that equality need not hold. If $S = \{a\}$, then $C_G(S)$ is denoted by $C_G(a)$.

4.4.22 Call a subgroup M of a group to be a **maximal subgroup** of a group (G, o) if (i) $M \neq G$, and (ii) there is no proper subgroup of G containing M properly. Show that $(\mathbb{Q}, +)$ has no maximal subgroup.

4.4.23 Show that every finitely generated group has a maximal subgroup.

4.4.24 Suppose that S is a set of generators of the group $(\mathbb{Q}, +)$ and F a finite subset of S. Show that $S - F$ is also a set of generators of $\mathbb{Q}$.

4.4.25 Show that the group P of roots of unity contains no maximal subgroup.

4.4.26 Show that if a group (G, o) is finitely generated, then every set of generators of G contains a finite set of generators.

4.4.27 Let S and T be subsets of a group (G, o) and H a subgroup such that $S \subseteq H$. Show that $(ST \cap H) = S(T \cap H)$. This identity is called the Dedekind law.

4.4.28 Let G be a finite group of order n. Show that $|Aut(G)| \leq n^{log_2(n)}$. Hint. If G has a set $X = \{a_1, a_2, \ldots, a_m\}$ of generators containing m elements, then any automorphism of G is completely determined by its effect on X and so it contains at most n^m elements. Consider the smallest m such that $< X > = G$. Let $H_i = < \{a_1, a_2, \ldots, a_i\} >$. Observe that $\frac{|H_i|}{|H_{i-1}|} > 2$. Conclude that $n \geq 2^m$.

4.4.29* Let (G, o) be finite group of order n. Let m be the number of prime divisors of n. Show that $|Aut(G)| \leq n^m$. Give an example to show that equality may hold.

4.4.30 Define a map f from $\mathbb{R}$ to S^1 by $f(x) = e^{2\pi ix}$. Show that f is a surjective homomorphism. Find the kernel of f. Also find $f^{-1}(P)$, where P is the subgroup of roots of unity.

4.4.31 Define a map f from $\mathbb{C}$ to $\mathbb{C}^*$ by $f(z) = e^z$, where $\mathbb{C}$ is the additive group of complex numbers and $\mathbb{C}^*$ the multiplicative group of nonzero complex numbers. (i) Is f surjective? (ii) Is f a homomorphism? If yes find the kernel.

4.4.32 Show that $\{m + n\sqrt{2} \mid m, n \in \mathbb{Z}\}$ is a subgroup of $(\mathbb{R}, +)$. Deduce that it is a dense subset of $\mathbb{R}$.

4.4.33 Show that $\{\frac{m}{2^n} \mid m, n \in \mathbb{Z}\}$ is a subgroup of $(\mathbb{R}, +)$. Deduce that it is a dense subset of $\mathbb{R}$.

4.4.34 Show that $\{e^{i\sqrt{2}m} \mid m \in \mathbb{Z}\}$ is a subgroup of S^1. Deduce that it is a dense subset of S^1.

4.4.35 Describe all finite subgroups of the circle group S^1.

4.4.36 Let G be a group all of whose proper subgroups are finitely generated. Can we conclude that G is also finitely generated. Support your claim.

4.4.37 Find the number of singleton generators of $\mathbb{Z}_{16}$.

4.4.38 Let $\phi_2(n)$ denote the number of pair (r_1, r_2) of integers such that $1 \leq r_1 < r_2 \leq n$ and the positive g.c.d. of r_1, r_2, and n is 1. Find $\phi_2(p^r)$, p is a prime and r is a positive integer. Find the number of doubleton generators of $\mathbb{Z}_9$.

4.5   Cyclic Groups

Order of an Element of a Group

Let (G, o) be a group and a an element of G. The subgroup $< \{a\} >$ generated by $\{a\}$ is denoted by $< a >$. Thus, $< a > = \{a^n \mid n \in \mathbb{Z}\}$.

Definition 4.5.1 The subgroup $< a >$ of (G, o) generated by $\{a\}$ is called the **cyclic subgroup generated by a**. A group (G, o) is said to be a **cyclic group** if it is generated by a single element.

Let (G, o) be a group and $a \in G$. Define a map f from $\mathbb{Z}$ to G by $f(n) = a^n$. Then, f is a homomorphism (follows from the law of exponents), and the image of f is the cyclic subgroup $< a >$. There are two cases:

(i) f is injective.
(ii) f is not injective.

In case (i), a is said to be of infinite order. Thus, a is of infinite order if and only if $a^n = a^m$ implies that $n = m$. This means that distinct integral powers of a are distinct.

Consider the case (ii) In this case, f is not injective, and so there exists a pair m, n of integers such that $m \neq n$ and $a^m = a^n$. But, then $a^{n-m} = a^n(a^m)^{-1} = e = a^m(a^n)^{-1} = a^{m-n}$. Thus, in this case, there exists a $l \in \mathbb{N}$ such that $a^l = e(n - m \in \mathbb{N}$ or $m - n \in \mathbb{N})$. Therefore, the subset $A = \{l \in \mathbb{N} \mid a^l = e\}$ of $\mathbb{N}$ is nonempty. By the well-ordering principle in $\mathbb{N}$, A has the least element m (say). This least element m is called the **order** of a and is denoted by $o(a)$. Thus, a natural number m is order of a if and only if $a^m = e$ and whenever $n \in \mathbb{N}$ and $a^n = e$, $m \leq n$.

Proposition 4.5.2 *Let (G, o) be a group and $a \in G$. Then a is of infinite order if and only if the kernel of the homomorphism f from $\mathbb{Z}$ to G defined by $f(n) = a^n$ is $\{0\}$.*

Proof Since f is a homomorphism, it is injective if and only if the kernel of f is $\{0\}$.   ♯

Proposition 4.5.3 *Let (G, o) be a group and a an element of finite order m. Then*

$$`a^n \; = \; e \; if \; and \; only \; if \; m \; divides \; n'$$

Proof Suppose that the order of a is $m > 0$. Then, by the definition of order of an element, m is the smallest positive integer such that $a^m \; = \; e$. If m divides n, then $n \; = \; mq$ for some $q \in Z$. But, then $a^n \; = \; a^{mq} \; = \; (a^m)^q \; = \; e$. Conversely, suppose that $a^n \; = \; e$. By the division algorithm, there are integers q, r such that

$$n \; = \; mq \; + \; r,$$

where $r \; = \; 0$ or else $0 \; < \; r \; < \; m$. Now,

$$e \; = \; a^n \; = \; a^{mq+r} \; = \; (a^m)^q o a^r \; = \; e o a^r \; = \; a^r$$

Hence, $r \; = \; 0$, for otherwise $0 \; < \; r \; < \; m$ and $a^r \; = \; e$, a contradiction to the supposition that m is the order of a. Thus, m divides n.   ♯

Corollary 4.5.4 *Let (G, o) be a group and $a \in G$. Then the order of a is m if and only if Kernel of f is $m\mathbb{Z}$, where f is the map given by $f(n) \; = \; a^n$.*

Proof $\ker f \; = \; \{n \in \mathbb{Z} \mid a^n \; = \; f(n) \; = \; e\}$. The result follows from the above proposition.   ♯

Example 4.5.5 The order of the identity element e is 1, for 1 is the smallest positive integer such that $e^1 \; = \; e$.

Example 4.5.6 Order of each nonidentity element a, b, c in V_4 is 2, for $a^2 \; = \; b^2 \; = \; c^2 \; = \; e$, and $a^1 \neq e, b^1 \neq e$ and $c^1 \neq e$.

Example 4.5.7 Order of -1 in Q_8 is 2, for $-1 \neq 1$ and $(-1)^2 \; = \; 1$. Order of i is 4, for $i \neq 1, i^2 \; = \; -1 \neq 1, i^3 \; = \; -i \neq 1$, and $i^4 \; = \; 1$. Similarly, orders of $j, k, -i, -j$ and $-k$ are 4.

Example 4.5.8 Order of each nonzero element in $(\mathbb{Z}, +)$ is infinite, for $m \neq 0$, and $n \cdot m \; = \; 0$ implies that $n \; = \; 0$.

Example 4.5.9 Order of $\overline{1}$ in $\mathbb{Z}_m$ is m, for $r\overline{1} \; = \; \overline{r} \; = \; \overline{0}$ if and only if m divides r.

Example 4.5.10 Consider the multiplicative group U_{10} of prime residue classes modulo 10. The order of $\overline{3}$ in U_{10} is 4, for $\overline{3} \neq \overline{1}, \overline{3}^2 \; = \; \overline{9} \neq \overline{1}, \overline{3}^3 \; = \; \overline{27} \; = \; \overline{7} \neq \overline{1}$, and $\overline{3}^4 \; = \; \overline{81} \; = \; \overline{1}$.

Example 4.5.11 Order of each nonidentity elements of $(\mathbb{Q}, +)$ and $(\mathbb{R}, +)$ is infinite. Such groups are called **torsion free**.

Example 4.5.12 Order of each element of the group P of roots of unity is finite. Such groups are called **torsion groups or periodic groups**. Observe that P is infinite.

Example 4.5.13 Order of some elements of S^1 (the circle group) is infinite and order of some elements of S^1 is finite. In fact, elements of finite order are those of P and those of $S^1 - P$ are of infinite orders. Such groups are called **mixed groups**. The subgroup P is the torsion part of S^1.

Proposition 4.5.14 *If (G, o) is a finite group, then order of each element of G is finite.*

Proof Let $a \in G$. Then, the map $n \rightsquigarrow a^n$ can not be injective, for otherwise G would become infinite.   ♯

Sometimes, we use notation $\mid a \mid$ also for the order of an element of a of a group.

Proposition 4.5.15 *Let (G, o) be a group and $a \in G$.*

 (i) *If a is of infinite order, then the cyclic subgroup $< a >$ generated by a is also infinite.*
 (ii) *If $o(a) = m$ is finite, then the cyclic subgroup $< a >$ generated by a is also finite and contains m elements. In fact, then*

$$< a > = \{e, a, a^2, \ldots, a^{m-1}\}$$

Proof (i) If a is of infinite order, then the map f from $\mathbb{Z}$ to $< a > = \{a^n \mid n \in \mathbb{Z}\}$ defined by $f(n) = a^n$ is a bijective homomorphism. Since $\mathbb{Z}$ is infinite, $< a >$ is infinite.
 (ii) Suppose that $o(a) = m$. Then, $a^n = e$ *if and only if* m/n. Let $a^n \in < a >$. By the division algorithm, there exist integers q and r such that $n = mq + r$, where $0 \le r < m$. But, then $a^n = a^{mq+r} = (a^m)^q o a^r = a^r$ belongs to $\{e, a, a^2, \ldots, a^{m-1}\}$. Thus,

$$< a > = \{e, a, a^2, \ldots, a^{m-1}\}$$

Observe further that all the elements of $\{e, a, a^2, \ldots, a^{m-1}\}$ are distinct.   ♯

The following corollary is immediate from the above proposition.

Corollary 4.5.16 $o(a) = o(< a >)$.   ♯

Proposition 4.5.17 *Let (G, o) be a group and $a, b \in G$ such that $aob = boa$. Suppose that $o(a) = m$, $o(b) = n$, and also $< a > \bigcap < b > = \{e\}$. Then, $o(aob) = [m, n]$ (the least common multiple of m and n).*

Proof Since $aob = boa$, $(aob)^r = a^r o b^r$ for all $r \in \mathbb{Z}$. Thus,

$$(aob)^r = e \text{ if and only if } a^r o b^r = e$$

But, then $a^r = b^{-r} \in < a > \cap < b > = \{e\}$. Hence, $a^r = e = b^{-r}$. Since $o(a) = m$ and $o(b) = n$, m and n both divide r. Hence, the least common multiple $[m, n]$ divides r. Further,

$$(aob)^{[m,n]} = a^{[m,n]}ob^{[m,n]} = eoe = e$$

This shows that order of $aob = [m, n]$. ♯

Remark 4.5.18 If we drop the condition $< a > \cap < b > = \{e\}$ from the above proposition, then all that we can say is that $o(aob)$ divides $[m, n]$. Further, if $aob \neq boa$, the $o(aob)$ is highly indeterminate. It can be of any finite number, or it can be infinite also.

Proposition 4.5.19 *Let (G, o) be a group. Let $a \in G$ and $o(a) = n$. Then,*

(i) $< a^r > = < a^{(n,r)} >$
(ii) $o(a^r) = \frac{n}{(n,r)}$,
       where (n, r) denotes the g.c.d of n and r.

Proof (i) Let $d = (n, r)$. Then, $r = dq$ for some q. But, then $a^r = (a^d)^q$ belongs to $< a^d >$. Hence, $< a^r > \subseteq < a^d >$. Further, by the Euclidean algorithm, *there exist $u, v \in \mathbb{Z}$ such that $un + vr = d$*. But, then

$$a^d = a^{un+vr} = (a^n)^u o(a^r)^v = (a^r)^v$$

Hence, $a^d \in < a^r >$, and so $< a^d > \subseteq < a^r >$. This proves (i).
(ii) From the Corollary 4.5.16, it follows that

$$o(a^r) = o(< a^r >) = o(< a^{(n,r)} >) = o(a^{(n,r)}) = \frac{n}{(n, r)} \quad ♯$$

Definition 4.5.20 A group (G, o) is said to be **cyclic** if there is an element $a \in G$ such that $G = < a >$.

Thus, a group (G, o) is cyclic if $G = \{a^n \mid n \in \mathbb{Z}\}$ for some $a \in G$.

Example 4.5.21 The additive group $(\mathbb{Z}, +)$ of integers is cyclic. In fact, $\mathbb{Z} = < 1 > = < -1 >$.

Example 4.5.22 For each $m > 0$, the group $(\mathbb{Z}_m, \oplus)$ is a cyclic group. In fact, $\mathbb{Z}_m = < \bar{1} >$. Also, $\mathbb{Z}_m = < \bar{a} >$ if and only if a and m are co-prime.

Proposition 4.5.23 *A group (G, o) is cyclic if and only if it is homomorphic image of $(\mathbb{Z}, +)$.*

Proof Suppose that (G, o) is cyclic and $G = < a >$. The map f from $\mathbb{Z}$ to G defined by $f(n) = a^n$ is surjective, for any element of G is of the form a^n for some $n \in \mathbb{Z}$. By the law of exponents, it is also a homomorphism. Conversely, suppose that f is a surjective homomorphism from the group $(\mathbb{Z}, +)$ to a group (G, o). Let $f(1) = a$. Then, $f(n) = a^n$ for all $n \in \mathbb{Z}$. Since f is surjective, $G = \{a^n \mid n \in \mathbb{Z}\}$. Thus, (G, o) is cyclic. ♯

Corollary 4.5.24 *A homomorphic image of a cyclic group is a cyclic group.*

Proof Since composition of surjective homomorphisms is surjective homomorphisms, it follows that homomorphic image of a homomorphic image of $(\mathbb{Z}, +)$ is a homomorphic image of $(\mathbb{Z}, +)$. ♯

The following theorem says that essentially there are two types of cyclic groups: (i) $(\mathbb{Z}, +)$, and (ii) $(\mathbb{Z}_m, \oplus)$ for some $m > 0$.

Theorem 4.5.25 *Any two infinite cyclic groups are isomorphic. Any two finite groups of same orders are isomorphic.*

Proof Since the relation of isomorphism is an equivalence relation on any set of groups, it is sufficient to show that any infinite cyclic group is isomorphic to the group $(\mathbb{Z}, +)$, and any finite cyclic group of order m is isomorphic to $(\mathbb{Z}_m, \oplus)$. Let (G, o) be an infinite cyclic group. Suppose that $G = < a >$. Then, a is of infinite order, for otherwise G will be a finite group (Proposition 4.5.15). Therefore, the map f from $\mathbb{Z}$ to G given by $f(n) = a^n$ is injective. It is already surjective homomorphism and so it is an isomorphism.

Next, suppose that $G = < a >$ contains m elements. Then, again by Proposition 4.5.15, a is of order m. Suppose that $\bar{r} = \bar{s}$ in $\mathbb{Z}_m$. Then, $m/r - s$. Since m is the order of a, $a^{r-s} = e$, and hence, $a^r = a^s$. This ensures that we have a map f from $\mathbb{Z}_m$ to G given by $f(\bar{r}) = a^r$. Clearly, f is a surjective homomorphism. Since both the groups contain the same number of elements, it is an isomorphism. ♯

The proofs of the following propositions are imitations of the arguments in Examples 4.4.11 and 4.4.12. One can prove these propositions using these examples and the above theorem. However, we repeat the arguments again.

Theorem 4.5.26 *Every subgroup of a cyclic group is cyclic.*

Proof Let (G, o) be a cyclic group and $G = < a >$. Let H be a subgroup of G. If $H = \{e\}$, then $H = < e >$ is a cyclic group generated by e. Suppose that $H \neq \{e\}$. Then, there exists an integer $n \neq 0$ such that $a^n \in H$. Since H is a subgroup, $a^{-n} = (a^n)^{-1}$ belongs to H. Hence, *there exists* $l \in \mathbb{N}$ such that $a^l \in H$. By the well-ordering principle in $\mathbb{N}$, we have the smallest $m \in \mathbb{N}$ with the property that $a^m \in H$. We show that $H = < a^m >$. Since H is a subgroup and $a^m \in H$, it follows that $< a^m > \subseteq H$. Let $a^n \in H$. By the division algorithm, *there exist* $q, r \in \mathbb{Z}$ such that $n = mq + r$, where $0 \leq r < m$. But, then $a^n = a^{mq+r} = (a^m)^q o a^r = a^r$. Thus, $a^r \in H$. Since m is the smallest positive integer such that $a^m \in H$, it follows that $r = 0$. This shows that $a^n = (a^m)^q$ belongs to $< a^m >$. Hence, $H = < a^m >$. ♯

Theorem 4.5.27 *Let (G, o) be an infinite cyclic group generated by a. Then, $<a^m> = <a^n>$ if and only if $m = \pm n$. In particular $<a^m> = G$ if and only if $m = \pm 1$.*

Proof Clearly, $<a^m> = <a^{-m}>$. Suppose that $<a^n> = <a^m>$. Then, *there exist* $r, s \in \mathbb{Z}$ such that $a^m = (a^n)^r$ and $a^n = (a^m)^s$. Since a is of infinite order, $m = nr$ and $n = ms$. This shows that $rs = 1$, and so $r = \pm 1$. Hence, $m = \pm n$.   ♯

Corollary 4.5.28 *There are infinitely many proper subgroups of an infinite cyclic group.*

Proof It follows from the above theorem that the map f from $\mathbb{N} \bigcup \{0\}$ to the set $S(G)$ of all subgroups of G given by $f(n) = <a^n>$ is bijective.   ♯

Proposition 4.5.29 *Let (G, o) be a finite cyclic group of order m generated by a. Then, $<a^r> = <a^s>$ if and only if $(m, r) = (m, s)$ (equivalently $\frac{[m,r]}{r} = \frac{[m,s]}{s}$).*

Proof Since the order of a is m the order of G, the result follows from Theorem 4.5.19. (Note that $(m, r) = \frac{mr}{[m,r]}$ and $(m, s) = \frac{ms}{[m,s]}$.)   ♯

Corollary 4.5.30 *Let (G, o) be a cyclic group of order m generated by a. Let d_1 and d_2 be divisors of m. Then $<a^{d_1}> = <a^{d_2}>$ if and only if $d_1 = \pm d_2$.*

Proof Follows from the fact that d is a divisor of m if and only if $(m, d) = \pm d$. ♯

Corollary 4.5.31 *Let (G, o) be a cyclic group of order m. Then the number of subgroups of G is $\tau(m)$, where τ is the divisor function ($\tau(m)$ is the number of positive divisors of m).*

Proof Let (G, o) be the cyclic group generated by a. Then, order of a is also m. Let D denote the set of positive divisors of m. Then, from the above results, it follows that the map f from D to the set $S(G)$ of subgroups of G defined by $f(d) = <a^d>$ is bijective.   ♯

Corollary 4.5.32 *Corresponding to every divisor of the order of a cyclic group there is a unique subgroup of that order.*

Proof Let (G, o) be a cyclic group of order m generated by a. Let r/m. Then, $<a^{\frac{m}{r}}>$ is the unique subgroup of order r (note that if H is a subgroup of G, then there exists a unique positive divisor d of m such that $H = <a^d>$). Further, the order of $<a^d>$ is $\frac{m}{d}$.   ♯

Corollary 4.5.33 *Let (G, o) be a cyclic group of order m generated by a. Then there are $\phi(m)$ singleton generators of G, where ϕ is the Euler's phi function.*

Proof We have already seen that $<a^r> = <a^{(m,r)}>$ and also if d_1 and d_2 are two positive divisors of m such that $<a^{d_1}> = <a^{d_2}>$, then $d_1 = d_2$. Thus, $<a^r> = <a>$ if and only if $(m, r) = 1$. Since $G = \{e, a, a^2, \ldots, a^{m-1}\}$, it follows that there are as many singleton generators of G as many positive integers less than m and co-prime to m. Thus, there are $\phi(m)$ singleton generators of G.   ♯

Proposition 4.5.34 *Let (G, o) be a finite group such that the subgroups of G form a chain in the sense that for any pair H, K of subgroups of G, $H \subseteq K$ or $K \subseteq H$. Then, G is a cyclic group of order p^n for some prime p and $n \geq 0$.*

Proof Since G is finite, order of each element of G is finite. Let a be an element of G of highest order. We show that $G = < a >$. Let $b \in G$. Since subgroups of G form a chain, $< b > \subseteq < a >$ or $< a > \subseteq < b >$. But $< b >$ can not contain $< a >$ properly, for otherwise $o(b) > o(a)$. Thus, $< b > \subseteq < a >$ and so $b \in < a >$. This shows that G is cyclic. Next, suppose that $o(G) = o(a) = m$ is not a power of a prime. Then, $m = m_1 m_2$ for some m_1, m_2 with $1 < m_1 < m$ and $1 < m_2 < m$ and $(m_1, m_2) = 1$. But, then $< a^{m_1} > = \{e, a^{m_1}, a^{2m_1}, \ldots, a^{(m_2-1)m_1}\}$ does not contain a^{m_2}, and similarly, $< a^{m_2} >$ does not contain a^{m_1}. This is contradiction to the hypothesis. ♯

Corollary 4.5.35 *Let (G, o) be a finite group such that union of any two subgroup of G is a subgroup. Then G is cyclic.* ♯

Remark 4.5.36 The above proposition and the corollary are not true if we do not assume G to be finite. For example, let p be a prime number and $G = \{z \in \mathbb{C} \mid z^{p^n} = 1 \text{ for some } n \in \mathbb{N}\}$. Then, G is a group with respect to complex multiplication which is not cyclic, but the family of subgroups of G form a chain (verify).

An Application To Number Theory

Theorem 4.5.37 *Let $n \in \mathbb{N}$. Then,*

$$\Sigma_{d/n}\phi(d) = n,$$

where ϕ is the Euler's phi function.

Proof Let (G, o) be a cyclic group of order n generated by a. Let X_d be the set of elements of order d of G. Since order of each element of G is a divisor (prove it) of n, $X_d = \emptyset$, if d does not divide n. Also,

$$G = \bigcup_{d/n} X_d$$

Further, if r and s are distinct divisors of n, then $X_r \cap X_s = \emptyset$. Hence, $o(G) = \Sigma_{d/n} o(X_d)$. Now, corresponding to each divisor d of n, there is a unique subgroup $< b >$ of order d, where $b = a^{\frac{n}{d}}$. Since each element of G of order d will generate a subgroup of order d, all the elements of X_d are elements of $< b >$ which generate $< b >$. Thus, $X_d = \{b^r \mid 1 \leq r \leq n \text{ and } (d, r) = 1\}$. Hence, $o(X_d) = \phi(d)$. This shows that $n = \Sigma_{d/n}\phi(d)$. ♯

Exercises

4.5.1 Find the order of $\overline{3}$ in U_{11}. Show that U_{11} is a cyclic group of order 10.

4.5.2 Find the order of $\overline{8}$ in $\mathbb{Z}_{60}$.

4.5.3 Find the number of subgroups of a cyclic group of order 200. Enumerate the subgroups of $\mathbb{Z}_{200}$.

4.5.4 Give an example of an infinite group all of whose elements are of finite order.

4.5.5 Show that every cyclic group is abelian.

4.5.6 Give an example of an abelian group which is not cyclic.

4.5.7 Find the subgroup of $\mathbb{Z}$ generated by $\{20, 45, 50\}$.

4.5.8 Let $m \in \mathbb{N}$. Suppose that $(10, m) = 1$. Show that r is the order of $\overline{10}$ in U_m if and only if the decimal representation of $\frac{1}{m}$ recurs at the rth step.

4.5.9 Find the subgroup of $\mathbb{Z}_{100}$ generated by $\{\overline{4}, \overline{6}, \overline{18}\}$.

4.5.10 Suppose that $o(a) = n$ and $o(b) = m$. Suppose that $a^r = b^{-r}$. Show that $[n, r] = [m, r]$.

4.5.11 Find the number of singleton generators of $\mathbb{Z}_{300}$.

4.5.12 Find the number of homomorphisms from a cyclic group of order 12 to a cyclic group of order 15. Also find them.

4.5.13 Find the number of homomorphisms from a cyclic group of order 16 to the circle group S^1 and also find them.

4.5.14 Give an example of a nonabelian group all of whose proper subgroups are cyclic.

4.5.15 Show that $(\mathbb{Q}, +)$, $(\mathbb{R}, +)$, and $(\mathbb{C}, +)$ are not cyclic groups.

4.5.16 Show that every proper closed subgroup of $(\mathbb{R}, +)$ is a cyclic group.

4.5.17 Show that if all proper subgroups of a nonabelian group G are abelian, then G is generated by two elements.

4.5.18 Show that all proper closed subgroups of the circle group S^1 are finite cyclic.

4.5.19 Characterize a group all of whose proper nontrivial subgroups are isomorphic to the group itself.

4.5.20 Let (G, o) be a group and $a, b \in G$. Suppose that $aob = boa$ $o(a) = 100$ $o(b) = 60$. Suppose that $a^{10} = b^6$. Find the order of aob if possible.

4.5.21 Suppose that $o(a) = 24$, $o(b) = 15$, and $aob = boa$ and also $< a > \cap < b > = \{e\}$. Find the order of aob.

4.5.22 Suppose that $o(a) = 18$ and $o(b) = 25$ and $ab = ba$. Find the order of aob.

4.5.23 Derive all the results about cyclic groups using Examples 4.4.12 and 4.4.13 and Theorem 4.5.25.

4.5.24 Let f be a homomorphism from a group (G, o) to a group (G', o'). Let $a \in G$ be an element of finite order. Show that $f(a)$ is also of finite order and $o(f(a))$ divides $o(a)$.

4.5.25 Let (G, o) be a group and $a, b \in G - \{e\}$. Let p be a prime such that $2^p - 1$ is also a prime. Suppose that $a^p = \{e\}$ and $aoboa^{-1} = b^2$. Find the $o(a)$ and $o(b)$.

4.5.26 Let (G, o) be a group and $a, b \in G - \{e\}$. Let p be a prime such that $a^p = e$ and $aoboa^{-1} = b^3$. Suppose that $3^p - 1 = 2q$, where q is a prime. Suppose that $aob \neq boa$. Show that $o(a) = p$ and $o(b)$ is q or $3^p - 1$.

4.5.27 Let (G, o) be a finite group of even order. Suppose that half of the elements of G are of order 2 and the rest of the elements of G form a subgroup H. Show that H is an abelian subgroup of odd order.

Hint. The number of elements of order 2 is odd. Thus, $o(G) = 2n$, where n is odd. Let b be an element of order 2 and $c \in H$. Then, $bc \notin H$. Hence, $bocob = c^{-1}$ for all $c \in H$. Deduce that $c^{-1}od^{-1} = d^{-1}oc^{-1}$ for all $c, d \in H$.

4.5.28 Let (G, o) be a group and $a, b \in G$ such that $aob = boa$. Suppose that $o(a) = m$ and $o(b) = n$. Show that G has an element of order $[m, n]$.

4.5.29 (G, o) be a finite abelian group. Let m be maximum among the orders of elements of G. Use the above exercise to show that $a^m = e$ for all $a \in G$.

4.5.30 Show that the semigroup $End(G, o)$ of endomorphisms of an infinite cyclic group (G, o) is isomorphic to the multiplicative semigroup $\mathbb{Z}$ of integers. Show further that the group $Aut(G, o)$ of automorphisms of (G, o) is isomorphic to the multiplicative group $\{1, -1\}$.

4.5.31 Let (G, o) be a cyclic group of order m. Show that the semigroup $End(G, o)$ of endomorphisms of (G, o) is isomorphic to the semigroup $(\mathbb{Z}_m, \star)$. Show further that the group $Aut(G, o)$ of automorphisms of (G, o) is isomorphic to the group U_m of prime residue classes modulo m.

Chapter 5
Fundamental Theorems

This chapter is devoted to some fundamental theorems such as Lagrange Theorem and Isomorphism Theorems. We also discuss the direct decomposition of groups into indecomposable groups.

5.1 Coset Decomposition, Lagrange Theorem

In this section, we partition a group into disjoint union of sets with the help of a subgroup such that each member of the partition contains the same number of elements as the subgroup, and thereby deduce that the order of a subgroup divides the order of the group. We also give some applications of this fundamental result, especially in number theory.

Unless specified otherwise, a binary operation of a group will be denoted by juxtaposition. Instead of saying that (G, o) is a group, we will simply say that G is a group.

Let H be a subgroup of a group G. Define relations R^l and R^r on G as follows:

$$R^l = \{(a, b) \in G \times G \mid a^{-1}b \in H\},$$

and

$$R^r = \{(a, b) \in G \times G \mid ab^{-1} \in H\}.$$

Here, l stands for left and r stands for right.

Proposition 5.1.1 R^l and R^r are equivalence relations with G.

Proof Reflexive: Since H is a subgroup $x^{-1}x = e$ belongs to H for all $x \in G$. This shows that $(x, x) \in R^l$ for all $x \in G$.
Symmetric: Suppose that $(x, y) \in R^l$. Then, $x^{-1}y \in H$. Since H is a subgroup,

© Springer Nature Singapore Pte Ltd. 2017
R. Lal, *Algebra 1*, Infosys Science Foundation Series in Mathematical Sciences,
DOI 10.1007/978-981-10-4253-9_5

$(x^{-1}y)^{-1} = y^{-1}x$ belongs to H. Hence, $(y, x) \in R^l$.

Transitive: Suppose that $(x, y) \in R^l$ and $(y, z) \in R^l$. Then, $x^{-1}y$ and $y^{-1}z$ both belong to H. Since H is a subgroup, $x^{-1}z = x^{-1}yy^{-1}z$ belongs to H. This shows that $(x, z) \in R^l$. Similarly, we can prove that R^r is an equivalence relation.   ♯

Example 5.1.2 If $H = \{e\}$, then $x^{-1}y \in H$ means that $x = y$. Hence, in this case $R^l = R^r$ is the diagonal relation Δ with G.

Example 5.1.3 The relations R^l and R^r in general may be different. For example, consider the symmetric group $S_3 = Sym(X)$, where $X = \{1, 2, 3\}$ of degree 3. Let $H = \{I, p\}$, where I is the identity map on X, and p is the map defined by $p(1) = 2$, $p(2) = 1$, and $p(3) = 3$. Clearly, H is a subgroup of S_3 ($pop = I$). Check that $(f, g) \in R^l$, where f and g are given by $f(1) = 3$, $f(2) = 1$, $f(3) = 2$, $g(1) = 1$, $g(2) = 3$, and $g(3) = 2$. Also check that (f, g) does not belong to R^r.

Example 5.1.4 Let H be a subgroup of a group G contained in the center of G. Let $(x, y) \in R^l$. Then, $x^{-1}y \in H$. Since H is contained in the center, $x^{-1}y = x(x^{-1}y)x^{-1} = yx^{-1}$. This means that $(y, x) \in R^r$. Since R^r is an equivalence relation, $(x, y) \in R^r$. Thus, $R^l \subseteq R^r$. Similarly, $R^r \subseteq R^l$. It follows that $R^l = R^r$.

For $a \in G$, the equivalence class R_a^l of G modulo the equivalence relation R^l is given by

$$R_a^l = \{b \in G \mid (a, b) \in R^l\} = \{b \in G \mid a^{-1}b \in H\}$$

$$= \{b \in G \mid b = ah \text{ for some } h \in H\} = \{ah \mid h \in H\}.$$

Definition 5.1.5 The set $\{ah \mid h \in H\}$ is denoted by aH and is called the **left coset of G modulo H determined by a**. The set $\{ha \mid h \in H\}$ denoted by Ha is called the **right coset of G modulo H determined by a**.

Thus, $R_a^l = aH$ is the left coset of G modulo H determined by a, and $R_a^r = Ha$ is the right coset of G modulo H determined by a. Following the properties of left and right cosets can be easily verified. Indeed, these properties are also a direct consequence of Proposition 2.3.4.

(i) $\bigcup_{a \in G} aH = G$.

(ii) $aH = bH$ *if and only if* $a^{-1}b \in H$.

(iii) $aH \neq bH$ *if and only if* $aH \bigcap bH = \emptyset$.

Also,

(i) $\bigcup_{a \in G} Ha = G$.

(ii) $Ha = Hb$ *if and only if* $ab^{-1} \in H$.

(iii) $Ha \neq Hb$ *if and only if* $Ha \bigcap Hb = \emptyset$.

The quotient set $G/R^l = \{aH \mid a \in G\}$ is the set of left cosets of G modulo H. This set is denoted by $G/^l H$. Similarly, $G/R^r = \{Ha \mid a \in G\}$ is the set of right cosets of G modulo H and is denoted by $G/^r H$. Recall that if R and S are equivalence relations on X, then

(i) $R = S$ if and only if $R_x = S_x$ for all $x \in X$.

(ii) $R = S$ if and only if $X/R = X/S$.

Thus,

(i) $R^l = R^r$ if and only if $aH = Ha$ for all $a \in G$.

(ii) $R^l = R^r$ if and only if $G/^l H = G/^r H$.

Proposition 5.1.6 *There is a bijective map from $G/^l H$ to $G/^r H$.*

Proof $a^{-1}b = a^{-1}(b^{-1})^{-1}$. Thus, to say that $a^{-1}b \in H$ is to say that $a^{-1}(b^{-1})^{-1} \in H$. Hence, $aH = bH$ if and only if $Ha^{-1} = Hb^{-1}$. This shows that we have an injective map f from $G/^l H$ to $G/^r H$ given by $f(aH) = Ha^{-1}$. Also any $Ha = f(a^{-1}H)$. Hence, f is surjective. ♮

Remark 5.1.7 The correspondence $aH \rightsquigarrow Ha$ need not be a map from $G/^l H$ to $G/^r H$. Consider Example 5.1.3. It can be checked that $fH = gH$ but $Hf \neq Hg$.

Corollary 5.1.8 *$G/^l H$ is finite if and only if $G/^r H$ is finite, and then they contain the same number of elements.* ♮

Definition 5.1.9 Let H be a subgroup of a group G. If $G/^l H$ (or equivalently $G/^r H$) is infinite, then we say that H is of **infinite index** in G. If $G/^l H$ is finite, then the number of elements in $G/^l H$ (which is the same as the number of elements in $G/^r H$) is called the **index** of H in G. The index of H in G is denoted by $[G : H]$.

Thus, $[G : H]$ is the number of left cosets of G modulo H (which is same as number of right cosets of G modulo H).

Example 5.1.10 Let $m \in \mathbb{N}$. Then, $m\mathbb{Z}$ is a subgroup of the additive group $\mathbb{Z}$ of integers. Since $\mathbb{Z}$ is abelian, every left coset of $\mathbb{Z}$ modulo $m\mathbb{Z}$ is also a right coset. Thus, $\mathbb{Z}/^l m\mathbb{Z} = \mathbb{Z}/^r m\mathbb{Z}$). The left coset $a + m\mathbb{Z}$ of $\mathbb{Z}$ modulo $m\mathbb{Z}$ determined by a is given by $a + m\mathbb{Z} = \{a + mr \mid r \in \mathbb{Z}\} = \{b \in \mathbb{Z} \mid m/a - b\}$. Thus, $a + m\mathbb{Z} = \overline{a}$ in $\mathbb{Z}_m$. This means that $\mathbb{Z}/^l m\mathbb{Z} = Z_m = \mathbb{Z}/^r m\mathbb{Z}$ and the index $[\mathbb{Z} : m\mathbb{Z}]$ of $m\mathbb{Z}$ in $\mathbb{Z}$ is m.

Example 5.1.11 Consider the Klein's four group V_4 and the subgroup $H = \{e, a\}$ of V_4. Then, $eH = \{e, a\}$, $aH = H = Ha$ (note that $aH = H$ if and only if $a \in H$). Also, $bH = \{b, c\} = cH = Hb = Hc$. Thus, $V_4/^l H = V_4/^r H = \{\{e, a\}, \{b, c\}\}$ and the index $[V_4 : H] = 2$.

Example 5.1.12 Consider the Quaternion group Q_8 and the subgroup $H = \{1, -1\}$ of Q_8. It is easy to observe that $Q_8/^l H = Q_8/^r H = \{\{1, -1\}, \{i, -i\}, \{j, -j\}, \{k, -k\}\}$, and so the index $[Q_8, H]$ of H in Q_8 is 4. Further, if $K = \{1, -1, i, -i\}$, then $Q_8/^l K = Q_8/^r K = \{\{1, -1, i, -i\}, \{j, -j, k, -k\}\}$, and so $[Q_8 : K] = 2$. Similarly, we observe that all other nontrivial proper subgroups of Q_8 are of index 2.

Theorem 5.1.13 (Lagrange) *Let H be a subgroup of a finite group G. Then,*

$$|G| = |H| [G : H]$$

Proof Since G is finite, the set $G/^l H$ of left cosets of G modulo H is finite. Suppose that $| G/^l H | = [G, H] = r$. Let

$$G/^l H = \{a_1 H = H, a_2 H, a_3 H, \ldots, a_r H\}$$

Since the union of left cosets of G modulo H is G, we have

$$G = a_1 H \bigcup a_2 H \bigcup \cdots \bigcup a_r H \tag{5.1}$$

Since distinct left cosets are disjoint, $a_i H \bigcap a_j H = \emptyset$ *for all* $i \neq j$. Thus, by counting the elements of G in the above equation we get

$$| G | = | a_1 H | + | a_2 H | + \cdots + | a_r H | \tag{5.2}$$

Next, let aH be any left coset of G modulo H. Then, the map $h \rightsquigarrow ah$ from H to aH is clearly surjective (clear from the definition of aH), and it is also injective, for by the cancellation law, $ah_1 = ah_2$ implies that $h_1 = h_2$. This shows that any left coset contains the same number of elements as H. Thus, from the above equation it follows that

$$| G | = | H | r = | H | [G, H] \qquad \natural$$

Corollary 5.1.14 *Order of a subgroup (index of a subgroup) H of a finite group G divides the order of the group, and*

$$[G, H] = \frac{| G |}{| H |} \qquad \natural$$

Remark 5.1.15 From Lagrange theorem, order of a subgroup divides the order of the group. Naturally, we have the following problem: Let G be a finite group and m divides the order of the group. Do we have a subgroup of G of order m? We shall see later that G need not have any subgroup of order m. If corresponding to every divisor of the order of a group G there is a subgroup of that order, then G is called a *C.L.T* group.

Corollary 5.1.16 *Let G be a group and $| G | = n$. Let $a \in G$. Then, $o(a)$ divides n (note that notation $o(G)$ also stands for order of G.).*

Proof Since $o(a) = o(< a >)$, the result follows from the Lagrange theorem. $\natural$

Corollary 5.1.17 *Let G be a group of order n and $a \in G$. Then, $a^n = e$*

Proof From the above corollary, $o(a)$ divides n. The result follows. $\natural$

Corollary 5.1.18 *Every group of prime order is cyclic. Indeed, in a group of prime order, every nonidentity element generates the group.*

Proof Let G be a group and $|G| = p$, where p is a prime. Let $a \in G$, $a \neq e$. Then, $<a> \neq \{e\}$, and so $o(<a>) > 1$. By the Lagrange theorem, $o(<a>)$ divides p. Hence, $o(<a>) = p$. But, then $G = <a>$ is a cyclic group.   ♯

The following theorem characterizes the cyclic groups of prime orders.

Theorem 5.1.19 *A group G has no nontrivial proper subgroups if and only if it is a cyclic group of prime order.*

Proof Suppose that $|G| = p$, where p is a prime. Let $H \neq \{e\}$ be a subgroup of G. Then, $|H| > 1$. By the Lagrange theorem, $|H| / |G|$. Hence, $|H| = p = |G|$, and so $H = G$. Conversely, suppose that G has no nontrivial proper subgroups. Let $a \in G$, $a \neq e$. Then, $<a>$ is a nontrivial subgroup of G. By our supposition, $<a> = G$. Thus, G is cyclic. If G is infinite cyclic group, then by Corollary 4.5.28, it has infinitely many nontrivial proper subgroups. Hence, $G = <a>$ is finite cyclic group. Suppose that $|G| = m$ is not prime. Then, $m = r \cdot s$ for some r, s, $1 < r < m$, $1 < s < m$. Then clearly $<a^r>$ is a subgroup of G of order s, and so it is a proper nontrivial subgroup of G. This is a contradiction to the supposition. Hence, m is prime.   ♯

Theorem 5.1.20 (Poincare Theorem) *Intersection of a family of finitely many subgroups of finite indexes is again a subgroup of finite index.*

Proof It is sufficient to show that the intersection of two subgroups of finite indexes is a subgroup of finite index. Let H and K be subgroups of G of finite indexes. Suppose that $a(H \bigcap K) = b(H \bigcap K)$. Then $a^{-1}b \in H \bigcap K$. In turn, $aH = bH$ and $aK = bK$. Thus, $(aH, aK) = (bH, bK)$. This gives us a map η from $G/^l(H \bigcap K)$ to $G/^l H \times G/^l K$ defined by $\eta(a(H \bigcap K)) = (aH, aK)$. Suppose that $(aH, aK) = (bH, bK)$. Then, $aH = bH$ and $aK = bK$. But, then $a^{-1}b \in H \bigcap K$. Hence, $a(H \bigcap K) = b(H \bigcap K)$. This shows that η is injective. Thus, if $G/^l H$ and $G/^l K$ are finite, then $G/^l(H \bigcap K)$ is also finite, and

$$|G/^l(H \bigcap K)| \leq |G/^l H| \cdot |G/^l K|$$

This proves that if $[G, H] < \infty$ and $[G, K] < \infty$, then

$$[G, H \bigcap K] \leq [G, H] \cdot [G, K] < \infty.$$ ♯

Proposition 5.1.21 *Let H and K be subgroups of a group G. Then, there is a bijective map from $H/^l(H \bigcap K)$ to $HK/^l K = \{aK \mid a \in HK\}$.*

Proof Suppose that $a(H \bigcap K) = b(H \bigcap K)$, where $a, b \in H$. Then, $a^{-1}b \in H \bigcap K$. In particular, $a^{-1}b \in K$, and so $aK = bK$. This shows the existence of a map η from $H/^l(H \bigcap K)$ to $HK/^l K$ defined by $\eta(a(H \bigcap K)) = aK$. Any element of $HK/^l K$ is of the form hkK for some $h \in H$ and $k \in K$. Clearly, $hkK = hK = \eta(h(H \bigcap K))$. This shows that η is surjective. Suppose that

$\eta(a(H \cap K)) = \eta(b(H \cap K))$. Then, $aK = bK$. Hence, $a^{-1}b \in K$. Since $a, b \in H$, it follows that $a^{-1}b \in H \cap K$. Hence, $a(H \cap K) = b(H \cap K)$. This shows that η is bijective. ♯

Corollary 5.1.22 *Let H and K be finite subgroups of a group G. Then, HK is also finite, and*

$$| HK | = \frac{| H | \cdot | K |}{| H \cap K |}.$$

Proof The map η from $H \times K$ to HK defined by $\eta((h, k)) = hk$ is surjective. Hence, if H and K are finite, then HK is also finite. Since $x \in HK$ implies that $xK \subseteq HK$, HK is a complete union of members of $HK/^l K$. Thus,

$$HK = \bigcup_{x \in HK} xK.$$

Since distinct left cosets are disjoint

$$| HK | = | K | \cdot | HK/^l L |$$

From the previous proposition and the Lagrange theorem,

$$| HK/^l K | = | H/^l H \cap K | = [H, H \cap K] = \frac{| H |}{| H \cap K |}$$

Thus,

$$| HK | = \frac{| K | \cdot | H |}{| H \cap K |}.$$

Applications to Number Theory

Theorem 5.1.23 (Euler–Fermat) *Let m be a positive integer and a an integer such that $(a, m) = 1$. Then*

$$a^{\phi(m)} \equiv 1 (mod\ m).$$

Equivalently,

$$\overline{a}^{\phi(m)} = \overline{1}$$

in $\mathbb{Z}_m$

Proof Consider the group U_m of prime residue classes modulo m. Since $(a, m) = 1$, $\overline{a} \in U_m$. Since $| U_m | = \phi(m)$, by Corollary 5.1.17, $\overline{a}^{\phi(m)} = \overline{1}$, or equivalently, $a^{\phi(m)} - 1$ is divisible by m. This means that $a^{\phi(m)} \equiv 1 (mod\ m)$. ♯

Remark 5.1.24 The above result asserts that if we divide $a^{\phi(m)}$ by m, the remainder obtained is 1.

Corollary 5.1.25 (Fermat) *Let p be a prime number which does not divide an integer a. Then*

$$a^{p-1} \equiv 1 (mod\ p).$$

Equivalently, $\overline{a}^{p-1} = \overline{1}$ in $\mathbb{Z}_p$.

Proof The result follows from the above Euler–Fermat's theorem, if we note that $\phi(p) = p - 1$ for all prime p (note that all positive integers less than p are co-prime to p). ♯

Corollary 5.1.26 *Let p be a prime number and a an integer. Then, $a^p \equiv a(mod\ p)$.*

Proof If p does not divide a, then from the above Fermat's theorem $a^{p-1} \equiv 1 (mod\ p)$, and so $a^p \equiv a(mod\ p)$. Next, if p divides a, then it also divides $a^p - a$, and therefore, again $a^p \equiv a(mod\ p)$. ♯

More generally, we have the following corollaries.

Corollary 5.1.27 *Let p be a prime number and it does not divide an integer a. Then*

$$a^{p^r - p^{r-1}} \equiv 1 (mod\ p^r).$$ ♯

Corollary 5.1.28 *If p is a prime which does not divide a, then*

$$a^{p^r} \equiv a^{p^{r-1}} (mod\ p^r).$$ ♯

Illustrations

1.1. Let G be a group, and H and K be subgroups of G such that $aH = bK$ for some $a, b \in G$. Then, $H = K$.

Proof Suppose that $aH = bK$. Then, $a \in aH = bK$. Hence, $a \in bK$. But, then $aK = bK = aH$. Since $ak = ah$ implies that $k = h$, we see that $K = H$. ♯

1.2. Let $G = < a >$ be a cyclic group (not necessarily finite) and H a subgroup of index m. Then, $H = < a^m >$.

Proof If $H = \{e\}$, then $m = [G, \{e\}] = |G|$ and so $a^m = e$ and $H = < a^m >$. Suppose that $H \neq \{e\}$ and that $H = < a^r >$, where r is the least positive integer such that $a^r \in H$. Then, it is easy to show that $a^n \in H$ if and only if r divides n. This means that $a^t H = a^s H$ if and only if r divides $s - t$. This shows that $\{eH, aH, a^2H, \ldots, a^{r-1}H\}$ is precisely the set $(G/H)^l$. Hence, $m = [G, H] = r$. ♯

1.3. Let G_1 and G_2 be groups of co-prime orders. Then, the only homomorphism from G_1 to G_2 is the trivial homomorphism.

Proof Let f be a homomorphism from G_1 to G_2. Let $a \in G_1$. Then, $| a |$ divides $| G_1 |$ and $| f(a) |$ divides $| G_2 |$. Also $| f(a) |$ divides $| a |$. Hence, $| f(a) |$ divides $| G_1 |$ and $| G_2 |$. Since $| G_1 |$ and $| G_2 |$ are co-prime, it follows that $| f(a) | = 1$ and so $f(a) = e$ for all $a \in G_1$.   ♯

1.4. Let G be a finite group. Suppose that for each divisor d of $| G |$, the equation $X^d = e$ has at most d solutions. Then, G is cyclic.

Proof Suppose that $| G | = n$. Let d be a divisor of n. If there exists an element a of order d, then all elements of $< a >$ are the solutions of the equation $X^d = e$. By our hypothesis, there cannot be more. In $< a >$, there are $\phi(d)$ elements of order d. This shows that either G contains $\phi(d)$ elements of order d or it contains no element of order d. Let $A = \{d \in \mathbb{N} \mid d \ divides \ n \ and \ G \ has \ an \ element \ of \ order \ d\}$. Since each element of G is of order a divisor of n, we have

$$n = | G | = \sum_{d \in A} \phi(d)$$

Since,

$$n = \sum_{d/n} \phi(d)$$

(by Theorem 4.5.37), it follows that A is the set of all divisors of n. In other words, corresponding to every divisor d of n, there is an element of that order. In particular, G has an element of order n. This proves that G is cyclic.   ♯

1.5. Let G be a group of order $p^r \cdot q^s$, where p and q are distinct primes. Let P be a subgroup of order p^r and Q a subgroup of order q^s. Then $G = P \cdot Q$.

Proof By Corollary 5.1.22, $| P \cdot Q | = \frac{|P| \cdot |Q|}{|P \cap Q|}$, and since $P \cap Q$ is a subgroup of P as well as Q, $| P \cap Q |$ divides $| P |$ as well as $| Q |$. Since p and q are distinct primes, $| P \cap Q | = 1$. Thus, $| P \cdot Q | = p^r q^s = | G |$, and so $G = P \cdot Q$.   ♯

1.6. Let G be a finite group and f an automorphism of G. Let $X = \{a \in G \mid f(a) = a^{-1}\}$. Suppose that $| X | > \frac{3}{4} | G |$. Then, G is abelian and $f(a) = a^{-1}$ for all $a \in G$.

Proof Let $b \in X$. We show that the centralizer $C_G(b)$ of b is the group G itself. Consider $b^{-1}X = \{b^{-1}a \mid a \in X\}$. Since the multiplication by any element of G is a bijective map from G to G, $| b^{-1}X | = | X |$. Now,

$$| b^{-1}X | + | X | - | b^{-1}X \bigcap X | = | b^{-1}X \bigcup X | \leq | G |$$

Since $| b^{-1}X | = | X | > \frac{3}{4} \cdot | G |$,

$$| b^{-1}X \bigcap X | \geq | b^{-1}X | + | X | - | G | > \frac{| G |}{2}$$

Let $z \in b^{-1}X \cap X$. Then, $z, bz \in X$. Hence, $z^{-1}b^{-1} = (bz)^{-1} = f(bz)$ (for $bz \in X$) $= f(b)f(z)$ (for f is a homomorphism) $= b^{-1}z^{-1}$ (for $b, z \in X$). Taking inverses $bz = zb$. Thus, each element of $b^{-1}X \cap X$ commutes with b. In other words, $b^{-1}X \cap X \subseteq C_G(b)$, the centralizer of b. Since $| b^{-1}X \cap X | > \frac{|G|}{2}$, $| C_G(b) | > \frac{|G|}{2}$ and so by the Lagrange theorem $C_G(b) = G$. Thus, every element of X commutes with every element of G and so $X \subseteq Z(G)$. Again, $| Z(G) | \geq | X | \geq \frac{3|G|}{4}$, and hence by the Lagrange theorem $Z(G) = G$. This shows that G is abelian. Further, then, X forms a subgroup(check it). Again, since $| X | > \frac{3}{4} | G |$, by the Lagrange theorem $X = G$.   ♯

Remark 5.1.29 The result in the above illustration is best possible. For example, consider Q_8 the Quaternion group of order 8. The map f from Q_8 to Q_8 given by $f(1) = 1$, $f(-1) = -1$, $f(i) = -i$, $f(j) = -j$, $f(-i) = i$, $f(-j) = j$, $f(k) = k$, $f(-k) = -k$ is an automorphism for which $| X | = \frac{3}{4} | Q_8 |$, but Q_8 is not abelian.

1.7. Let a be an integer greater than 1. Then, n divides $\phi(a^n - 1)$ *for all* $n \geq 1$, where ϕ is the Euler's phi function.

Proof Consider the group U_{a^n-1} of prime residue classes modulo $a^n - 1$. Then, $| U_{a^n-1} | = \phi(a^n - 1)$. Since $a > 1$ and $n \geq 1$, $(a, a^n - 1) = 1$. Hence, $\bar{a} \in U_{a^n-1}$. Also $\bar{a}^n = \bar{1}$, for $a^n - 1$ divides $a^n - 1$. Further, $a^m - 1$ is not divisible by $a^n - 1$ for any $m < n$. Hence, the order of $\bar{a}$ in U_{a^n-1} is n. The result follows from Corollary 5.1.16.   ♯

1.8. The remainder obtained when $42^{61} + 18^{31} + 7$ is divided by 31 is 5, for using Fermat's theorem, we have

$$\overline{42^{61} + 18^{31} + 7} = \overline{42}^{61} + \overline{18}^{31} + \overline{7} = (\overline{42}^{30})^2 \cdot \overline{42} + \overline{18} + \overline{7} = \overline{42} + \overline{25} = \overline{5}$$

Exercises

5.1.1 Let R be a relation on a group G such that R is a subgroup of $G \times G$ (recall that $G \times G$ is a group with respect to coordinate-wise operation). Show that $R_e = \{x \in G \mid (x, e) \in R\}$ is a subgroup of G. Let R and S be subgroups of G such that $R_e = S_e$. Can we conclude that $R = S$? Let R be a reflexive relation on G which is a subgroup of $G \times G$. Show that $xhx^{-1} \in R_e$ for all $x \in G$ and $h \in R_e$. Let R be a symmetric relation on G which is a subgroup of $G \times G$. Show that if $x \in R_e$, then $x^3 \in R_e$. Show that a map f from G to itself considered as a subset of $G \times G$ is a subgroup if and only if f is an endomorphism of G.

5.1.2 Let G be a group of order 80. Show that it does not contain any subgroup of order 3.

5.1.3 Give an example of an infinite group in which all nontrivial subgroups are of finite indexes.

5.1.4 Give an example of an infinite group in which all proper subgroups are of finite orders.

5.1.5 Let H and K be subgroups of a finite group G such that $H \subseteq K$. Show that $[G, H] = [G, K][K, H]$.

5.1.6 Suppose that $[G, H] = m$ and $[G, K] = n$. Show that $[m, n]$ divides $[G, H \cap K]$. When does equality hold?

5.1.7 Let f be a homomorphism from a finite group G_1 to a finite group G_2. Show that $| f(G_1) |$ divides $| G_2 |$ and $| \ker f |$ divides $| G_1 |$.

5.1.8 Show that every finite cyclic group is a $C.L.T$ group.

5.1.9 Let S be a nonempty subset of a group G. Define a relation $\sim$ on G by $a \sim b$ *if and only if* $a^{-1}b \in S$. Show that $\sim$ is an equivalence relation if and only if S is a subgroup.

5.1.10 Let H be a subgroup of G. Show that R^l is a subgroup of $G \times G$ if and only if $R^l = R^r$ (equivalently, $aH = Ha$ *for all* $a \in G$).

5.1.11 Let H and K be subgroups of a finite group G. Suppose that $| H | > \sqrt{| G |}$ and $| K | > \sqrt{| G |}$. Show that $H \cap K \neq \{e\}$.

5.1.12 Suppose that $| G | = p \cdot q$, where p and q are distinct primes and $p > q$. Use the above exercise to show that there can be at most one subgroup of G of order p.

5.1.13 Show that $(a + b)^p \equiv (a^p + b^p)(mod\ p)$ *for all* $a, b \in \mathbb{N}$, where p is a prime number. Hint: Use the Binomial theorem.

5.1.14 Let H and K be subgroups of G. Show that $a(H \cap K) = aH \cap aK$ for all $a \in G$.

5.1.15 Suppose that $[G, H]$ and $[G, K]$ are finite and co-prime. Show that $[G, H \cap K] = [G, H] \cdot [G, K]$. Suppose further that G is finite. Show that $G = HK$.

5.1.16 Let G be a group having just one proper nontrivial subgroup. Show that G is a cyclic group of order p^2 for some prime p.

5.1.17 Find the remainders when
  (i) $28^{16} + 5$ is divided by 17.
  (ii) $14^{24} + 5$ is divided by 13.
  (iii) $(42^{20} + 6)^{10}$ is divided by 11.
  (iv) 3^{47} is divided by 23.
  (v) $(18! + 7)^{36} + 2$ is divided by 19.

5.1.18 Show that $(10! + 5)^{20}$ is of the form $1 + 11k$ for some $k > 0$.

5.1.19 Show by means of an example that $a^{m-1} \equiv 1(mod\ m)$ need not imply that m is prime.

5.1.20 Find $\phi(1000)$.

5.1.21 Let G be a group and H a subgroup of G. A subset S obtained by selecting one and only one member from each right(left) coset of G modulo H with the choice e from the coset H is called a **right(left) transversal** of G modulo H. Let H be a subgroup of K and K a subgroup of G. Let T be a right transversal of K in G and S a right transversal of H in K. Show that $S \cdot T$ is a right transversal to H in G.

5.1.22 Let G be a finite group and H a subgroup of order m and index r. Find the number of right transversals to H in G.

5.2 Product of Groups and Quotient Groups

Let $G_1, G_2, \ldots, G_n$ be groups and $G = G_1 \times G_2 \times \cdots \times G_n$ the Cartesian product of the sets $G_1, G_2, \ldots, G_n$. Define a binary operation $\star$ on G by

$$a \star b = (a_1b_1, a_2b_2, \ldots, a_nb_n),$$

where $a = (a_1, a_2, \ldots, a_n)$ and $b = (b_1, b_2, \ldots, b_n)$. The element $(e_1, e_2, \ldots, e_n)$ is the identity, and $(a_1^{-1}, a_2^{-1}, \ldots, a_n^{-1})$ is the inverse of $(a_1, a_2, \ldots, a_n)$. This group is called the **Cartesian product** or the **external direct product** of $G_1, G_2, \ldots, G_n$, and it is denoted by $\prod_{i=1}^{n} G_i$.

Theorem 5.2.1 *Let $G = \prod_{k=1}^{n} G_k$ be the external direct product of $G_1, G_2, \ldots, G_n$. Let*

$$H_k = \{(e_1, e_2, \ldots, e_{k-1}, a, e_{k+1}, \ldots, e_n) \mid a \in G_k\},\ k = 1, 2, \ldots, n.$$

Then, the following holds:
   (i) H_k is a subgroup of G, for all k.
   (ii) H_k is isomorphic to G_k for all k.
   (iii) Every element of H_k commutes with every element of H_l for all $k \neq l$.
   (iv) Every element $g \in G$ has a unique representation as

$$g = h_1h_2 \cdots h_n, h_k \in H_k.$$

   (v) $(h_1h_2 \cdots h_n) \cdot (h_1'h_2' \cdots h_n') = h_1h_1'h_2h_2' \cdots h_nh_n'.$ ♯

The proof of the above theorem is straight forward, and it is left as an exercise. The map η_k from G_k to H_k given by $\eta_k(a) = (e_1, e_2, \ldots, e_{k-1}, a, e_{k+1}, \ldots, e_n)$ is the required isomorphism.

Definition 5.2.2 A group G is said to be the **internal direct product** of its subgroups $H_1, H_2, \ldots, H_n$ if
(i) each element of H_k commutes with each element of H_l for all $k \neq l$, and
(ii) every element g of G can be expressed uniquely as

$$g = h_1 h_2 \cdots h_n, \ h_k \in H_k.$$

Thus, the external direct product $G = \prod_{i=1}^{n} G_i$ is the internal direct product of its subgroups $H_1, H_2, \ldots, H_n$ described in the previous theorem. Conversely, we have the following proposition.

Proposition 5.2.3 *If G is the internal direct product of its subgroups $H_1, H_2, \ldots, H_n$, then G is isomorphic to the external direct product $\prod_{i=1}^{n} H_i$.*

Proof The map η from $\prod_{i=1}^{n} H_i$ to G defined by $\eta((h_1, h_2, \ldots, h_n)) = h_1 h_2 \cdots h_n$ is easily be seen to be an isomorphism.   ♯

Proposition 5.2.4 *Let G be a group. Then, G is the internal direct product of its subgroups $H_1, H_2, \ldots, H_n$ if and only if the following holds.*
(i) $G = H_1 H_2 \cdots H_n$.
(ii) $h_k h_l = h_l h_k$ for all $h_k \in H_k$ and $h_l \in H_l, k \neq l$.
(iii) $H_k \cap (H_1 H_2 \cdots H_{k-1} H_{k+1} H_{k+2|} \cdots H_n) = \{e\}$ for all k.

Proof Suppose that G is the internal direct product of its subgroups $H_1, H_2, \ldots, H_n$. The conditions (i) and (ii) follow from the definition of the internal direct product. To prove (iii), let $g \in H_k \cap (H_1 H_2 \cdots H_{k-1} H_{k+1} \cdots H_n)$. Then, $g = e_1 e_2 \cdots e_{k-1} g e_{k+1} \cdots e_n$, where each $e_i = e$ the identity of G. Again, since $g \in (H_1 H_2 \cdots H_{k-1} H_{k+1} \cdots H_n)$, $g = h_1 h_2 \cdots h_{k-1} e h_{k+1} \cdots h_n$. By the uniqueness of the representation, $g = e$. This proves the condition (iii).

Conversely, assume that the conditions (i), (ii), and (iii) of the proposition hold. By (i), every element g of G can be expressed as

$$g = h_1 h_2 \cdots h_n, \ h_i \in H_i$$

For the uniqueness of the representation, suppose that

$$g = h_1 h_2 \cdots h_n = h_1' h_2' \cdots h_n', \ h_k, h_k' \in H_k.$$

From (ii), it follows that $H_1 H_2 \cdots H_{k-1} H_{k+1} \cdots H_n$ is a subgroup of G for all k. Again, by (ii),

$$h_k^{-1} h_k' = (h_1 h_2 \cdots h_{k-1} h_{k+1} \cdots h_n) \cdot (h_1' h_2' \cdots h_{k-1}' h_{k+1}' \cdots h_n')^{-1}.$$

This shows that $h_k^{-1} h_k'$ belongs to $H_k \cap H_1 H_2 \cdots H_{k-1} H_{k+1} \cdots H_n$ for all k. By the condition (iii) of the proposition, $h_k^{-1} h_k' = e$ for all k. Hence, $h_k = h_k'$ for all i. This proves the uniqueness of the representation.   ♯

In view of the above discussion, we need not to distinguish internal and external direct products.

Proposition 5.2.5 *The direct product of any two nontrivial cyclic groups is cyclic if and only if they are finite and of co-prime orders.*

Proof Let G be a group which is a direct product of a nontrivial cyclic group $H = <a>$ and a nontrivial cyclic group $K = <b>$. G cannot be infinite cyclic group. For, if $G = <c>$ is infinite cyclic, then $a = c^l$ and $b = c^m$ for some $l \neq 0$ and $m \neq 0$. But, then $c^{lm} \in H \cap K$. Since G, and so c is of infinite order $c^{lm} \neq e$. This contradicts the supposition that G is a direct product of H and K. Thus, if the direct product of H and K is cyclic, then both H and K should be finite. Now suppose that $|H| = |a| = r$ and $|K| = |b| = s$. Then, G is cyclic if and only if $|a^i b^j| = rs$ for some i, j. Since the elements of H and K commute and also $H \cap K = \{e\}$, by Proposition 4.5.17, $|a^i b^j| = [|a^i|, |b^j|]$ (the least common multiple of the order of a^i and the order of b^j). Now, by Proposition 4.5.19, $|a^i| = \frac{r}{(r,i)}$ and $|b^j| = \frac{s}{(s,j)}$. Thus, G is cyclic if and only if *there exist* i, j such that $|a^i b^j| = [\frac{r}{(r,i)}, \frac{s}{(s,j)}] = rs$. But, the above equality holds if and only if $(r, i) = 1 = (s, j)$ and $(r, s) = 1$. (Note that $[\frac{r}{(r,i)}, \frac{s}{(s,j)}] \leq \frac{r}{(r,i)} \frac{s}{(s,j)}$). Hence, the above conclusion. ♯

Definition A group G is said to be **indecomposable** if it cannot be written as a direct product of two nontrivial proper subgroups.

Example 5.2.6 Since any two nontrivial subgroups of an infinite cyclic group intersect nontrivially, an infinite cyclic group cannot be a direct product of two nontrivial proper subgroups. Thus, an infinite cyclic group is always indecomposable.

Example 5.2.7 A finite cyclic group is indecomposable if and only if it is of prime power order. For, if G is a cyclic group of prime power order, then no two nontrivial proper subgroups of G are of co-prime orders and so from Proposition 5.2.5, G is indecomposable. Next, suppose that $G = <a>$ is not of prime power order. Suppose that $|G| = m$. Then, $m = m_1 m_2$, where $1 < m_1$, $1 < m_2$ and $(m_1, m_2) = 1$. Take $H = <a^{m_1}>$ and $K = <a^{m_2}>$. Then, $|H| = m_2$ and $|K| = m_1$ are of co-prime orders, and so by the Lagrange theorem $H \cap K = \{e\}$. But, then $G = HK$. Clearly, the elements of H and K commute. Suppose that $h_1 k_1 = h_2 k_2$. Then, $h_1^{-1} h_2 = k_2^{-1} k_1 \in H \cap K = \{e\}$. Hence, $h_1 = h_2$ and $k_1 = k_2$. This shows that every element g of G has a unique representation as $g = hk$. This means that G is a direct product of H and K, and so it cannot be indecomposable.

Example 5.2.8 The additive group $\mathbb{Q}$ of rational numbers is indecomposable. For, let H and K be nontrivial subgroups of $\mathbb{Q}$. Let $r = \frac{m}{n} \neq 0$ be a member of H and $s = \frac{k}{l} \neq 0$ be a member of K. Then, $mk \in H \cap K$ and so $H \cap K \neq \{0\}$. Hence, $\mathbb{Q}$ cannot be a direct product of H and K.

Based on the following analogy, the language of group theory can be developed as the language of set theory was developed. We shall illustrate it very briefly, and the reader may complete the details as exercises.

Sets $\longleftrightarrow$ Groups.

Subsets $\longleftrightarrow$ Subgroups.

Cartesian products $\longleftrightarrow$ Direct products.

Relation R on a set X $\longleftrightarrow$ Relation R on a group G which is a subgroup of $G \times G$.

Equivalence relation on a set X $\longleftrightarrow$ Equivalence relation on a group G which is also a subgroup of $G \times G$.

Map from a set X to a set Y $\longleftrightarrow$ Homomorphisms from a group G to a group G'.

Note that a map f from a group G to G' is a homomorphism if and only if f considered as a subset of $G \times G'$ is a subgroup of $G \times G'$.

Definition 5.2.9 Let G be a group. An equivalence relation R on G is called a **congruence** on G if it is a subgroup of $G \times G$.

Now, we discuss the relationship between the congruences on G and subgroups of G.

Proposition 5.2.10 *Let R be a congruence on a group G. Then,*

(i) $R_e = \{a \in G \mid (a, e) \in R\}$ is a subgroup of G.

(ii) $aR_e = R_a = R_e a$ for all $a \in G$.

(iii) R is uniquely determined by the subgroup R_e. In fact, $(a, b) \in R$ if and only if $a^{-1}b \in R_e$ (if and only if $ab^{-1} \in R_e$).

Proof (i) R, being congruence, is an equivalence relation on G. Hence $(e, e) \in R$, and so $e \in R_e$. Thus, $R_e \neq \emptyset$. Let $a, b \in R_e$. Then $(a, e), (b, e) \in R$. Since R is a subgroup of $G \times G$, $(a, e)(b, e)^{-1} = (ab^{-1}, e) \in R$. This means that $ab^{-1} \in R_e$. Thus, R_e is a subgroup of G.

(ii) Let $x \in R_a$. Then, $(a, x) \in R$. Since R is an equivalence relation, $(a^{-1}, a^{-1}) \in R$. Since R is a congruence, it is a subgroup of $G \times G$, and hence $(a^{-1}, a^{-1}) \cdot (a, x) = (e, a^{-1}x) \in R$. This shows that $a^{-1}x \in R_e$, or equivalently $x \in aR_e$. Conversely, if $h \in R_e$, then $(ah, a) = (a, a) \cdot (h, e)$ belongs to R, for R is a congruence. This shows that $ah \in R_a$.

(iii) Since R is a congruence, to say that $(a, b) \in R$ is equivalent to say that $(e, a^{-1}b) = (a^{-1}, a^{-1}) \cdot (a, b)$ belongs to R. Thus, $(a, b) \in R$ if and only if $a^{-1}b \in R_e$. Similarly, one can show that $(a, b) \in R$ if and only if $ab^{-1} \in R_e$.   ♯

Remark 5.2.11 The above proposition says that $R^l = R^r$ is a necessary condition for a subgroup H to be R_e for some congruence R on G. The following proposition says that the condition is also sufficient.

Proposition 5.2.12 *Let H be a subgroup of a group G. Then, R^l is a congruence if and only if $R^l = R^r$. Further, then $H = R_e^l$.*

Proof Suppose that R^l is a congruence. To say that $(a, b) \in R^l$ is equivalent to say that $(a^{-1}, b^{-1}) = (a, b)^{-1} \in R^l$. This is equivalent to say that $ab^{-1} = (a^{-1})^{-1}b^{-1} \in H$. Hence, $R^l = R^r$. Thus, if R^l is a congruence, then $R^l = R^r$.

Conversely, suppose that $R^l = R^r$. Then, we have to show that R^l is a congruence. It is already an equivalence relation. Thus, it is sufficient to show that it is a subgroup. Since $R^l = R^r$ and $a^{-1}ah \in H$ for all $h \in H$, it follows that $aha^{-1} \in H$ for all $a \in G$ and $h \in H$. Now, let $(a, b), (c, d) \in R^l$. Then, $a^{-1}b$ and $c^{-1}d$ belong to H. Hence, $(ac)^{-1}bd = c^{-1}a^{-1}bd = c^{-1}dd^{-1}(a^{-1}b)d$ belongs to H. This shows that $(ac, bd) \in R^l$. Already we have seen that the inverses of the elements of R^l are the elements of $R^r = R^l$. It follows that R^l is a congruence. Finally, it is easily observed that $R^l_e = H$.   ♯

For further analogy between set theory and group theory, consider a group G and congruence R on G. Consider the quotient set

$$G/R = \{R_a \mid a \in G\} = \{aR_e \mid a \in G\} = G/^l R_e = G/^r R_e$$

If $(a, c), (b, d) \in R$, then $(ab, cd) \in R$, for R is a subgroup. Thus, $R_a = R_c$ and $R_b = R_d$ implies that $R_{ab} = R_{cd}$. This shows that there is a unique binary operation $\star$ on $G/R = G/^l R_e = G/^r R_e$ given by

$$R_a \star R_b = R_{ab}$$

It is easily seen that $(G/R, \star)$ is a group. R_e is the identity and the inverse of R_a is $R_{a^{-1}}$. This group is called the **quotient group of G modulo R**.

Normal Subgroups and Quotient Groups

Definition 5.2.13 A subgroup H of a group G is called a **normal subgroup** or **invariant subgroup** if $aH = Ha$ for all $a \in G$.

Before having some examples, let us have some necessary and sufficient conditions for a subgroup to be a normal subgroup.

Theorem 5.2.14 *Let H be a subgroup of a group G. Then, the following conditions are equivalent:*

1. H is a normal subgroup of G.
2. $R^l = R^r$.
3. $G/^l H = G/^r H$.
4. $R^l (R^r)$ is a congruence on G.
5. The correspondence which associates aH with Ha is a map from $G/^l H$ to $G/^r H$.
6. $(a, b) \in R^l$ if and only if $(ag, bg) \in R^l$ for all $g \in G$.
7. $aha^{-1} \in H$ for all $a \in G$ and $h \in H$.
8. $aHa^{-1} = H$ for all $a \in G$.
9. The binary relation $\star$ on $G/^l H$ given by $aH \star bH = abH$ is a binary operation on $G/^l H$.

Proof $1 \Longleftrightarrow 2$. We know that $R^l_a = aH$ and $R^r_a = Ha$. Thus, to say that H is normal is to say that $R^l_a = R^r_a$ for all $a \in G$. This is equivalent to say that $R^l = R^r$.

$2 \Longleftrightarrow 3$. This follows if we observe that two equivalence relations on a set X are same if and only if the corresponding quotient sets are same.

$3 \Longleftrightarrow 4$. This is precisely Proposition 5.2.12.

$1 \Longleftrightarrow 5$. Clearly, 1 implies that $aH \rightsquigarrow Ha$ is the identity map. Assume 5. Then, $aH = bH$ implies that $Ha = Hb$. Equivalently, $a^{-1}b \in H$ *implies that* $ab^{-1} \in H$. This means that $R^l \subseteq R^r$. Since the inverses of the elements of R^l in $G \times G$ are precisely the elements of R^r, and those of elements of R^r are precisely the elements of R^l, it follows that $R^r \subseteq R^l$. Thus, $R^l = R^r$. This completes the proof of the fact that $1 \Longleftrightarrow 5$.

$4 \Longrightarrow 6$. Assume 4. Since R^l is a congruence, it is a subgroup, and $(g, g) \in R^l$ for all $g \in G$. Hence, $(a, b) \in R^l$ *if and only if* $(a, b) \cdot (g, g) \in R^l$.

$6 \Longrightarrow 2$. Assume 6. Then, $(a, b) \in R^l$ *if and only if* $(ab^{-1}, bb^{-1}) \in R^l$. In other words, $a^{-1}b \in H$ *if and only if* $ab^{-1} \in H$. This shows that $R^l = R^r$.

$1 \Longrightarrow 7$. Assume 1. Then, $aH = Ha$ for all $a \in G$. Let $a \in G$ and $h \in H$. Then, $ah \in aH = Ha$. Hence, $ah = ka$ for some $k \in H$. But, then $aha^{-1} = k \in H$.

$7 \Longrightarrow 1$. Assume 7. Let $ah \in aH$, where $h \in H$. Then, $aha^{-1} \in H$ and so $ah = aha^{-1}a \in Ha$. Thus, $aH \subseteq Ha$. Further, $ha = a(a^{-1}ha) \in aH$ for all $a \in G$ and $h \in H$. This shows that $aH = Ha$.

$7 \Longrightarrow 8$. Assume 7. Then, aha^{-1} and $a^{-1}h(a^{-1})^{-1}$ belong to H for all $a \in G$ and $h \in H$. This shows that $aHa^{-1} = H$.

$8 \Longrightarrow 7$ is obvious.

Finally, we prove that $7 \Longleftrightarrow 9$. Assume 7. Then, $ghg^{-1} \in H$ for all $g \in G$ and $h \in H$. Suppose that $aH = bH$ and $cH = dH$. Then, $a^{-1}b \in H$ and $c^{-1}d \in H$. But, then $(ac)^{-1}bd = c^{-1}a^{-1}bd = c^{-1}(a^{-1}b)c(c^{-1}d)$ belongs to H. This shows that $acH = bdH$. Thus, $\star$ defined by

$$aH \star cH = acH$$

is a binary operation, and so 7 *implies* 9. Assume 9. Let $a \in G$ and $h \in H$. Then, $hH = eH$. Since $\star$ is a binary operation

$$haH = hH \star aH = eH \star aH = aH.$$

This shows that $a^{-1}ha \in H$ *for all* $a \in G$ and *for all* $h \in H$. Thus, $9 \Longrightarrow 7$.   ♮

Remark 5.2.15 We have shown the existence of a natural bijective map $\mathfrak{R}$ from the set $C(G)$ of congruences on G to the set $NS(G)$ of all normal subgroups of G defined by $\mathfrak{R}(R) = R_e$.

Notation. We use the notation $H \trianglelefteq G$ to say that H is a normal subgroup of G.

Let G be a group. Then, G and $\{e\}$ are always normal subgroups of G (verify). The normal subgroup G of G is called the **improper** normal subgroup and $\{e\}$ is called the **trivial** normal subgroup of G. Other normal subgroups are called **proper** normal subgroups. A group G is called **Simple** if it has no proper normal subgroups.

Remark 5.2.16 A simple group is not really simple to study.

Notation. The notation $H \lhd G$ is used to say that H is a proper normal subgroup of G. Thus,

$$H \lhd G \text{ if and only if } (H \unlhd G \text{ and } H \neq G, \; H \neq \{e\})$$

If H is a subgroup of a group G which is contained in the center of G, then it is normal in G, for $aH = \{ah \mid h \in H\} = \{ha \mid h \in H\} = Ha$. In particular, every subgroup of an abelian group is normal. We may ask the following converse of the above statement: Let G be a group all of whose subgroups are normal. Can we infer that G is abelian? The answer is no. Consider the Quaternion group Q_8. $\{1\}$ and Q_8 are clearly normal. The subgroup $\{1, -1\}$ is the center, and so it is normal. All other subgroups are of order 4, and so of index 2. They are all normal because of the following proposition:

Proposition 5.2.17 *Every subgroup of index 2 is normal.*

Proof Let G be a group and H a subgroup of index 2. We have to show that $aH = Ha$ for all $a \in G$. If $a \in H$, then $aH = H = Ha$. Suppose that $a \notin H$. Then, $aH \neq H \neq Ha$. Since H is of index 2, $G/^l H = \{H, aH\}$ and $G/^r H = \{H, Ha\}$. Since Left(right) cosets form a partition of G, $aH = G - H = Ha$. Thus, $aH = Ha$ for all $a \in G$, and so H is normal in G. ♯

Remark 5.2.18 We observed that Q_8 is a nonabelian group all of whose subgroups are normal in G. Baer (Situation der untergruppen und struktur der group, S.B. Heidelberg Akad. Mat. Nat, Klasse 2, 1933, 12–17) in 1933 proved that a nonabelian group has all its subgroups normal if and only if it is a direct product of Q_8, several copies of $\mathbb{Z}_2$, and an abelian group all of whose elements are of odd order. In particular, all subgroups of $Q_8 \times \mathbb{Z}_2$ are normal.

Remark 5.2.19 Every subgroup of index 2 is normal. The result is not true for any other prime. However, the result can be generalized as follows: If p is the smallest prime dividing the order of G, then every subgroup of index p is normal. The proof of this fact will be given later.

Proposition 5.2.20 *An abelian group is simple if and only if it is prime cyclic.*

Proof Since every subgroup of an abelian group is normal, an abelian group is simple if and only if it has no nontrivial proper subgroups. But, then, it is necessarily prime cyclic (Theorem 5.1.19). ♯

Proposition 5.2.21 *Let f be a homomorphism from a group G_1 to a group G_2. Let $H_2 \unlhd G_2$. Then, $f^{-1}(H_2) \unlhd G_1$. Suppose further that f is surjective and $H_1 \unlhd G_1$. Then, $f(H_1) \unlhd G_2$.*

Proof We already know that $f^{-1}(H_2)$ is a subgroup (Proposition 4.4.25). Let $h \in f^{-1}(H_2)$ and $g \in G_1$. Then $f(ghg^{-1}) = f(g)f(h)f(g)^{-1}$. Since $h \in f^{-1}(H_2)$, $f(h) \in H_2$, and again since $H_2 \trianglelefteq G_2$, $f(g)f(h)f(g)^{-1} \in H_2$. Thus, $f(ghg^{-1}) \in H_2$. But, then $ghg^{-1} \in f^{-1}(H_2)$. Since g and h are arbitrary, $f^{-1}(H_2) \trianglelefteq G_1$.

Next, suppose that f is surjective and $H_1 \trianglelefteq G_1$. Let $f(h) \in f(H_1)$, where $h \in H_1$. Let $x \in G_2$. Since f is surjective, *there exists* $g \in G_1$ such that $f(g) = x$. But, then $xf(h)x^{-1} = f(g)f(h)f(g)^{-1} = f(ghg^{-1})$ (for f is a homomorphism). Since $H_1 \trianglelefteq G_1$, $ghg^{-1} \in H_1$, and so $xf(h)x^{-1} \in f(H_1)$. This shows that $f(H_1)$ is a normal subgroup of G_2. ♯

Corollary 5.2.22 *Under the map ϕ defined in Theorem 4.4.29 (Correspondence Theorem), normal subgroups correspond.* ♯

Corollary 5.2.23 *Kernel of a homomorphism is a normal subgroup.* ♯

Proof Let f be a homomorphism from a group G_1 to a group G_2. Since $\{e_2\}$ is a normal subgroup of G_2, it follows that the *ker* $f = f^{-1}(\{e_2\})$ is a normal subgroup of G_1. ♯

Proposition 5.2.24 *Let H and K be normal subgroups of a group G. Then, the following holds.*
  (i) HK is a normal subgroup.
  (ii) If $H \bigcap K = \{e\}$, then $hk = kh$ for all $h \in H$ and $k \in K$.

Proof (i) Since K is normal $HK = \bigcup_{h \in H} hK = \bigcup_{h \in H} Kh = KH$. This shows that HK is a subgroup of G. Further, let $hk \in HK$, where $h \in H$ and $k \in K$. Since H and K are normal in G, $ghg^{-1} \in H$ and also $gkg^{-1} \in K$. Hence, $ghkg^{-1} = ghg^{-1}gkg^{-1}$ also belongs to HK. This shows that HK is normal in G.

  (ii) Let $h \in H$ and $k \in K$. Since H is normal in G, $khk^{-1} \in H$, and so $hkh^{-1}k^{-1} \in H$. Similarly, $hkh^{-1}k^{-1} \in K$. Since $H \bigcap K = \{e\}$, it follows that $hkh^{-1}k^{-1} = e$. This means that $hk = kh$. ♯

Corollary 5.2.25 *Let G be a group. Then, G is the internal direct product of its subgroups $H_1, H_2, \ldots, H_n$ if and only if the following hold.*
(i) $G = H_1 H_2 \cdots H_n$.
(ii) H_k is normal for each k.
(iii) $H_k \bigcap (H_1 H_2 \ldots H_{k-1} H_{k+1} \ldots H_n) = \{e\}$ for all k.

Proof Since $H_k \bigcap (H_1 H_2 \ldots H_{k-1} H_{k+1} \ldots H_n) = \{e\}$ for all k, $H_k \bigcap H_l = \{e\}$ for all $k \neq l$. From the above proposition, for $k \neq l$, each element of H_k commutes with each element of H_l. The result follows from Proposition 5.2.4. ♯

Theorem 5.2.26 *Let $H \trianglelefteq G$. Then, we have a binary operation $\star$ on G/H defined by*

$$aH \star bH = abH$$

with respect to which G/H is a group.

Proof By Theorem 5.2.14(9), $\star$ is a binary operation on G/H. The verification of the associativity of $\star$ is straightforward. The identity is $eH = H$, and the inverse of aH is $a^{-1}H$. ♯

Definition 5.2.27 The group $(G/H, \star)$ described in the above theorem is called the **quotient group** or the **factor group** of G modulo H.

Check that $\mathbb{Z}/m\mathbb{Z}$ is the group $\mathbb{Z}_m$ of residue classes modulo m.

Definition 5.2.28 Let $H \trianglelefteq G$. The map ν from G to G/H defined by $\nu(a) = aH$ is called the **quotient map** from G to G/H.

It follows from the definition of the binary operation in G/H that the map ν is a surjective homomorphism. Further, the kernel of ν is $\{a \in G \mid \nu(a) = H\}$ (note that the identity of G/H is H) = $\{a \in G \mid aH = H\}$. Now $aH = H$ if and only if $a \in H$. This shows that kernel of ν is H. This also says that normal subgroups are precisely kernels of homomorphisms.

Proposition 5.2.29 *Let $H \trianglelefteq G$. Then any subgroup of G/H is of the form K/H, where K is a subgroup of G containing H. Further $K_1/H = K_2/H$ if and only if $K_1 = K_2$. The subgroup K/H is normal in G/H if and only if K is normal in G.*

Proof The quotient map ν is a surjective homomorphism from G to G/H whose kernel is H. Thus, by correspondence theorem, ν induces a bijective map from the set $S(G)$ of subgroups of G containing H (the kernel of ν) to the set of all subgroups of G/H given by $K \rightsquigarrow K/H$ under which normal subgroups correspond. The result follows. ♯

Example 5.2.30 Consider the quotient map ν from $\mathbb{Z}$ to $\mathbb{Z}/m\mathbb{Z} = \mathbb{Z}_m$. It follows from the above proposition that every subgroup of $\mathbb{Z}_m$ is of the form $r\mathbb{Z}/m\mathbb{Z}$, where $m\mathbb{Z} \subseteq r\mathbb{Z}$. Now $m\mathbb{Z} \subseteq r\mathbb{Z}$ if and only if r divides m. Clearly, $r\mathbb{Z}/m\mathbb{Z} = \{\overline{0}, \overline{2r}, \ldots, \overline{(q-1)r}\}$, where $qr = m$.

Theorem 5.2.31 (Cauchy Theorem for Abelian groups) *Let G be a finite abelian group. Let p be a prime dividing the order of G. Then, G has an element (and so also a subgroup) of order p.*

Proof The proof is by induction on $|G|$. If $|G| = 1$, the statement is vacuously true. Assume that the result is true for all those groups whose order is less than the order of G. We have to prove the result for G. If p does not divide the order of G, then the result is vacuously true. Suppose that p divides the order of G. Let $a \in G$, $a \neq e$ and order of a is m. If p divides m, then order of $a^{\frac{m}{p}}$ is p and we are done. Suppose that p does not divide m. Then, $(p, m) = 1$. Since G is abelian $< a >$ is normal in G. Further,

(i) $|G/< a >| = \frac{|G|}{|<a>|} < |G|$.

(ii) p divides $|G/< a >|$, for p divides $|G|$ and $(p, |a|) = 1$.

By the induction hypothesis, there is an element $b < a >$ of $G/< a >$ of order p. Thus,

$$b < a > \neq < a > \text{ and } (b < a >)^p = < a > .$$

Equivalently, $b \notin < a >$ and $b^p \in < a >$. If $b^p = e$, b is an element of order p, and we are done. Suppose that $b^p = a^r \neq e$. Then

$$(b^p)^m = (a^r)^m = a^{mr} = e.$$

Hence, $(b^m)^p = e$. Suppose that $b^m = e$. Then

$$(b < a >)^m = b^m < a > = < a >$$

the identity of $G / < a >$. Since $b < a >$ is of order p, p divides m, a contradiction to the supposition that $(p, m) = 1$. Thus, b^m is a nonidentity element which is of order p.   ♯

Theorem 5.2.32 *Let G be a group such that $G/Z(G)$ is cyclic. Then, G is abelian and $G/Z(G)$ is the trivial group.*

Proof Suppose that $G/Z(G)$ is cyclic and is generated by $aZ(G)$. Then, $G/Z(G) = \{(aZ(G))^n \mid n \in \mathbb{Z}\} = \{a^n Z(G) \mid n \in Z\}$. Since G is the union of its cosets,

$$G = \bigcup_{n \in \mathbb{Z}} a^n Z(G)$$

Now, let $x, y \in G$. Then, $x = a^n u$ and $y = a^m v$ for some $m, n \in \mathbb{Z}$ and $u, v \in Z(G)$. Since u and v are in the center of G,

$$xy = a^n u a^m v = a^{n+m} vu = a^{m+n} vu = a^m a^n vu = a^m v a^n u = yx$$

This shows that G is abelian, and so $G/Z(G)$ is the trivial group.   ♯

Definition 5.2.33 Let G be a group. An element of the form $aba^{-1}b^{-1}$ is denoted by (a, b) and is called a **commutator**. The subgroup G' of G generated by all commutators of G is called the **commutator subgroup** or the **derived subgroup** of G.

Theorem 5.2.34 *The commutator subgroup G' of G is a normal subgroup of G such that G/G' is abelian. Further, let H be a normal subgroup of G. Then, G/H is abelian if and only if $G' \subseteq H$. Also if H is any subgroup of G containing G', then it is normal in G.*

Proof Let $h \in G'$ and $g \in G$. Then, $ghg^{-1}h^{-1}$ being a commutator, belongs to G'. But, then $ghg^{-1} = ghg^{-1}h^{-1}h$ belongs to G'. Hence, G' is a normal subgroup of G. The above argument also shows that if H is any subgroup containing G', then it is normal in G. Now, $abH = aH \star bH = bH \star aH = baH$ for all $a, b \in G$ if and only if $(ba)^{-1}ab = a^{-1}b^{-1}ab$ belongs to H for all $a, b \in G$. It follows that G/H is abelian if and only if all commutators of G are in H. Since commutators generate G', G/H is abelian if and only if $G' \subseteq H$.   ♯

Remark 5.2.35 It follows from the above theorem that G' is the smallest normal subgroup of G such that G/G' is an abelian group. The quotient group G/G' is the largest quotient group of G which is abelian. The group G/G' is called the **abelianizer** of G and is denoted by G_{ab}.

Group with operator

Definition 5.2.36 Let Ω be a set and G a group. We say that Ω **operates** on G through a map $\star$ from $G \times \Omega$ to G if

$$ab \star \omega = (a \star \omega)(b \star \omega)$$

for all $\omega \in \Omega$ and $a, b \in G$, where $a \star \omega$ denotes the image of (a, ω) under $\star$. We also say that G is a Ω group through the operation $\star$. Sometimes we also say that the pair (G, Ω) is a **group with operator**.

Let (G, Ω) be a group with operator. Then for each $\omega \in \Omega$, we have a map f_ω from G to G given by $f_\omega(g) = g \star \omega$. Since

$$f_\omega(ab) = ab \star \omega = (a \star \omega)(b \star \omega) = f_\omega(a) f_\omega(b)$$

for all $a, b \in G$, it follows that the map f_ω is an endomorphism of G. Thus, we have a map f from Ω to the set $End(G)$ of endomorphisms of G defined by $f(\omega) = f_\omega$. Conversely, given a map f from Ω to the set $End(G)$, Ω operates on G through $\star$ given by $a \star \omega = f(\omega)(a)$. Let G_1 and G_2 be two Ω groups with the operations $\star_1$ and $\star_2$, respectively. A homomorphism ϕ from G_1 to G_2 is called a $\Omega-$ homomorphism if

$$\phi(a \star_1 \omega) = \phi(a) \star_2 \omega$$

for all $a \in G_1$ and $\omega \in \Omega$. Equivalently, a homomorphism ϕ from G_1 to G_2 is a $\Omega-$ homomorphism if $\phi o f_\omega = f_\omega o \phi$ for all $\omega \in \Omega$. Clearly, the composite of any two $\Omega-$ homomorphisms is a $\Omega-$ homomorphism. As usual an injective $\Omega-$ homomorphism is called a $\Omega-$ monomorphism and a surjective $\Omega-$ homomorphism is called a $\Omega-$ epi morphism. A bijective $\Omega-$ homomorphism is called a $\Omega-$ isomorphism. A subgroup H of a $\Omega-$ group G is called a $\Omega-$ subgroup of G if $h \star \omega \in H$ for all $h \in H$ and $\omega \in \Omega$. In turn, a $\Omega-$ subgroup H of G is a $\Omega-$ group at its own right such that the inclusion map is a $\Omega-$ monomorphism. A normal $\Omega-$ subgroup H of G is called a $\Omega-$ normal subgroup. If H is a $\Omega-$ normal subgroup of a $\Omega-$ group G, then the quotient group G/H also becomes a $\Omega-$ group with respect to the $\Omega-$ operation $\star$ given by $aH \star \omega = (a \star \omega)H$ such that the quotient map ν from G to G/H given by $\nu(a) = aH$ is a $\Omega-$ homomorphism. This group is called the $\Omega-$ quotient group. All the earlier relevant results hold good if we replace groups by $\Omega-$ groups, subgroups by $\Omega-$ subgroups, homomorphisms by $\Omega-$ homomorphisms, normal subgroups by $\Omega-$ normal subgroups and quotient groups by $\Omega-$ quotient groups. The verification to this effect is left as exercise.

Let G be a group and $\Omega \subseteq End(G)$. Then, G is a $\Omega-$ group in a natural way. A $EndG$ subgroup of G is called a **fully invariant** subgroup of G. More precisely, a subgroup H of G is called a fully invariant subgroup of G if $\eta(h) \in H$ for all $\eta \in End(G)$ and $h \in H$. An $Aut(G)$ subgroup of G is called a **characteristic** subgroup of G. Thus, a subgroup H of G is a characteristic subgroup if $\eta(h) \in H$ for all $\eta \in Aut(G)$ and $h \in H$. A subgroup H is a $Inn(G)$ subgroup if $f_g(h) \in H$ for all $h \in H$ and $g \in G$ (f_g being the inner automorphism determined by g). Thus, $Inn(G)$ subgroups are precisely normal subgroups of G. It is evident that a fully invariant subgroup is a characteristic subgroup, and a characteristic subgroup is a normal subgroup. However, reverse implication is not true.

Example 5.2.37 The subgroup $\{e, a\}$ of the Klein's four group $V_4 = \{e, a, b, c\}$ is a normal subgroup. However, it is not a characteristic subgroup, for it is not invariant under the automorphism ϕ of V_4 which maps e to e, a to b, b to c, and c to a. Indeed, V_4 is **characteristically simple** in the sense that it has no nontrivial proper characteristic subgroup.

Example 5.2.38 Let $Z(G)$ denote the center of G and η an automorphism of G. Let $a \in Z(G)$ and $x \in G$. Then, $x = \eta(y)$ for a unique $y \in G$. Now, $\eta(a)x = \eta(a)\eta(y) = \eta(ay) = \eta(ya) = \eta(y)\eta(a) = x\eta(a)$. This shows that $\eta(a) \in Z(G)$. Thus, the center of a group is always a characteristic subgroup of the group. However, the center of a group need not be fully invariant. Consider, for example, the group $Q_8 \times \mathbb{Z}_4$, where Q_8 denotes the Quaternion group. Clearly,

$$Z(Q_8 \times \mathbb{Z}_4) = \{1, -1\} \times \mathbb{Z}_4$$

Take an isomorphism τ from $\mathbb{Z}_4$ to the subgroup $\{1, i, -1, -i\} \times \{\overline{0}\}$ of $Q_8 \times \mathbb{Z}_4$. Let p_2 denote the second projection. Then, $\tau o p_2$ is an endomorphism of $Q_8 \times \mathbb{Z}_4$ which maps the center $\{1, -1\} \times \mathbb{Z}_4$ to the subgroup $\{1, i, -1, -i\} \times \{\overline{0}\}$ of $Q_8 \times \mathbb{Z}_4$ which is not contained in the center. This shows that center of a group need not be fully invariant.

Example 5.2.39 The commutator subgroup $G' = [G, G]$ of a group is generated by the set $\{(a, b) = a^{-1}b^{-1}ab \mid a, b \in G\}$ of commutators of G. If η is an endomorphism of G, then $\eta((a, b)) = (\eta(a))^{-1}(\eta(b))^{-1}\eta(a)\eta(b) = (\eta(a), \eta(b))$ is again a commutator. This shows that the commutator subgroup G' of G is fully invariant in G. More generally, if H and K are fully invariant subgroups of G, then the subgroup $[H, K] = < \{h^{-1}k^{-1}hk \mid h \in H, k \in K\} >$ is also fully invariant.

Example 5.2.40 All subgroups of a cyclic group are fully invariant (verify). Are there more groups all of whose subgroups are fully invariant?

Let $\Omega \subseteq End(G)$. Recall that an endomorphism η of G is a $\Omega-$ endomorphism if η commutes with all members of Ω. A $Inn(G)-$ endomorphism of G is called a **normal** endomorphism. Thus, an endomorphism η of G is a normal endomorphism if $\eta o f_g = f_g o \eta$ for all $g \in G$. More precisely, an endomorphism η of G is a normal endomorphism if and only if

$$\eta(gxg^{-1}) = g\eta(x)g^{-1}$$

for all $g, x \in G$

Proposition 5.2.41 *An endomorphism σ of G is a normal endomorphism if and only if $g^{-1}\sigma(g) \in C_G(\sigma(G))$ for all $g \in G$. Indeed, then the map η from G to $C_G(\sigma(G))$ defined by $\eta(g) = \sigma(g^{-1})g$ is a homomorphism.*

Proof Let σ be an endomorphism of G. Suppose that σ is a normal endomorphism. Then by the definition,

$$\sigma(gxg^{-1}) = g\sigma(x)g^{-1} \text{ for all } g, x \in G.$$

In turn,

$$\sigma(g)\sigma(x)(\sigma(g))^{-1} = g\sigma(x)g^{-1} \text{ for all } g, x \in G.$$

Hence,

$$g^{-1}\sigma(g)\sigma(x) = \sigma(x)\, g^{-1}\sigma(g) \text{ for all } g, x \in G.$$

This shows that $g^{-1}\sigma(g) \in C_G(\sigma(G))$ for all $g \in G$. Finally, consider the map η from G to $C_G(\sigma(G))$ defined by $\eta(g) = (\sigma(g))^{-1}g$. Then
$\eta(g_1 g_2)$
$= (\sigma(g_1 g_2))^{-1}g_1 g_2$
$= (\sigma(g_2))^{-1}(\sigma(g_1))^{-1}g_1 g_2$
$= (\sigma(g_1))^{-1}g_1(\sigma(g_2))^{-1}g_2$ (for $(\sigma(g_1))^{-1}g_1 \in C_G(\sigma(G))$)
$= \eta(g_1)\eta(g_2)$.
This shows that η is a homomorphism. ♯

Definition 5.2.42 An endomorphism σ of G is called a **central endomorphism** if σ induces identity automorphism on $G/Z(G)$, or equivalently $g^{-1}\sigma(g) \in Z(G)$ *for all $g \in G$.*

consequently, observe that a central endomorphism maps center to itself.

Corollary 5.2.43 *All central endomorphisms are normal endomorphisms. Also a surjective normal endomorphism is central.* ♯

Corollary 5.2.44 *Let σ be an automorphism of G. Then $\sigma \in C_{Aut(G)}(Inn(G))$ if and only if the map τ defined by $\tau(g) = g^{-1}\sigma(g)$ is a homomorphism from G to its center $Z(G)$. Conversely, let τ be a homomorphism from G to its center $Z(G)$. Then, the map σ defined by $\sigma(g) = g\tau(g)$ is a normal endomorphism of G.*

Proof To say that $\sigma \in C_{Aut(G)}(Inn(G))$ is to say that σ is a normal automorphism of G. By Proposition 5.2.41, this is equivalent to say that the map η defined by $\eta(g) = \sigma(g^{-1})g$ is a homomorphism from G to $C_G(\sigma(G)) = Z(G)$. Since $Z(G)$ is abelian, the map τ defined by $\tau(g) = g^{-1}\sigma(g) = (\eta(g))^{-1}$ is a homomorphism from G to $Z(G)$. Conversely, let τ be a homomorphism from G to its center $Z(G)$.

Consider the map σ from G to G defined by $\sigma(g) = g\tau(g)$. Then

$\sigma(g_1 g_2)$

$= g_1 g_2 \tau(g_1 g_2)$

$= g_1 g_2 \tau(g_1) \tau(g_2)$

$= g_1 \tau(g_1) g_2 \tau(g_2)$ (for $\tau(g_1) \in Z(G)$)

$= \sigma(g_1) \sigma(g_2)$

Thus, σ is an endomorphism of G. Further,

$\sigma(gxg^{-1})$

$= gxg^{-1} \tau(gxg^{-1})$

$= gxg^{-1} \tau(g) \tau(x) (\tau(g))^{-1}$

$= gxg^{-1} \tau(x)$ (for $\tau(x) \in Z(G)$)

$= gx\tau(x)g^{-1}$

$= g\sigma(x)g^{-1}.$

This shows that the map σ defined by $\sigma(g) = g\tau(g)$ is a normal endomorphism of G.   ♯

The following corollary is an immediate consequence of Proposition 5.2.41.

Corollary 5.2.45 *Let G be a center less group in the sense that $Z(G)$ is trivial. Then, the only normal automorphism of G is the identity map. In particular, $Z(Aut(G))$ is also trivial. In turn, for any nonabelian simple group G, $Z(Aut(G))$ is trivial.* ♯

Remark 5.2.46 Indeed, for any simple group G, $Aut(G)$ is a complete group in the sense that $Z(Aut(G)) = \{I_G\}$ and every automorphism of $Aut(G)$ is an inner automorphism of $Aut(G)$.

Example 5.2.47 Every endomorphism of an abelian group is normal endomorphism. We determine normal endomorphisms of the Quaternion group Q_8. The trivial endomorphism and the identity automorphism are obviously normal endomorphisms as they commute with all inner automorphisms. Let η be another endomorphism of Q_8. If the image $\eta(Q_8) = \{1, -1\}$, then since $C_{Q_8}(\{1, -1\}) = Q_8$, it follows that $g^{-1}\eta(g) \in C_{Q_8}(\{1, -1\})$ for all $g \in Q_8$. By Proposition 5.2.41, η is a normal endomorphism. The image of an endomorphism cannot be a subgroup of order 4 (justify). Suppose that η is a nonidentity automorphism. Since $Z(Q_8) = \{1, -1\}$, η is a normal endomorphism if and only if $\eta(g) = \pm g$. There are 3 such automorphisms. There are no more automorphisms. This shows that all endomorphisms of Q_8 are normal endomorphisms.

Example 5.2.48 Trivial subgroup, Q_8, and $\{1, -1\}$ are characteristic subgroups of Q_8. However, the rest of the subgroups are not characteristic.

Definition 5.2.49 Let G be a direct product of its subgroups $H_1, H_2, \ldots, H_n$. Then for each k, we have the map p_k from G to G defined by $p_k(g) = g_k$, where $g = g_1 g_2 \cdots g_n$ is the unique representation g as product of elements of $H_1, H_2, \ldots, H_n$. The map p_k, thus obtained, is called the **k_{th} Projection**.

Proposition 5.2.50 *Let G be a direct product of its subgroups $H_1, H_2, \ldots, H_n$. Then, we have the following:*
(i) Each p_k is a normal endomorphism which is idempotent in the sense that $(p_k)^2 = p_k$. The image $p_k(G) = H_k$.
(ii) Each pair p_k, p_l, $k \neq l$ of distinct projections is summable in the sense that the map $p_k + p_l$ defined by $(p_k + p_l)(g) = p_k(g)p_l(g)$ is again a normal endomorphism (indeed, a projection on $H_k H_l$).
(iii) The composition $p_k o p_l$ is the trivial endomorphism for all $k \neq l$.
(iv) $p_1 + p_2 + \cdots + p_n = I_G$, the identity automorphism on G.
Conversely, if G is a group which has endomorphisms p_k satisfying conditions (i), (ii), (iii), and (iv), then G is a direct product of its subgroups $p_1(G)$, $p_2(G)$, $\ldots$, $p_n(G)$.

Proof (i) Let g and h be elements of G with their unique representations $g = g_1 g_2 \cdots g_n$ and $h = h_1 h_2 \cdots h_n$, respectively. Then, $gh = g_1 h_1 g_2 h_2 \cdots g_n h_n$. Hence, $p_k(gh) = g_k h_k = p_k(g) p_k(h)$. This shows that p_k is an endomorphism of G. By the definition $p_k(p_k(g)) = p_k(g_k) = g_k = p_k(g)$. Again, $ghg^{-1} = g_1 h_1 g_1^{-1} g_2 h_2 g_2^{-1} \cdots g_k h_k g_k^{-1} \cdots g_n h_n g_n^{-1}$. Thus, $p_k(ghg^{-1}) = g_k h_k g_k^{-1} = g h_k g^{-1} = g p_k(h) g^{-1}$. This proves (i).
(ii) Let g and h be elements of G with their unique representations $g = g_1 g_2 \cdots g_n$ and $h = h_1 h_2 \cdots h_n$, respectively. Clearly, then, for $k \neq l$, $(p_k + p_l)(gh) = p_k(gh) p_l(gh) = g_k h_k g_l h_l = g_k g_l h_k h_l = p_k(g) p_l(g) p_k(h) p_l(h) = (p_k + p_l)(g)(p_k + p_l)(h)$. This proves (ii).
(iii) An element $a \in H_l$ has unique representation as $a = e_1 e_2 \cdots e_{l-1} a e_{l+1} \cdots e_n$, where $e_i = e$ the identity of G, and which belongs to H_i for all i. Thus, for $k \neq l$, $p_k(a) = e$ for all $a \in H_l$. Again, since $p_l(G) = H_l$, it follows that $p_k(p_l(g)) = e$ for all $g \in G$. This proves (iii).
(iv) Let g be an element of G with unique representations $g = g_1 g_2 \cdots g_n$. Then by the definition, $g_i = p_i(g)$ for all i. Thus, $g = p_1(g) p_2(g) \cdots p_n(g) = (p_1 + p_2 + \cdots + p_n)(g)$ for all $g \in G$. This proves (iv).

Conversely, suppose that G is a group which has endomorphisms p_k satisfying conditions (i), (ii), (iii), and (iv) of the proposition. Put $p_k(G) = H_k$. Since the image of a normal endomorphism is a normal subgroup (verify), H_k is a normal subgroup of G for each k. This shows that the condition (ii) of Corollary 5.2.25 holds. By (iv), every element $g \in G$ is expressible as $g = (p_1 + p_2 + \cdots + p_n)(g) = p_1(g) p_2(g) \cdots p_n(g)$. This shows that $G = H_1 H_2 \cdots H_n$ and so condition (i) of Corollary 5.2.25 also holds. Let $a \in H_k \cap H_1 H_2 \cdots H_{k-1} H_{k+1} \cdots H_n$. Then $a = a_1 a_2 \cdots a_{k-1} e_k a_{k+1} \cdots a_n$. Since $a \in H_k = p_k(G)$ and $(p_k)^2 = p_k$ (by (i)), it follows that $e = e_k = p_k(a) = a$. This proves condition (iii) of Corollary 5.2.25. The result follows from Corollary 5.2.25. ♯

Proposition 5.2.51 *Let G be a group and η an idempotent normal endomorphism of G. Then, G is a direct product of $\eta(G)$ and Ken η.*

Proof Since η is normal endomorphism $\eta(G)$ is a normal subgroup. Already, *ker* η is a normal subgroup. Let $g \in G$. Then, since η is idempotent, $(\eta(g))^{-1} g$ is in the *ker* η. Further, $g = \eta(g)(\eta(g))^{-1} g$. This shows that $G = \eta(G) ker \eta$. Suppose that

$g \in \eta(G) \cap ker\ \eta$. Then, since η is idempotent, $g = \eta(g) = e$. This shows that $\eta(G) \cap ker\ \eta = \{e\}$. The result follows from Corollary 5.2.25.   ♯

Corollary 5.2.52 *Let G be an indecomposable group and η a normal idempotent endomorphism. Then, η is either trivial or a normal automorphism of G.*

Proof It follows from the above proposition that G is a direct product of $\eta(G)$ and $ker\ \eta$. Since G is indecomposable, $\eta(G) = \{e\}$, or else $\eta(G) = G$ and $ker\ \eta = \{e\}$. This shows that η is trivial or else an automorphism.   ♯

Proposition 5.2.53 *Every finite group is a direct product of indecomposable groups.*

Proof The proof is by induction on $\mid G \mid$. If $\mid G \mid = 1$, then there is nothing to prove. Assume that every group whose order is less than that of G is a direct product of indecomposable groups. We show that G is also a direct product of indecomposable groups. If G itself is indecomposable, there is nothing to do. If not, then G is a direct product of two nontrivial subgroups H and K of G. Clearly, then $\mid H \mid < \mid G \mid$ and $\mid K \mid < \mid G \mid$. By the induction hypothesis, H and K are direct products of indecomposable subgroups. Hence, G is also a direct product of indecomposable subgroups.   ♯

Definition 5.2.54 A subgroup H of a group G is called a **direct factor** of G if there is a subgroup K of G such that G is a direct product of H and K.

Thus, all direct factors are normal subgroups. Normal subgroups need not be direct factors. Indeed, all subgroups of the additive group $\mathbb{Z}$ of integers are normal but no nontrivial proper subgroups are direct factors (verify).

The following theorem known as Krull–Remak–Schmidt Theorem will be proved in Chap. 10.

Theorem 5.2.55 (Krull–Remak–Schmidt) *If a group G is a direct product of indecomposable subgroups $H_1, H_2, \ldots, H_r$ and also a direct product of indecomposable subgroups $K_1, K_2, \ldots, K_s$, then*

(i) $r = s$, and

(ii) there is a bijective correspondence between the sets $\{H_1, H_2, \ldots, H_r\}$ and $\{K_1, K_2, \ldots, K_r\}$ such that the corresponding subgroups are isomorphic.   ♯

Remark 5.2.56 The above results reduce the problem of classification of finite groups to the problem of classification of indecomposable groups. The solution to this problem is beyond a dream to mathematicians.

Exercises

5.2.1 Show that the direct product of abelian groups is abelian.

5.2.2 Let G be the internal direct product of groups $G_1, G_2, \ldots, G_n$. Show that each G_i is normal in G.

5.2.3 Let $H \trianglelefteq G$ and K a subgroup of G containing H. Show that $H \trianglelefteq K$.

Remark 5.2.57 We shall see that $H \trianglelefteq K$ and $K \trianglelefteq G$ need not imply that $H \trianglelefteq G$.

5.2.4 Let $H \trianglelefteq G$ and K a subgroup of G. Show that HK is a subgroup of G. We shall see that this subgroup need not be normal in G.

5.2.5 Show that the Klein's four group $V_4 = \{e, a, b, c\}$ is the internal direct product of its subgroups $H = \{e, a\}$ and $K = \{e, b\}$. Deduce that it is isomorphic to $Z_2 \times Z_2$.

5.2.6 (i) Show that the group P of roots of unity is indecomposable.
(ii) Show that Q_8 is indecomposable.
(iii) Show that Direct factor of a direct factor is itself a direct factor.
(iv) Characterize groups all of whose subgroups are direct factors.

5.2.7 Is the additive group $\mathbb{R}$ of reals indecomposable?

5.2.8 Show that the multiplicative group of positive rational numbers is indecomposable.

5.2.9 Express V_4 as a direct product of indecomposable groups.

5.2.10 Characterize subgroups H of a group G for which the relation R^l is a normal subgroup of $G \times G$.

5.2.11 Show that all subgroups of $Q_8 \times \mathbb{Z}_2$ are normal. Find all of its direct factors.

5.2.12 Show that all subgroups of $Q_8 \times \mathbb{Z}_3$ are normal. Find all of its direct factors.

5.2.13 Show that every abelian group of order pq, where p and q are distinct primes, is cyclic.

5.2.14 Is it true that all subgroups of $Q_8 \times \mathbb{Z}_4$ are normal? Support your claim.

5.2.15 Show that the direct image of a normal subgroup under a homomorphism need not be a normal subgroup.

5.2.16 Show that the set $I(G)$ of inner automorphisms of G is a normal subgroup of $Aut(G)$.

5.2.17 Let H be a subgroup of G such that $x^2 \in H \; \forall x \in G$. Show that $H \trianglelefteq G$.

5.2.18 Let G be a finite simple group and f a nontrivial endomorphism of G. Show that f is an automorphism of G.

5.2.19 Let G be a group which has a unique subgroup H of order $m > 1$. Show that $H \trianglelefteq G$.

5.2.20 Show that the intersection of a family of normal subgroups of a group G is normal in G.

5.2.21 Characterize the elements of the normal subgroup generated by a set S. Show that, in general, it is bigger than the subgroup generated by S.

5.2.22 Let S be a normal set in the sense that $ghg^{-1} \in S$ for all $g \in G$ and $h \in S$. Show that $< S >$ is normal in G.

5.2.23 Write explicitly the elements of the group $Q_8/\{1, -1\}$. Show that it is isomorphic to V_4.

5.2.24 Call a normal subgroup H of G to be a **maximal normal subgroup**, if it is a proper normal subgroup which is not properly contained in any proper normal subgroup. Show that a normal subgroup H is maximal normal if and only if G/H is simple.

5.2.25 Show that the additive group $\mathbb{Q}$ of rationals, the additive group $\mathbb{R}$ of reals and the circle group do not contain any maximal normal subgroups.

5.2.26 Determine all maximal normal subgroups of the additive group $\mathbb{Z}$ of integers.

5.2.27 Let G be a finite abelian group and m divides $|G|$. Show that G has a subgroup of order m.
Hint. Prove it by induction on $|G|$, using Cauchy theorem.

5.2.28 Let G be a finite abelian group all of whose elements are of orders a power of p, where p is a prime. Show that $|G| = p^n$ for some n.

5.2.29 Let H be a subgroup of G. Let $N_G(H) = \{g \in G \mid gH = Hg\}$. Show that $N_G(H)$ is a subgroup of G containing H as a normal subgroup. Show further that $N_G(H)$ is the largest subgroup of G in which H is normal. This subgroup is called the **normalizer** of H in G.

5.2.30 Let H be a subgroup of G. Show that $Core_G(H) = \bigcap_{g \in G} gHg^{-1}$ is the largest normal subgroup of G contained in H. This subgroup is called the **core** of H in G.

5.2.31 Let G be a group and $H = < a >$ a cyclic subgroup which is normal in G. Show that every subgroup of H is normal in G.

5.2.32 Let H, K, L be subgroups of G, where L is normal in G. Suppose that $HL = KL$ and $H \bigcap L = K \bigcap L$. Show that $H = K$.

5.2.33 Let H be a subgroup of G. Show that $C_G(H) \trianglelefteq N_G(H)$, where $C_G(H)$ is the centralizer of H in G.

5.2.34 Find the commutator subgroup of the Quaternion group Q_8.

5.2.35 Find the commutator subgroups of D_4.

5.2.36 Find all characteristic subgroups and also all fully invariant subgroups of a Dihedral group.

5.2.37 Let σ and τ be a pair of normal summable idempotent endomorphisms of a group G such that $\sigma + \tau$ is an automorphism. Show that one of σ and τ is an automorphism.

5.2.38 Let η be a normal endomorphism of G. Show that all normal subgroups of $\eta(G)$ are also normal subgroups of G.

5.2.39** Let G be a group and H a subgroup of G. Let S be right transversal to H in G (S is obtained by selecting one and only one member from each right coset of H in G with the choice e for the coset H). Define a binary operation o on S by

$$\{xoy\} = Hxy \bigcap S$$

Show that (S, o) is a right quasigroup in the sense that equations $Xoa = b$ have unique solutions for all $a, b \in S$ (Conversely, it is proved in [Ramji Lal, 'Transversals in Groups,' Journal of Algebra 1996] that every right quasigroup with identity turns out to be a right transversal to a subgroup H in group G with G, H universal in certain sense). Show that if $H \lhd G$, then right quasigroups determined by all right transversals are isomorphic to G/H (Conversely, it is shown (Ramji Lal— R.P. Shukla, 'Perfectly Stable Subgroups of a Finite Group,' Communications in Algebra 1996), using classification of finite simple groups, that if all right transversals to a subgroup H of a finite group G determine isomorphic right quasigroups, then $H \trianglelefteq G$).

5.3 Fundamental Theorem of Homomorphism

In the last section, we noticed that a quotient group G/H is homomorphic image of G. In this section, we show that every homomorphic image of G is isomorphic to a quotient group of G.

Theorem 5.3.1 (Fundamental Theorem of Homomorphism) *Let f be a homomorphism from a group G to a group G'. Let K be the kernel of f. Let $H \lhd G$. Then, there exists a homomorphism $\overline{f}$ from G/H to G' making the diagram*

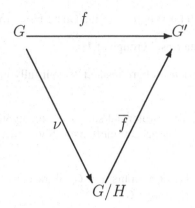

commutative ($\overline{f} \circ \nu = f$) *if and only if* $H \subseteq K$. *Further, then,*
   (i) such a $\overline{f}$ *is unique.*
   (ii) $\overline{f}$ *is injective if and only if* $H = K$.
   (iii) $\overline{f}$ *is surjective if and only if* f *is so.*

Proof Suppose that such a homomorphism $\overline{f}$ exists. Let $h \in H$. Since H is the identity of G/H and $\overline{f}$ is a homomorphism,

$$e' = \overline{f}(H) = \overline{f}(hH) = \overline{f}(\nu(h)) = f(h)$$

Hence, $h \in K$. Thus, $H \subseteq K$. Conversely, suppose that $H \subseteq K$. Suppose further that $aH = bH$. Then, $a^{-1}b \in H$. Since $H \subseteq K$, $a^{-1}b \in K$. In turn, $e' = f(a^{-1}b) = f(a)^{-1}f(b)$, and so $f(a) = f(b)$. This ensures that we have a map $\overline{f}$ from G/H to G' defined by $\overline{f}(aH) = f(a)$. Now,

$$\overline{f}(aH \star bH) = \overline{f}(abH) = f(ab) = f(a)f(b) = \overline{f}(aH)\overline{f}(bH).$$

Thus, $\overline{f}$ is a homomorphism. By the definition of $\overline{f}$, $\overline{f} \circ \nu = f$.

   Next, suppose that such a homomorphism $\overline{f}$ exists. If g is also a homomorphism such that $\overline{f} \circ \nu = f = g \circ \nu$, then since ν is surjective, $\overline{f} = g$. This proves (i). We know that $\overline{f}$ is injective if and only if kernel of $\overline{f}$ is the trivial subgroup $\{H\}$ of G/H. Now,

$$ker\, \overline{f} = \{aH \in G/H \mid e' = \overline{f}(aH) = f(a)\} = \{aH \mid a \in K\} = K/H$$

Clearly, $K/H = \{H\}$ if and only if $K = H$. This proves (ii).

   Finally, since ν is surjective, $\overline{f} \circ \nu = f$ is surjective if and only if $\overline{f}$ is surjective.   ♯

Corollary 5.3.2 *Let* f *be a surjective homomorphism from* G *to* G'. *Then,* $G/ker\, f$ *is isomorphic to* G'.   ♯

Remark 5.3.3 The last corollary is used very frequently and is also termed as the fundamental theorem of homomorphism.

Applications of the fundamental theorem of homomorphism

3.1. $\mathbb{R}/\mathbb{Z}$ is isomorphic to the circle group S^1.

Proof Consider the map f from the additive group $\mathbb{R}$ of real numbers to the circle group S^1 defined by $f(a) = e^{2\pi i a}$. By the law of exponents, it follows that f is a homomorphism which is surjective. Further, $e^{2\pi i a} = 1$ if and only if $a \in \mathbb{Z}$. Thus, the kernel of f is $\mathbb{Z}$. From the fundamental theorem of homomorphism, the result follows. ♯

3.2. $\mathbb{Q}/\mathbb{Z}$ is isomorphic to the group P of roots of unity.

Proof $e^{2\pi i a} \in P$ if and only if *there exists* $n \in \mathbb{Z}$ such that $(e^{2\pi i a})^n = e^{2\pi i n a} = 1$. This is equivalent to say that $a \in \mathbb{Q}$. Thus, we have a surjective homomorphism f from $\mathbb{Q}$ to P given by $f(a) = e^{2\pi i a}$ whose kernel is $\mathbb{Z}$. The result follows from the fundamental theorem of homomorphism. ♯

3.3. Suppose that n divides m. Then $m\mathbb{Z} \subseteq n\mathbb{Z}$ and $n\mathbb{Z}/m\mathbb{Z}$ is isomorphic to $\mathbb{Z}_{\frac{m}{n}}$.

Proof Clearly, $m\mathbb{Z} \subseteq n\mathbb{Z}$ if and only if every integral multiple of m is also an integral multiple of n. This means n divides m. Define a map ϕ from $\mathbb{Z}$ to $n\mathbb{Z}/m\mathbb{Z}$ by $\phi(r) = nr + m\mathbb{Z}$. ϕ is clearly a surjective homomorphism. Now $nr + m\mathbb{Z} = m\mathbb{Z}$ if and only if $nr \in m\mathbb{Z}$. This is equivalent to say that r is a multiple of $\frac{m}{n}$. Thus, the kernel of ϕ is $\frac{m}{n}\mathbb{Z}$. The result follows from the fundamental theorem of homomorphism. ♯

3.4. Let G be a group. Then, $G/Z(G)$ is isomorphic to the group $I(G)$ of inner automorphisms of G.

Proof The map f from G to $I(G)$ defined by $f(g) = f_g$, where f_g is the inner automorphism determined by g ($f_g(x) = gxg^{-1}$) is a surjective map which is easily seen to be a homomorphism. Now,
$$ker\ f = \{g \in G \mid f_g = I_G\} = \{g \in G \mid gxg^{-1} = x \forall x \in G\} = \{g \in G \mid gx = xg \forall x \in G\} = Z(G).$$ By the fundamental theorem of homomorphism the result follows. ♯

Theorem 5.3.4 (1st Isomorphism Theorem) *Let f be a surjective homomorphism from G to G' and $H \trianglelefteq G$ such that $ker\ f \subseteq H$. Let $H' = f(H)$. Then $H' \trianglelefteq G'$, and there exists a unique isomorphism $\overline{f}$ from G/H to G'/H' such that the following diagram is commutative ($\overline{f} \circ \nu_G = \nu_{G'} \circ f$), where ν_G and $\nu_{G'}$ are the corresponding quotient maps).*

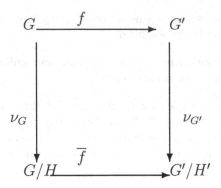

Proof Consider the map $\phi = \nu_{G'} \circ f$ from G to G'/H'. Since $\nu_{G'}$ and f are surjective homomorphisms, ϕ is a surjective homomorphism. Now,

$$ker\, \phi = \{a \in G \mid \phi(a) = H'\}.$$
$$= \{a \in G \mid \nu_{G'}(f(a)) = H'\}.$$
$$= \{a \in G \mid f(a)H' = H'\}.$$
$$= \{a \in G \mid f(a) \in H'\}.$$
$$= f^{-1}(H').$$
$$= f^{-1}(f(H)).$$

Since H contains the kernel of f, by Proposition 4.4.28, $f^{-1}(f(H)) = H$. Hence, the kernel of ϕ is H. The result follows from the fundamental theorem of homomorphism.   ♯

Corollary 5.3.5 *Let H and K be normal subgroups of G such that $H \subseteq K$. Then, $K/H \trianglelefteq G/H$ and $(G/H)/(K/H)$ is isomorphic to G/K.*

Proof Consider the quotient map ν from G to G/H. Then, ν is surjective homomorphism, and $ker\, \nu = H$. Also $\nu(K) = K/H$. The result follows from the first isomorphism theorem.   ♯

Theorem 5.3.6 (Noether 2nd Isomorphism Theorem) *Let H and K be subgroups of G and $K \trianglelefteq G$. Then, $H \cap K \trianglelefteq H$ and $H/H \cap K$ is isomorphic to HK/K.*

Proof Since $K \trianglelefteq G$, $HK = KH$, and so HK is a subgroup of G. Since $K \trianglelefteq G$, $K \trianglelefteq HK$. Thus, we can consider the quotient group HK/K. The map ϕ from H to HK/K defined by $\phi(h) = hK$ is clearly a homomorphism. Further, any element of HK/K is of the form hkK for some $h \in H$ and $k \in K$. Since $h^{-1}hk = k$ belongs to K, $hkK = hK = \phi(h)$. This shows that ϕ is surjective homomorphism. Now,

$$ker\, \phi = \{h \in H \mid \phi(h) = K\} = \{h \in H \mid hK = K\} = \{h \in H \mid h \in K\}.$$

Thus, $ker\, \phi = H \cap K$. Since kernel of a homomorphism is a normal subgroup $H \cap K \trianglelefteq H$. Also by the fundamental theorem of homomorphism, $H/H \cap K$ is isomorphic to HK/K.   ♯

Remark 5.3.7 The assumption that $K \unlhd G$ is essential, for otherwise HK/K or even HK need not be groups.

Exercises

5.3.1 Show that $Q_8/\{1, -1\}$ is isomorphic to V_4. Deduce that $I(Q_8)$ is isomorphic to V_4.

5.3.2 Let G be a group such that $Z(G) = \{e\}$. Show that G is isomorphic to $I(G)$. Deduce that any nonabelian simple group is isomorphic to its group of inner automorphisms.

5.3.3 Show that $G/Z(G)$ and $I(G)$ cannot be a nontrivial cyclic group.

5.3.4 Can $|I(G)| = 29$? Support.

5.3.5 Let $G = <a>$ be a cyclic group of order m. Define a map f from $\mathbb{Z}$ to G by $f(n) = a^n$. Show that f is surjective homomorphism with kernel $m\mathbb{Z}$. Deduce that G is isomorphic to $\mathbb{Z}_m$.

5.3.6 Show that $\mathbb{Z}_n$ is isomorphic to $(\mathbb{Z}/m\mathbb{Z})/(n\mathbb{Z}/m\mathbb{Z})$, where n divides m.

5.3.7 Show that $\mathbb{R}/\mathbb{Q}$ is isomorphic to S^1/P.
Hint. Consider the map $x \leadsto e^{2\pi\, ix}$, and use the first isomorphism theorem.

5.3.8 Show that $m\mathbb{Z}/m\mathbb{Z} \cap n\mathbb{Z}$ is isomorphic to $m\mathbb{Z} + n\mathbb{Z}/n\mathbb{Z}$. Deduce that $\mathbb{Z}_{\frac{[m,n]}{m}}$ is isomorphic to $\mathbb{Z}_{\frac{n}{(m,n)}}$. Deduce further that $[m, n](m, n) = mn$.

5.3.9 Show that $\mathbb{Q} + 2\pi\mathbb{Z}/2\pi\mathbb{Z}$ is isomorphic to $\mathbb{Q}$. Deduce that S^1 contains an isomorphic copy of $\mathbb{Q}$.
Hint. Observe that $\mathbb{Q} \cap 2\pi\mathbb{Z} = \{0\}$. Use the Noether isomorphism theorem.

5.3.10 Show that $\mathbb{Q} + e\mathbb{Z}/e\mathbb{Z}$ is isomorphic to $\mathbb{Q}$.

5.3.11 Show that $2\pi\mathbb{Z} + \mathbb{Q}/\mathbb{Q}$ is isomorphic to $\mathbb{Z}$.

5.3.12 Let H, K be subgroups of G and $L \unlhd G$ such that $HL = KL$, $H \cap L = \{e\} = K \cap L$. Show that H is isomorphic to K.

5.3.13 Show by means of an example that $H_1 \approx H_2$ and $G_1/H_1 \approx G_2/H_2$ need not imply that $G_1 \approx G_2$.

5.3.14 Show that $N_G(H)/C_G(H)$ is isomorphic to a subgroup of $Aut(H)$.
Hint. Consider the map $g \leadsto f_g$, where $f_g(h) = ghg^{-1}$.

5.3.15 Use the fundamental theorem of maps in set theory to prove the fundamental theorem of homomorphism and Isomorphism theorems.

Chapter 6
Permutation Groups and Classical Groups

The two main sources of groups are the permutation groups and the matrix groups. This chapter is devoted to introduce these groups, and to study some of their fundamental and elementary properties. In Algebra 2 and in Algebra 3, we shall study the representations of abstract groups as permutation groups, and also the representations of abstract groups as matrix groups.

6.1 Permutation Groups

In this section, we shall introduce the permutation groups (also called the symmetric groups) and representations of abstract groups in these groups.

Let X be a set. Let $Sym(X)$ denote the set of all bijective maps from X to X. Then, $Sym(X)$ is a group with respect to the composition of maps. This group is called **the Symmetric group or the Transformation group** or **the Permutation group** on X. The proof of the following proposition is an easy verification.

Proposition 6.1.1 *If f is a bijective map from X to Y, then it induces an isomorphism $Sym(f)$ from $Sym(X)$ to $Sym(Y)$ defined by*

$$Sym(f)(g) = fogof^{-1}.$$ ♯

Thus, the group $Sym(X)$ depends only on the cardinality of X. In particular if X is finite, then $Sym(X)$ depends (up to isomorphism) on the number of elements in X. In this section, we shall be interested in symmetric groups on finite sets only. If X contains n elements, then without any loss we can take $X = \{1, 2, \ldots, n\}$.

The symmetric group $Sym(X)$, where $X = \{1, 2, \ldots, n\}$ will be denoted by S_n and will be called **the symmetric group** or **the permutation group of degree n**.

© Springer Nature Singapore Pte Ltd. 2017
R. Lal, *Algebra 1*, Infosys Science Foundation Series in Mathematical Sciences,
DOI 10.1007/978-981-10-4253-9_6

Thus, S_n is the group of all bijective maps from $\{1, 2, \ldots, n\}$ to itself, the operation being the composition of maps. We may represent an element $f \in S_n$ (without any ambiguity) by

$$\begin{pmatrix} 1 & 2 & \ldots & n \\ f(1) & f(2) & \ldots & f(n) \end{pmatrix}$$

Since f is a bijective map, the second row is just the rearrangement (permutation) of $1, 2, \ldots, n$. Thus, any $f \in S_n$ gives a unique permutation described above. Conversely, if we have a rearrangement of $1, 2, \ldots, n$, then it gives rise to a unique bijective map from $\{1, 2, \ldots, n\}$ to itself by putting the rearrangement below $12 \ldots n$ as above. For example, if $n = 4$, the rearrangement 2314 of 1234 gives rise to a bijective map α from $\{1, 2, 3, 4\}$ to itself given by $\alpha(1) = 2$, $\alpha(2) = 3$, $\alpha(3) = 1$ and $\alpha(4) = 4$. In the above notation

$$\alpha = \begin{pmatrix} 1\,2\,3\,4 \\ 2\,3\,1\,4 \end{pmatrix}$$

Thus, the members of S_n can be viewed as permutations. The product gf of permutations

$$f = \begin{pmatrix} 1 & 2 & \ldots & n \\ f(1) & f(2) & \ldots & f(n) \end{pmatrix}$$

and

$$g = \begin{pmatrix} 1 & 2 & \ldots & n \\ g(1) & g(2) & \ldots & g(n) \end{pmatrix}$$

is given by

$$gf = \begin{pmatrix} 1 & 2 & \ldots & n \\ g(f(1)) & g(f(2)) & \ldots & g(f(n)) \end{pmatrix}$$

Example 6.1.2 If

$$\alpha = \begin{pmatrix} 1\,2\,3\,4 \\ 3\,4\,1\,2 \end{pmatrix}$$

and

$$\beta = \begin{pmatrix} 1\,2\,3\,4 \\ 2\,1\,3\,4 \end{pmatrix}$$

then

$$\alpha\beta = \begin{pmatrix} 1\,2\,3\,4 \\ 4\,3\,1\,2 \end{pmatrix}$$

and

$$\beta\alpha = \begin{pmatrix} 1\,2\,3\,4 \\ 3\,4\,2\,1 \end{pmatrix}$$

Thus, $\alpha\beta \neq \beta\alpha$

Example 6.1.3 Take $n = 3$,

$$\alpha = \begin{pmatrix} 1 & 2 & 3 \\ 2 & 1 & 3 \end{pmatrix}$$

and

$$\beta = \begin{pmatrix} 1 & 2 & 3 \\ 1 & 3 & 2 \end{pmatrix}$$

Then

$$\alpha\beta = \begin{pmatrix} 1 & 2 & 3 \\ 2 & 3 & 1 \end{pmatrix}$$

and

$$\beta\alpha = \begin{pmatrix} 1 & 2 & 3 \\ 3 & 1 & 2 \end{pmatrix}$$

Again $\alpha\beta \neq \beta\alpha$. This shows that S_3 is nonabelian.

Let $n \leq m$. Every permutation in S_n can be viewed as a permutation in S_m by fixing all $i > n$. This defines an embedding of S_n in to S_m. In other words S_n can be thought of as a subgroup of S_m. It follows that S_n is abelian if and only if $n \leq 2$.

Since the number of permutations on n symbols is $n_{P_n} = n!$, we have

Theorem 6.1.4 $|S_n| = n!$. ♯

Corollary 6.1.5 $p^{n!} = I$ for all $p \in S_n$, where I is the identity permutation.

Proof Follows from the Corollary 5.1.17. ♯

Cycles and Transpositions

Now, we consider special types of permutations, for example, the permutation

$$\alpha = \begin{pmatrix} 1 & 2 & 3 & 4 & 5 & 6 \\ 2 & 5 & 3 & 1 & 6 & 4 \end{pmatrix}$$

α takes 1 to 2, 2 to 5, 5 to 6, 6 to 4 and 4 to 1. The remaining symbol 3 is fixed. We can faithfully represent the permutation α by the row (1 2 5 6 4) with the understanding that each symbol goes to the following symbol, the last symbol is mapped to the first symbol, and the symbol not appearing in the row is kept fixed. Thus, the permutation

$$\begin{pmatrix} 1 & 2 & 3 & 4 & 5 & 6 & 7 \\ 1 & 5 & 2 & 4 & 7 & 6 & 3 \end{pmatrix}$$

can be represented by (2 5 7 3) whereas

$$\begin{pmatrix} 1 & 2 & 3 & 4 \\ 2 & 1 & 4 & 3 \end{pmatrix}$$

cannot be represented in this form.

Definition 6.1.6 A permutation $\alpha \in S_n$ is called a **cycle** of length $r \geq 1$ if there exists a subset $\{i_1, i_2, \ldots, i_r\}$ of $\{1, 2, \ldots, n\}$ containing r distinct elements such that $\alpha(i_1) = i_2, \alpha(i_2) = i_3, \ldots, \alpha(i_{r-1}) = i_r, \alpha(i_r) = i_1$ and $\alpha(j) = j$ for all $j \notin \{i_1, i_2, \ldots, i_r\}$. The cycle α is denoted by $(i_1 i_2 \ldots i_r)$. A cycle of length 2 is called a **transposition**. Thus, a transposition is represented by $(i\ j)$ which interchanges i and j and keeps the rest of the symbols fixed.

Thus, the identity permutation is not treated as a cycle.

Cycles $(i_1 i_2 \ldots i_r)$ and $(j_1 j_2 \ldots j_s)$ are said to be disjoint if

$$\{i_1, i_2, \ldots, i_r\} \bigcap \{j_1, j_2, \ldots j_s\} = \emptyset$$

Proposition 6.1.7 *Any two disjoint cycles commute.*

Proof Let $\alpha = (i_1 i_2 \ldots i_r)$ and $\beta = (j_1 j_2 \ldots j_s)$ be disjoint cycles. If $l \notin \{i_1, i_2, \ldots, i_r, j_1, j_2, \ldots, j_s\}$, then $\alpha(l) = \beta(l) = l$ and so $\alpha(\beta(l)) = \beta(\alpha(l))$. If $l \in \{i_1, i_2, \ldots, i_r\}$, then β fixes l and also $\alpha(l)$. This means that $\alpha(\beta(l)) = \alpha(l) = \beta(\alpha(l))$. Similarly, if $l \in \{j_1, j_2, \ldots, j_s\}$, then also $(\alpha\beta)(l) = \beta(l) = (\beta\alpha)(l)$. This shows that $\beta\alpha = \alpha\beta$.   ♯

Proposition 6.1.8 *If α and β are disjoint cycles, then $<\alpha> \bigcap <\beta> = \{I\}$.*

Proof If α and β are disjoint cycles, then the symbols changed by any power of α is fixed by any power of β and the symbols changed by any power of β is fixed by any power of α. Thus, $\alpha^l = \beta^m$ if and only if $\alpha^l = I = \beta^m$.   ♯

Remark 6.1.9 If p and q are permutations such that the symbols changed by p are fixed by q and the symbols changed by q are fixed by p, then $pq = qp$ and $<p> \bigcap <q> = \{I\}$, and so in this case $o(pq) = [o(p), o(q)]$.

Proposition 6.1.10 *Let $\alpha \in S_n$ be a cycle of length r. Then, $o(\alpha) = r$.*

Proof Let $\alpha = (i_1 i_2 \ldots i_r)$. One observes inductively that $\alpha^t(i_1) = i_{1+t}$ for $t \leq r - 1$. Hence, $\alpha^t \neq I$ for $t \leq r - 1$. Next, $\alpha^r(i_1) = \alpha(\alpha^{r-1}(i_1)) = \alpha(i_r) = i_1$. Also $\alpha^r(i_t) = \alpha^r(\alpha^{t-1}(i_1)) = \alpha^t(\alpha^{r-t}(\alpha^{t-1}(i_1))) = \alpha^t(\alpha^{r-1}(i_1)) = \alpha^{t-1}(\alpha(i_r)) = \alpha^{t-1}(i_1) = i_t$. Hence, $\alpha^r = I$. This shows that $o(\alpha) = r$.   ♯

Proposition 6.1.11 *Let $\{\alpha_1, \alpha_2, \ldots, \alpha_r\}$ be a set of pairwise disjoint cycles of lengths $m_1, m_2, \ldots m_r$ respectively. Let $\alpha = \alpha_1 \alpha_2 \ldots \alpha_r$. Then, $o(\alpha) = [m_1, m_2, \ldots, m_r]$ the least common multiple of $m_1, m_2, \ldots m_r$.*

Proof Let $m = [m_1, m_2, \ldots m_r]$. By the above proposition, $o(\alpha_i) = m_i$ *for all* $i \in$ $\{1, 2, \ldots, r\}$. Since $\alpha_1, \alpha_2, \ldots, \alpha_r$ are pairwise disjoint, they also commute pairwise and so $\alpha^m = \alpha_1^m \alpha_2^m \ldots \alpha_r^m = I$. Suppose further that $\alpha^t = \alpha_1^t \alpha_2^t \ldots \alpha_r^t = I$. Since the symbols changed by α_i^t is kept fixed by α_j^t, $j \neq i$, it follows that $\alpha_i^t = I$ for all i. Since $o(\alpha_i) = m_i$, it follows that each m_i divides t. Hence, m divides t. This shows that $o(\alpha) = m$. ♯

Example 6.1.12 The order of the permutation $(1\ 3\ 4)(2\ 5)$ is $[3, 2] = 6$.

Theorem 6.1.13 *Every nonidentity permutation can be written as a product of disjoint cycles. Further, any two representations of a nonidentity permutation as product of disjoint cycles is same up to rearrangement of cycles.*

Proof Let α be a nonidentity permutation in S_n. Then, *there exists* i_1 such that $\alpha(i_1) \neq i_1$. Since $o(S_n) = n!$, $\alpha^{n!} = I$. Hence, $\alpha^{n!}(i_1) = i_1$. Let l_1 be the least positive integer such that $\alpha^{l_1}(i_1) = i_1$. Given $m \in \mathbb{Z}$, by the division algorithm, *there exist* q, r such that $m = l_1 q + r$, where $0 \leq r \leq l_1 - 1$. But, then $\alpha^m(i_1) = \alpha^r(i_1)$. It is clear from the above observation that the effect of the permutation α on symbols in $\{i_1, \alpha(i_1), \alpha^2(i_1), \ldots, \alpha^{l_1-1}(i_1)\}$ is same as that of the cycle $C_1 = (i_1\ \alpha(i_1)\ \alpha^2(i_2)\ldots\alpha^{l_1-1}(i_1))$. If $\alpha = C_1$, there is nothing to do. If not, *there exists* $i_2 \notin \{i_1, \alpha(i_1), \alpha^2(i_1), \ldots \alpha^{l_1-1}(i_1)\}$ such that $\alpha(i_2) \neq i_2$. As before consider the cycle $C_2 = (i_2\ \alpha(i_2)\ \alpha^2(i_2)\ldots\alpha^{l_2-1}(i_2))$, where l_2 is the smallest positive integer such that $\alpha^{l_2}(i_2) = i_2$. Clearly, C_1 and C_2 are disjoint cycles. If $\alpha = C_1 C_2$, there is nothing to do. If not proceed. This process stops after finitely many steps giving α as product of disjoint cycles, because the symbols are finitely many.

Finally, we prove the uniqueness. Suppose that $\alpha \neq I$ and

$$\alpha = C_1 C_2 \ldots C_r = C_1' C_2' \ldots C_s',$$

where C_i and C_j are disjoint for $i \neq j$, and also C_k' and C_l' are disjoint for $k \neq l$. Suppose that $\alpha(p) \neq p$. Then, *there exist* i, k such that $C_i(p) \neq p$ and also $C_k'(p) \neq p$. We may assume that $C_1(p) \neq p$ and $C_1'(p) \neq p$. But, then, using the arguments of the previous paragraph, we find that $C_1 = C_1'$. Canceling C_1 and C_1', using induction and the fact that products of nonidentity disjoint cycles can never be identity, we find that $r = s$ and $C_i = C_i'$ for all i. ♯

Remark 6.1.14 The proof of the above theorem is algorithmic and gives an algorithm to express a permutation as product of disjoint cycles. Thus, using the above propositions we can find the order of any permutation.

Example 6.1.15 Consider the permutation α given by

$$\begin{pmatrix} 1\ 2\ 3\ 4\ 5\ 6\ 7 \\ 4\ 5\ 2\ 7\ 3\ 6\ 1 \end{pmatrix}$$

$\alpha(1) = 4, \alpha^2(1) = \alpha(4) = 7, \alpha^3(1) = \alpha(7) = 1$. Thus, $C_1 = (1\ 4\ 7)$. Now $2 \notin \{1, 4, 7\}$ and $\alpha(2) = 5 \neq 2$ $\alpha^2(2) = 3, \alpha^3(2) = 2$. Thus, $C_2 = (2\ 5\ 3)$.

Further, 6 is the only symbol left which is fixed by α. Hence, $\alpha = C_1 \cdot C_2$. Further, by the Proposition 5.4.11, $o(\alpha) = [3, 3] = 3$.

Example 6.1.16 The number of elements of order 6 in S_5 is 20. For, an element in S_5 is of order 6 if and only if it is product of two disjoint cycles one of length 3 and the other of length 2. Once a cycle of length 3 is picked up there is a unique cycle of length 2 which is disjoint from the selected cycle of length 3. Thus, there are as many elements of order 6 in S_5 as many distinct cycles of length 3. Further, corresponding to a choice $\{\alpha_1, \alpha_2, \alpha_3\} \subseteq \{1, 2, 3, 4, 5\}$ of 3 elements, there are exactly 2 distinct cycles of length 3, namely $(\alpha_1\alpha_2\alpha_3)$ and $(\alpha_1\alpha_3\alpha_2)$. Thus, there are $2 \cdot {}^5C_3 = 20$ distinct cycles of length 3, and as such there are 20 elements of order 6.

Example 6.1.17 S_7 contains no elements of order 8. For, a permutation is of order 8 if and only if it can be written as product of disjoint cycles such that the least common multiple of lengths of these cycles equals 8. We also observe that l.c.m of certain numbers equals 8 only if at least one of them is 8. Thus, we cannot have any such permutation in S_7. Similarly, S_7 contains no element of order 15, for a permutation of order 15 contains at least one cycle of length 5 and at least one cycle of length 3, or it should contain a cycle of length 15 in its decomposition as product of disjoint cycles.

Definition 6.1.18 Let α and β be two nontrivial permutations in S_n. We say that α and β are of same form if

(i) the number of cycles in the representation of α as product of disjoint cycles is same as that in the representation of β as product of disjoint cycles, and
(ii) there is a bijective correspondence between the set of cycles used in the representation of α and that in the representation of β so that the lengths of the corresponding cycles are same.

Thus, $(1\ 2\ 3)(4\ 7)(5\ 6\ 9\ 8)$ and $(4\ 5\ 2\ 3)(8\ 9)(1\ 6\ 7)$ are permutations of same form where as $(1\ 2\ 3)(4\ 7)$ and $(4\ 5\ 2\ 3)(8\ 9)$ are not of the same form.

Theorem 6.1.19 *Let $\alpha, \beta \in S_n$. Then, α and β are conjugate in S_n if and only if they have same form.*

Proof Let $\alpha = \alpha_1\alpha_2 \ldots \alpha_r$, where $\{\alpha_1, \alpha_2, \ldots, \alpha_r\}$ is a set of pairwise disjoint cycles. Let p be a permutation. Then,

$$p\alpha p^{-1} = p\alpha_1 p^{-1} \cdot p\alpha_2 p^{-1} \ldots p\alpha_r p^{-1}.$$

Let $\alpha_i = (\xi_1^i \xi_2^i \ldots \xi_{n_i}^i), i = 1, 2, \ldots, r$. Then,

$$p\alpha_i p^{-1} = (p(\xi_1^i)p(\xi_2^i) \ldots p(\xi_{n_i}^i))$$

Thus, $p\alpha p^{-1} = \beta_1\beta_2 \ldots \beta_r$, where $\beta i = (p(\xi_1^i)p(\xi_2^i) \ldots p(\xi_{n_i}^i))$ is a cycle of length n_i. Note that $\{\beta_1, \beta_2, \ldots \beta_r\}$ is a set of pairwise disjoint cycles. This shows that α and $p\alpha p^{-1}$ are of the same form.

Conversely, suppose that $\alpha = \alpha_1 \alpha_2 \ldots \alpha_r$ and $\beta = \beta_1 \beta_2 \ldots \beta_r$ are of the same form, where $\{\alpha_1, \alpha_2, \ldots, \alpha_r\}$ and $\{\beta_1, \beta_2, \ldots, \beta_r\}$ are sets of pairwise disjoint cycles with $l(\alpha_i) = l(\beta_i)$ *for all* i. Suppose that

$$\alpha_i = (\xi_1^i \xi_2^i \ldots \xi_{n_i}^i)$$

and

$$\beta_i = (\eta_1^i \eta_2^i \ldots \eta_{n_i}^i)$$

Take a permutation p which takes ξ_j^i to η_j^i and maps the set $\{1, 2, \ldots n\} - \{\xi_j^i \mid 1 \leq j \leq n_i, 1 \leq i \leq r\}$ bijectively to the set $\{1, 2, \ldots, n\} - \{\eta_j^i \mid 1 \leq j \leq n_i, 1 \leq i \leq r\}$. Then, it is evident that $p \alpha p^{-1} = \beta$. ♯

Following corollary is immediate from the definition of a normal subgroup.

Corollary 6.1.20 *A subgroup H of S_n is normal if and only if $\alpha \in H$ implies that all permutations which are of the same form as of α belongs to H. More precisely, normal subgroups of S_n are subgroups which are unions of classes of permutations of same forms.* ♯

Example 6.1.21 To find all normal subgroups of S_3, we divide S_3 into the classes of permutations of same forms:

$C_1 = \{I\}$.
$C_2 = \{(123), (132)\}$.
$C_3 = \{(12), (23), (13)\}$.

Apart from S_3 and $C_1 = \{I\}$, $C_1 \cup C_2 = \{I, (123), (132)\}$ is a subgroup of S_3 (verify) which is therefore normal. Observe that $C_1 \cup C_3$ cannot be a subgroup because it contains 4 elements and 4 does not divide 6. Thus, we have only one nontrivial proper normal subgroup, viz $\{I, (123), (132)\}$.

Example 6.1.22 To find normal subgroups of S_4, we divide it into the classes of permutations of same forms:

$C_1 = \{I\}$.
$C_2 = \{(1234), (1324), (1423), (1243), (1342), (1432)\}$ the sets of cycles of length 4.
$C_3 = \{(123), (132), (234), (243), (134), 143), (124), (142)\}$ the set of cycles of length 3.
$C_4 = \{(12)(34), (14)(23), (13)(24)\}$ the set of products of two disjoint transposition.
$C5 = \{(12), (13), (14), (23), (24), (34)\}$ the set of transpositions.

Let H be a nontrivial proper normal subgroup of S_4. Then, clearly $C_1 \subseteq H$. Further, H must be complete union of some of the classes C_i. By the Lagrange theorem, $|H|$ divides 24. The only possibilities, therefore are $H = C_1 \cup C_3 \cup C_4$ and $H = C_1 \cup C_4$ (it can be checked easily that these are, indeed, subgroups). Thus, nontrivial proper normal subgroups of S_4 are

$$\{I, (123), (132), (234), (243), (134), (143), (124), (142), (12)(34), (13)(24), (14)(23)\},$$

and
$$\{I, (12)(34), (13)(24), (14)(23)\}.$$

Remark 6.1.23 The subgroup $H = \{I, (12)(34), (13)(24), (14)(23)\}$ of S_4 which is normal in S_4 is isomorphic to the Klein's four group V_4 (the map $e \rightsquigarrow I, a \rightsquigarrow (12)(34), b \rightsquigarrow (13)(24), c \rightsquigarrow (14)(23)$ is an isomorphism). This subgroup is also denoted by V_4 and called the **Klein's four** group. Thus, $V_4 \lhd S_4$.

Definition 6.1.24 Let $n \in \mathbb{N}$. An ordered set $\{i_1 \leq i_2 \leq \cdots \leq i_r\}$ of natural numbers is called a **partition** of n if $i_1 + i_2 + \cdots + i_r = n$. The partition of n in which each $i_j = 1$ is called the trivial partition of n. Let $p(n)$ denote the number of partitions of n. The function p thus obtained is called the **partition function**.

Thus, $p(1) = 1$, for $1 = 1$, $p(2) = 2$, for $2 = 2, 2 = 1 + 1$, $p(3) = 3$, for $3 = 3, 3 = 2 + 1, 3 = 1 + 1 + 1$, and $p(4) = 5$, for $4 = 4, 4 = 3 + 1, 4 = 2 + 2, 4 = 2 + 1 + 1$, and $4 = 1 + 1 + 1 + 1$.

Proposition 6.1.25 *The number of conjugacy classes of S_n is $p(n)$.*

Proof The trivial partition of n corresponds to the trivial conjugacy class $\{I\}$ consisting of the identity permutation only. Given a nontrivial partition $\{i_1 \leq i_2 \leq \cdots \leq i_r\}$ of n, there is a unique $s < r$ such that $i_j = 1$ for all $j \leq s$ and $i_j \geq 2$ for all $j \geq s + 1$. This partition determines uniquely a class of permutations which are products of $r - s$ disjoint cycles of lengths $i_{s+1}, i_{s+2}, \ldots i_r$, and conversely, every class of permutation determines uniquely a nontrivial partition.   ♯

Example 6.1.26 $V_4 \lhd S_4$ and $\{I, (12)(34)\} \lhd V_4$, but $\{I, (12)(34)\}$ is not normal in S_4. Thus, $H \lhd K$ and $K \lhd G$ need not imply that $H \lhd G$.

Proposition 6.1.27 $S_4/V_4 \approx S_3$.

Proof $V_4 = \{I, (12)(34), (13)(24), (14)(23)\}$ is a normal subgroup of S_4 and $S_3 = \{I, (123), (132), (12), (13), (23)\}$ is a subgroup of S_4 such that $V_4 \cap S_3 = \{I\}$. By the Noether isomorphism theorem,

$$S_3 V_4 / V_4 \approx S_3/(V_4 \cap S_3) \approx S_3/\{I\} \approx S_3$$

Also since $\mid V_4 \cap S_3 \mid = 1$, it follows that
$$\mid S_3 V_4 \mid = \frac{|S_3| \cdot |V_4|}{|S_3 \cap V_4|} = 24.$$
Thus, $S_3 V_4 = S_4$. Hence, $S_4/V_4 \approx S_3$.   ♯

Example 6.1.28 The number of distinct cycles of length r in S_n is $\frac{1}{r} \cdot \frac{n!}{(n-r)!}$. For, every arrangement of r elements from $\{1, 2, \ldots, n\}$ determines a cycle of length r and two cycles $(i_1 i_2 \ldots i_r)$ and $(j_1 j_2 \ldots j_r)$ of length r are same if the arrangement $j_1 j_2 \ldots j_r$ can be obtained by cyclically permuting $i_1 i_2 \ldots i_r$. Thus, r distinct arrangements define same cycles. Since there are $\frac{n!}{(n-r)!}$ arrangements of r elements the result follows.

Proposition 6.1.29 *Every cycle is product of transpositions.*

Proof $(i_1 i_2 \ldots i_r) = (i_1 i_r)(i_1 i_{r-1}) \ldots (i_1 i_2).$ ♯

Since every permutation is product of disjoint cycles, we have

Corollary 6.1.30 *Every permutation is a product of transpositions.* ♯

Remark 6.1.31 Representation of a permutation as product of transpositions is not unique. For example,

$$(1234) = (14)(13)(12) = (14)(13)(12)(24)(24) = (14)(23)(13).$$

6.2 Alternating Maps and Alternating Groups

Let $\alpha \in S_n$. Consider the following rational number:

$$\frac{\alpha(1) - \alpha(2)}{1 - 2} \frac{\alpha(1) - \alpha(3)}{1 - 3} \cdots \frac{\alpha(1) - \alpha(n)}{1 - n} \cdot \frac{\alpha(2) - \alpha(3)}{2 - 3} \cdots \frac{\alpha(2) - \alpha(n)}{2 - n} \cdots \frac{\alpha(n - 1) - \alpha(n)}{(n - 1) - n}.$$

The above expression in short is denoted by

$$\prod_{1 \leq i < j \leq n} \frac{\alpha(i) - \alpha(j)}{i - j}.$$

Proposition 6.2.1 $\prod_{1 \leq i < j \leq n} \frac{\alpha(i) - \alpha(j)}{i - j} = \pm 1$ *for all* $\alpha \in S_n$.

Proof Since α is a permutation, for all pair (k, l) there is a unique pair (p, q), $\alpha(p) = k$ and $\alpha(q) = l$. If $p < q$, then $k - l$ appears once and only once in the numerator of the expression, and if $q < p$, then $l - k$ appears once and only once in the numerator of the expression. Also $k - l$ or $l - k$ appears once and only once in the denominator according as $k < l$ or $l < k$. This proves the result. ♯

Definition 6.2.2 The map χ from S_n to $\{1, -1\}$ defined by

$$\chi(\alpha) = \prod_{1 \leq i < j \leq n} \frac{\alpha(i) - \alpha(j)}{i - j}.$$

is called the **alternating map** of degree n.

Theorem 6.2.3 *The alternating map* χ *is a surjective homomorphism from* S_n *to* $\{1, -1\}$ *which takes any transposition to* -1.

Proof We first show that χ is a homomorphism.

$$\chi(\alpha\beta) = \prod_{1 \le i < j \le n} \frac{\alpha\beta(i) - \alpha\beta(j)}{i - j} =$$

$$\prod_{1 \le i < j \le n} \frac{\alpha(\beta(i)) - \alpha(\beta(j))}{\beta(i) - \beta(j)} \prod_{1 \le i < j \le n} \frac{\beta(i) - \beta(j)}{i - j} = \prod_{1 \le i < j \le n} \frac{\alpha(\beta(i)) - \alpha(\beta(j))}{\beta(i) - \beta(j)} \cdot \chi(\beta).$$

Since β is a permutation of $1, 2, \ldots n$, it follows that

$$\prod_{1 \le i < j \le n} \frac{\alpha(\beta(i)) - \alpha(\beta(j))}{\beta(i) - \beta(j)} = \chi(\alpha).$$

This shows that χ is a homomorphism. Since $\{1, -1\}$ is abelian, it follows that χ is constant on each conjugacy class. Since any two transpositions (being permutations of same form) are conjugate to each other, it suffices now to show that $\chi(\tau) = -1$, where $\tau = (12)$. Since $\tau(i) = i$ *for all $i \ge 3$*, we have

$$\chi(\tau) = \frac{2 - 1}{1 - 2} \cdot \frac{2 - 3}{1 - 3} \cdots \frac{2 - n}{1 - n} \cdot \frac{1 - 3}{2 - 3} \cdot \frac{1 - 4}{2 - 4} \cdots \frac{1 - n}{2 - n} = -1$$

Thus, χ takes any transposition to -1. Clearly, it takes I to 1.   ♯

Corollary 6.2.4 *Let $\alpha \in S_n$. Suppose that*

$$\alpha = \sigma_1 \sigma_2 \ldots \sigma_r = \tau_1 \tau_2 \ldots \tau_s,$$

where σ_i and τ_j are transpositions. Then, $r \equiv s \,(mod\,2)$, i.e., 2 divides $r - s$ (equivalently r and s both are simultaneously even or simultaneously odd).

Proof From the above theorem, it follows that

$$\chi(\alpha) = \chi(\sigma_1)\chi(\sigma_2)\ldots\chi(\sigma_r) = \chi(\tau_1)\chi(\tau_2)\ldots\chi(\tau_s)$$

Since χ takes a transposition to -1, $(-1)^r = (-1)^s$. Hence, $r - s$ is even.   ♯

Remark 6.2.5 From the above corollary, it follows that if we can write a permutation as a product of even number of transpositions, then we cannot write it as a product of odd number of transpositions, and if we can write it as a product of odd number of transpositions, then we cannot write it as a product of even number of transpositions.

Definition 6.2.6 A permutation α is called an **even permutation** if it can be expressed as a product of even number of permutations, or equivalently $\chi(\alpha) = 1$. It is said to be an **odd permutation** if it can be expressed as a product of odd number of transpositions.

Thus, the set of all even permutations is the kernel of χ and we have the following:

Corollary 6.2.7 *The set of all even permutations of degree n is a normal subgroup of S_n.* ♯

Definition 6.2.8 The group of all even permutations of degree n is denoted by A_n, and it is called the **alternating group** of degree n. Thus, $A_n \lhd S_n$.

By the fundamental theorem of homomorphism, we have

Corollary 6.2.9 $S_n/A_n \approx \{1, -1\}$. ♯

By the Lagrange theorem, $|S_n| = |A_n| \cdot |S_n/A_n| = 2|A_n|$. Thus, we have the following.

Corollary 6.2.10 $|A_n| = \frac{n!}{2}$. ♯

Remark 6.2.11 There are $\frac{n!}{2}$ even permutations and the rest $\frac{n!}{2}$ permutations are odd. Product of any two even or product of any two odd permutations are even and product of an even and an odd permutation is odd.

Proposition 6.2.12 *A cycle of length r is even if and only if r is odd.*

Proof Let $\alpha = (i_1 i_2 \ldots i_r)$ be cycle of length r. Then,

$$\alpha = (i_1 i_r)(i_1 i_{r-1}) \ldots (i_1 i_2).$$

Hence, α is product of $r - 1$ transpositions. The result now follows from the definition. ♯

In general, we have the following:

Proposition 6.2.13 *If α is product of cycles of lengths $n_1, n_2, \ldots, n_r$, then it is even if and only if $n_1 + n_2 + \cdots + n_r - r$ is even.* ♯

Proposition 6.2.14 *If $\alpha \in S_n$ is of odd order, then $\alpha \in A_n$.*

Proof If α is product of disjoint cycles of lengths $m_1, m_2, \ldots, m_r$, then its order is the least common multiple of $m_1, m_2, \ldots m_r$. This is odd if and only if each m_i is odd. But, then each $m_i - 1$ is even. The result follows from the above proposition. ♯

Corollary 6.2.15 *All odd ordered subgroups of S_n are also subgroups of A_n.*

Proof Let H be a subgroup of odd order. Then, by the Lagrange theorem, order of every element (which is the order of the subgroup generated by that element) divides the order of the group, and so it is of odd order. From the above proposition, it follows that all elements of H are in A_n. Thus, H is a subgroup of A_n. ♯

Proposition 6.2.16 *Let H be a subgroup of S_n which has no subgroup of index 2 (in particular if H is simple). Then, $H \subseteq A_n$.*

Proof Suppose that $H \nsubseteq A_n$. Then, HA_n is a subgroup of S_n which contains A_n properly. Since there is no proper divisor of $n!$ which is greater that $\frac{n!}{2}$, it follows by the Lagrange theorem that $HA_n = S_n$. By the Noether 2_{nd} isomorphism theorem,

$$\{1, -1\} \approx S_n/A_n = HA_n/A_n \approx H/H \bigcap A_n.$$

This shows that $H \bigcap A_n$ is a subgroup of H of index 2. This is a contradiction to the supposition that H has no subgroup of index 2.   ♯

Example 6.2.17 A_5 contains no element of order 6 whereas S_5 contains 20 elements of order 6. For, an element of S_5 is of order 6 if and only if it is product of two disjoint cycles one of length 2 and another of length 3. These permutations are odd (being product of 3 transpositions). Thus, there is no element of order 6 in A_5. It is already seen (Example 5.4.16) that S_5 contains 20 elements of order 6.

Example 6.2.18 A_8 contains no element of order 8 whereas S_8 contains 7! elements of order 8. For, a permutation α may be of order 8 only if in the representation of α as product of disjoint cycles, one of the cycle is of length 8. Thus, in S_8 cycles of length 8 and they are the only elements of order 8. They are obviously odd and $\frac{{}^8P_8}{8} = 7!$ in number.

Example 6.2.19 A_8 contains $2 \cdot \frac{1}{5}\frac{8!}{3!}$ elements of order 15. Indeed, all elements of order 15 in S_8 (being of odd order) are in A_8. Further, an element of S_8 is of order 15 if and only if it can be written as product of two disjoint cycles, one of length 5 and the other of length 3. For each choice of cycle of length 5, there are exactly 2 distinct cycles of length 3 which are disjoint to the chosen cycle of length 5. Thus, there are $2 \cdot \frac{1}{5} \cdot \frac{8!}{3!}$ elements of A_8 of order 15.

Example 6.2.20 In this example, we give an example of a group in which converse of Lagrange theorem is not true. Consider the group A_4 which is of order 12. We show that A_4 contains no subgroup of order 6. Suppose the contrary. Let H be a subgroup of A_4 of order 6. Then, H is of index 2 (Lagrange theorem) and so it is normal in A_4. Since H is of even order, it contains an element of order 2. The elements of order 2 in A_4 are $(12)(34), (13)(24), (14)(23)$. Suppose that $(12)(34) \in H$. Since H is normal $(14)(23) = (123)(12)(34)(132) = (123)(12)(34)(123)^{-1}$ belongs to H. Similarly, $(13)(24) \in H$ (in fact all elements of order 2 in A_4 are conjugates in A_4). This shows that V_4 is a subgroup of H. This is a contradiction to the Lagrange theorem (4 does not divide 6). Thus, there is no subgroup of order 6 in A_4.

Our next aim is to represent an abstract group as a subgroup of a permutation group.

Let G be a group and X a set. A homomorphism ρ from G to $Sym(X)$ is called a **permutation representation** of G on X. It is said to be **faithful** if ρ is injective $(ker\, \rho = \{e\}))$.

The first result on permutation representation is the following theorem of Cayley.

Theorem 6.2.21 (Cayley's Theorem) *Every group is isomorphic to a permutation (also called transformation) group (i.e., every group has a faithful representation on some set.). Also a finite group of order n is isomorphic to a subgroup of S_n.*

Proof Let G be a group and consider the symmetric group $Sym(G)$ on G. Let $g \in G$. Define a map L_g from G to G by $L_g(x) = gx$. The map L_g is called the left multiplication by g. Suppose that $L_g(x_1) = L_g(x_2)$. Then, $gx_1 = gx_2$. By the cancellation law, $x_1 = x_2$. Thus, L_g is an injective map. Further, given any $y \in G, L_g(g^{-1}y) = y$. Thus, L_g is also surjective. Hence, $L_g \in Sym(G)$. We have a map f from G to $Sym(G)$ defined by $f(g) = L_g$. Suppose that $f(g_1) = f(g_2)$. Then, $L_{g_1} = L_{g_2}$, and so $L_{g_1}(e) = L_{g_2}(e)$. But, then, $g_1 = g_1e = g_2e = g_2$. This shows that f is injective. Next, we show that f is a homomorphism. Now

$$f(g_1g_2)(x) = L_{g_1g_2}(x) = (g_1g_2)x = g_1(g_2x) = L_{g_1}(L_{g_2}(x)) =$$
$$f(g_1)(f(g_2)(x)) \text{ for all } x \in G.$$

This shows that $f(g_1g_2) = f(g_1)of(g_2)$ for all $g_1, g_2 \in G$. Thus, f is injective homomorphism from G to $Sym(G)$. In turn, it follows that G is isomorphic to the subgroup $f(G)$ of $Sym(G)$. Lastly, if G is finite containing n elements, then $Sym(G)$ is isomorphic to S_n, and hence, in this case G is isomorphic to a subgroup of S_n. ♯

Remark 6.2.22 If G is a group of order p, where p is prime, then G cannot be isomorphic to a subgroup of S_m for any $m < p$. This is because p does not divide $m!$. Thus, in general, G is the smallest set in the sense of cardinality such that G is isomorphic to a subgroup of $Sym(G)$.

Corollary 6.2.23 *Let G be a group of order n which has no subgroup of index 2. Then, G is isomorphic to a subgroup of A_n.*

Proof From the Cayley's theorem, we have an injective homomorphism ρ from G to S_n. Consider $\chi o\rho$, where χ is the alternating homomorphism from S_n to $\{1, -1\}$. Now, $\chi o\rho$ cannot be surjective, for otherwise, by the fundamental theorem of homomorphism, $G/ker\ \chi o\rho$ is isomorphic $\{1, -1\}$ and $ker\ \chi o\rho$ becomes a subgroup of G of index 2. Thus, $\chi(\rho(g)) = 1$ *for all* $g \in G$. This means that $\rho(g) \in A_n$ *for all* $g \in G$. It follows that $\rho(G) \subseteq A_n$, and so G is isomorphic to a subgroup of A_n. ♯

Since a group of odd order has no subgroup of index 2 (Lagrange theorem), and so also a simple group of order > 2 we have the following two corollaries.

Corollary 6.2.24 *Every group of odd order n is isomorphic to a subgroup of A_n.* ♯

Corollary 6.2.25 *If G is a simple group of order $n, n > 2$, then G is isomorphic to a subgroup of A_n.*

Every permutation of degree n can be thought of as a permutation of degree m for each $m \geq n$ which keeps $n + 1, n + 2, \ldots, m$ fixed. Thus, S_n is a subgroup of S_m in a natural way for all $m > n$.

Proposition 6.2.26 S_n *is isomorphic to a subgroup of* A_{n+2}.

Proof Define a map ϕ from S_n to A_{n+2} by $\phi(\alpha) = \alpha$ if $\alpha \in A_n$, and $\phi(\alpha) = \alpha \cdot (n+1\ n+2)$ if $\alpha \notin A_n$. Clearly, ϕ is an injective homomorphism.   ♮

The following corollary follows from the Cayley's theorem and the above proposition.

Corollary 6.2.27 *Every group of order n is isomorphic to a subgroup of* A_{n+2}.   ♮

Proposition 6.2.28 *Let G be a group of order* $2l$, *where* l *is odd.Then, G has a subgroup of index 2 (i.e., G has subgroup of order* l).

Proof Since G is of even order, it has an element $a \neq e$ such that $a^{-1} = a$. Thus, left multiplication L_a by a will interchange b to ab and ab to b. Therefore, the permutation determined by L_a in S_{2l} is product of l transpositions which is odd. Hence, if ψ represents the injective homomorphism from G to S_{2l} as given in the Cayley's theorem, then $\psi(a)$ is an odd permutation which is product of l transpositions. Thus, $\chi o \psi(a) = -1$, where χ is the alternating map. This shows that $\chi o \psi$ is a surjective homomorphism from G to $\{1, -1\}$. By the fundamental theorem of homomorphism, $G/ker\ \chi o \psi$ is isomorphic to $\{1, -1\}$, and hence, $ker\ \chi o \psi$ is a subgroup of G of index 2.   ♮

Corollary 6.2.29 *Let G be a group of order* $4n+2, n \geq 1$. *Then, G cannot be simple.*   ♮

Since the number of subgroups of S_n is finite, and any group of order n is isomorphic to a subgroup of S_n, we have

Proposition 6.2.30 *There are only finitely many nonisomorphic group of order n.*♮

Corollary 6.2.31 *There are only countably many nonisomorphic finite groups.*

Proof Since countable union of finite sets is countable, the result follows from the above proposition.   ♮

Proposition 6.2.32 *The group* S_n *is generated by* $\{(12), (23), \ldots, (n-1\ n)\}$.

Proof Since every permutation is product of transpositions, it suffices to show that (ij), $i < j$ belongs to H, where H is the subgroup generated by $\{(12), (23), \ldots, (n-1\ n)\}$. We show that $(i\ i+k) \in H$ *for all* $k > 1$. The proof is by induction on k. If $k = 1$, there is nothing to do. Assume that $(i\ i+k) \in H$ *for all* i. Then, $(i\ i+k+1) = (i\ i+1)(i+1\ i+1+k)(i\ i+1)$ also belongs to H. The result follows.   ♮

Proposition 6.2.33 *The group* S_n *is generated by* $\{(12), (12\ldots n)\}$.

Proof Using the above proposition, it is sufficient to show that $(i\ i+1) \in H$ for all i, $1 \le i < n$, where H is the subgroup generated by $\{(12), (12\ldots n)\}$. Now

$$(12\ldots n)(12)(12\ldots n)^{-1} = (23).$$

and

$$(12\ldots n)(23)(12\ldots n)^{-1} = (34).$$

Proceeding inductively, it follows that

$$(12\ldots n)^{i-1}(12)(12\ldots)^{-(i-1)} = (i\ i+1).$$

This shows that $(i\ i+1) \in H$, and so $H = S_n$. ♯

From the Cayley's theorem, we have,

Corollary 6.2.34 *Every finite group can be embedded as a subgroup into a group which is generated by 2 elements.* ♯

Proposition 6.2.35 A_n *is generated by cycles of length 3.*

Proof Since permutations in A_n are products of even number of transpositions, it is sufficient to show that product of any two transpositions is a product of cycles of length 3. Consider $(\alpha\beta)(\gamma\delta)$. It is $I = (\alpha\beta\gamma)(\alpha\gamma\delta)$ if $\{\alpha, \beta\} = \{\gamma, \delta\}$. Next, suppose that $\beta = \gamma$ and $\alpha \ne \delta$. Then, $(\alpha\beta)(\gamma\delta) = (\alpha\beta)(\beta\delta) = (\alpha\beta\delta)$. Now suppose that $\{\alpha, \beta\} \bigcap \{\gamma, \delta\} = \emptyset$. Then, $(\alpha\beta)(\gamma\delta) = (\gamma\alpha\delta)(\alpha\beta\gamma)$. ♯

Proposition 6.2.36 *All cycles of length 3 in A_n, $n \ge 5$ are conjugate to each other.*

Proof Let $(\alpha\beta\gamma)$ and $(\alpha'\beta'\gamma')$ be two cycles of length 3 in A_n, $n \ge 5$. We can find a permutation $p \in S_n$ which takes α' to α, β' to β and γ' to γ. We may assume that p is an even permutation, for if not $p(\mu\nu)$, where $\{\mu, \nu\} \bigcap \{\alpha', \beta', \gamma'\} = \emptyset$ will belong to A_n, and it will serve the purpose. Such a pair μ, ν exists, for $n \ge 5$. It is clear that $p^{-1}(\alpha\beta\gamma)p = (\alpha'\beta'\gamma')$. ♯

Remark 6.2.37 The above proposition is not true for $n = 4$. Not all cycles of length 3 are conjugate in A_4. For example (123) and (132) are not conjugate in A_4 (verify).

Theorem 6.2.38 A_n, $n \ne 4$ *is simple.*

Proof A_1 and A_2 are trivial groups whereas A_3 is a cyclic group of prime order 3, and so simple. Assume that $n \ge 5$. Let H be a nontrivial normal subgroup of A_n. We need to show that $H = A_n$. Since A_n is generated by cycles of length 3 (above proposition), it is sufficient to show that H contains all cycles of length 3. Since H is assumed to be normal, it contains all conjugates of its elements. Since all cycles of length 3 are conjugates, it is sufficient to show that H contains a cycle of length 3.

Since $H \neq \{I\}$, it contains a nontrivial permutation. Let p be a permutation in H which is different from identity, and keeps maximum number of symbols fixed. We shall show that p is a cycle of length 3. Suppose the contrary. Then, there are two cases:

(i) p is product of even number of disjoint transpositions, i.e.,

$$p = (\alpha\beta)(\gamma\delta)\ldots$$

(ii) In the representation of p as product of disjoint cycles, there is a cycle of length ≥ 3, i.e.,

$$p = (\alpha\beta\gamma\ldots)\ldots$$

Consider the case (i). Since $n \geq 5$, *there exists* $\rho \notin \{\alpha, \beta, \gamma, \delta\}$. Since $H \trianglelefteq A_n$, $q = (\gamma\delta\rho)p(\gamma\delta\rho)^{-1} = (\gamma\delta\rho)p(\gamma\rho\delta)$ belongs to H. Clearly, $p \neq q$ for p takes γ to δ but q takes ρ to δ. Thus, $p^{-1}q \in H$, $p^{-1}q \neq I$. Now $p^{-1}q$ fixes all symbols fixed by p except possibly ρ. But $p^{-1}q(\alpha) = \alpha$ and $p^{-1}q(\beta) = \beta$. This means that $p^{-1}q$ is a nontrivial permutation in H which keeps more symbols fixed than p does. This is a contradiction to the choice of p. Thus, case (i) cannot occur.

Consider the case (ii). In this case, if p is not a cycle of length 3, then it must change two more symbols δ, μ (say) other than α, β and γ, for a cycle of length 4 is not an even permutation. Since $H \trianglelefteq A_n$, $q = (\gamma\delta\mu)p(\gamma\delta\mu)^{-1} = (\gamma\delta\mu)p(\gamma\mu\delta)$ belongs to H. Now q is a permutation in H which fixes all symbols fixed by p. Also $p \neq q$, for $p(\beta) = \gamma$ and $q(\beta) = \delta$. Thus, $q^{-1}p$ is a nontrivial permutation in H which fixes all symbols fixed by p and it fixes one more symbol (α) which is changed by p. Thus, $q^{-1}p$ is a nontrivial member of H which fixes more symbols than p does. This is a contradiction to the choice of p. Hence, p is a cycle of length 3.   ♯

Since every group of order n is isomorphic to a subgroup of A_{n+2}, we have following corollary.

Corollary 6.2.39 *Every finite group is isomorphic to subgroup of a finite simple group.*   ♯

Example 6.2.40 A_5 has no subgroup of order 30 whereas 30 divides order of A_5. For, a subgroup of order 30, being a subgroup of index 2 (Lagrange Theorem) will be a proper nontrivial normal subgroup of A_5, a contradiction to the above theorem.

Example 6.2.41 In this example, we find all normal subgroups of S_n. Normal subgroups of S_3 and S_4 have already been determined. Assume that $n \geq 5$. Let H be a nontrivial proper normal subgroup of S_n. Then, $H \cap A_n$ is a normal subgroup of A_n. Since A_n is simple $H \cap A_n = A_n$ or $H \cap A_n = \{I\}$. If $H \cap A_n = A_n$, then $A_n \subseteq H$. Since there is no proper subgroup of S_n containing A_n properly, $H = A_n$. Next, suppose that $H \cap A_n = \{I\}$. Since $H \neq \{I\}$, $HA_n = S_n$. Since $H \cap A_n = \{I\}$, H is isomorphic to $H/H \cap A_n$. By the Noether isomorphism theorem, $H/H \cap A_n$ is isomorphic to $HA_n/A_n = S_n/A_n$. Since S_n/A_n is isomorphic to $\{1, -1\}$, it follows that $H \approx \{1, -1\}$. Thus, H is a subgroup of order 2. Since every nonidentity permutation in S_n has more than one permutations of same form, this is impossible.

Example 6.2.42 Let G be a group of odd order and ϕ a homomorphism from S_n to $G, n \geq 5$. From the above example, it follows that $ker \; \phi$ is $\{I\}$ or A_n or S_n. If $ker \; \phi = \{I\}$, then by the fundamental theorem of homomorphism, S_n is isomorphic to a subgroup of G. But, then, by the Lagrange theorem, $n!$ divides the order of G. Since $n!$ is even, this is not possible. If $ker \; \phi = A_n$, then again S_n / A_n is isomorphic to a subgroup of G. This is also impossible, for G is of odd order. Thus, $ker \; \phi = S_n$ and so ϕ is the trivial map.

S_n can be considered as a subgroup of S_m *for all* $m \geq n$. Similarly, A_n can be considered as a subgroup of A_m *for all* $m \geq n$. Thus, we get two chains

$$S_1 \subseteq S_2 \subseteq \cdots \subseteq S_n \subseteq S_{n+1} \subseteq \cdots .$$

and

$$A_1 \subseteq A_2 \subseteq \cdots \subseteq A_n \subseteq A_{n+1} \subseteq \cdots .$$

of groups. Let $S_\infty = \bigcup_{n=1}^\infty S_n$ and $A_\infty = \bigcup_{n=1}^\infty A_n$. Then, S_∞ is a group which contains all S_n as subgroups and A_∞ is a group such that all A_n are its subgroups. Clearly, A_∞ is a subgroup of S_∞. Further, since S_n can be considered as a subgroup of A_{n+2} in natural manner, S_∞ can also be considered as a subgroup of A_∞. In turn, S_∞ is isomorphic to one of its proper subgroup and also A_∞ is isomorphic to one of its proper subgroup. By the Cayley's theorem, all finite groups are subgroups (upto isomorphism) of S_∞ and also of A_∞.

Proposition 6.2.43 A_∞ *is a simple group.*

Proof Let H be a nontrivial proper normal subgroup of A_∞. Since $A_n \subseteq A_{n+1}$ *for all* n, $A_\infty = \bigcup_{n=m}^\infty A_n$ *for all* m. Since H is nontrivial $H \cap A_m \neq \{I\}$ *for some* $m \geq 5$. But, then, $H \cap A_n \neq \{I\}$ for all $n \geq m$. Since $H \trianglelefteq A_\infty$, $H \cap A_n \trianglelefteq A_n \; \forall n$. Since A_n is simple for all $n \geq 5$, $H \cap A_n = A_n$ *for all* $n \geq m$. But then, $A_n \subseteq H$ *for all* $n \geq m$. Hence, $A_\infty \subseteq H$. This shows that $H = A_\infty$. ♯

Remark 6.2.44 Since A_∞ is a normal subgroup of S_∞ and A_∞ is simple, S_∞ is not isomorphic to A_∞.

Exercises

6.2.1 Let f be an injective map from X to Y. Let $g \in Sym(X)$. Define a map $E(f)(g)$ from Y to Y by $E(f)(g)(y) = f(g(x))$ if $y = f(x)$ and y if $y \notin f(X)$. Show that $E(f)(g) \in Sym(Y)$. Show further that $E(f)$ defines an injective homomorphism from $Sym(X)$ to $Sym(Y)$. Show also that if g is another injective map from Y to Z, then $E(gof) = E(g)oE(f)$.

6.2.2 Find the product $\alpha\beta$ of permutations

$$\alpha = \begin{pmatrix} 1\,2\,3\,4\,5\,6 \\ 3\,5\,2\,6\,1\,4 \end{pmatrix}$$

and

$$\beta = \begin{pmatrix} 1\,2\,3\,4\,5\,6 \\ 2\,3\,1\,6\,5\,4 \end{pmatrix}$$

6.2.3 Express the following permutations as product of disjoint cycles and find their orders. Determine which of them are even and which of them are odd.

$$\alpha = \begin{pmatrix} 1\,2\,3\,4\,5\,6\,7 \\ 4\,3\,2\,6\,7\,1\,5 \end{pmatrix}$$

$$\beta = \begin{pmatrix} 1\,2\,3\,4\,5\,6\,7\,8\,9 \\ 9\,4\,5\,7\,6\,3\,8\,2\,1 \end{pmatrix}$$

$$\gamma = \begin{pmatrix} 1\,2\,3\,4\,5\,6\,7\,8\,9\,10\,11 \\ 4\,6\,9\,8\,10\,2\,5\,1\,11\,7\,3 \end{pmatrix}$$

$$\delta = \begin{pmatrix} 1\,2\,3\,4\,5\,6\,7\,8 \\ 2\,5\,4\,3\,1\,7\,8\,6 \end{pmatrix}$$

6.2.4 Express the permutations in Exercise 6.2.3 as product of transpositions.

6.2.5 Show that S_{11} contains no elements of order 35. Find the number of elements of order 35 in S_{12}. Show that they are all in A_{12}. Show that they are all conjugates. Find the number of elements in S_{12} and also in A_{12} which commute with a given permutation of order 35.

6.2.6 Show that S_{15} contains no element of order 16 whereas S_{16} contains several elements of order 16. Find the number of elements of order 16 in S_{16}. Does A_{16} contain an element of order 16. Find the number of elements in S_{16} which commute with a given element of order 16.

6.2.7 Find the number of elements of order 55 in S_{16}.

6.2.8 Find the number of permutations in S_{10} commuting with a cycle of length 5.

6.2.9 Find the number of even permutations in S_n which commute with a cycle of length r.

6.2.10 Let H be a subgroup of S_n all of whose nonidentity permutations are odd. Show that $| H | = 1$ or 2.

6.2.11 Find the number of conjugacy classes of A_4 and also of A_5.

6.2.12 Do A_n and S_n contain same number of conjugacy classes? Support.

6.2.13 Is every conjugacy class in A_n also a conjugacy class in S_n? Support.

6.2.14 Show that S_{11} contains an element of order 18. Find the number of such elements.

6.2.15 Is the converse of Lagrange theorem true in A_6? Support.

6.2.16 Find the number of elements of order 21 in A_{10}.

6.2.17 Show that A_5 and in fact A_n, $n \geq 5$ is generated by 2 elements.

6.2.18 Show that A_4 is generated by 2 elements.

6.2.19 Show that the center $Z(S_n) = \{I\}$, $n \geq 3$.

6.2.20 Show that $Z(A_n) = \{I\}$ for all $n \geq 4$.

6.2.21 Show that the group $Inn(S_n)$ of inner automorphisms of S_n is isomorphic to S_n.

6.2.22 Show that the group $Inn(A_n)$, $n \geq 4$ is isomorphic to A_n.

6.2.23 Let $\alpha \in S_n$ be cycle of length n and τ a transposition in S_n. Show that S_n is generated by $\{\alpha, \tau\}$.

6.2.24 Find a subgroup of S_8 isomorphic to Q_8 the Quaternion group of degree 8.

6.2.25 Obtain a subgroup of S_4 isomorphic to $\mathbb{Z}_4$. Can $\mathbb{Z}_4$ be isomorphic to a subgroup of A_4?

6.2.26 Let G be a group and $g \in G$. Let R_g denote the right multiplication by g ($x \rightsquigarrow xg$). Show that $g \rightsquigarrow R_g$ defines an anti isomorphism from G to a subgroup of $Sym(G)$ (a bijective map f from a group to a group is said to be an anti isomorphism if $f(ab) = f(b)f(a) \; \forall a, b$). Show also that $g \rightsquigarrow R_{g^{-1}}$ defines an isomorphism from G to a subgroup of $Sym(G)$.

6.2.27 Find the a subgroup of S_8 which is isomorphic (and also a subgroup which is anti isomorphic) to the dihedral group D_4.

6.2.28 Let $n = n_1 + n_2 + \cdots + n_r$ be a partition of n. Show that $S_{n_1} \times S_{n_2} \times \cdots \times S_{n_r}$ is isomorphic to a subgroup of S_n. Are subgroups corresponding to different partitions isomorphic? Support.

6.2.29 Show that V_4 is the only nontrivial proper normal subgroup of A_4.

6.2.30 Show that A_6 contains no element of order 180.

6.2.31 Show that every nonabelian simple subgroup of S_n is also a subgroup of A_n.

6.2.32 Let H be a nontrivial subgroup of S_n such that $H \not\subseteq A_n$. Show that H contains a subgroup of index 2.

6.2.33 Find $\mid C_{A_n}(\alpha) \mid$, where α is a cycle of odd length r.

6.2.34 Show that every homomorphism from A_n $n \geq 5$ to a group G is either injective or trivial homomorphism.

6.2.35 Show that every nontrivial endomorphism of A_n, $n \geq 5$ is an automorphism.

6.2.36 Show that there are at most $^{n!-1}C_{n-1}$ nonisomorphic groups of order n.

6.2.37 Show that no group of order $4n + 2$, $n \geq 1$ is simple.

6.2.38 Show that the number of elements of order 21 in A_{10} is double the number of elements of order 7.

6.2.39 Give an example of an infinite simple group.

6.2.40 Show that all finite groups are isomorphic to a subgroup of S_∞ and also to a subgroup of A_∞. Deduce that there are countably many finite nonisomorphic groups. Show that S_∞ is isomorphic to one of its proper subgroup.

6.2.41 Is $Sym(\mathbb{N}) \approx S_\infty$? Support.

6.2.42 Is $S_\infty \lhd Sym(\mathbb{N})$? Support.

6.2.43 Is $Sym(\mathbb{N})$ simple? Support.

6.2.44 Show that $S_\infty \not\approx A_\infty$.

6.2.45 Show that the order of each element of A_∞ is finite. Show also that it contains elements of any finite order.

6.2.46 Show that the additive group $\mathbb{Q}$ of rational numbers is isomorphic to a subgroup of $Sym(\mathbb{N})$.

6.2.47 Is the additive group $\mathbb{Q}$ of rationals isomorphic to a subgroup of S_∞? Support.

6.2.48 Is the group P of roots of unity isomorphic to a subgroup of A_∞? Support.

6.2.49 Let $\alpha \in S_n$ such that it is product of m_1 cycles of length r_1, m_2 cycles of length $r_2, \ldots, m_t$ cycles of length r_t, $r_j \geq 2$ *for all j*. Show that the number of permutations which have same form as α is

$$\frac{n!}{(n - m_1 r_1 - m_2 r_2 - \cdots - m_t r_t)! \cdot (m_1)! r_1^{m_1} \cdot (m_2)! r_2^{m_2} \ldots (m_t)! r_t^{m_t}}.$$

Hence, find the order of $C_{S_n}(\alpha)$.

6.2.50 What is the normal subgroup of S_n generated by a transposition?

6.2.51* Let G be a group and H a subgroup of G. A set S obtained by selecting one and only one member from each right coset of G modulo H with $e \in S$ is called a right transversal of H in G. Determine a right transversal of (i) S_{n-1} in S_n, (ii) $S_r \times S_s, r + s = n$ in S_n, (iii) S_n in A_{n+2} and (iv) A_{n-1} in A_n. Determine also, if possible, right transversals which generate the corresponding groups.

6.2.52 Show that $S_r \times S_s, r + s = n$ is a maximal subgroup of S_n. Give another type of maximal subgroups of S_n.

6.2.53* Show that on every nonempty set X there is a group structure.
Hint. If X contains n elements, then there is a bijective map from X to the cyclic group $\mathbb{Z}_n$. If X is infinite, then there is a subgroup of $Sym(X)$ which has the same cardinality as X.

6.2.54* Use the above exercise to show that there is no set containing all groups.

6.2.55* Show that there is no set X of groups such that every group is isomorphic to a member of X.

6.2.56* Show that there is no set containing all groups isomorphic to a given group.

6.2.57 Find the centralizer of (12) in S_n.

6.2.58 Find all normal endomorphisms of S_n and also of A_n.

6.2.59* Let p be an element of order 2 in $A_n, n \geq 5$. Show that there is an element $q \in A_n$ such that $\{p, q\}$ generate A_n.

6.2.60 Determine the number of involutions in S_n, and also in A_n.

6.3 General Linear Groups

The purpose of this section and the following section is to introduce some important classical matrix groups, and study some of their fundamental properties. We shall not discuss any of their topological properties. Indeed, many algebraic properties follow from topological considerations. Further properties of these groups will be discussed in Algebra 2 and Algebra 3.

A $n \times m$ matrix with entries in a set X is an arrangement of n rows and m columns. Thus, a $n \times m$ matrix A with entries in X is given by

$$A = \begin{bmatrix} a_{11} & a_{12} & \cdots\cdots & a_{1m} \\ a_{21} & a_{22} & \cdots\cdots & a_{2m} \\ \cdot & \cdot & \cdots\cdots & \cdot \\ \cdot & \cdot & \cdots\cdots & \cdot \\ \cdot & \cdot & \cdots\cdots & \cdot \\ \cdot & \cdot & \cdots\cdots & \cdot \\ \cdot & \cdot & \cdots\cdots & \cdot \\ a_{n1} & a_{n2} & \cdots\cdots & a_{nm} \end{bmatrix},$$

where $a_{ij} \in X$ for all i, j. We denote this matrix by $[a_{ij}]$, where a_{ij} denote the ith row jth column entry of the matrix, and call it a matrix of order $n \times m$ with entries in X. If $n = m$, then it is said to be a square matrix of order n. The set of $n \times m$ matrices with entries in X is denoted by $M_{nm}(X)$. The set of all square matrices of order n is denoted by $M_n(X)$.

Consider the set $M_{nm}(\mathbb{R})$ of $n \times m$ matrices with entries in $\mathbb{R}$. All the matrices in this section will have their entries in $\mathbb{R}$, unless stated otherwise. Let $A = [a_{ij}]$ and $B = [b_{ij}]$ be members of $M_{nm}(\mathbb{R})$. We define the sum $A + B$ of A and B to be the matrix $C = [c_{ij}]$, where $c_{ij} = a_{ij} + b_{ij}$ for all i, j. This defines a binary operation $+$ on $M_{nm}(\mathbb{R})$ such that $(M_{nm}(\mathbb{R}), +)$ is an abelian group. $O_{n\times m}$ denotes the $n \times m$ matrix all of whose entries are 0. This matrix is called the zero matrix, and it is the additive identity of the group $(M_{nm}(\mathbb{R}), +)$. The negative of the matrix $A = [a_{ij}]$ is the matrix $-A = [b_{ij}]$, where $b_{ij} = -a_{ij}$ for all i, j.

Let $A = [a_{ij}]$ be a $n \times m$ matrix and $B = [b_{jk}]$ a $m \times p$ matrix with entries in $\mathbb{R}$. The product $A \cdot B$ of A and B is defined to be the $n \times p$ matrix $C = [c_{ik}]$, where

$$c_{ik} = \Sigma_{j=1}^{j=n} a_{ij} b_{jk}$$

for all i, k. The following properties can be easily verified:

(i) The matrix multiplication $\cdot$ is associative, i.e, $(A \cdot B) \cdot C = A \cdot (B \cdot C)$ for all matrices A, B and C, where the relevant products are defined.
(ii) Matrix multiplication $\cdot$ distributes over $+$, i.e., $A \cdot (B + C) = (A \cdot B) + (A \cdot C)$, and $(A + B) \cdot C = (A \cdot C) + (B \cdot C)$ for all matrices A, B and C, where the relevant sums and products are defined.

Evidently, $A \cdot O_{m\times p} = O_{n\times p}$, and $O_{p\times n} \cdot A = O_{p\times m}$ for all $n \times m$ matrices A. For each n, we have the square $n \times n$ matrix $I_{n\times n}$ all of whose diagonal entries are 1 and all off diagonal entries are 0. The matrix $I_{n\times n}$ is called the **identity matrix** of order n. Clearly, $A \cdot I_{m\times m} = A = I_{n\times n} \cdot A$ for all $A \in M_{nm}(\mathbb{R})$.

General linear group: The pair $(M_n(\mathbb{R}), \cdot)$ is a semigroup with identity. However, $(M_n(\mathbb{R}), \cdot)$ is not a group, for $O_{n\times n}$ has no inverse. A matrix $A \in M_n(\mathbb{R})$ is said to be an invertible matrix (also called a nonsingular matrix) if $AB = I_{n\times n} = BA$ for some $B \in M_n(\mathbb{R})$. Let A and B be invertible matrices with entries in $\mathbb{R}$. Then, there is a matrix C and a matrix D such that $AC = I_{n\times n} = CA$ and $BD = I_{n\times n} = DB$. But, then $(AB)(DC) = I_{n\times n} = (DC)(AB)$. This shows that the product of any two invertible matrices is an invertible matrix. Thus, the multiplication of matrices induces a multiplication on the set $GL(n, \mathbb{R})$ of invertible matrices, and $GL(n, \mathbb{R})$ is a group with respect to this multiplication. This group is called the **General linear group**.

Diagonal subgroup: A matrix is said to be a **diagonal matrix** if all its off diagonal entries are 0. A diagonal matrix whose diagonal entries are $\lambda_1, \lambda_2, \ldots, \lambda_n$ is denoted by $Diag(\lambda_1, \lambda_2, \ldots, \lambda_n)$. Observe that the effect of multiplying $Diag(\lambda_1, \lambda_2, \ldots, \lambda_n)$ from left to a matrix $A = [a_{ij}]$ is replacing the ith row of A by λ_i times the ith row of A. More explicitly, $Diag(\lambda_1, \lambda_2, \ldots, \lambda_n) \cdot A = B = [b_{ij}]$, where $b_{ij} = \lambda_i a_{ij}$. Similarly, $A \cdot Diag(\lambda_1, \lambda_2, \ldots, \lambda_m) = [C = c_{ij}]$, where $c_{ij} = \lambda_j a_{ij}$. In particular, $Diag(\lambda_1, \lambda_2, \ldots, \lambda_n) \cdot Diag(\mu_1, \mu_2, \ldots, \mu_n) = Diag(\lambda_1\mu_1, \lambda_2\mu_2, \ldots, \lambda_n\mu_n)$. It follows that a diagonal matrix is invertible if and only if all its diagonal entries are nonzero, and then the inverse of $Diag(\lambda_1, \lambda_2, \ldots, \lambda_n)$ is $Diag(\lambda_1^{-1}, \lambda_2^{-1}, \ldots, \lambda_n^{-1})$. Let $D_n(\mathbb{R})$ denote the set of all diagonal matrices with nonzero entries. The above discussion ensures that $D_n(\mathbb{R})$ is a subgroup of $GL(n, \mathbb{R})$ called the diagonal subgroup. Clearly, $D_n(\mathbb{R})$ is isomorphic the direct product of $n-$ copies of the multiplicative group $\mathbb{R}^*$ of nonzero real numbers. For $n \geq 2$, the subgroup $D_n(\mathbb{R})$ is not a normal subgroup of $GL(n, \mathbb{R})$. For, consider the diagonal matrix $Diag(\lambda_1, \lambda_2, \ldots, \lambda_n)$, where $\lambda_1 \neq \lambda_2$. Let E_{12}^1 denote the matrix all of whose diagonal entries are 1, the first row second column entry is also 1, and the rest of the entries are 0. Then, E_{12}^1 is invertible, and its inverse is the matrix E_{12}^{-1} all of whose diagonal entries are 1, the first row second column entry is -1, and the rest of the entries are 0. Check that $E_{12}^1 \cdot Diag(\lambda_1, \lambda_2, \ldots, \lambda_n) \cdot E_{12}^{-1}$ is a nondiagonal matrix whose first row second column entry is $-\lambda_1 + \lambda_2$. Describe the normalizer of $D_n(\mathbb{R})$, and also the centralizer of $D_n(\mathbb{R})$.

Borel subgroup $B(n, \mathbb{R})$: A square $n \times n$ matrix $A = [a_{ij}]$ is called an upper triangular matrix if all below diagonal entries are 0, i.e., $a_{ij} = 0$ for all $i > j$. Let $B(n, \mathbb{R})$ denote the set of all invertible upper triangular matrices. Observe that the product of upper triangular matrices are upper triangular. Suppose that $A = [a_{ij}]$ is an invertible upper triangular matrix. Then, $a_{ij} = 0$ for all $i > j$, and there is a matrix $B = [b_{ij}]$ such that $B \cdot A = I_{n \times n}$. Equating the first row first column entries of both sides, we obtain that $b_{11}a_{11} = 1$. Hence, $a_{11} \neq 0$ and $b_{11} = a_{11}^{-1} \neq 0$. Next, equating the second row first column entries of both sides, we get that $b_{21}a_{11} = 0$, and so $b_{21} = 0$. Again, equating the second row second column entries, we get that $b_{21}a_{12} + b_{22}a_{22} = 1$. Since $b_{21} = 0, b_{22}a_{22} = 1$. Hence, $a_{22} \neq 0$ and $b_{22} = a_{22}^{-1} \neq 0$. Further, equating the third row first column entries, we obtain that $b_{31}a_{11} = 0$. This means that $b_{31} = 0$. Now, equating the third row second column entries, we get that $b_{31}a_{12} + b_{32}a_{22} = 0$. But, then $b_{32}a_{22} = 0$. This shows that $b_{32} = 0$. Proceeding inductively, we find that $a_{ii} \neq 0$, $b_{ii} = a_{ii}^{-1} \neq 0$, and $b_{ij} = 0$ for all $i > j$. This shows that the inverse of an invertible upper triangular matrix is again an invertible upper triangular matrix. Thus, $B(n, \mathbb{R})$ is a subgroup of $GL(n, \mathbb{R})$. This subgroup and all its conjugates are called the **Borel subgroups**.

Unipotent subgroup $U(n, \mathbb{R})$: Let $U(n, \mathbb{R})$ denote the set of all matrices in $B(n, \mathbb{R})$ with all diagonal entries 1. It can be checked that $U(n, \mathbb{R})$ is a subgroup of $GL(n, \mathbb{R})$. This subgroup is called a unipotent subgroup of $GL(n, \mathbb{R})$. Indeed, any conjugate of $U(n, \mathbb{R})$ is called a unipotent subgroup of $GL(n, \mathbb{R})$.

Let A be a $n \times n$ matrix. The matrix obtained by interchanging the rows in to the corresponding columns is called the **transpose** of A, and it is denoted by A^t. Thus, if $A = [a_{ij}]$, then $A^t = [b_{ij}]$, where $b_{ij} = a_{ji}$. It is easy to observe the following:

(i) $(A + B)^t = A^t + B^t$.

(ii) $(A \cdot B)^t = B^t \cdot A^t$.

(iii) If A is invertible, then A^t is also invertible, and $(A^{-1})^t = (A^t)^{-1}$.

A $n \times n$ matrix A is said to be a **symmetric** matrix if $A^t = A$. A matrix A is said to be a **skew symmetric matrix** if $A^t = -A$.

Elementary Matrices, Permutation Matrices

For $i \neq j$ and $\lambda \in \mathbb{R}$, consider the matrix E_{ij}^{λ} all of whose diagonal entries are 1, the ith row jth column entry is λ, and the rest of the entries are 0. The matrices of the form E_{ij}^{λ} are called **transvections** or **shear matrices**. Clearly, $E_{ij}^{\lambda} = I_{n \times n} + e_{ij}^{\lambda}$, where e_{ij}^{λ} is the matrix all of whose entries are 0 except the ith row jth column entry which is λ. It is easy to observe that $e_{ij}^{\lambda} \cdot e_{ij}^{\mu} = 0_{n \times n}$. In turn, it follows that $E_{ij}^{\lambda} \cdot E_{ij}^{\mu} = E_{ij}^{\lambda + \mu}$. Consequently, $E_{ij}^{\lambda} \cdot E_{ij}^{-\lambda} = I_{n \times n} = E_{ij}^{-\lambda} E_{ij}^{\lambda}$. Thus, $E_{ij}^{\lambda} \in GL(n, \mathbb{R})$ for all natural numbers $i, j \leq n$, and $\lambda \in \mathbb{R}$. Also check that E_{12}^1 and E_{21}^1 do not commute. Thus, $GL(n, \mathbb{R})$ is an infinite noncommutative group for all $n \geq 2$ (note that $GL(1, \mathbb{R})$ is the multiplicative group $\mathbb{R}^*$ of nonzero real numbers). Clearly, for each i, j; $i \neq j$, there is an embedding μ_{ij} of the additive group $\mathbb{R}$ of real numbers to $GL(n, \mathbb{R})$ given by $\mu_{ij}(\lambda) = E_{ij}^{\lambda}$. The subgroup $E(n, \mathbb{R})$ of $GL(n, \mathbb{R})$ generated by the transvections is called the **elementary subgroup** of $GL(n, \mathbb{R})$ (we shall have another interpretation of $E(n, \mathbb{R})$).

Let p be a permutation of degree n. Let M_p denote the matrix obtained by permuting the rows of the identity matrix $I_{n \times n}$ through the permutation p, and M^p denote the matrix obtained by permuting the columns of the identity matrix through the permutation p. These matrices are called the **permutation matrices**. It can be easily observed that for any matrix $A = [a_{ij}]$, $M_p \cdot A$ is the matrix obtained by permuting the rows of the matrix A through the permutation p. In particular, $M_p \cdot M_q = M_{pq}$ for all permutations p and q. This defines an embedding M of the symmetric group S_n in to the group $GL(n, \mathbb{R})$ given by $M(p) = M_p$. Consequently, using the Cayley's theorem, we have the following proposition.

Proposition 6.3.1 *Every group of order n is isomorphic to a subgroup of* $GL(n, \mathbb{R})$. ♮

Definition 6.3.2 A square $n \times n$ matrix A is called an **elementary matrix** if it is one of the following types:

(i) A transvection E_{ij}^{λ}.

(ii) A permutation matrix M_{τ}, where $\tau = (ij)$ is a transposition.

(iii) A diagonal matrix D_i^{λ} all of whose diagonal entries are 1 except the ith diagonal entry, and the ith diagonal entry is $\lambda \neq 0$.

Clearly, all the elementary matrices are invertible matrices. Indeed, the inverse of E_{ij}^λ is $E_{ij}^{-\lambda}$, the inverse of M_τ is M_τ itself, and the inverse of D_i^λ is $D_i^{-\lambda}$. Hence, they are all members of $GL(n, \mathbb{R})$. We shall show that the set of all elementary matrices generate $GL(n, \mathbb{R})$.

Definition 6.3.3 The following operations on matrices are called the elementary row (column) operations:

  (i) Add λ times a row (column) of the matrix to another row (column) of the matrix.
 (ii) Interchange any two distinct rows (columns) of the matrix.
(iii) Multiply a row (column) by a nonzero real number.

Remark 6.3.4 It is easily observed that multiplying E_{kl}^λ from left to a matrix A is equivalent to adding λ times the lth row of A to the kth row of A, and multiplying it from right to a matrix A is equivalent to adding λ times the kth column of A to the lth column of A. Also, multiplying M_τ from left to a matrix A is equivalent to interchanging the kth row and lth row of A, where τ is the transposition (k, l). Further, multiplying M_τ from right is equivalent to interchanging the kth and lth columns. Finally, multiplying the kth row (column) of A by λ is equivalent to multiplying the elementary matrix D_k^λ from left (right).

Definition 6.3.5 A $m \times n$ matrix $A = [a_{ij}]$ is said to be a matrix in **reduced row (column) echelon form**, or it is said to be a **reduced row echelon matrix** if the following hold:

  (i) The first nonzero entry in each row (column) is 1. This entry is called a **pivot entry**, and the corresponding columns (rows) are called the **pivot column(row)** of the matrix. The columns (rows) which are not pivot columns (rows) are called **free** columns (rows).
 (ii) The pivot entry in any row (column) is toward right (bottom) side to the pivot in the previous row (column).
(iii) All of the rest of the entries in a pivot column (row) are 0.
(iv) All the zero rows (columns) are toward bottom (right).

Example 6.3.6 The matrix

$$A = \begin{bmatrix} 1 & 2 & 0 & 0 & 2 \\ 0 & 0 & 1 & 0 & 1 \\ 0 & 0 & 0 & 1 & 2 \\ 0 & 0 & 0 & 0 & 0 \end{bmatrix}$$

is in reduced row echelon form. The 1st row 1st column, 2nd row 3rd column and 3rd row 4th column entries are pivot entries, 2nd and 5th columns are free columns.

Proposition 6.3.7 *A reduced row (column) echelon matrix is invertible if and only if it is the identity matrix.*

Proof The result follows if we observe that a matrix having a zero row (column) is not invertible.   ♯

Proposition 6.3.8 *Using elementary row/column operations, every matrix can be reduced to a matrix in reduced row/column echelon form.*

Proof Let A be a $m \times n$ matrix. If A is the zero matrix, then it is already in reduced row echelon form. Suppose that A is nonzero matrix. Let j_1 be the least number such that the j_1th column is a nonzero column. Further, let i_1 be the smallest number such that $a_{i_1 j_1} \neq 0$. Interchanging the i_1th row and the first row, we may assume that $a_{1 j_1} \neq 0$, and $a_{ik} = 0$ for all $k < j_1$, and for all i. Multiplying the first row by $a_{1 j_1}^{-1}$, we may assume that $a_{1 j_1} = 1$, and $a_{ik} = 0$ for all i and $k < j_1$. Next, adding $-a_{i j_1}$ times the first row to the ith row for each $i \geq 2$, we reduce A to a matrix $[a_{ij}]$, where $a_{ik} = 0$ for all $k \leq j_1 - 1$, $a_{1 j_1} = 1$, and $a_{i j_1} = 0$ for all $i \geq 2$. If in this reduced matrix $a_{ij} = 0$ for all $i \geq 2$ and j, then it is already in reduced row echelon form. If not, let j_2 be the smallest number such that $a_{i j_2} \neq 0$ for some $i \geq 2$. Further, let i_2 be the smallest number greater than 2 such that $a_{i_2 j_2} \neq 0$. Note that $j_2 > j_1$. Interchanging the i_2th row and the second row, we may assume that $a_{2 j_2} \neq 0$. Then, multiplying the second row by $a_{2 j_2}^{-1}$, we may assume that $a_{2 j_2} = 1$. In turn, adding $-a_{i j_2}$ times the second row to the ith row for each $i \neq 2$, A may have been reduced to a matrix in reduced row echelon form. If not, proceed as before. This process reduces A in to reduced row echelon form after finitely many steps (if worst comes, at the nth step). Similarly, using elementary column operations, every matrix can be reduced to a matrix in reduced column echelon form.   ♮

Corollary 6.3.9 *Let A be $n \times n$ matrix A. Then, there is a finite sequence $E_1, E_2, \ldots, E_r$ of elementary matrices such that*

$$E_1 \cdot E_2 \cdots \cdot E_r \cdot A$$

is in reduced row echelon form, and also there is a finite sequence $E_1', E_2', \ldots, E_s'$ of elementary matrices such that

$$A E_1' \cdot E_2' \cdots \cdot E_s'$$

is in reduced column echelon form. Further, A is invertible if and only if

$$E_1 \cdot E_2 \cdots \cdot E_r \cdot A = I_n = A E_1' \cdot E_2' \cdots \cdot E_s'$$

Proof Let A be a $n \times n$ matrix. From the Remark 6.3.4 and Proposition 6.3.8, there is a finite sequence $E_1, E_2, \ldots, E_r$ of elementary matrices such that

$$E_1 \cdot E_2 \cdots \cdot E_r \cdot A$$

is a matrix in reduced row echelon form. Since all elementary matrices are invertible, A is invertible if and only if

$$E_1 \cdot E_2 \cdots \cdot E_r \cdot A$$

is invertible. Again, by Proposition 6.3.7, a reduced row echelon matrix is invertible if and only if it is the identity matrix. Similarly, the remaining part also follows. ♮

Since the inverse of an elementary matrix is an elementary matrix, we have the following corollaries.

Corollary 6.3.10 *Every invertible matrix is a product of elementary matrices. Further, a square $n \times n$ matrix A is invertible if and only if $A \cdot \overline{X}^t = \overline{0}^t$ implies that $\overline{X} = \overline{0}$, where $\overline{X}$ denote the unknown row vector and $\overline{0}$ is the zero row vector.*

Proof Since the inverse of an elementary matrix is an elementary matrix, the first assertion is immediate from the above corollary. Now, assume that A is invertible and $A \cdot \overline{X}^t = \overline{0}^t$. Then, $\overline{X}^t = A^{-1} \cdot \overline{0}^t = \overline{0}^t$. Conversely, suppose that A is singular. Then, the reduced column echelon form of A will have its last column as zero column vector $\overline{0}^t$. From the above corollary, there is a finite sequence $E_1, E_2, \ldots, E_r$ of elementary matrices such that

$$A \cdot E_1 \cdot E_2 \cdots \cdot E_r$$

is a matrix in reduced column echelon form. Since the last column of this reduced column echelon form is $\overline{0}^t$,

$$A \cdot E_1 \cdot E_2 \cdots \cdot E_r \cdot \overline{e_n}^t = \overline{0}^t,$$

where $\overline{e_n}$ is the nonzero vector whose all entries are 0 except the nth column entry which is 1. Since the elementary matrices are invertible, $E_1 \cdot E_2 \cdots \cdot E_r \cdot \overline{e_n}^t$ is a nonzero column vector. The result follows. ♮

Corollary 6.3.11 *The general linear group $GL(n, \mathbb{R})$ is generated by the set of elementary matrices.* ♮

Determinant map

Definition 6.3.12 The **determinant** is a map det from $M_n(\mathbb{R})$ to $\mathbb{R}$ defined by

$$det(A) = \Sigma_{p \in S_n} signp \prod_{i=1}^{n} a_{p(i)i},$$

where $A = [a_{ij}]$. The image $det(A)$ is called the determinant of A.

Consider the $n \times m$ matrix $A = [a_{ij}]$. This matrix can be faithfully represented by a column

$$\begin{bmatrix} \overline{r_1} \\ \overline{r_2} \\ \cdot \\ \cdot \\ \cdot \\ \overline{r_n} \end{bmatrix}$$

of n rows, where $\overline{r}_i$ denote the ith row vector

$$[a_{i1}, a_{i2}, \ldots, a_{im}]$$

of the matrix A.

Similarly, the $n \times m$ matrix $A = [a_{ij}]$ can be faithfully represented by a row

$$[\overline{c_1}^t, \overline{c_2}^t, \ldots, \overline{c_m}^t].$$

of m columns, where $\overline{c_j}^t$ denote the jth column vector

$$\begin{bmatrix} a_{1j} \\ a_{2j} \\ \cdot \\ \cdot \\ \cdot \\ a_{nj} \end{bmatrix}$$

of the matrix A.

Proposition 6.3.13 *The determinant map det from $M_n(\mathbb{R})$ to $\mathbb{R}$ satisfies the following properties:*

(i) $det(I_{n \times n}) = 1$.

(ii) det is a linear map on the rows of the matrices in the sense that

$$det\begin{bmatrix} \overline{r}_1 \\ \overline{r}_2 \\ \cdot \\ \cdot \\ a\overline{r}_i + b\overline{r}_i' \\ \cdot \\ \cdot \\ \overline{r}_n \end{bmatrix} = a\,det\begin{bmatrix} \overline{r}_1 \\ \overline{r}_2 \\ \cdot \\ \cdot \\ \overline{r}_i \\ \cdot \\ \cdot \\ \overline{r}_n \end{bmatrix} + b\,det\begin{bmatrix} \overline{r}_1 \\ \overline{r}_2 \\ \cdot \\ \cdot \\ \overline{r}_i' \\ \cdot \\ \cdot \\ \overline{r}_n \end{bmatrix}$$

for each i.

(iii) Let $A = [a_{ij}]$ be a $n \times n$ matrix. Let $B = [bij]$ be the matrix obtained by interchanging the kth row and the lth row of A, i.e., $b_{kj} = a_{lj}, b_{lj} = a_{kj}$ and $b_{ij} = a_{ij}$ for all $i \notin \{k, l\}$. Then, $det(B) = -Det(A)$. In other words, if we interchange the rows of a matrix, then the determinant changes its sign.

Proof (i) Since all off diagonal entries of the identity matrix $I_{n \times n}$ is 0, for each nonidentity permutation p, $a_{p(i)i} = 0$ for some i. Also sign of identity permutation is 1. Hence, $Det(I_{n \times n}) = \prod_{i=1}^{n} a_{ii} = 1$.

(ii) Let

$$\overline{r}_i = [a_{i1}, a_{i2}, \ldots, a_{in}],$$

$1 \leq i \leq n$, and for a fix k,

$$\overline{r_k}' = [a'_{k1}, a'_{k2}, \ldots, a'_{kn}].$$

Then,

$$det \begin{bmatrix} \overline{r_1} \\ \overline{r_2} \\ . \\ . \\ . \\ \alpha\overline{r_k} + \beta\overline{r_k}' \\ . \\ . \\ . \\ \overline{r_n} \end{bmatrix}$$

$$= \Sigma_{p \in S_n} signp(\alpha a_{p(p^{-1}(k))p^{-1}(k)} + \beta a'_{p(p^{-1}(k))p^{-1}(k)}) \prod_{i \neq p^{-1}(k)} a_{p(i)i}$$

$$= \alpha\Sigma_{p \in S_n} signp \prod_{i=1}^{n} a_{p(i)i} + \beta\Sigma_{p \in S_n} signp a'_{p(p^{-1}(k))p^{-1}(k)}) \prod_{i \neq p^{-1}(k)} a_{p(i)i}$$

$$= \alpha det \begin{bmatrix} \overline{r_1} \\ \overline{r_2} \\ . \\ . \\ \overline{r_k} \\ . \\ . \\ \overline{r_n} \end{bmatrix} + \beta det \begin{bmatrix} \overline{r_1} \\ \overline{r_2} \\ . \\ . \\ \overline{r_k}' \\ . \\ . \\ \overline{r_n} \end{bmatrix}$$

(iii) Let $B = [b_{ij}]$ be the matrix obtained by interchanging the kth row and the lth row of $A = [a_{ij}]$. Let τ denote the transposition (k, l). Then,

$$det(B) = \Sigma_{p \in S_n} signp \prod_{i=1}^{n} b_{p(i)i} = \Sigma_{p \in S_n} signp \prod_{i=1}^{n} a_{\tau(p(i))i} =$$

$$- \Sigma_{p \in S_n} sign\tau p \prod_{i=1}^{n} a_{\tau(p(i))i} = -\Sigma_{q \in S_n} signq \prod_{i=1}^{n} a_{q(i)i} = -det(A). \natural$$

Conversely, we shall see that the map *det* as given by the Definition 6.3.12 is the only map which satisfies the conditions stated in the Proposition 6.3.13.

Corollary 6.3.14 *If $A = [a_{ij}]$ is a $n \times n$ matrix whose two distinct rows are same, then $det(A) = 0$.*

Proof Suppose that the kth row and the lth row of A are same. Interchanging the kth row and lth row, the matrix remains the same, and so, from the Proposition 6.3.13, $det(A) = -det(A)$. Hence, $det(A) = 0$. $\natural$

Proposition 6.3.15 $det(E \cdot A) = det(E)det(A)$ *for all elementary matrices* E, *and for all matrices* A. *In fact,*

(i) $det(E_{kl}^\lambda) = 1$, *and* $det(E_{kl}^\lambda \cdot A) = det(A) = det(E_{kl}^\lambda)det(A)$ *for all transvections* E_{kl}^λ, *and for all matrices* A.

(ii) $det(M_\tau) = -1$, *and*

$$det(M_\tau \cdot A) = -det(A) = det(M_\tau)det(A)$$

for all transpositions τ, *and for all matrices* A.

(iii) $det(D_k^\lambda) = \lambda$, *and*

$$det(D_k^\lambda \cdot A) = \lambda det(A) = det(D_k^\lambda)det(A)$$

for all elementary matrices D_k^λ, *and for all matrices* A.

Proof (i) From the Proposition 6.3.13 (ii), det is a linear map on the rows of the matrix. Hence,

$$det(E_{kl}^\lambda) = det(I_{n \times n}) + \lambda det([b_{ij}]),$$

where $[b_{ij}]$ is a matrix whose kth row and lth rows are same. By the Corollary 6.3.14, $det(E_{kl}^\lambda) = 1$. The rest of the identities in (i) follow from the same observation.

(ii) Follows from Proposition 6.3.13 (iii).

(iii) This follows from the fact that the det is a linear map on rows.   ♯

Proposition 6.3.16 *A* $n \times n$ *matrix* A *is invertible if and only if* $det(A)$ *is non zero. Further, if* A *is invertible and* $A = E_1 E_2 \dots E_r$, *then* $det(A) = det(E_1)det(E_2) \dots det(E_r)$.

Proof Let A be a $n \times n$ matrix. Then, there is a finite sequence $E_1, E_2, \dots, E_r$ of elementary matrices such that $E_1 \cdot E_2 \cdot \dots \cdot E_r \cdot A$ is in reduced row echelon form. From the above proposition, $det(E_1 \cdot E_2 \cdot \dots \cdot E_r \cdot A) = det(E_1)det(E_2) \dots det(E_r)det(A)$. If A is a noninvertible matrix, then there is a zero row in its reduced row echelon form, and since the determinant of a matrix with a zero row is zero, it follows that $det(A) = 0$. Further, if A is invertible, then $E_1 \cdot E_2 \cdot \dots \cdot E_r \cdot A = I_{n \times n}$. Hence, from the Proposition 6.3.15, $1 = det(E_1 \cdot E_2 \cdot \dots \cdot E_r \cdot A) = det(E_1)det(E_2) \dots det(E_r)det(A)$. This shows that $det(A) \neq 0$. The final assertion also follows from the Proposition 6.3.15.   ♯

The following is an immediate corollary.

Corollary 6.3.17 $GL(n, \mathbb{R}) = \{A \in M_n(\mathbb{R}) \mid det(A) \neq 0\}$.   ♯

Proposition 6.3.18 $det(A \cdot B) = det(A)Det(B)$ *for all matrices* A *and* B.

Proof Suppose that A (or B) is noninvertible, then both sides are zero, and so the equality holds. Suppose that both the matrices are invertible. Suppose that $A = E_1 \cdot E_2 \cdots \cdot E_r$, where E_i is elementary matrix for each i. Then, $det(A \cdot B) = det(E_1 \cdot E_2 \cdots \cdot E_r \cdot B)$. Further, from the Proposition 6.3.15, $det(A \cdot B) = det(E_1)det(E_2)\ldots det(E_r)det(B)$. Again, from Proposition 6.3.16, $det(A) = det(E_1)det(E_2)\ldots det(E_r)$. This shows that $det(A \cdot B) = det(A)det(B)$. ♯

Corollary 6.3.19 *The determinant map det is a surjective homomorphism from $GL(n, \mathbb{R})$ to the multiplicative group $\mathbb{R}^*$ of nonzero real numbers.*

Proof The fact that det defines a map from $GL(n, \mathbb{R})$ to $\mathbb{R}^*$ follows from the Proposition 6.3.16. The above proposition ensures that the map det is a homomorphism. If $\lambda \in \mathbb{R}^*$, then $det(D_i^\lambda) = \lambda$. Hence, it is a surjective homomorphism. ♯

The kernel of the determinant map is the set $SL(n, \mathbb{R})$ of all $n \times n$ matrices with determinant 1. Thus, $SL(n, \mathbb{R})$ is a normal subgroup of $GL(n, \mathbb{R})$. This group is called the **special linear group**. From the fundamental theorem of homomorphism, we have the following corollary.

Corollary 6.3.20 $GL(n, \mathbb{R})/SL(n, \mathbb{R})$ *is isomorphic to the multiplicative group $\mathbb{R}^*$ of nonzero real numbers.* ♯

Following result is an immediate consequence of the Corollary 6.3.10 and the Proposition 6.3.16.

Corollary 6.3.21 *Let A be a $n \times n$ matrix. Then, $det(A) = 0$ if and only if there is a nonzero vector $\bar{a}$ such that $A \cdot \bar{a} = \bar{0}$.* ♯

6.4 Classical Groups

Proposition 6.4.1 *Let A be a $n \times n$ skew symmetric matrix which is invertible. Then, n is even. More explicitly, if $n = 2m + 1$ is odd, and A is skew symmetric $n \times n$ matrix, then $det(A) = 0$.*

Proof Observe, by the definition, that $det(-A) = (-1)^n det(A)$ for any $n \times n$ matrix. Suppose that A is a skew symmetric $n \times n$ matrix. Then, $det(A) = det(A^t) = det(-A) = (-1)^n det(A)$. Thus, if n is odd, then $det(A) = 0$. ♯

Proposition 6.4.2 *Let P be a symmetric matrix (skew symmetric) in $GL(n, \mathbb{R})$. Then, the set*

$$O(P) = \{A \in M_n(\mathbb{R}) \mid A \cdot P \cdot A^t = P\}$$

is a subgroup of $GL(n, \mathbb{R})$.

Proof Let $A \in O(P)$. Then, $A \cdot P \cdot A^t = P$. Since P is invertible, $det(A)det(P)$
$det(A^t) \neq 0$. Hence, $det(A) \neq 0$, and so $A \in GL(n, \mathbb{R})$. Let $A, B \in O(P)$. Then,

$$(A \cdot B) \cdot P \cdot (A \cdot B)^t = A \cdot B \cdot P \cdot B^t \cdot A^t = A \cdot P \cdot A^t = P.$$

This shows that $A \cdot B \in O(P)$. Further, suppose that $A \in O(P)$. Then, $A \cdot P \cdot A^t = P$. In turn, $P = A^{-1} \cdot P \cdot (A^t)^{-1} = A^{-1} \cdot P \cdot (A^{-1})^t$. This shows that $A^{-1} \in O(P)$.   ♯

Definition 6.4.3 Two $n \times n$ matrices P and Q are said to be **congruent** to each other if there exists a nonsingular matrix A such that $APA^t = Q$.

The relation of being congruent to is an equivalence relation on the set of all $n \times n$ matrices.

Proposition 6.4.4 *Let P and Q be nonsingular symmetric (skew symmetric) matrices in $GL(n, \mathbb{R})$ which are congruent. Then, $O(P)$ and $O(Q)$ are conjugate in $GL(n, \mathbb{R})$. In particular, they are isomorphic.*

Proof Let L be a nonsingular matrix such that $LQL^t = P$. We show that $L^{-1}O(P)$ $L = O(Q)$. Let $A \in O(P)$. Then, $APA^t = P$, and so $ALQL^tA^t = P$. This implies that $L^{-1}ALQL^tA^t(L^t)^{-1} = L^{-1}P(L^t)^{-1}$. In turn, $L^{-1}ALQ(L^{-1}AL)^t = Q$. This shows that $L^{-1}AL \in O(Q)$. Thus, $L^{-1}O(P)L \subseteq O(Q)$. Similarly, $LO(Q)L^{-1} \subseteq O(P)$. Hence, $L^{-1}O(P)L = O(Q)$.   ♯

The following proposition classifies the congruent classes of symmetric (skew symmetric) matrices. The proof of the proposition can be found in Algebra 2.

Proposition 6.4.5 *For each pair (p, q) of nonnegative integers with $p + q = n$, let $J(p, q)$ denote the $n \times n$ diagonal matrix whose first p diagonal entries are -1 and the rest of the q diagonal entries are 1. Then, any nonsingular symmetric $n \times n$ matrix is congruent to a unique matrix $J(p, q)$. Also there is a unique congruence class of nonsingular $2m \times 2m$ skew symmetric matrices.*   ♯

Different choices of nonsingular symmetric (skew symmetric) matrices P give rise to different groups $O(P)$ which play crucial roles in the respective geometries.

Orthogonal Groups

Orthogonal groups play a very crucial role in Euclidean and spherical geometries. Indeed, they are the group of isometries of the Euclidean and spherical spaces.

Consider the symmetric matrix $P = I_{n \times n}$. The group $O(I_{n \times n})$ is denoted by $O(n)$. Thus, $O(n) = \{A \in GL(n, \mathbb{R}) \mid AA^t = I_{n \times n}\}$. This group is called the **orthogonal group**. The members of $O(n)$ are called the orthogonal matrices. If A is an orthogonal matrix, then $(det(A))^2 = 1$. Thus, det induces a homomorphism from $O(n)$ to the two element group $\{1, -1\}$. The diagonal matrix $diag(-1, 1, \ldots, 1)$ is an orthogonal matrix with determinant -1. Hence, det is a surjective homomorphism from $O(n)$ to $\{1, -1\}$. The kenel of this homomorphism is the normal subgroup

$\{A \in O(n) \mid Det(A) = 1\}$ of $O(n)$. This subgroup is denoted by $SO(n)$, and it is called the **special orthogonal group** (also called the rotation group). From the fundamental theorem of homomorphism, we get the following corollary:

Corollary 6.4.6 $O(n)/SO(n)$ is isomorphic to $\{1, -1\}$. ♯

Let $\overline{X}$ be a row vector with n columns. Then, $\overline{X} \cdot \overline{X}^t$ is a real number. Let

$$A = [\overline{X_1}^t, \overline{X_2}^t, \ldots, \overline{X_n}^t]$$

be a $n \times n$ matrix, where $\overline{X_j}^t$ represent the jth column of the matrix. Then, $A^t A = [\overline{X_i} \cdot \overline{X_j}^t]$. Thus, A is orthogonal if and only if $\overline{X_i} \cdot \overline{X_i}^t = 1$, and $\overline{X_i} \cdot \overline{X_j}^t = 0$ for $i \neq j$. Similarly, if we represent A as a column of row vectors whose ith vector is $\overline{R_i}$, then $AA^t = [\overline{R_i} \cdot \overline{R_j}^t]$. In turn, A is orthogonal if and only if $\overline{R_i} \cdot \overline{R_i}^t = 1$ and $\overline{R_i} \cdot \overline{R_j}^t = 0$ for $i \neq j$.

Proposition 6.4.7 A $n \times n$ matrix A is an orthogonal matrix if and only if $(A\overline{X}^t)^t \cdot A\overline{Y}^t = \overline{X} \cdot \overline{Y}^t$ for all vectors $\overline{X}$ and $\overline{Y}$.

Proof Suppose that A is orthogonal. Then, $A^t \cdot A = I_n$. Hence,

$$(A\overline{X}^t)^t \cdot A\overline{Y}^t = \overline{X} \cdot A^t \cdot A \cdot \overline{Y}^t = \overline{X} \cdot I_n \cdot \overline{Y}^t = \overline{X} \cdot \overline{Y}^t$$

for all vectors $\overline{X}$ and $\overline{Y}$. Conversely, suppose that $(A\overline{X})^t \cdot A\overline{Y} = \overline{X}^t \cdot \overline{Y}$ for all vectors $\overline{X}$ and $\overline{Y}$. Let $\overline{e_i}$ denote the vector with ith column 1 and the rest of the entries 0. Then, $A \cdot \overline{e_i}^t$ is the ith column $\overline{X_i}^t$ of the matrix A. Hence, $\overline{X_i} \cdot \overline{X_j}^t = (A \cdot \overline{e_i}^t)^t \cdot A \cdot \overline{e_j}^t = \overline{e_i} \cdot \overline{e_j}^t$. Clearly, $\overline{e_i} \cdot \overline{e_i}^t = 1$, and $\overline{e_i} \cdot \overline{e_j}^t = 0$ for $i \neq j$. It follows that A is orthogonal. ♯

Proposition 6.4.8 Let A be a $n \times n$ orthogonal matrix, where $n = 2m + 1$ is odd. If $det(A) = \pm 1$, then there is a vector $\overline{X}$ with $\| \overline{X} \|^2 = \overline{X} \cdot \overline{X}^t = 1$ such that $A \cdot \overline{X}^t = \pm \overline{X}^t$.

Proof Suppose that A is orthogonal matrix with $det(A) = \pm 1$. Then,

$$det(A - \pm I_n) = det(A^t)det(A - \pm I_n) = det(A^t(A - \pm I_n)) =$$
$$det(I_n - \pm A^t) = det(I_n - \pm A) = (-1)^n det(A - \pm I_n).$$

Since n is odd, $det(A - \pm I_n) = 0$. From the Corollary 6.3.21, there is a nonzero vector $\overline{Y}$ such that $(A - \pm I_n) \cdot \overline{Y}^t = \overline{0}$. This means that $A \cdot \overline{Y}^t = \pm \overline{Y}^t$. Take $\overline{X} = \frac{1}{\sqrt{\overline{Y} \cdot \overline{Y}^t}} \overline{Y}$. ♯

Let $P(n, \mathbb{R})$ denote the set of all $n \times n$ permutation matrices. Then, $P(n, \mathbb{R})$ is a subgroup of $GL(n, \mathbb{R})$ isomorphic to the symmetric group S_n. Recall that the map M from S_n to $GL(n, \mathbb{R})$ given by $M(p) = M_p$ is an injective homomorphism whose image is $P(n, \mathbb{R})$ (by definition). For each transposition τ, M_τ is an orthogonal matrix of determinant -1. Indeed, M_τ is symmetric, and $(M_\tau)^2 = I_{n \times n}$. Since $P(n, \mathbb{R})$ is generated by the matrices of the type M_τ, it follows that $P(n, \mathbb{R})$ is a subgroup of $O(n)$. The following corollary is consequence of the Cayley's theorem and the above observation.

Corollary 6.4.9 *Every finite group of order n is isomorphic to a subgroup of $O(n)$.* ♯

Corollary 6.4.10 *Let G be a finite group of order n which has no subgroup of index 2. Then, G is isomorphic to a subgroup of $SO(n)$.*

Proof $det \ o \ M$ is a homomorphism from G to the two element group $\{1, -1\}$. Since G has no subgroup of order of index 2, $(det \ o \ M)(g) = 1$ for all $g \in G$. Hence, $det(M(g)) = 1$. This shows that M is an injective homomorphism from G to $SO(n)$. ♯

Example 6.4.11 Let

$$A = \begin{bmatrix} a & b \\ c & d \end{bmatrix}$$

be a member of $O(2)$. Then, the condition that $AA^t = I_{2 \times 2}$ implies the following identities:

(i) $a^2 + b^2 = 1 = c^2 + d^2$.
(ii) $ac + bd = 0$.

Indeed, then we also have

(iii) $a^2 + c^2 = 1 = c^2 + d^2$, and
(iv) $ab + cd = 0$

Hence, for a, there is a unique angle θ, $0 \leq \theta \leq \pi$ such that $a = cos\theta$. Then, $b = \pm sin\theta$. If $b = sin\theta$, then the conditions (ii) and (iii) imply that $(c = sin\theta \ and \ d = -cos\theta)$ or $(c = -sin\theta \ and \ d = cos\theta)$. Again, if $b = -sin\theta$, then the conditions (ii) and (iii) imply that $(c = sin\theta \ and \ d = cos\theta)$ or $(c = -sin\theta \ and \ d = -cos\theta)$. Thus, there is a unique angle θ, $0 \leq \theta \leq \pi$ such that

$$A = \begin{bmatrix} cos\theta & sin\theta \\ \pm sin\theta & \mp cos\theta \end{bmatrix} \ or \ A = \begin{bmatrix} cos\theta & -sin\theta \\ \pm sin\theta & \pm cos\theta \end{bmatrix}.$$

In particular, $A \in SO(2)$ if and only if there is a unique angle θ, $0 \leq \theta \leq \pi$ such that

$$A = \begin{bmatrix} cos\theta & \pm sin\theta \\ \mp sin\theta & cos\theta \end{bmatrix}$$

The correspondence $A \mapsto \cos\theta + i\sin\theta$ gives an isomorphism from $SO(2)$ to the circle group S^1.

Example 6.4.12 Consider the group $SO(3)$. Let $A \in SO(3)$. From Proposition 6.4.8, there is a unit vector $\overline{X}$ in the Euclidean space $\mathbb{R}^3$ such that $A \cdot \overline{X}^t = \overline{X}^t$. Let $\overline{Y}, \overline{Z}$ be a pair of mutually orthogonal unit vectors in the plane orthogonal to $\overline{X}$. More explicitly, $\overline{X} \cdot \overline{X}^t = \overline{Y} \cdot \overline{Y}^t = \overline{Z} \cdot \overline{Z}^t = 1$, and $\overline{X} \cdot \overline{Y}^t = \overline{X} \cdot \overline{Z}^t = \overline{Y} \cdot \overline{Z}^t = 0$. Thus, the matrix $P = [\overline{X}^t, \overline{Y}^t, \overline{Z}^t]$ is also an orthogonal matrix in $O(3)$. Interchanging the second and the third vector, if necessary, we may suppose that $P \in SO(3)$. Further,

$$\overline{Y} \cdot A^t \cdot \overline{X}^t = \overline{Y} \cdot A^t \cdot A \cdot \overline{X}^t = \overline{Y} \cdot \overline{X}^t = 0.$$

Hence, $\overline{Y} \cdot A^t$ is orthogonal to the vector $\overline{X}$, and so it lies in the plane determined by $\overline{Y}$ and $\overline{Z}$. Similarly, $\overline{Z} \cdot A^t$ also lies in the plane determined by $\overline{Y}$ and $\overline{Z}$. Suppose that

$$\overline{Y} \cdot A^t = a\overline{Y} + b\overline{Z},$$

and

$$\overline{Z} \cdot A^t = c\overline{Y} + d\overline{Z}.$$

Since A^t is an orthogonal matrix, $\overline{Y} \cdot A^t$ and $\overline{Z} \cdot A^t$ are pairwise orthogonal unit vectors. Hence,

$$a^2 + b^2 = 1 = c^2 + d^2, \tag{6.4.1}$$

and

$$ac + bd = 0. \tag{6.4.2}$$

Further,

$$P^t \cdot A \cdot P = P^t \cdot [A \cdot \overline{X}^t, A \cdot \overline{Y}^t, A \cdot \overline{Z}^t] = P^t \cdot [\overline{X}^t, a\overline{Y}^t + b\overline{Z}^t, c\overline{Y}^t + d\overline{Z}^t].$$

Thus,

$$P^t \cdot A \cdot P = \begin{bmatrix} \overline{X} \cdot \overline{X}^t & \overline{X} \cdot (a\overline{Y}^t + b\overline{Z}^t) & \overline{X} \cdot (c\overline{Y}^t + d\overline{Z}^t) \\ \overline{Y} \cdot \overline{X}^t & \overline{Y} \cdot (a\overline{Y}^t + b\overline{Z}^t) & \overline{Y} \cdot (c\overline{Y}^t + d\overline{Z}^t) \\ \overline{Z} \cdot \overline{X}^t & \overline{Z} \cdot (a\overline{Y}^t + b\overline{Z}^t) & \overline{Z} \cdot (c\overline{Y}^t + d\overline{Z}^t) \end{bmatrix} = \begin{bmatrix} 1 & 0 & 0 \\ 0 & a & c \\ 0 & b & d \end{bmatrix}$$

Since $P^t = P^{-1}$, equating the determinant of both sides, we get that $det(A) = ad - bc$. Thus,

$$ad - bc = 1 \tag{6.4.3}$$

From the Eqs. (6.4.1)–(6.4.3), we obtain a unique $\theta, 0 \leq \theta \leq \pi$ such that $a = \cos\theta, c = \pm\sin\theta, b = \mp\sin\theta, and\ d = \cos\theta$. Thus, A is conjugate to a unique matrix of the type

$$\begin{bmatrix} 1 & 0 & 0 \\ 0 & cos\theta & \pm sin\theta \\ 0 & \mp sin\theta & cos\theta \end{bmatrix},$$

where $0 \leq \theta \leq \pi$.

Lorentz Groups

Lorentz groups play a very crucial role in hyperbolic geometry. Indeed, they are the group of isometries of the hyperbolic spaces.

Let J denote the $(n + 1) \times (n + 1)$ diagonal matrix $diag(-1, 1, 1, \ldots, 1)$. Then, J is nonsingular symmetric matrix. We denote the group $O(J)$ by $O(1, n)$. Thus,

$$O(1, n) = \{A \in GL(n + 1, \mathbb{R}) \mid A \cdot J \cdot A^t = J\}.$$

The group $O(1, n)$ is called the **Lorentz orthogonal group**. The members of $O(1, n)$ are called the **Lorentz matrices**.

If A is a Lorentz matrix, then $det(A)^2 = 1$. Clearly, J is a Lorentz matrix whose determinant is -1. Thus, det induces a surjective homomorphism from $O(1, n)$ to the two element group $\{1, -1\}$. We denote the kernel of this map by $SO(1, n)$ and call it the **special Lorentz group**. Thus, $SO(1, n) = \{A \in O(1, n) \mid det(A) = 1\}$. By the fundamental theorem of homomorphism, $O(1, n)/SO(1, n)$ is isomorphic to the group $\{1, -1\}$.

A Lorentz matrix $A = [a_{ij}]$ is called a **positive Lorentz matrix** if $a_{11} > 0$. It can be checked with a little effort that the product of any two positive Lorentz matrix is a positive Lorentz matrix. Thus, the set $PO(1, n)$ of positive Lorentz matrices is a subgroup of $O(1, n)$. This group is called the **positive Lorentz group**. Let $PSO(1, n)$ denote the set of all positive Lorentz matrices with determinant 1. Then, $PSO(1, n)$ is a subgroup of $PO(1, n)$, and it is called **positive special Lorentz group**.

Example 6.4.13 We try to describe $PSO(1, 1)$, $PO(1, 1)$ and $O(1, 1)$. Let

$$A = \begin{bmatrix} a & b \\ c & d \end{bmatrix}$$

be a member of $PSO(1, 1)$. Then, the conditions that $AJA^t = J$, $a > 0$, and $det(A) = 1$ imply the following identities:

(i) $a^2 - b^2 = 1 = c^2 - d^2$.
(ii) $ac = bd$.
(iii) $ad - bc = 1$.

Since $a > 0$, $a \geq 1$. Hence, there is a unique $x \geq 0$ such that $a = coshx$ and $b = sinhx$. The condition (ii) implies that $c = sinhx$ and $d = coshx$. Thus,

$$PSO(1, 1) = \{\begin{bmatrix} coshx & sinhx \\ sinhx & coshx \end{bmatrix} \mid x \in \mathbb{R}^+\}.$$

Further, it can be observed that $SO(1, 1) = \{\pm A \mid A \in PSO(1, 1)\}$ and $O(1, 1) = SO(1, 1) \bigcup D_1^{-1} SO(1, 1)$.

Symplectic Group

So far, we discussed examples of the groups $O(P)$, where P is a non singular symmetric matrix. Now, we consider the case where P is a nonsingular skew symmetric matrix. Observe that all nonsingular $2m \times 2m$ skew symmetric matrices are congruent to each other. Hence, we shall get the unique group upto isomorphism. Consider the matrix J_{2n} given by

$$J_{2n} = \begin{bmatrix} 0_{n \times n} & I_{n \times n} \\ -I_{n \times n} & 0_{n \times n} \end{bmatrix}.$$

The group $O(J_{2n})$ is denoted by $SP(2n, \mathbb{R})$, and it is called the **symplectic group** of $2n \times 2n$ matrices. Thus,

$$SP(2n, \mathbb{R}) = \{A \in GL(2n, \mathbb{R}) \mid A J_{2n} A^t = J_{2n}\}.$$

Example 6.4.14 Let

$$A = \begin{bmatrix} a & b \\ c & d \end{bmatrix}$$

be a 2×2 matrix. Observe that $A J_2 A^t = J_2$ if and only if $ad - bc = 1$. Thus, $SP(2, \mathbb{R}) = SL(2, \mathbb{R})$. However, in general, $SP(2m, \mathbb{R}) \subseteq SL(2m, \mathbb{R})$.

We may replace $\mathbb{R}$ by $\mathbb{C}$ or by $\mathbb{Q}$ in most of the discussions in this section to get other groups. Thus, $GL(n, \mathbb{C})$ denote general linear group with entries in $\mathbb{C}$ and $SL(n, \mathbb{C})$ denote the special linear group of $n \times n$ matrices with entries in $\mathbb{C}$.

Unitary Groups

Let $A = [a_{ij}]$ be a $n \times n$ matrix with entries in $\mathbb{C}$. The conjugate $\overline{A}$ of A is defined to be the matrix $[b_{ij}]$, where $b_{ij} = \overline{a_{ij}}$. The matrix $\overline{A^t} = \overline{A}^t$ is called the **tranjugate** or **hermitian conjugate** of A, and it is denoted by $A^\star$. Clearly, (i) $(A + B)^\star = A^\star + B^\star$, (ii) $(AB)^\star = B^\star A^\star$, and (iii) $det(A^\star) = \overline{det(A)}$. A $n \times n$ matrix $A \in GL(n, \mathbb{C})$ is called a **unitary matrix** if $A^\star = A^{-1}$. It is easy to observe that the set $U(n)$ of all unitary matrices in $GL(n, \mathbb{C})$ form a subgroup of $GL(n, \mathbb{C})$. The group $U(n)$ is called the **unitary group**. Let $A \in U(n)$. Then,

$$1 = det(I_{n \times n}) = det(AA^\star) = det(A)\overline{det(A)} = |\, det(A)\,|^2 .$$

Thus, det induces a homomorphism from $U(n)$ to the circle group. Clearly, $diag$ $(z, 1, 1, \ldots, 1) \in U(n)$ for all $z \in S^1$, and then, $det((diag(z, 1, 1, \ldots, 1)) = z \in S^1$. This shows that det is a surjective homomorphism from $U(n)$ to S^1. The kernel of det is denoted by $SU(n)$, and it is called the **special unitary group**. By the fundamental theorem of homomorphism,

$$U(n)/SU(n) \approx S^1.$$

Example 6.4.15 Let

$$A = \begin{bmatrix} a & b \\ c & d \end{bmatrix}$$

be a member of $SU(2)$. Then, $AA^\star = I_{2\times2}$ and $det(A) = 1$. Equating the corresponding entries of the matrices, we get the following identities:

$$|a|^2 + |b|^2 = 1 = |c|^2 + |d|^2,$$

$$a\bar{c} + b\bar{d} = 0, \text{ and } ad - bc = 1.$$

This shows that $d = \bar{a}$, and $c = -\bar{b}$. Thus, $SU(2)$ consists of the matrices of the type

$$A = \begin{bmatrix} a & b \\ -\bar{b} & \bar{a} \end{bmatrix},$$

where $|a|^2 + |b|^2 = 1$.

Exercises

6.4.1 Show that $\{A \in GL(n, \mathbb{R}) \mid det(A) \in \mathbb{Q}\}$ is a normal subgroup of $GL(n, \mathbb{R})$. Is this the same as the subgroup $GL(n, \mathbb{Q})$ of $GL(n, \mathbb{R})$?

6.4.2 Show that $GL(n, \mathbb{R})$ $(SL(n, \mathbb{R}))$ can viewed as a subgroup of $GL(n + 1, \mathbb{R})$ $(SL(n + 1, \mathbb{R}))$ through the embedding i_n given by

$$i_n(A) = \begin{bmatrix} A & 0_{n\times1} \\ 0_{1\times n} & 1 \end{bmatrix},$$

6.4.3 Recall that a group G is the semi direct product $K \prec H$ if there is a normal subgroup H and a subgroup K of G such that $G = HK$, and $H \cap K = \{e\}$. Prove the following:

(i) Let G be a group and β a surjective homomorphism from G to a group K. Let t be a homomorphism from K to G such that $\beta \circ t = I_K$. Show that G is the semi direct product $K \prec ker\ \beta$.
(ii) $GL(n, \mathbb{R})$ is the semi direct product $\mathbb{R}^\star \prec SL(n, \mathbb{R})$.
(iii) $O(n)$ is the semi direct product $\{1, -1\} \prec SO(n)$.
(iv) $U(n)$ is the semi direct product $S^1 \prec SU(n)$.

6.4.4 Show that $GL(n, \mathbb{R})$ and $SL(n, \mathbb{R})$ are not finitely generated.

6.4.5 Let A be a nonsingular matrix in $GL(n, \mathbb{R})$. Show that A can be reduced to the diagonal matrix $D_n^{det(A)}$ by using the elementary row and column operations of the type (i) as described in the Definition 5.5.3, i.e., adding a nonzero multiple of a row/column to an other row/column.

Hint. See the proof of the Proposition 5.5.8 and modify it suitably.

6.4.6 Use the above exercise to show that every nonsingular matrix A is expressible as

$$A = E_{i_1 j_1}^{\lambda_1} E_{i_2 j_2}^{\lambda_2} \cdots E_{i_r j_r}^{\lambda_r} D_n^{det(A)} E_{k_1 l_1}^{\mu_1} E_{k_2 l_2}^{\mu_2} \cdots E_{k_s l_s}^{\mu_s}.$$

6.4.7 Use the above exercise to show that the group $SL(n, \mathbb{R})$ is generated by the set of all transvections.

6.4.8 Show that $GL(n_1, \mathbb{R}) \times GL(n_2, \mathbb{R})$ is isomorphic to a subgroup of $GL(n_1 + n_2, \mathbb{R})$.

6.4.9 Is $GL(n, \mathbb{R})$ isomorphic to $GL(m, \mathbb{R})$ for $n \neq m$? Support.

6.4.10 Describe the centralizer of a transvection E_{ij}^{λ}.

6.4.11 Describe the centralizer of M_τ.

6.4.12 Describe the centralizer of D_k^{λ}.

6.4.13 Describe the centers of the groups $GL(n, \mathbb{R})$, $SL(n, \mathbb{R})$, $O(n)$ and $U(n)$.

6.4.14 Find the commutator subgroups of $GL(n, \mathbb{R})$, $SL(n, \mathbb{R})$, and $O(n)$.

6.4.15 Describe the conjugacy classes of matrices in $O(3)$ having the determinant -1.

6.4.16 Interpret the members of $SO(3)$ as rotations in $\mathbb{R}^3$ about an axis passing through the origin.

6.4.17 Interpret the matrices of determinant -1 in $O(3)$ as reflections in $\mathbb{R}^3$ about a plane passing through origin.

6.4.18 Describe the conjugacy classes of subgroups of $SO(3)$.

6.4.19 Let $A \in SO(1, n)$, where n is even. Show that there is a nonzero vector $\overline{X} \in \mathbb{R}^{n+1}$ such that $A \cdot \overline{X}^t = \overline{X}^t$.

6.4.20 Describe the conjugacy classes in $SO(1, 2)$, and also the conjugacy classes of subgroups of $SO(1, 2)$.

6.4.21 Show that every element of $SU(2)$ is uniquely expressible as

$$a_0 \begin{bmatrix} 1 & 0 \\ 0 & 1 \end{bmatrix} + a_1 \begin{bmatrix} i & 0 \\ 0 & -i \end{bmatrix} + a_2 \begin{bmatrix} 0 & 1 \\ -1 & 0 \end{bmatrix} + a_3 \begin{bmatrix} 0 & i \\ i & 0 \end{bmatrix},$$

where (a_0, a_1, a_2, a_3) belongs to the unit 3-sphere S^3. Deduce that $SU(2)$ is isomorphic to the group S^3 described in Exercise 4.1.26.

Chapter 7
Elementary Theory of Rings and Fields

Ring is an another important algebraic structure with two compatible binary operations whose intrinsic presence in almost every discipline of mathematics is frequently noticed. The theory of rings, in the beginning, will be developed on the pattern the theory of groups was developed.

7.1 Definition and Examples

Definition 7.1.1 A **ring** is a triple $(R, +, \cdot)$, where R is a set, $+$ and $\cdot$ are binary operations on R such that the following three conditions hold.

1 $(R, +)$ is an abelian group.
2 $(R, \cdot)$ is a semigroup.
3 The binary operation $\cdot$ distributes over $+$ from left as well as from right. Thus,

(i) $a \cdot (b + c) = a \cdot b + a \cdot c$
(ii) $(a + b) \cdot c = a \cdot c + b \cdot c$
    for all $a, b, c \in R$.

Remark 7.1.2 The condition 3 in the above definition is the compatibility condition between the two binary operations.

Example 7.1.3 $(\mathbb{Z}, +, \cdot)$, $(\mathbb{Q}, +, \cdot)$, $(\mathbb{R}, +, \cdot)$, $(\mathbb{C}, +, \cdot)$ are all rings($+$ and $\cdot$ are the usual addition and multiplication).

Example 7.1.4 $(\mathbb{Z}_m, \oplus, \star)$ (see Chap. 3) is a ring.

Example 7.1.5 $\mathbb{Z}[\sqrt{2}] = \{a + b\sqrt{2} \mid a, b \in \mathbb{Z}\}$ is a ring with respect to the addition and multiplication induced by those in $\mathbb{R}$. We can replace 2 by any integer in this example and get other rings.

© Springer Nature Singapore Pte Ltd. 2017
R. Lal, *Algebra 1*, Infosys Science Foundation Series in Mathematical Sciences,
DOI 10.1007/978-981-10-4253-9_7

Example 7.1.6 $\mathbb{Z}[i] = \{a + b\sqrt{-1} \mid a, b \in \mathbb{Z}\}$ is a ring with respect to the addition and multiplication induced by the addition and the multiplication of complex numbers. This ring is called the ring of **Gaussian integers**.

Example 7.1.7 $\mathbb{Z}[\sqrt{-5}] = \{a + b\sqrt{-5} \mid a, b \in \mathbb{Z}\}$ is also a ring with respect to the usual addition and multiplication induced by those in $\mathbb{C}$.

Example 7.1.8 The set $M_n(\mathbb{Z})$ of $n \times n$ matrices with entries in $\mathbb{Z}$ is a ring with respect to the addition $+$ and multiplication $\cdot$ of matrices

In the above example, we can replace $\mathbb{Z}$ by $\mathbb{Q}$, $\mathbb{R}$ or $\mathbb{C}$ to get other matrix rings. In fact, we can replace $\mathbb{Z}$ by any ring to get a matrix ring $M_n(R)$ with entries in the ring R.

Example 7.1.9 Let **H** denote the set of all 2×2 matrices with entries in the field $\mathbb{C}$ of complex numbers which are of the form

$$\begin{bmatrix} a & b \\ -\bar{b} & \bar{a} \end{bmatrix},$$

where $\bar{a}$ denotes the complex conjugate of a. Then, H is a ring with respect to the addition and multiplication of matrices (verify). This ring is called the ring of **Quaternions**.

Example 7.1.10 Let $(M, +)$ be an abelian group and $End(M)$ denote the set of all group endomorphisms of M. Then, $End(M)$ is a ring with respect to the operations $\oplus$ and o, where $\oplus$ and o are defined by

$$(\eta \oplus \rho)(m) = \eta(m) + \rho(m),$$

and

$$(\eta o \rho)(m) = \eta(\rho(m))$$

for all $m \in M$ (verify).

Example 7.1.11 Every abelian group $(M, +)$ can be made a ring by defining the multiplication $\cdot$ in M by $a \cdot b = 0$. This ring is called the **zero ring** on M.

Example 7.1.12 Let $(R_1, \oplus_1, \star_1)$ and $(R_2, \oplus_2, \star_2)$ be rings, and $R = R_1 \times R_2$. Define the operations $\oplus$ and $\star$ on R by

$$(x_1, x_2) \oplus (y_1, y_2) = (x_1 \oplus_1 y_1, x_2 \oplus_2 y_2),$$

and

$$(x_1, x_2) \star (y_1, y_2) = (x_1 \star_1 y_1, x_2 \star_2 y_2).$$

Then, R is a ring with respect to these operations. This ring is called the **directproduct** of R_1 and R_2.

Example 7.1.13 Let $(R, +, \cdot)$ be a ring. Define operations $\oplus$ and $\star$ on $\mathbb{Z} \times R$ by

$$(n, a) \oplus (m, b) = (n + m, a + b),$$

and

$$(n, a) \star (m, b) = (nm, nb + ma + ab).$$

Then, $\mathbb{Z} \times R$ is a ring with respect to these operations (verify).

Example 7.1.14 Let $(R, +, \cdot)$ be a ring, and X be a set. Let R^X denote the set of all mappings from X to R. Then, R^X is a ring with respect to the binary operations $\oplus$ and $\star$ defined by

$$(f \oplus g)(x) = f(x) + g(x)$$

for all $x \in X$, and

$$(f \star g)(x) = f(x) \cdot g(x)$$

for all $x \in X$.

7.2 Properties of Rings

Proposition 7.2.1 *Let $(R, +, \cdot)$ be a ring. Let 0 denote the additive identity (the zero of the ring) of the ring and $-a$ be the inverse of a in $(R, +)$. Then,*

(i) $a \cdot 0 = 0 = 0 \cdot a$
(ii) $a \cdot (-b) = -(a \cdot b) = a \cdot (-b)$
(iii) $(-a) \cdot (-b) = a \cdot b.$
  for all $a, b \in R$.

Proof (i) $0 + a \cdot 0 = a \cdot 0 = a \cdot (0 + 0) = a \cdot 0 + a \cdot 0$ (by the distributive condition). By the cancellation law in the group $(R, +)$, we get that $a \cdot 0 = 0$. Similarly, using the right cancellation law, we get that $0 \cdot a = 0$ *for all $a \in R$*.
(ii) $a \cdot (-b) + a \cdot b = a \cdot (-b + b)$ (by the distributive condition) $= a \cdot 0 = 0$ (by (i)). Hence $a \cdot (-b) = -(a \cdot b)$ *for all $a, b \in R$*. Similarly, $(-a) \cdot b = -(a \cdot b)$.
(iii) Follows by applying (ii) twice. ♯

Remark 7.2.2 In the proof of the above proposition, we have not used the commutativity of $+$. Only the cancellation law and the distributive condition is used. Thus, the above proposition is also true for an algebraic structure $(R, +, \cdot)$ satisfying all properties of a ring except (possibly) the commutativity of $+$.

The following example illustrates the impact of the multiplication on the addition through the distributive condition.

Example 7.2.3 Let $(R, +, \cdot)$ be an algebraic structure which satisfies all the postulates of a ring except (possibly) the commutativity of the addition. Suppose that there is a $c \in R$ which can be left canceled in the sense that

$$ca = cb \text{ implies that } a = b.$$

Then, $+$ is necessarily commutative, and so $(R, +, \cdot)$ becomes a ring.

Proof Using Proposition 7.2.1(ii), and the distributivity, we have

$$c \cdot (-(a + b)) = (-c) \cdot (a + b) = (-c) \cdot a + (-c) \cdot b =$$
$$c \cdot (-a) + c \cdot (-b) = c \cdot ((-a) + (-b)).$$

Canceling c, we obtain that $-(a + b) = (-a) + (-b)$. Taking the inverses, we get $a + b = b + a$ for all $a, b \in R$.

Remark 7.2.4 The cancellation of an element c in the above proposition is a condition on $(R, \cdot)$ which controls (through the distributive condition) the addition by making it commutative. Similarly, addition also has control over multiplication.

A ring $(R, +, \cdot)$ is called a **commutative** ring if $a \cdot b = b \cdot a$ for all $a, b \in R$. It is said to be with **identity** if *there exists an element* $1 \neq 0$ *in R such that* $1 \cdot a = a = a \cdot 1$ *for all* $a \in R$. This element 1 is called the **identity** of R. Examples 7.1.3–7.1.7 are all commutative rings with identities. Examples 7.1.8 and 7.1.9 are noncommutative rings with identities. The identity of $M_n(\mathbb{Z})$ is the identity matrix $I_{n \times n}$. It can be seen easily that $E_{ij}^\alpha \cdot E_{jk}^\beta \neq E_{jk}^\beta \cdot E_{ij}^\alpha$, $\alpha \neq 0$, $\beta \neq 0$, $i \neq k$. Thus, $M_n(\mathbb{Z})$ is a noncommutative ring for $n \geq 2$. The ring R^X is commutative ring if and only if R is commutative (verify). R^X has identity if and only if R has identity (prove it). The set $2\mathbb{Z}$ of even integers with respect to the usual addition and multiplication of integers is a commutative ring without identity. The ring $\mathbb{Z} \times \mathbb{Z}$ is a commutative ring with identity $(1, 1)$ with respect to coordinate-wise addition and multiplication.

Example 7.2.5 Let $(R, +, \cdot)$ be ring such that $a^2 = a$ for all $a \in R$. Then, the ring is commutative, and $a + a = 0$ for all $a \in R$.

Proof $a + a = (a + a)^2 = (a + a) \cdot (a + a) = a^2 + a^2 + a^2 + a^2 = a + a + a + a$ for all $a \in R$. By the cancellation law, $a + a = 0$ for all $a \in R$. Next,

$$(a + b) = (a + b)^2 = (a + b)(a + b) = a^2 + ab + ba + b^2 = a + ab + ba + b.$$

By the cancellation property in $(R, +)$, $ab + ba = 0$. In turn, $ab + ba = 0 = ab + ab$. By the cancellation law, $ab = ba$.

Theorem 7.2.6 (Binomial Theorem) *Let $(R, +, \cdot)$ be a commutative ring. Then,*

$$(a + b)^n = a^n + {}^nC_1 a^{n-1}b + \cdots + {}^nC_r a^{n-r} b^r + \cdots + b^n = \Sigma_{r=0}^{n} {}^nC_r a^{n-r} b^r.$$

for all $a, b \in R$ and $n \in \mathbb{N}$.

Proof The proof is by induction on n. If $n = 1$, then there is nothing to do. Assume that the result is true for n. Then,

$$(a + b)^n = \Sigma_{r=0}^{n} {}^nC_r a^{n-r} b^r.$$

Now,

$$(a + b)^{n+1} = (a + b)^n \cdot (a + b) = (\Sigma_{r=0}^{n} {}^nC_r a^{n-r} b^r)(a + b).$$

By the distributive law and the commutativity of $\cdot$,

$$(a + b)^{n+1} = \Sigma_{r=0}^{n} {}^nC_r a^{n-r+1} b^r + \Sigma_{r=0}^{n} {}^nC_r a^{n-r} b^{r+1}.$$

Thus,

$$(a + b)^{n+1} = a^{n+1} + \Sigma_{r=1}^{n} {}^nC_r a^{n+1-r} b^r + \Sigma_{r=1}^{n} {}^nC_{r-1} a^{n-r+1} b^r + b^{r+1}.$$

Since ${}^nC_r + {}^nC_{r-1} = {}^{n+1}C_r$, we have

$$(a + b)^{n+1} = \Sigma_{r=0}^{n+1} {}^{n+1}C_r a^{n+1-r} b^r.$$

♯

Example 7.2.7 Let $(R, +, \cdot)$ be a ring with identity 1. Let a be an element of R which has more that one left inverse with respect to the multiplication in R. Then, it has infinitely many left inverses. In particular, the ring R is infinite. Thus, in any finite ring with identity, if left inverse (right inverse) of an element exists, then it is unique.

Proof Let X be the set of all left inverses of a. Then, since a has more than 1 left inverse, none of them are right inverses of a. Let $a_0 \in X$. Then, $a_0 \cdot a = 1$. If $x \in X$, then $(ax - 1 + a_0)a = axa - a + a_0 a = a \cdot 1 - a + 1 = 1$, and so $(ax - 1 + a_0) \in X$. Thus, we have a map f from X to X defined by $f(x) = ax - 1 + a_0$. Suppose that $f(x) = f(y)$. Then, $ax - 1 + a_0 = ay - 1 + a_0$. Multiplying by x from left and using the fact that x is a left inverse of a, we get that $x = y$. Thus, f is injective. We show that $f(x) \neq a_0$ for all $x \in X$. If $f(x) = ax - 1 + a_0 = a_0$, then $ax = 1$, and so x will also be a right inverse, a contradiction. It follows that f is not surjective. Hence, X is an infinite set.

♯

7.3 Integral Domain, Division Ring, and Fields

Let $(R, +, \cdot)$ be a ring. An element $a \in R$ is called a **left(right) zero divisor** if there exists a nonzero element b in R such that $ab = 0$ $(ba = 0)$. Thus, 0 is always a zero (left as well as right) divisor in any nontrivial ring. In $\mathbb{Z}_6$, $\bar{2}$ is a left as well as right zero divisor, for $\bar{2} \star \bar{3} = \bar{0} = \bar{3} \star \bar{2}$ whereas $\bar{3} \neq \bar{0}$. In $M_n(\mathbb{Z})$, the matrix e_{ij}^1, $i \neq j$ is a zero divisor, for $(e_{ij}^1)^2$ is the zero matrix.

Proposition 7.3.1 *Let $(R, +, \cdot)$ be a ring. Then, the following conditions are equivalent.*
1. *R has no nonzero left zero divisors.*
2. *R has no nonzero right zero divisors.*
3. *If $a \cdot b = 0$, then $a = 0$ or $b = 0$.*
4. *If a and b are nonzero elements of R, then $a \cdot b \neq 0$.*
5. *Restricted cancellation law holds in R in the sense that*

$$[a \neq 0 \text{ and } a \cdot b = a \cdot c] \text{ implies that } b = c,$$

and

$$[a \neq 0 \text{ and } b \cdot a = c \cdot a] \text{ implies that } b = c$$

6. *The multiplication $\cdot$ induces a binary operation in $R^* = R - \{0\}$ with respect to which R^* is a semigroup with cancellation law.*

Proof $1 \implies 2$. Assume 1. Let b be a right zero divisor. Then, there is a nonzero element $a \in R$ such that $a \cdot b = 0$. But then $b = 0$, for otherwise a will be a nonzero left zero divisor. Thus, the ring has no nonzero right zero divisors.

$2 \implies 3$. Assume 2. Suppose that $a \cdot b = 0$. If $b = 0$, then there is nothing to do. If $b \neq 0$, then since R has no nonzero right zero divisors, b cannot be a right zero divisor, and so $a = 0$.

$3 \iff 4$. This equivalence is a tautology.

$4 \implies 5$. Assume 4 (and so 3 also). Let $a \neq 0$ and $a \cdot b = a \cdot c$. Then, $0 = a \cdot b - a \cdot c = a \cdot b + a \cdot (-c) = a \cdot (b + (-c))$. Since $a \neq 0$, $b + (-c) = 0$. Hence $b = c$.

$5 \implies 4$. Assume 5. Suppose that $a \cdot b = 0$. If $a = 0$, then there is nothing to do. If $a \neq 0$, then $a \cdot b = 0 = a \cdot 0$. By the restricted cancellation law, $b = 0$.

$5 \implies 6$. Assume 5 (and so 3 and 4 also follows). Because of 4, the multiplication $\cdot$ induces a binary operation on $R^* = R - \{0\}$, and because of 5, R^* is a semigroup with cancellation law.

$6 \implies 1$. Assume 6. If there is a nonzero left zero divisor a, then there is a nonzero element b such that $a \cdot b = 0$. But then $a, b \in R^*$, and $a \cdot b \notin R^*$. This contradicts 6. ♯

Definition 7.3.2 A ring $(R, +, \cdot)$, $R \neq \{0\}$ is called an **integral domain** if it satisfies any one (and hence all) of the six conditions in the above proposition.

Examples 7.1.3, 7.1.5, 7.1.6, and 7.1.7 are all integral domains.

Proposition 7.3.3 *The ring* $(\mathbb{Z}_m, \oplus, \star)$ *is an integral domain if and only if m is prime.*

Proof Suppose that $m = p$ is a prime and $\overline{0} = \overline{a} \star \overline{b} = \overline{ab}$. Then, p divides ab. Since p is prime, it divides a or b. But then $\overline{a} = \overline{0}$ or $\overline{b} = \overline{0}$. Thus, $\mathbb{Z}_p$ is an integral domain. Suppose that m is not prime, and so $m = m_1 \cdot m_2$ for some m_1, m_2 such that $1 < m_1 < m$ and $1 < m_2 < m$. But then $\overline{m_1} \neq \overline{0} \neq \overline{m_2}$, whereas $\overline{m_1} \star \overline{m_2} = \overline{m_1 m_2} = \overline{m} = \overline{0}$. Hence, in this case, $\mathbb{Z}_m$ is not an integral domain. ♯

For example, $\mathbb{Z}_6$ is not an integral domain, whereas $\mathbb{Z}_7$ is an integral domain. The ring $M_n(\mathbb{Z})$ is not an integral domain for e_{12}^2 is a zero divisor.

Example 7.3.4 Let $C[0, 1]$ denote the set of all real-valued continuous functions from the closed unit interval $[0, 1]$. Since the sum and product of continuous functions are continuous, $C[0, 1]$ is a commutative ring with identity(the constant function which takes every member of $[0, 1]$ to 0 is the additive identity and the constant function which takes every member to 1 is the multiplicative identity). It is not an integral domain: Let f be the function on $[0, 1]$ defined by $f(x) = o$ if $x \in [0, \frac{1}{2}]$ and $f(x) = (x - \frac{1}{2})^2$ if $x \in [\frac{1}{2}, 1]$ and g the function given by $g(x) = (x - \frac{1}{2})^2$ if $x \in [0, \frac{1}{2}]$ and 0 at rest of the places. Then, f and g are nonzero elements of $C[0, 1]$ whose product is zero function. What can you say about the ring of differentiable functions? What can we say about the ring of analytic functions on a domain in the complex plane?

Remark 7.3.5 On every abelian group $(M, +)$, we can define a multiplication · by $a \cdot b = 0$ for all $a, b \in M$. Then, $(M, +, \cdot)$ becomes a ring(called the zero ring on M). Can we always define addition $+$ on a semigroup $(R, \cdot)$ so that $(R, +, \cdot)$ becomes a ring? Obviously, the answer to this question is in negative. For example, $(R, \cdot)$ can never be a group. There should be an element $0 \in R$ such that $a \cdot 0 = 0 = 0 \cdot a$ for all $a \in R$. Even on such semigroup, we may not be able to define an addition $+$ so that $(R, +, \cdot)$ becomes a ring. The characterization of multiplicative semigroup of a ring may be a difficult problem.

Proposition 7.3.6 *Let* $(R, +, \cdot)$ *be a ring containing p elements, where p is a prime. Then, it is a zero ring, or a commutative integral domain(compare with the corresponding result in groups).*

Proof Since R contains p elements, $(R, +)$ is a cyclic group of order p in which every nonzero element is a generator of the group $(R, +)$. Suppose that $(R, +, \cdot)$ is not an integral domain. Then, we have a pair of nonzero elements $a, b \in R$ such that $a \cdot b = 0$. Let $x, y \in R$. Since a as well as b generate $(R, +)$, $x = na$ and $y = mb$ for some $n, m \in \{1, 2, \ldots, p - 1\}$. Hence, $xy = na \cdot mb = nma \cdot b$ (by the distributive condition)$= 0$. Thus, if $(R, +, \cdot)$ is not an integral domain, then it is a zero ring. Further, let $a \in R^*$. Then, the group $(R, +)$ is generated by a. Let $x, y \in R$. Then, $x = na$ and $y = ma$ for some $n, m \in \{1, 2, \ldots, p - 1\}$. Clearly, $xy = na \cdot ma = nma^2 = ma \cdot na = yx$. Thus, the ring is commutative. ♯

Let $(R, +, \cdot)$ be a ring. An element $e \in R$ is called an **idempotent** element if $e^2 = e$. An element $a \in R$ is called a **nilpotent** element if $a^n = 0$ for some $n > 0$. If $(R, +, \cdot)$ is with identity 1, then an element u is called a **unit** (or **invertible**) if there is an element $v \in R$ such that $u \cdot v = 1 = v \cdot u$. The set of units of R is denoted by $U(R)$. An element u is called **unipotent** if $u - 1$ is nilpotent, or equivalently if $u = 1 + b$ for some nilpotent element b.

Example 7.3.7 Let Y be a set and X the power set of Y. Then, X is a ring with respect to the addition as symmetric difference and multiplication as the intersection. Every element of this ring is idempotent (verify). It is a commutative ring with identity Y. There is no unit except the identity. There is no nilpotent element except the zero (i.e., $\emptyset$). There is no unipotent element except the identity (i.e., Y) element. It is not an integral domain (why?).

Example 7.3.8 In $\mathbb{Z}_6$, $\bar{3}$ is an idempotent which is neither a unit nor a nilpotent element. This is also not a unipotent element (verify).

Example 7.3.9 In the ring $M_n(\mathbb{Z})$ of $n \times n$ matrices with entries in $\mathbb{Z}$, all strictly upper triangular matrices are nilpotent (verify). All upper triangular with diagonal entries 1 (such matrices are called the unitriangular) are unipotent (verify).

Proposition 7.3.10 *Let $(R, +, \cdot)$ be an integral domain. Let $e \in R$, and u be a nonzero element of R such that $eu = u = ue$. Then, e is the identity.*

Proof Let $a \in R$. Then, $u \cdot (ea - a) = ua - ua = 0$. Since R is an integral domain and $u \neq 0$, $ea = a$. Similarly, $ae = a$ for all $a \in R$.   ♯

Corollary 7.3.11 *The only nonzero idempotent element in an integral domain is the identity of the integral domain.*   ♯

Proposition 7.3.12 *Let $(R, +, \cdot)$ be a ring with identity, and u be a unipotent element of the of the ring. Then, u is a unit of R.*

Proof Suppose that $u = 1 - b$, where $b^n = 0$ for some $n \in \mathbb{N}$. Then,

$$(1-b)(1+b+b^2+\cdots+b^{n-1}) = 1-b+b-b^2+b^2-\cdots-b^{n-1}+b^{n-1}-b^n = 1$$

Similarly, $(1 + b + \cdots + b^{n-1})(1 - b) = 1$. This shows that $u = 1 - b$ is a unit, and its inverse is $1 + b + \cdots + b^{n-1}$.   ♯

Remark 7.3.13 If a and b are nilpotent elements of R which commute with each other, then $a + b$ is also nilpotent: Suppose that $a^n = 0 = b^m$. Then, by the Binomial theorem(note that in the proof of the binomial theorem, we only need the fact that a and b commute), we get that $(a + b)^{n+m} = 0$. Thus, in particular, if $u = 1 - b$ is a unipotent element, then its inverse $1 + b + \cdots + b^{n-1}$ is also a unipotent element. Let us denote the set of unipotent elements of R by $Uni(R)$. If $u \in Uni(R)$, then $u^{-1} \in Uni(R)$. In general, product of unipotent elements need not be unipotent. The matrices E_{ij}^1 and E_{ji}^1 are unipotent in $M_n(\mathbb{Z})$, but there product is not (verify). Clearly, $Uni(R) \subseteq U(R)$.

Proposition 7.3.14 *Let $(R, +, \cdot)$ be a ring with identity 1. The multiplication in R induces a multiplication in the set $U(R)$ of units with respect to which $U(R)$ is a group.*

Proof Clearly $1 \in U(R)$. Let $a, b \in U(R)$. Then, *there exist $c, d \in R$ such that $ac = ca = bd = db = 1$.* But, then $abdc = ac = 1$. Also, $dcab = 1$. Hence, $ab \in U(R)$. Further, if $a \in U(R)$, then *there exists $c \in R$ such that $ac = 1 = ca$.* Clearly, $c \in U(R)$. This shows that $U(R)$ is a group with respect to the induced multiplication. ♯

A maximal subgroup of $U(R)$ consisting of unipotent elements is called a **unipotent subgroup** of $U(R)$. The group of unitriangular matrices in $M_n(\mathbb{C})$ is a unipotent subgroup of $U(M_n(\mathbb{C}))$.

Example 7.3.15 The group $U((\mathbb{Z}, +, \cdot))$ is the cyclic group $\{1, -1\}$ of order 2. $a \in \mathbb{Z}$ is a unit if there is a $b \in \mathbb{Z}$ such that $ab = 1$. But then $\mid a \mid \cdot \mid b \mid = 1$. Hence, $\mid a \mid = 1$, and so $a = 1$ or $a = -1$.

Example 7.3.16 The group $U((\mathbb{Z}_m, \oplus, \star))$ is the group U_m of prime residue classes modulo m: $\bar{a}$ is a unit in $\mathbb{Z}_m$ if there is an element $\bar{b}$ in $\mathbb{Z}_m$ such that $\bar{a} \star \bar{b} = \bar{1}$. But then $ab - 1$ is divisible by m. This means that a and m are co-prime, and so $\bar{a} \in U_m$.

Example 7.3.17 Consider the ring $\mathbb{Z}[i] = \{a + bi \mid a, b \in \mathbb{Z}\}$ of Gaussian integers. Then, $U(\mathbb{Z}[i]) = \{1, -1, i, -i\}$ is the cyclic group of order 4: Suppose that $a + bi$ is a unit. Then, there is an element $c + di \in \mathbb{Z}[i]$ such that $(a + bi) \cdot (c + di) = 1$. Taking square of the modulus of both sides, we got that $a^2 + b^2 = 1$. This means that $a = \pm 1$ and $b = 0$ or $b = \pm 1$ and $a = 0$. Hence, $a + bi = 1$ or -1 or i or $-i$.

Definition 7.3.18 A ring $(R, +, \cdot)$ is called a **division ring** or a **skew field** if the binary operation $\cdot$ induces a binary operation on $R^\star = R - \{0\}$ with respect to which it is a group. A commutative division ring is called a **field**.

Thus, a division ring is always an integral domain. The integral domain $\mathbb{Z}$ of integers is not a division ring, for $\mathbb{Z}^\star$ is not a group (2 has no inverse). The rings $(\mathbb{Q}, +, \cdot)$, $(\mathbb{R}, +, \cdot)$ and $(\mathbb{C}, +, \cdot)$ are all fields, for $(\mathbb{Q}^\star, \cdot)$, $(\mathbb{R}^\star, \cdot)$, $(\mathbb{C}^\star, \cdot)$ are all commutative groups.

Proposition 7.3.19 *A ring $(R, +, \cdot)$ with identity $1 \neq 0$ is a division ring if and only if for all nonzero element a in R, there is an element a^{-1} in R such that $a^{-1} \cdot a = 1 = a \cdot a^{-1}$.*

Proof Suppose that $(R, +, \cdot)$ is a division ring, and $a \in R^\star$. Then, since $\cdot$ induces a binary operation on $R^\star$ with respect to which it is a group, there is an element b in $R^\star$ such that $a \cdot b = b \cdot a = 1$. Conversely, suppose that for all $a \in R^\star$, there is an element $a^{-1} \in R$ such that $a \cdot a^{-1} = 1 = a^{-1} \cdot a$. Then, $U(R) = R^\star$. ♯

Example 7.3.20 The Quaternion ring $(\mathbf{H}, +, \cdot)$ (Example 7.1.9) is a non commutative ring. We show that it is a division ring (which is not a field). Let

$$\begin{bmatrix} a & b \\ -\overline{b} & \overline{a} \end{bmatrix}$$

be a nonzero element of $\mathbf{H}$. Then, $\mid a \mid^2 + \mid b \mid^2 \neq 0$. Let us denote $\mid a \mid^2 + \mid b \mid^2 \neq 0$ by δ. Then, it can be checked that

$$\begin{bmatrix} \frac{\overline{a}}{\delta} & -\frac{b}{\delta} \\ \frac{\overline{b}}{\delta} & \frac{a}{\delta} \end{bmatrix}$$

is the inverse of the above element, and it belongs to $\mathbf{H}$.

Proposition 7.3.21 *Every finite integral domain is a division ring.*

Proof Let $(R, +, \cdot)$ be a finite integral domain. Then, $\cdot$ induces a binary operation in $R^\star$ with respect to which it is a finite semigroup in which cancellation law holds. Since a finite semigroup in which cancellation law holds is a group, $R^\star$ is a group with respect to the induced operation. This shows that $(R, +, \cdot)$ is a division ring. ♯

Remark 7.3.22 The finiteness assumption in the above proposition is essential, for $(\mathbb{Z}, +, \cdot)$ is an integral domain which is not a division ring.

The following is an immediate corollary of the above proposition.

Corollary 7.3.23 *Every finite commutative integral domain is a field.* ♯

Remark 7.3.24 In fact, every finite integral domain (without the assumption of commutativity) is a field. A more general result is that a division ring in which every nonzero element is of finite multiplicative order is a field. The proof of this assertion can be found in Algebra 2.

Corollary 7.3.25 $(\mathbb{Z}_m, \oplus, \star)$ *is a field if and only if m is prime.*

Proof By proposition 7.3.3, $(\mathbb{Z}_m, \oplus, \star)$ is a finite commutative ring which is an integral domain if and only if m is prime. The result follows from the above corollary. ♯

Corollary 7.3.26 *A ring containing prime number of elements is either a zero ring or a field.*

Proof By Proposition 7.3.6, a ring containing prime number of elements is either a zero ring or a commutative integral domain. The result follows from the fact that every finite commutative integral domain is a field. ♯

Remark 7.3.27 Let $(R, +, \cdot)$ be a ring such that $(R, +)$ has no proper subgroup, then $\mid R \mid$ is prime, and so it is a zero ring or a field. This illustrates that a condition on $(R, +)$ puts a restriction on $(R, \cdot)$.

Proposition 7.3.28 *Let $(D, +, \cdot)$ be a division ring containing q elements. Then,* $a^q = a$ *for all* $a \in D$.

Proof If $a = 0$, then there is nothing to do. Suppose that $a \neq 0$. Then, $a \in D^*$, and since D^* is a group with respect to the induced multiplication containing $q - 1$ elements, it follows that $a^{q-1} = 1$. Thus, $a^q = a$. ♯

Characteristic of an Integral Domain

Proposition 7.3.29 *Let $(R, +, \cdot)$ be an integral domain. Then, additive order of any two nonzero elements are same.*

Proof It is sufficient to show that $ma = 0$ *if and only if* $mb = 0$ *for all* $a, b \in R^*$. Let $a, b \in R^*$. Suppose that $ma = 0$. Then, $0 = (ma) \cdot b = a \cdot (mb)$ (by the distributive law). Since $a \neq 0$, and R is an integral domain, $mb = 0$. Similarly, $mb = 0$ implies that $ma = 0$. ♯

Definition 7.3.30 Let $(R, +, \cdot)$ be an integral domain. If additive order of a nonzero element (and so of all nonzero elements) is infinite, then we say that R is of **characteristic 0**. If order of a nonzero element (and so of all nonzero elements) is finite and m, then we say that R is of **characteristic m**. The characteristic of R is denoted by **charR**.

Proposition 7.3.31 *The characteristic of an integral domain is either 0 or else a prime number.*

Proof Suppose that $charR = m \neq 0$, and m is not prime. Then, $m = m_1 m_2$ for some m_1, m_2 with $1 < m_1 < m, 1 < m_2 < m$. Let a be a nonzero element of R. Since R is integral domain, $a^2 \neq 0$. Hence m is additive order of a as well as a^2. Now, $0 = ma^2 = m_1 m_2 a^2 = m_1 a \cdot m_2 a$. Since R is an integral domain, $m_1 a = 0$ or $m_2 a = 0$. This is a contradiction to the supposition that m is order of a. ♯

Every division ring (field) is an integral domain, and hence, characteristic of a division ring or a field is zero or a prime number.

Corollary 7.3.32 *If $(R, +, \cdot)$ is a finite integral domain, then the characteristic of R is some prime p such that p divides $|R|$.*

Proof Since R is finite, no element can be of infinite order. Thus, characteristic of R is a prime p which is additive order of any nonzero element of R. Since order of an element divides the order of the group, p divides $|R|$. ♯

Corollary 7.3.33 *Order of a finite integral domain (division ring or a field) is p^n for some prime p and $n > 0$.*

Proof Since R is finite, the characteristic of R is some prime p. Thus, the order of each nontrivial element of $(R, +)$ is p. No other prime q will divide the order of R, for otherwise, by Cauchy theorem for abelian groups, $(R, +)$ will have an element of order q. Hence, $|R| = p^n$ for some $n > 0$. ♯

In particular, there is no integral domain of order 6.

Characteristics of $(\mathbb{Z}, +, \cdot)$, $(\mathbb{Q}, +, \cdot)$, $(\mathbb{R}, +, \cdot)$, and $(\mathbb{C}, +, \cdot)$ are all 0. The characteristic of $(\mathbb{Z}_p, \oplus, \star)$ is p.

Remark 7.3.34 An infinite integral domain may also have characteristic p. An example will follow later.

Proposition 7.3.35 *If $(R, +, \cdot)$ is a commutative integral domain of characteristic p, then $(a + b)^p = a^p + b^p$ for all $a, b \in R$.*

Proof We know that p divides ^pC_r for all $r, 1 \leq r \leq p - 1$, and so $^pC_r c = o$ for all $c \in R$, $1 \leq r \leq p - 1$. Applying the binomial theorem the result follows. ♯

Corollary 7.3.36 *If R is a finite field of characteristic p, then given any $b \in R$, there is an element $a \in R$ such that $a^p = b$.*

Proof Suppose that $a^p = b^p$. Then, $(a - b)^p = a^p - b^p = 0$. Since R is an integral domain, $a - b = 0$, and so $a = b$. Thus, the map $a \rightsquigarrow a^p$ is an injective map. Since R is finite, it is surjective. Hence, given any $b \in R$, there is an element $a \in R$ such that $b = a^p$. ♯

Integral Domains of Orders 4 and 8

Let $(R, +, \cdot)$ be an integral domain of order 4. Since its characteristic is prime and divides 4, it is 2. In turn, the additive order of each nonzero element is 2. Hence, it is the Klein's four group. Since it is finite, it is a division ring. Thus, R^* is a cyclic group of order 3. This also shows that it is a field. Let $R = \{0, 1, \alpha, \alpha^2\}$. The operations in R are obvious. One can check that with these operations, it is a ring.

Let $(R, +, \cdot)$ be an integral domain of order 8. Again, it is of characteristic 2 and R^* is a cyclic group of order 7. Let $R = \{0, 1, \alpha, \alpha^2, \alpha^3, \alpha^4, \alpha^5, \alpha^6\}$. The multiplication is clear. We try to give the addition $+$ in R. Since R is of characteristic 2, $a + a = 0$ for all $a \in R$. Consider $1 + \alpha$. $1 + \alpha \neq 1$, for otherwise $\alpha = 0$. $1 + \alpha \neq \alpha$, for otherwise $1 = 0$. Suppose that $1 + \alpha = \alpha^2$. Then, $\alpha^4 = (1 + \alpha)^2 = 1 + \alpha^2 = 1 + 1 + \alpha = \alpha$. But then $\alpha^3 = 1$. This is a contradiction, since $|R^*| = 7$. Thus, $1 + \alpha \neq \alpha^2$. Next, suppose that

$$1 + \alpha = \alpha^3 \tag{7.3.1}$$

Then,

$$1 + \alpha^2 = (1 + \alpha)^2 = \alpha^6 \tag{7.3.2}$$

$$1 + \alpha^3 = 1 + 1 + \alpha = \alpha \tag{7.3.3}$$

$$1 + \alpha^4 = (1 + \alpha^2)^2 = \alpha^{12} = \alpha^5 \tag{7.3.4}$$

$$1 + \alpha^5 = 1 + 1 + \alpha^4 = \alpha^4 \tag{7.3.5}$$

$$1 + \alpha^6 = 1 + 1 + \alpha^2 = \alpha^2. \tag{7.3.6}$$

Since $\alpha^i + \alpha^j = \alpha^i(1 + \alpha^{j-i})$ for $i < j$, we can find $\alpha^i + \alpha^j$ for all i, j. For example, $\alpha^3 + \alpha^5 = \alpha^3(1 + \alpha^2) = \alpha^3\alpha^6 = \alpha^2$. The addition in R, therefore, is given by the following table:

$+$	0	1	α	α^2	α^3	α^4	α^5	α^6
0	0	1	α	α^2	α^3	α^4	α^5	α^6
1	1	0	α^3	α^6	α	α^5	α^4	α^2
α	α	α^3	0	α^4	1	α^2	α^6	α^5
α^2	α^2	α^6	α^4	0	α^5	α	α^3	1
α^3	α^3	α	1	α^5	0	α^6	α^2	α^4
α^4	α^4	α^5	α^2	α	α^6	0	1	α^3
α^5	α^5	α^4	α^6	α^3	α^2	1	0	α
α^6	α^6	α^2	α^5	1	α^4	α^3	α	0

The verification of the correctness of the table is left as an exercise. Also find other possibilities for $1 + \alpha$.

Exercises

7.3.1 Let $(R, +, \cdot)$ be a ring. Show that $\mathbb{Z} \times R$ is a ring with respect to the coordinate-wise addition and multiplication $\star$ given by $(n, a) \star (m, b) = (nm, nb + ma + ab)$. Show that it is a ring with identity. Find the identity.

7.3.2 Let $D[0, 1]$ denote the set of all real-valued differentiable functions on $[0,1]$. Show that it is a commutative ring with respect to pointwise addition and pointwise multiplication. Is it an integral domain?

7.3.3 Let $(R, +, \cdot)$ be a ring. Show that $(na) \cdot b = n(a \cdot b) = a \cdot (nb)$ for all $a, b \in R$, and $n \in \mathbb{Z}$.

7.3.4 Let $M_n(R)$ denote the set of all $n \times n$ matrices with entries in a ring $(R, +, \cdot)$ with identity. Show that this is a ring with identity with respect to the usual matrix addition and matrix multiplication. Show that for $n \geq 2$, this is non commutative, and also it is not an integral domain.

7.3.5 Let $(R, +, \cdot)$ be an integral domain. Let $e \in R^*$ such that $e^n = e$ for some $n > 1$. Show that e is the identity of R.

7.3.6 Let $(R_1, +_1, \star_1)$ and $(R_2, +_2, \star_2)$ be rings. Define operations $\oplus$ and $\star$ in $R = R_1 \times R_2$ by $(a, b) \oplus (c, d) = (a +_1 c, b +_2 d)$ and $(a, b) \star (c, d) = (a \star_1 c, b \star_2 d)$. Show that R is a ring which is commutative if and only if R_1 and R_2 are commutative. It is with identity if and only if both the rings are with identities. Show also that $U(R_1 \times R_2) = U(R_1) \times U(R_2)$.

7.3.7 Let $(R, +, \cdot)$ be a commutative ring with identity. Let $Uni(R)$ denote the set of unipotent elements of R. Show that it is a subgroup of $U(R)$. Is this result true for non commutative rings?

7.3.8 Let $(R, +, \cdot)$ be an integral domain of order 32. Show that it is a field. Let $\alpha \in R - \{0, 1\}$. Show that (i) $1 + \alpha \neq \alpha^2$ (ii) $1 + \alpha \neq \alpha^3$. Find the possible values of i for which $1 + \alpha = \alpha^i$.

7.3.9 Do the above problem for an integral domain of order 16.

7.3.10 Can we define multiplication on the additive group $\mathbb{Z}_4$ to make it an integral domain? Support.

7.3.11 Is the ring $\mathbb{Z}_8$ an integral domain?

7.3.12 Is $\mathbb{R}^3$ a ring with respect to vector addition and vector product?

7.3.13 Let X be a set containing more than 1 elements. Define a product $\cdot$ on X by $a \cdot b = b$. Can we define addition $+$ on X so that $(X, +, \cdot)$ becomes a ring? Support (observe that $(X, \cdot)$ is a semigroup).

7.3.14 Let $(R, +, \cdot)$ be an structure, where $(R, +)$ and $(R, \cdot)$ are groups and $+$ distributes over $\cdot$. Show that R is a singleton.

7.3.15 Show that every ring of order 15 is commutative. Can it be an integral domain?

7.3.16 A formal expression of the type $a_0 + a_1 i + a_2 j + a_3 k$, $a_i \in \mathbb{R}$ is denoted by a and is called a **Quaternion**. The above expression is written as 0 if and only if $a_i = 0 \ \forall i$. The Quaternion $a_0 - a_1 i - a_2 j - a_3 k$ is called the conjugate of a and is denoted by $\bar{a}$. Also, the norm $\mid a \mid$ is defined to be $+\sqrt{a_0^2 + a_1^2 + a_2^2 + a_3^2}$. It is easy to see that the usual property of norm is satisfied. Let $\mathbb{R}^4$ denote the set of all Quaternions. Define $\oplus$ and $\cdot$ in $\mathbb{R}^4$ by

$$a + b = (a_0 + b_0) + (a_1 + b_1)i + (a_2 + b_2)j + (a_3 + b_3)k,$$

and

$$a \cdot b = (a_0 b_0 - a_1 b_1 - a_2 b_2 - a_3 b_3) + (a_0 b_1 + a_1 b_0 + a_2 b_3 - a_3 b_2)i + (a_0 b_2 + a_2 b_0 - a_1 b_3 + a_3 b_1)j + (a_0 b_3 + a_3 b_0 + a_1 b_2 - a_2 b_1)k$$

Show that it is a division ring (check that if $a \neq 0$, then $\frac{\bar{a}}{|a|}$ is the inverse of a).

7.3.17 Let $R = \{0, e, a, b, c\}$. Define a binary operation $\cdot$ on R by $0 \cdot x = 0 = x \cdot 0$, $e \cdot x = x = x \cdot e$ for all $x \in R$, and the product of any two of a, b, c is the third. Show that there does not exist any operation $+$ on R such that $(R, +, \cdot)$ is a ring.

7.3.18 Show that the ring of analytic functions on a domain is an integral domain.

7.3.19 Show that there is no set containing all rings.

7.4 Homomorphisms and Isomorphisms

From now onward, the binary operations of rings will not be written unless necessary. They will usually be denoted by $+$ and $\cdot$. Thus, we shall say that R is a ring instead of saying that $(R, +, \cdot)$ is a ring.

Definition 7.4.1 Let R_1 and R_2 be rings. A map f from R_1 to R_2 is called a **ring homomorphism** or simply a **homomorphism** if

(i) $f(a + b) = f(a) + f(b)$.
(ii) $f(a \cdot b) = f(a) \cdot f(b)$.
  for all $a, b \in R_1$.

An injective homomorphism is called a **monomorphism**, and a surjective homomorphism is called an **epimorphism**. A bijective homomorphism is called an **isomorphism**. As in case of groups, composite of homomorphisms are homomorphism. Inverse of an isomorphism is an isomorphism (verify).

Proposition 7.4.2 *Let f be a homomorphism from a ring R_1 to a ring R_2. Then, the following hold:*

(i) $f(0) = 0$.
(ii) $f(-a) = -f(a)$.
(iii) $f(a - b) = f(a) - f(b)$.
(iv) $f(na) = nf(a)$
  for all $a, b \in R_1$ and $n \in \mathbb{Z}$.

Proof A ring homomorphism from R_1 to R_2 is also a group homomorphism from the group $(R_1, +)$ to $(R_2, +)$. The result follows from the corresponding results in groups.   ♯

Remark 7.4.3 A group homomorphism from $(R_1, +)$ to $(R_2, +)$ need not be ring homomorphism. For example, $n \rightsquigarrow 2n$ is a group homomorphism from $(\mathbb{Z}, +)$ to $(\mathbb{Z}, +)$ but it is not a ring homomorphism from $(\mathbb{Z}, +, \cdot)$ to itself.

We say that a ring R_1 is isomorphic to R_2 if there is an isomorphism from R_1 to R_2. The notation $R_1 \approx R_2$ stands to say that R_1 is isomorphic to R_2. The relation of isomorphism is an equivalence relation on a set of rings. Let R_1 and R_2 be rings. The map which takes every element of R_1 to the zero of R_2 is a homomorphism called the zero homomorphism. The identity map I_R on a ring R is an isomorphism.

Example 7.4.4 Let f be a ring homomorphism from the ring $\mathbb{Z}$ of integers to itself. Then, f is also a group homomorphism from $(\mathbb{Z}, +)$ to itself. Thus, there exists $m \in \mathbb{Z}$ such that $f(a) = ma$ for all $a \in \mathbb{Z}$. Since it is also a ring homomorphism, $m \cdot 1 = f(1) = f(1 \cdot 1) = f(1) \cdot f(1) = m \cdot 1 \cdot m \cdot 1 = m^2 \cdot 1$. Hence, $m^2 = m$. This shows that $m = 0$ or $m = 1$, and so f is the zero homomorphism or the identity homomorphism.

Example 7.4.5 The map $a \rightsquigarrow \bar{a}$ is a surjective ring homomorphism from the ring $\mathbb{Z}$ of integers to the ring $\mathbb{Z}_m$ of integers modulo m.

Example 7.4.6 Let X be a set and R a ring. Consider the ring R^X of functions from X to R. Let $x \in X$. Define a map e_x from R^X to R by $e_x(f) = f(x)$. Then, e_x is a surjective homomorphism (verify). This map is called the evaluation map at x.

Remark 7.4.7 A ring homomorphism need not take identity to identity, even if the homomorphism is injective homomorphism. The ring $\mathbb{Z} \times \mathbb{Z}$ is a ring with identity $(1, 1)$. The map $a \rightsquigarrow (a, 0)$ is a ring homomorphism from $\mathbb{Z}$ to $\mathbb{Z} \times \mathbb{Z}$. It does not take the identity 1 of $\mathbb{Z}$ to that of $\mathbb{Z} \times \mathbb{Z}$.

Proposition 7.4.8 *Let f be a surjective homomorphism from a ring R_1 to a ring R_2. Let e_1 be the identity of R_1. Then, $f(e_1)$ is the identity of R_2.*

Proof Let $y \in R_2$. Since f is surjective, there is an element $x \in R_1$ such that $y = f(x)$. But then $f(e_1)y = f(e_1)f(x) = f(e_1x) = f(x) = y$. Similarly, $yf(e_1) = y$ for all $y \in R_2$. ‡

Proposition 7.4.9 *Let f be a nonzero ring homomorphism from a ring R_1 with identity e_1 to an integral domain R_2. Then, $f(e_1)$ is the identity of R_2.*

Proof Suppose that $f(e_1) = 0$. Then, $f(x) = f(xe_1) = f(x)f(e_1) = 0$. This is a contradiction to the supposition that f is a nonzero homomorphism. Thus, $f(e_1) \neq 0$. Now, $f(e_1)f(e_1) = f(e_1e_1) = f(e_1) = e_2f(e_1)$. By the restricted cancellation in $R_2, f(e_1) = e_2$. ‡

The above proposition also shows that the only nonzero ring homomorphism from $\mathbb{Z}$ to itself is the identity map, for it will take 1 to 1.

An injective homomorphism from a ring R_1 to a ring R_2 is called an **embedding**. If there is an embedding from R_1 to R_2, then we say that R_1 is embedded in R_2.

Proposition 7.4.10 *Every ring can be embedded in a ring with identity.*

Proof Let R be a ring. Consider the ring $\mathbb{Z} \times R$ of the Example 7.1.13 which is with identity $(1,0)$. Define a map f from R to $\mathbb{Z} \times R$ by $f(a) = (0, a)$. Then, f is an injective homomorphism, and so it is an embedding. ‡

Let $(M, +)$ be an abelian group. Then, $End(M)$ is a ring with respect to the pointwise addition and composition of maps.

Proposition 7.4.11 *Every ring can be embedded in a ring of endomorphisms of an abelian group.*

Proof Every ring can be embedded in a ring with identity, and since composite of embeddings are embeddings, it is sufficient to show that every ring with identity can be embedded in a ring of endomorphisms of an abelian group. Let R be a ring with identity. Consider the ring $End(R, +)$ of endomorphisms of $(R, +)$. Let $a \in R$. The left multiplication L_a from R to R given by $L_a(x) = ax$ is an element of $End(R, +)$. Define a map ϕ from R to $End(R, +)$ by $\phi(a) = L_a$. It is easy to check that ϕ is a ring homomorphism. Suppose that $\phi(a) = \phi(b)$. Then, $L_a = L_b$, and so $a = a \cdot 1 = L_a(1) = L_b(1) = b \cdot 1 = b$. This shows that ϕ is an embedding. ‡

The proof of the following theorem is an imitation of the process of embedding the ring $\mathbb{Z}$ of integers into the field $\mathbb{Q}$ of rational numbers as described in the Sect. 3.6 of the Chap. 3.

Theorem 7.4.12 *Every commutative integral domain can be embedded in a field.*

Proof Let $R \neq \{0\}$ be a commutative integral domain. Define a relation $\sim$ on $R \times R^*$ by

$$(a, b) \sim (c, d) \text{ if and only if } ad = bc.$$

Then, $\sim$ is an equivalence relation: The reflexivity and the symmetry follow from the fact that R is commutative. Suppose that $(a, b) \sim (c, d)$ and $(c, d) \sim (e, f)$. Then, $ad = bc$ and $cf = de$. Hence, $adf = bcf = bde$. Since $d \neq 0$, $af = be$. Thus, $(a, b) \sim (e, f)$.

Let us denote the equivalence class determined by (a, b) by $\frac{a}{b}$. Thus, $\frac{a}{b} = \{(c, d) \mid ad = bc\}$. Further,

$$\frac{a}{b} = \frac{c}{d} \text{ if and only if } ad = bc. \qquad (7.4.1)$$

Let F denote the quotient set $R \times R^* / \sim$. Thus,

$$F = \{\frac{a}{b} \mid a \in R, b \in R^*\}.$$

Suppose that $\frac{a}{b} = \frac{u}{v}$ and $\frac{c}{d} = \frac{x}{y}$. Then, $av = bu$ and $cy = dx$. But then $(ad + bc)vy = bd(uy + vx)$. Hence $\frac{ad+bc}{bd} = \frac{uy+vx}{vy}$ (by (7.4.1)). This ensures that we have a binary operation $+$ on F defined by

$$\frac{a}{b} + \frac{c}{d} = \frac{ad + bc}{bd}. \qquad (7.4.2)$$

Similarly, we have a binary operation $\cdot$ on F defined by

$$\frac{a}{b} \cdot \frac{c}{d} = \frac{ac}{bd}. \qquad (7.4.3)$$

We show that $(F, +, \cdot)$ is a field. The verification of associativity and commutativity of $+$ and $\cdot$, and the distributivity of $\cdot$ over $+$ is straightforward. $\frac{0}{d} = \frac{0}{b}$ for all $b, d \in R^*$. The equivalence class $\frac{0}{d}$ is the zero of F (verify). The negative of $\frac{a}{b}$ is $\frac{-a}{b}$ (check it). Also, $\frac{b}{b} = \frac{d}{d}$ for all $b, d \in R^*$. The element $\frac{d}{d}$ is the identity of F (verify). Let $\frac{a}{b}$ be a nonzero element of F. Then, $a \neq 0$. It is easy to observe that $\frac{b}{a} \cdot \frac{a}{b} = \frac{ab}{ba} = \frac{d}{d}$ the identity of F. Hence, every nonzero element of F has the inverse. This completes the proof of the fact that F is a field.

Define a map f from R to F by $f(a) = \frac{ad}{d}$, $d \neq 0$. It is easy to verify that f is a homomorphism. Suppose that $f(a) = f(b)$. Then, $\frac{ad}{d} = \frac{bd}{d}$. But then $ad^2 = bd^2$. Since $d^2 \neq 0$, $a = b$. This shows that f is an embedding. ♯

The field F together with the homomorphism f introduced in the above theorem can be characterized by the following universal property.

Theorem 7.4.13 *Let R, F, and f be as in the above theorem. Let F' be a field, and ϕ an injective homomorphism from R to F'. Then, there exists unique homomorphism ψ from F to F' such that ψ of $= \phi$. Indeed, if the pair (F', ϕ) also satisfies the same property, then ψ is an isomorphism.*

Proof Since ϕ is injective, $b \neq 0$ implies that $\phi(b) \neq 0$. Define a binary relation ψ from F to F' by

$$\psi(\frac{a}{b}) = \phi(a)\phi(b)^{-1}.$$

It is straightforward verification that ψ is a map which is a homomorphism (in fact an injective homomorphism) with the required property ψ of $= \phi$. Next, observe that $f(R) = \{\frac{ad}{d} \mid a \in R, d \in R^\star\}$ generates the field F, for $\frac{a}{b} = \frac{ad}{d} \cdot \frac{d}{bd} = f(a)f(b)^{-1}$. Thus, if ψ' is also a homomorphism with ψ' of $= \phi$, then $\psi = \psi'$.

Further, if (F', ϕ) also satisfies the same universal property, then there is a unique homomorphism χ from F' to F such that $\chi o\phi = f$. Clearly, then $\chi o\psi$ is the identity on $f(R)$, and so it is the identity on F. Similarly, $\psi o\chi$ is the identity on F'. ♯

Definition 7.4.14 The field F described in the above theorems is called the **field of fractions** or the **quotient field** of R.

Remark 7.4.15 It is clear from the above results that the field of fractions of R can be viewed as the smallest field containing R.

The field of fractions of the ring $\mathbb{Z}$ of integers is the field $\mathbb{Q}$ of rational numbers. Interpret the field of fractions of $\mathbb{Z}[i]$. The field of fractions of a field is the field itself.

Exercises

7.4.1 Find all ring homomorphisms from $\mathbb{Z}_6$ to itself.

7.4.2 Find all ring homomorphisms from $\mathbb{Q}$ to $\mathbb{Q}$.

7.4.3 Find all continuous ring homomorphisms from $\mathbb{R}$ to $\mathbb{R}$ and also from $\mathbb{C}$ to $\mathbb{C}$.

7.4.4 Find all ring homomorphisms from the ring of Gaussian integers to itself.

7.4.5 Let f be a nonzero ring homomorphism from an integral domain R_1 of characteristic p to an integral domain R_2. Show that R_2 is also of characteristic p.

7.4.6 Show that a homomorphic image of a commutative ring is commutative.

7.4.7 Let R be a ring with identity. Define new binary operations $\oplus$ and $\star$ on R by $a \oplus b = a + b + 1$ and $a \star b = ab + a + b$. Show that $(R, \oplus, \star)$ is a ring, and the map f defined by $f(a) = a - 1$ is an isomorphism from $(R, +, \cdot)$ to $(R, \oplus, \star)$.

7.4.8 Let F be a field and f a surjective ring homomorphism from F to a ring $R \neq \{0\}$. Show that R is also a field and f is an isomorphism.

7.4.9 Show that the field $\mathbb{R}$ of real numbers is not isomorphic to the field $\mathbb{C}$ of complex numbers.

Hint. Look at the images of -1 and i.

7.4.10 Show that there is no surjective homomorphism from $\mathbb{R}$ to $\mathbb{C}$ (or from $\mathbb{C}$ to $\mathbb{R}$).

7.4.11 Let F be a field of characteristic 0. Show that there is an injective homomorphism from $\mathbf{Q}$ to F.

Hint. Consider the map f from $\mathbb{Z}$ to F given by $f(n) = ne$, where e is the identity of F. Show that f is an injective homomorphism. Observe that $\mathbb{Q}$ is the field of fractions of $\mathbb{Z}$.

7.4.12 Let F be a field of characteristic $p \neq 0$. Show that there is an injective homomorphism from $\mathbb{Z}_p$ to F.

Hint. Show that $\bar{n} \rightsquigarrow ne$ is the required injective homomorphism.

7.4.13 What is the field of fractions of $2\mathbb{Z}$?

7.4.14 What is the field of fractions of $\mathbb{Z}[i]$?

7.4.15 Show that any two fields of order 4 are isomorphic.

7.4.16 Show that any two fields of order 8 are isomorphic.

7.4.17 Let F be a finite field of characteristic p. Show that the map f defined by $f(a) = a^p$ is an automorphism of F. Deduce that any equation of the form $x^p = a$ is solvable in F. How many solutions are there if $a \neq 0$?

7.4.18 Let R be a ring with identity 1 and S a subset of R such that

(i) $1 \in S$.

and

(ii) $a, b \in S$ implies that $ab \in S$.

Define a relation $\sim$ on $R \times S$ by

$$(a, s) \sim (b, t) \text{ if and only if there is an element } s' \in S \text{ such that}$$
$$s'(at - bs) = 0.$$

Show that $\sim$ is an equivalence relation. Let $S^{-1}R$ denote the quotient set, and $\frac{a}{s}$ denote the equivalence class determined by (a, s). Show that we have binary operations $+$ and $\cdot$ on $S^{-1}R$ defined by

$$\frac{a}{s} + \frac{b}{t} = \frac{at + bs}{st}$$

and

$$\frac{a}{s} \cdot \frac{b}{t} = \frac{ab}{st}$$

with respect to which it is a ring.

7.4.19 Show that if $o \in S$, then $S^{-1}R = \{0\}$.

7.4.20 What happens if S contains a zero divisor?

7.4.21 Suppose that R is an integral domain and $0 \notin S$. Show that the map f given by $f(a) = \frac{a}{1}$ is an embedding such that each element of $f(S)$ is invertible. What is $S^{-1}R$ if $S = R^*$?

7.4.22 Find $S^{-1}\mathbb{Z}$, where $S = \{p^n \mid n \in \mathbb{N} \bigcup \{0\}\}$ and p is prime.

7.5 Subrings, Ideals, and Isomorphism Theorems

Definition 7.5.1 Let R be a ring. A subset S of R is called a **subring** if the binary operations $+$ and $\cdot$ induce binary operations on S with respect to which it is a ring.

Proposition 7.5.2 *A subset S of a ring R is a subring if and only if*

(i) $S \neq \emptyset$.
(ii) $a - b \in S$ for all $a, b \in S$.
(iii) $a \cdot b \in S$ for all $a, b \in S$.

Proof Suppose that S is a subring. Then, $+$ and $\cdot$ induce a binary operation on S with respect to which it is a ring. Since every ring is nonempty (contains at least 0), (i) is satisfied. Since $\cdot$ induces a binary operation, (iii) is satisfied. Since $+$ induces a binary operation with respect to which S is a subgroup, (ii) follows from the corresponding result in group theory.

Conversely, suppose that (i), (ii), and (iii) holds. From (i) and (ii), and the corresponding result in group theory, it follows that $+$ induces a binary operation on S with respect to which S is a subgroup of $(R, +)$. Further, by (iii), $\cdot$ induces a binary operation in S. That S is a ring with respect to the induced operations is a consequence of the corresponding properties in the ring R.   ♯

Remark 7.5.3 If S is a subring, then $0 \in S$ and $ma \in S$ for all $m \in \mathbb{Z}$ and $a \in S$.

Example 7.5.4 For each $m \in \mathbb{Z}$, $m\mathbb{Z}$ is a subring of $\mathbb{Z}$. Since every subgroup of $(\mathbb{Z}, +)$ is of the form $m\mathbb{Z}$, it follows that every subring of $(\mathbb{Z}, +)$ is of the form $m\mathbb{Z}$.

$\mathbb{Z}$ is a subring of $\mathbb{Q}$, $\mathbb{Q}$ is a subring of $\mathbb{R}$, and $\mathbb{R}$ is a subring of $\mathbb{C}$.

Example 7.5.5 Consider the ring $\mathbb{Z} \times \mathbb{Z}$ with coordinate-wise addition and multiplication. $(1, 1)$ is the identity. The subset $\mathbb{Z} \times \{0\}$ is a subring. The identity of $\mathbb{Z} \times \{0\}$ is $(1,0)$ which is different from the identity of the ring. Thus, subring of a ring with identity may have identity which is different from the identity of the ring.

Example 7.5.6 Subring of a ring with identity may not have identity. For example, $2\mathbb{Z}$ is a subring of $\mathbb{Z}$ which is without identity.

Example 7.5.7 A ring may be without identity, but a subring may have identity. For example, the ring $\mathbb{Z} \times 2\mathbb{Z}$ with coordinate-wise operations is a ring without identity, but $\mathbb{Z} \times \{0\}$ is a subring with identity.

Proposition 7.5.8 *Let R be an integral domain with identity, and S be a subring of R which is also with identity. Then, the identities of R and S are same.*

Proof Let e be the identity of R and e' be that of S. Then, $e' \cdot e' = e' = e' \cdot e$. From the restricted cancellation law in an integral domain, $e' = e$.   ♯

Example 7.5.9 Let S be a subring of the field $\mathbb{R}$ of real numbers. Then, S is also a subgroup of the additive group $\mathbb{R}$. Hence, it is cyclic subgroup of $\mathbb{R}$, or it is dense in $\mathbb{R}$. Suppose that it is a nontrivial cyclic subgroup generated by a, where a is the smallest positive real number in S. Then, a^2 also belongs to S, and hence, $a^2 = na$ for some $n \in \mathbb{N}$. Thus, a is a natural number. This shows that every non-dense subring of $\mathbb{R}$ is also a subring of $\mathbb{Z}$. In particular, $\mathbb{Z}[\sqrt{2}]$ is dense in $\mathbb{R}$. Also, $\mathbb{Z}$ is the only subring of $\mathbb{R}$ which is with identity and which is not dense.

The proofs of the following two propositions are straightforward (similar to the proofs of the corresponding propositions in groups) and are left as exercises.

Proposition 7.5.10 *Intersection of a family of subrings is a subring.*   ♯

Proposition 7.5.11 *Union of two subrings is a subring if and only if one of them is contained in the other.*   ♯

Let S_1 and S_2 be subrings of a ring R. The sum $S_1 + S_2 = \{a+b \mid a \in S_1, b \in S_2\}$ is a subgroup of $(R, +)$, but, as in case of groups, it need not be a subring. Consider the subring

$$\mathbb{Z}[\pi] = \{a_0 + a_1\pi + a_2\pi^2 + \cdots + a_n\pi^n \mid a_i \in \mathbf{Z}, \, n \geq 0\},$$

and the subring $\mathbb{Q}$ of $\mathbb{R}$. We show that $\mathbb{Z}[\pi] + \mathbb{Q}$ is not a subring. We use the fact that π is a transcendental number in the following sense:

"*If* $u_0 + u_1\pi + u_2\pi^2 + \cdots + u_r\pi^r = 0$, $u_i \in \mathbb{Q}$, *then* $u_i = 0$ *for all i*".

Clearly, $\pi \in \mathbb{Z}[\pi] + \mathbb{Q}$ and also $\frac{1}{2} \in \mathbb{Z}[\pi] + \mathbb{Q}$. Suppose that $\frac{1}{2}\pi \in \mathbb{Z}[\pi] + \mathbb{Q}$. Then,

$$\frac{1}{2}\pi = a_0 + a_1\pi + \cdots + a_n\pi^n + r,$$

where $a_i \in \mathbb{Z}$ and $r \in \mathbb{Q}$. Using the transcendence of π, we find that $a_1 = \frac{1}{2}$. This is a contradiction. Hence, $\frac{1}{2}\pi \notin \mathbb{Z}[\pi] + \mathbb{Q}$. This shows that $\mathbb{Z}[\pi] + \mathbb{Q}$ is not a subring.

Subring Generated by a Subset

Let R be a ring and X a subset of R. The intersection of all subrings of R containing X is a subring of R, and it is the smallest subring of R containing X. This subring is called the **subring generated by** X and is denoted by $< X >$. The elements of $< X >$ are precisely finite sums of integral multiples of finite products of nonnegative integral powers of elements of X. In particular, if $X = \{a\}$, then $< X >$ denoted by $< a >$ is given by

$$< a > = \{\alpha_0 + \alpha_1 a + \alpha_2 a^2 + \cdots + \alpha_r a^r \mid \alpha_i \in \mathbb{Z}, \ r \geq 0\}.$$

Proposition 7.5.12 *Image of a subring under a homomorphism is a subring. Inverse image of a subring under a homomorphism is also a subring.*

Proof Let f be a homomorphism from a ring R_1 to a ring R_2. Let S_1 be a subring of R_1. Then, $0 \in S_1$, and so $0 = f(0) \in f(S_1)$. Thus, $f(S_1) \neq \emptyset$. Let $f(a), f(b) \in f(S_1)$, where $a, b \in S_1$. Since S_1 is a subring, $a - b \in S_1$ and also $a \cdot b \in S_1$. Since f is a homomorphism $f(a) - f(b) = f(a - b)$ and $f(a) \cdot f(b) = f(a \cdot b)$. Hence, $f(a) - f(b)$ and $f(a) \cdot f(b)$ both belong to $f(S_1)$. This shows that $f(S_1)$ is a subring. Similarly, if S_2 is a subring of R_2, it can be shown that $f^{-1}(S_2)$ is a subring of R_1. ♯

Corollary 7.5.13 $f^{-1}(\{0\})$ *is a subring of* R_1.

Proof Since $\{0\}$ is a subring of R_2, the result follows from the above proposition.

Definition 7.5.14 $f^{-1}(\{0\})$ is called the **kernel** of the homomorphism f, and it is denoted by $ker f$. Thus,

$$ker f = \{a \in R_1 \mid f(a) = 0\}.$$

is a subring of R_1.

Since image of a subring under a homomorphism is a subring, the image of an embedding is also a subring. Thus, if f is an embedding of R_1 into R_2, then R_1 is isomorphic to the subring $f(R_1)$ of R_2. The following three results follow from Propositions 6.4.10, 6.4.11 and Theorem 6.4.12.

Proposition 7.5.15 *Every ring is isomorphic to a subring of a ring with identity.* ♯

Proposition 7.5.16 *Every ring is isomorphic to a ring of endomorphism.* ♯

Proposition 7.5.17 *Every commutative integral domain is isomorphic to a subring of a field.* ♯

Since the images and the inverse images of subrings under homomorphisms are subrings, the following correspondence theorem for rings follows from the correspondence theorem for groups.

Theorem 7.5.18 (Correspondence Theorem) *Let f be a surjective homomorphism from a ring R_1 to a ring R_2. Let $S(R_1)$ denote the set of all subrings of R_1 containing ker f and $S(R_2)$ the set of all subrings of R_2. Then, f induces a bijective map ϕ from $S(R_1)$ to $S(R_2)$ defined by $\phi(S) = f(S)$. Also $\phi(S \bigcap T) = \phi(S) \bigcap \phi(T)$.* ♯

Example 7.5.19 Lagrange theorem holds for finite rings: If S is a subring of a finite ring R, then S is a subgroup of $(R, +)$. By the Lagrange theorem for groups, $|S|$ divides $|R|$. Here also the converse of Lagrange theorem is not true: Let F be a field of order 8. We show that F has no subring of order 4. If R is subring of F of order 4, then it being a subring of a field is an integral domain. Since a finite commutative integral domain is field, R is a field. But then $R^\star$ is a subgroup of $F^\star$ of order 3. Since order of $F^\star$ is 7, this is a contradiction to the Lagrange theorem.

Let A be a subring of a ring R. Then, A is a subgroup of $(R, +)$. Consider the quotient group

$$R/A = \{x + A \mid x \in R\}.$$

Since $(R, +)$ is an abelian group, R/A is also abelian. The addition in R/A is given by

$$(x + A) + (y + A) = (x + y) + A.$$

We want to make R/A a ring. Our temptation would be to define a multiplication $\cdot$ in R/A by

$$(x + A) \cdot (y + A) = x \cdot y + A. \tag{7.5.1}$$

But $\cdot$ defined above need not be a binary operation. For example, consider $\mathbb{R}/\mathbb{Z}$. We have $\sqrt{2} + \mathbb{Z} = 1 + \sqrt{2} + \mathbb{Z}$ and $\sqrt{3} + \mathbb{Z} = 1 + \sqrt{3} + \mathbb{Z}$. But $\sqrt{2}\sqrt{3} + \mathbb{Z} \neq (1 + \sqrt{2})(1 + \sqrt{3}) + \mathbb{Z}$, for $(1 + \sqrt{2})(1 + \sqrt{3}) - \sqrt{2}\sqrt{3}$ is not in $\mathbb{Z}$.

Suppose that $\cdot$ defined by 1 is a binary operation. Then, given $a \in A$ and $x \in R$, $a + A = 0 + A$ and so $(a + A) \cdot (x + A) = (0 + A)(x + A)$. This means that $ax + A = 0 + A$. Hence $ax \in A$. Thus, $ax \in A$ for all $a \in A$ and $x \in R$. Also $(x + A)(a + A) = (x + A)(0 + A)$, and so $xa + A = 0 + A$. It also follows that $xa \in A$ for all $a \in A$ and $x \in R$.

Conversely, suppose that S is a subring of R such that ax and xa belong to A for all $a \in A$ and $x \in R$. Suppose that $a + A = b + A$ and $c + A = d + A$. Then, $(a - b) \in A$ and $(c - d) \in A$. Hence, from our hypothesis, $(a - b) \cdot c \in A$ and $b \cdot (c - d) \in A$. Since A is a subring, $ac - bd = (a - b)c + b(c - d)$ belongs to A. This shows that $ac + A = bd + A$, and so $\cdot$ defined by 1 is indeed a binary operation. We have proved the following theorem.

Theorem 7.5.20 *Let R be a ring and A a subgroup of $(R, +)$. Then, we have a binary operation $\cdot$ on R/A defined by*

$$(a + A) \cdot (b + A) = ab + A.$$

if and only if ax and xa belong to A for all $x \in R$ and $a \in A$. ♯

Definition 7.5.21 A subring A of R is called a **left(right) ideal** if xa (ax) belongs to A for all $x \in R$ and for all $a \in A$. We say that A is an **ideal** of R if it is both-sided ideal.

Let A be an ideal of R. Then, it is easily seen that R/A is a ring with respect to the binary operations $+$ and $\cdot$ on R/A defined by

$$(x + A) + (y + A) = (x + y) + A.$$

and

$$(x + A) \cdot (y + A) = xy + A.$$

This ring is called the **difference ring** or **quotient ring** of R modulo A.

The map ν from R to R/A defined by $\nu(x) = x + A$ is a surjective homomorphism whose kernel is A. Thus, every ideal is a kernel of a homomorphism. The map ν is called the **quotient map** modulo A.

It can be seen as before that the image of a left ideal (right ideal) under a surjective homomorphism is a left (right) ideal. Also, in the correspondence theorem, ideals correspond. Inverse image of a left (right) ideal under a homomorphism is a left (right) ideal. In particular, kernel of a homomorphism is an ideal.

As in case of groups, any subring of R/A is of the form B/A, where B is a subring containing A. It is a left (right) ideal if and only if B is a left (right) ideal of R.

The results such as the fundamental theorem of homomorphism, 1st isomorphism theorem, and 2nd isomorphism theorem are true in case of rings also. In the corresponding results for groups, replace subgroups by subrings and normal subgroups by ideals to get the corresponding result for rings. The proofs are also on the same lines. As an illustration, we prove the 2nd isomorphism theorem.

Theorem 7.5.22 (Noether 2nd Isomorphism theorem). *Let A and B be subrings of R and B an ideal of R. Then, $A \cap B$ is an ideal of A, and $A/(A \cap B) \approx (A + B)/B$.*

Proof Since B is an ideal, $A + B$ is a subring (verify and compare with the corresponding result in groups) and B is an ideal of $A + B$. Define a map f from A to $(A + B)/B$ by $f(a) = a + B$. Then, f is clearly a ring homomorphism. Further, any element of $(A + B)/B$ is of the form $a + b + B$ for some $a \in A$ and $b \in B$. Since $(a + b) - a = b$ belongs to B, $a + b + B = a + B = f(a)$. Thus, f is surjective homomorphism. Also,

$$ker\, f = \{a \in A \mid f(a) = B\} = \{a \in A \mid a + B = B\} = \{a \in A \mid a \in B\}$$

Thus, $ker\, f = A \cap B$. Since kernel of a homomorphism is an ideal, $A \cap B$ is an ideal of A. By the fundamental theorem of homomorphism, $A/(A \cap B)$ is isomorphic to $(A + B)/B$.   ♯

Proposition 7.5.23 *A ring R is without proper left (right) ideals if and only if it is a zero ring of prime order, or it is a division ring.*

Proof Suppose that R is without proper left ideals. Let $a \in R$, $a \neq 0$. Then,

$$Ra = \{xa \mid x \in R\}.$$

is a left ideal of R. There are two cases:

(i) $Ra = \{0\}$ for some $a \neq 0$.
(ii) $Ra = R$ for all $a \in R^* = R - \{0\}$.

Consider the case (i). Let $a \neq 0$ be such that $Ra = \{0\}$. Consider the left ideal $< a >$ generated by a. Then,

$$< a > = \{na + xa \mid n \in \mathbb{Z} \text{ and } x \in R\}.$$

Since $< a > \neq \{0\}$ (note that $a \in < a >$), $< a > = R$. Also, since $Ra = \{0\}$, $xa = 0$ for all $x \in R$. Hence, $R = < a > = \{na \mid n \in \mathbb{Z}\}$. Since $Ra = \{0\}$, $na \cdot ma = nma \cdot a = 0$. Thus, R is a zero ring. But then every subgroup of $(R, +)$ will be a left ideal, and since R is supposed to be without proper left ideals, $(R, +)$ is without proper subgroups. Hence, in this case, R is a zero ring of prime order.

Consider the case (ii). In this case $Ra = R$ for all $a \neq 0$. Suppose that $a \neq 0 \neq b$. Then, $Rab = (Ra)b = Rb = R$. Hence, $a \neq 0, b \neq 0$ implies that $ab \neq 0$. This shows that R is an integral domain. Let $a \in R, a \neq 0$. Since $Ra = R$, there is an element $e \in R$, $e \neq 0$ such that $ea = a$. Also $(ae - a)a = a^2 - a^2 = 0$. Since $a \neq 0$, and R is an integral domain, $ae = a$. Let b be another nonzero element of R. Then, $a \cdot (eb - b) = ab - ab = 0$. Since $a \neq 0$, $eb = b$ for all $b \in R$. Also, $(be - b) \cdot b = b^2 - b^2 = 0$. Since $b \neq 0$, $be = b$ for all $b \in R$. Hence, e is the identity of R. If $a \neq 0$, then $Ra = R$, and so there is an element $a' \in R$, $a' \neq 0$ such that $a' \cdot a = e$. This shows that R^* is a group, and so R is a division ring.

Conversely, we show that a zero ring of prime order or a division ring will have no proper left ideals. If R is zero ring of prime order, then it has no proper subgroups of $(R, +)$, and so it has no proper left ideals. Next, suppose that R is a division ring. Let A be a nonzero left ideal. Let $a \in A$, $a \neq 0$. Since R is a division ring, there is an element $a' \in R$ such that $a' \cdot a = 1$. Since A is a left ideal, $1 \in A$. But then $x = x \cdot 1$ belongs to A for all $x \in R$. This shows that $A = R$. Thus, R has no proper left ideals. The result for right ideals follows on the same lines. ♯

Since a ring with identity cannot be a zero ring, the following result is immediate from the above proposition.

Corollary 7.5.24 *A ring with identity is without proper left (right) ideals if and only if it is a division ring.* ♯

Corollary 7.5.25 *A commutative ring with identity is a field if and only if it is without proper ideals.* ♯

Corollary 7.5.26 *An ideal M of a commutative ring R with identity is a maximal ideal if and only if R/M is a field.*

Proof Since every ideal of R/M is of the form B/M, where B is an ideal containing M, it follows that M is a maximal ideal if and only if R/M is without proper ideals. The result follows from the above corollary.   ♯

Proposition 7.5.27 *A ring R is without proper subrings if and only if it is a zero ring of prime order or a field of prime order.*

Proof If R has no proper subrings, then it has no proper left ideals also. Hence, R is a zero ring of prime order or a division ring. Suppose that R is a division ring, and 1 is the identity of R. The map f from $\mathbb{Z}$ to R defined by $f(n) = n1$ is a ring homomorphism, and so $f(\mathbb{Z}) \neq \{0\}$ is a subring of R. Hence, $f(\mathbb{Z}) = R$. Since R is a division ring and $\mathbb{Z}$ is not, f cannot be injective. It follows that $ker\, f = m\mathbb{Z}$ for some $m \neq 0$, and $\mathbb{Z}/m\mathbb{Z} = \mathbb{Z}_m \approx R$. Since $\mathbb{Z}_m$ is a division ring if and only if m is prime, the result follows.   ♯

Compare the results with the corresponding result in groups.

Example 7.5.28 Let F be a field. Consider the ring $M_2(F)$ of 2×2 matrices with entries in the field F. Let A be the set of all matrices whose 2nd column is the zero column. Then, A is a left ideal (verify). It is not a right ideal, for $e_{11} \in A$, and $e_{11} \cdot M_2(F) = B$ is the set of all matrices whose 2nd row is zero (verify). Note that B is a right ideal. More generally, consider the ring $M_n(F)$ of all $n \times n$ matrices. Let C_i denote the set of all matrices having all its columns zero except the ith column. Then, C_i is a left ideal but not a right ideal. It may also be observed that each C_i is minimal left ideal of $M_n(F)$. Let R_i denote the set of all matrices having all its rows zero except ith row. Then, R_i is a (minimal) right ideal which is not a left ideal.

Example 7.5.29 In this example, we show that the ring $M_2(F)$ of 2×2 matrices with entries in a field F is without proper two sided ideals. Let $\mho$ be a nonzero two-sided ideal of $M_2(F)$. Let A be a nonzero element of $\mho$. It is an elementary fact of matrix theory (see Algebra 2 or any book on linear algebra) that there are nonsingular matrices P and Q such that PAQ is the identity matrix I or e_{11} (depending on whether the rank is 2 or 1), where e_{ij} denote the matrix whose ith row and jth column entry is 1 and the rest of the entry is 0. Thus, $I \in \mho$ or $e_{11} \in \mho$. If $I \in \mho$, then since $\mho$ is an ideal, every element of $M_2(F)$ is in $\mho$ and so $\mho = M_2(F)$. Suppose that $e_{11} \in \mho$. Then, $e_{22} = e_{21}e_{11}e_{11}e_{12}$ belongs to $\mho$. But then $I = e_{11} + e_{22}$ belongs to $\mho$. Hence, in this case, also $\mho = M_2(F)$.

Remark 7.5.30 Propositions 7.5.23 and Corollary 7.5.24 are not true if we replace left ideals by ideals.

Theorem 7.5.31 (Krull). *Every proper ideal (left ideal/right ideal) of a ring with identity is contained in a maximal ideal(left ideal/right ideal).*

Proof Let R be a ring with identity and A an ideal (a left ideal/a right ideal) of R, $A \neq R$. Then, $1 \notin A$. Let X be the set of all ideals (left ideal/ right ideal) of R containing A but not 1. Then, $A \in X$, and so $X \neq \emptyset$. Thus, $(X, \subseteq)$ is a nonempty partially ordered set. Let $\{B_\alpha \mid \alpha \in \lambda\}$ be a chain in X. Then, the union of this chain is also an ideal (left ideal/right ideal) containing A but not 1. Thus, every chain in X has an upper bound. By the Zorn's Lemma, X has a maximal member M (say). Then, M is a maximal ideal (left ideal/right ideal), for if M is properly contained in N, then $N \notin X$. Since $A \subseteq N$, $1 \in N$. But then $N = R$.   ♯

Definition 7.5.32 Let R be a ring. An ideal $\wp$ of R is called a **prime ideal** of R if $R/\wp$ is an integral domain.

Proposition 7.5.33 *An ideal $\wp$ of R is a prime ideal if and only if 'ab $\in \wp$ implies that $a \in \wp$ or $b \in \wp$'.*

Proof Suppose that $\wp$ is a prime ideal. Then, $R/\wp$ is an integral domain. Suppose that $ab \in \wp$. Then, $(a + \wp)(b + \wp) = ab + \wp = 0 + \wp = \wp$ (the zero of $R/\wp$). Since $R/\wp$ is supposed to be an integral domain, $a + \wp = \wp$ or $b + \wp = \wp$. This means that $a \in \wp$ or $b \in \wp$.

Conversely, suppose that '$ab \in \wp$ implies that $a \in \wp$ or $b \in \wp$'. Further, suppose that $(a+\wp)(b+\wp) = \wp$ (the zero of $R/\wp$). Then, $ab \in \wp$. Hence, by the supposition, $a \in \wp$ or $b \in \wp$. But then $a + \wp = \wp$ or $b + \wp = \wp$.   ♯

Corollary 7.5.34 *Every maximal ideal of a commutative ring with identity is a prime ideal.*

Proof If M is a maximal ideal of a commutative ring with identity, then R/M is a field, and so an integral domain. Hence, M is a prime ideal.   ♯

Remark 7.5.35 The above corollary is not true for a noncommutative rings. For example, $M_2(F)$, where F is a field, is a ring without proper ideals, and so $\{0\}$ is the maximal ideal. This is not a prime ideal, for $M_2(F)$ is not an integral domain. We also observe that the result is not true if the ring is without identity. For example, the zero ring on a prime cyclic group has $\{0\}$ as maximal ideal, but it is not a prime ideal.

Example 7.5.36 $\{0\}$ is a prime ideal of $\mathbb{Z}$ which is not a maximal ideal. Since $\mathbb{Z}/m\mathbb{Z} = \mathbb{Z}_m$ is an integral domain if and only if m is prime, it follows that $m\mathbb{Z}$ is a prime ideal if and only if m is prime. In this case, it is also a maximal ideal.

Proposition 7.5.37 *Let R be a commutative ring and A an ideal of R. Then,*

$$\sqrt{A} = \{a \in R \mid a^n \in A \text{ for some } n \geq 1\}$$

is an ideal of R.

Proof Clearly $0 \in \sqrt{A}$, for $0 \in A$. Thus, $\sqrt{A} \neq \emptyset$. Let a, $b \in \sqrt{A}$. Then, a^n, $b^m \in A$ for some $m, n \in \mathbb{N}$. If $r + s = n + m$, then $r \geq n$ or $s \geq m$. But then $a^r b^s \in A$. Since A is an ideal, applying the binomial theorem, we see that $(a - b)^{n+m} \in A$. Hence, $(a - b) \in \sqrt{A}$. Next, suppose that $a \in \sqrt{A}$ and $x \in R$. Then, $a^n \in A$ for some $n \in \mathbb{N}$. But then $(xa)^n = x^n a^n \in A$. Hence, $xa \in \sqrt{A}$. This shows that $\sqrt{A}$ is an ideal of R.   ♯

Definition 7.5.38 The ideal $\sqrt{A}$ is called the **radical** of A and $\sqrt{R}$ is called the **nil radical** of R.

Proposition 7.5.39 *Let R be a commutative ring with identity. Let A be an ideal of R. Then, $\sqrt{A}$ is the intersection of all prime ideals containing A.*

Proof Let $x \in \sqrt{A}$, and $\wp$ a prime ideal containing A. Then, $x^n \in A \subseteq \wp$ for some $n \in \mathbb{N}$. Since $\wp$ is a prime ideal, and $x^n \in \wp$, $x \in \wp$. Hence, $\sqrt{A} \subseteq \wp$. Thus, $\sqrt{A}$ is contained in the intersection of all prime ideals containing A. Suppose that $x \notin \sqrt{A}$. Then, we shall show the existence of a prime ideal $\wp$ such that $A \subseteq \wp$ but $x \notin \wp$. Consider the subset $S = \{x^n \mid n \in \mathbb{N}\}$. Since $x \notin \sqrt{A}$, $S \bigcap A = \emptyset$. Let

$$X = \{B \mid B \text{ is an ideal of } R \text{ and } B \bigcap S = \emptyset\}.$$

Clearly, $A \in X$, and so $X \neq \emptyset$. Thus, $(X, \subseteq)$ is a nonempty partially ordered set. If $\{B_\alpha \mid \alpha \in \lambda\}$ is a chain in X, then its union is an ideal containing A whose intersection with S is empty set. Thus, the union is an upper bound of the chain, and so every chain in X has an upper bound. By the Zorn's lemma, X will have a maximal element $\wp$ (say). Then, $A \subseteq \wp$ and $\wp \bigcap S = \emptyset$. We show that $\wp$ is a prime ideal. Suppose that $a \notin \wp$ and $b \notin \wp$. Then, the ideals $< \wp \bigcup \{a\} > = \wp + Ra$ and $< \wp \bigcup \{b\} > = \wp + Rb$ do not belong to X. Thus, $\wp + Ra$ and $\wp + Rb$ intersect S non trivially. Hence,

$$x^n = \alpha + \beta a$$

and

$$x^m = \gamma + \delta b$$

for some $n, m \in \mathbb{N}$, α, $\gamma \in \wp$ and β, $\delta \in R$. But then

$$x^{n+m} = \beta\delta ab + \alpha\delta b + \gamma\beta u + \alpha\gamma = \beta\delta ab + \mu,$$

where $\mu \in \wp$. Since $x^{n+m} \notin \wp$, $ab \notin \wp$. This shows that $\wp$ is a prime ideal containing A but not x.   ♯

Example 7.5.40 Consider an ideal $m\mathbb{Z}$ of $\mathbb{Z}$, where $m = p_1^{\alpha_1} p_2^{\alpha_2} \cdots p_n^{\alpha_n}$, $p_i \neq p_j$ for $i \neq j$. Then, $\{p_1\mathbb{Z}, p_2\mathbb{Z}, \ldots, p_n\mathbb{Z}\}$ is the set of distinct prime ideals containing $m\mathbb{Z}$ (verify). Hence

$$\sqrt{m\mathbb{Z}} = p_1\mathbb{Z} \bigcap p_2\mathbb{Z} \bigcap \cdots \bigcap p_n\mathbb{Z} = p_1 p_2 \cdots p_n \mathbb{Z}.$$

Definition 7.5.41 An ideal A of a commutative ring R with identity is called a **radical** ideal if $\sqrt{A} = A$.

Thus, every prime ideal is a radical ideal. The ideal $6\mathbb{Z}$ of $\mathbb{Z}$ is a radical ideal, but it is not a prime ideal. It can be checked that in the correspondence theorem, prime ideals and radical ideals correspond.

Definition 7.5.42 A ring R is called a **reduced** ring if it is without nonzero nilpotent elements.

Thus, every integral domain is a reduced ring. $\mathbb{Z}_m$ is a reduced ring if and only if m is product of distinct primes. In particular, $\mathbb{Z}_{10}$ is a reduced ring (it is not an integral domain).

Proposition 7.5.43 *An ideal A of a commutative ring R with identity is a radical ideal if and only if R/A is a reduced ring.*

Proof Note that $x^n \in A$ is equivalent to say that $(x + A)^n = x^n + A = A$. Thus, $x \in \sqrt{A}$ if and only if $x + A$ is nilpotent in R/A. The result follows from the definition of a radical ideal and that of a reduced ring.   ♯

Exercises

7.5.1 Give two examples of subrings which are not ideals.

7.5.2 Find conditions on subrings A and B so that $A + B$ is a subring. Show that if A or B is an ideal, then $A + B$ is a subring. Show that sum of any two ideals is an ideal.

7.5.3 Find all subrings of $\mathbb{Z}_{16}$. Are they all ideals?

7.5.4 Show that if S is a non-dense subring of the field $\mathbb{R}$ of real numbers, then it is of the form $m\mathbb{Z}$ for some $m \in \mathbb{Z}$. Deduce that the set $\{a + b\sqrt{2} \mid a, b \in \mathbb{Z}\}$ is a dense subset of $\mathbb{R}$.

7.5.5 Characterize rings with identities which are generated by their identities.

7.5.6 Prove the Correspondence theorem and the 1st isomorphism theorem for rings.

7.5.7 Characterize fields in which all subrings are subfields.

7.5.8 Show that union of chain of subrings (left ideals, ideals) is a subring (left ideals, ideals).

7.5.9 Let F be a field of order p^n, and R a subring of F. Show that the order of R is p^m, where m divides n. The converse of this is also true, and the proof of this fact can be found in Chap. 9 of Algebra 2.
Hint. Observe that $p^m - 1$ divides $p^n - 1$ if and only if m divides n.

7.5.10 Let R be a ring. Let $Z(R) = \{a \in R \mid ax = xa \text{ for all } x \in R\}$. Show that $Z(R)$ is a subring of R. This subring is called the **center** of R.

7.5.11 Show that the center of a division ring is a field.

7.5.12 Show that $Z(M_n(F))$ is a subring isomorphic to the field F.

7.5.13 Let A be the subset of $M_n(F)$ consisting of matrices all of whose columns are zero except the 1st column which is arbitrary. Show that A is a minimal nonzero left ideal of $M_n(F)$.

7.5.14 For any field F, show that $M_n(F)$ has no proper two-sided ideal.

7.5.15 Show that the intersection of a maximal ideal of a ring with a subring need not be maximal ideal of the subring, whereas it remains a prime ideal.

7.5.16 Let R be a ring, and S a subring. Let $\wp$ be a prime ideal of R. Show that $\wp \cap S$ is a prime ideal of S.

7.5.17 Let R be a ring, and e an idempotent element of R. Show that eRe is a subring of R with e as identity.

7.5.18 Suppose that e is idempotent. Show that $1 - e$ is also idempotent.

7.5.19 Let R be a ring with identity. Let $a, \ b \in R$ be such that $1 - ab$ is a unit. Show that $1 - ba$ is also a unit, and

$$(1 - ba)^{-1} = 1 + b(1 - ab)^{-1}a.$$

7.5.20 Show by means of an example that radical of distinct ideals may be same.

7.5.21 Show that under the correspondence theorem radical ideals correspond.

7.5.22 Show that radical of radical of A is radical of A.

7.5.23 Show that the ideal generated by an element a of the ring R is $\{na + xa \mid n \in \mathbb{Z} \text{ and } x \in R\}$.

7.5.24 Let R be a ring. Consider the ring $R \times R$ with coordinate-wise addition and multiplication. Let T be an equivalence relation on R which is a subring of $R \times R$ (such an equivalence relation is called a congruence). Show that T_0 is an ideal. Conversely, suppose that B is an ideal of R. Consider the relation T on R given by '$T = \{(a, b) \in R \times R \mid (a - b) \in B$'. Show that T is a congruence such that $T_0 = B$.

7.5.25 Let f be a map from a ring R_1 to a ring R_2. Show that f is a homomorphism if and only if $\{(x, f(x)) \mid x \in R_1\}$ is a subring of $R_1 \times R_2$.

7.5.26 Describe the notion of composition series for rings. Establish the Jordan–Holder theorem for rings.

7.5.27 Describe the concept of indecomposable rings. Give some examples of indecomposable rings.

7.5.28 Discuss the Krull–Remauk–Schmidt theorem for rings.

7.5.29 Let A be an ideal of R. Let $r(A) = \{x \in R \mid xa = 0 \text{ for all } a \in A\}$. Show that $r(A)$ is a left ideal of R.

7.5.30* Let $C[a, b]$ denote the ring of all real-valued continuous functions on $[a, b]$. Let A be a proper ideal of $C[a, b]$. Show that there exists an element $x \in [a, b]$ such that all members of A vanish at x. Deduce that the maximal ideals of $C[a, b]$ are of the form $M_x = \{f \in C[a, b] \mid f(x) = 0\}$. Show that the map $x \rightsquigarrow M_x$ defines a bijective map from $[a, b]$ to the set $Max(C[a, b])$ of maximal ideals. Interpret the induced topology on $Max(C[a, b])$.

Ordered Integral Domain and Fields

Definition 7.5.44 Let R be a commutative integral domain. An **order structure** on R is a subset P of R which satisfies the following conditions:

(i) $x + y \in P$ for all $x, y \in P$.
(ii) $x \cdot y \in P$ for all $x, y \in P$.
(iii) For any $x \in R$, one and only one of the following holds:

(a) $x = 0$.
(b) $x \in P$.
(c) $-x \in P$.

Thus, $0 \notin P$. The set P is called the set of **positive** elements of R.

7.5.31 Let R be an ordered commutative integral domain with the order structure P on it. Show that

(i) $1 \in P$.
(ii) $-1 \notin P$.
(iii) If $a \neq 0$, then $a^2 \in P$.
(iv) If $a_1^2 + a_2^2 + \cdots + a_n^2 = 0$, then $a_i = 0$ for all i.

Deduce that the field $\mathbb{C}$ of complex numbers cannot be given an order structure.

7.5.32 Show that no finite integral domain can be given an order structure.

7.5.33 Show that every ordered integral domain is of characteristic 0.

7.5.34 Show that $\mathbb{N}$ is an order structure on $\mathbb{Z}$.

7.5.35 Show that every commutative ordered integral domain contains a copy of $\mathbb{Z}$. Deduce that $\mathbb{Z}$ is the smallest ordered integral domain.

7.5.36 Show that $\mathbb{Q}$ is an ordered field.

7.5.37 Let P be an order structure on a commutative integral domain R. Define a relation $<$ and $\leq$ on R as follows:

$$x < y \text{ if and only if } y - x \in P,$$

and

$$x \leq y \text{ if and only if } y - x \in P \bigcup \{0\}.$$

Show that $<$ is nonreflexive, antisymmetric, and transitive. Show that $(R, \leq)$ is a totally ordered set.

7.5.38 Let Γ denote the set of all Cauchy sequences in $\mathbb{Q}$ (see Sect. 3.7 for the definition). Show that Γ is a commutative ring with identity with respect to the pointwise addition and multiplication.

7.5.39 Define a map ϕ from $\mathbb{Q}$ to Γ by $\phi(r)(n) = r$ *for all* $n \in \mathbb{N}$. Show that ϕ is an embedding.

7.5.40 Let $\aleph$ denote the set of all null sequences (see Sect. 3.7). Show that $\aleph$ is an ideal of Γ.

7.5.41 Let $f \in \Gamma - \aleph$. Show that *there exist* $r \in \mathbb{Q}$, $r > 0$ and $n_0 \in \mathbb{N}$ such that $|f(n)| > r$ *for all* $n \geq n_0$.

7.5.42 Use the above exercise to show that $\aleph$ is a maximal ideal of Γ. Deduce that $\Gamma/\aleph$ is a field. Show that this is the field $\mathbb{R}$ of real numbers as introduced in Sect. 3.7. The field $\Gamma/\aleph$ is called the field $\mathbb{R}$ of real numbers.

7.5.43 Show that the map Φ from $\mathbb{Q}$ to $\mathbb{R}$ defined by $\Phi(r) = \phi(r) + \aleph$ is an embedding.

7.5.44 Let P denote the set

$$\{f + \aleph \mid \text{there exist } r \in \mathbb{Q}, r > 0 \text{ and } n_0 \in \mathbb{N} \text{ such that } n \geq n_0 \text{ implies that } f(n) > r\}.$$

Show that P is an order structure on $\mathbb{R}$.

7.5.45 Recall that a partially ordered set $(X, \leq)$ is said to an order complete set if every nonempty set which has an upper bound has a least upper bound. Show that $\mathbb{R}$ is an order complete field with respect to the order induced by the order structure P on $\mathbb{R}$.

7.6   Polynomial Ring

Let R' be a ring with identity containing a ring R with identity as a subring. We assume that the identity of R is same as that of R'. We also call R' a ring extension of R. Let S be a subset of R' which commute with R element wise. Let $< S >$ denote the sub-semigroup of the multiplicative semigroup R' which contains the identity of R' and which is generated by S. More explicitly, $< S >$ denotes the set of finite products of nonnegative integral powers of elements of S. Then, the subring $R(S)$ of R' generated by $R \bigcup S$ is given by

$$\{\Sigma_{g\in<S>}\alpha_g g \mid \alpha_g \in R \text{ and } g \in< S > \text{ with } \alpha_g =$$
$$0 \text{ for all but finitely many } g\}.$$

More generally, and more formally, let R be a ring with identity 1, and G be a semigroup with identity e. Let $R(G)$ denote the set of all maps from G to R which are zero at all but finitely many members of G. Thus, $f \in R(G)$ means that f is a map from G to R for which there is a finite subset F of G such that $f(x) = 0$ for all $x \in G - F$. Let $f, g \in R(G)$. Define a map $f + g$ from G to R by $(f + g)(x) = f(x) + g(x)$. If $f(x) = 0$ for all $x \in G - F_1$, and $g(x) = 0$ for all $x \in G - F_2$, where F_1 and F_2 are finite sets, then $F_1 \bigcup F_2$ is a finite set, and $(f + g)(x) = 0$ for all $x \in G - (F_1 \bigcup F_2)$. Thus, $(f + g) \in R(G)$. This defines a binary operation $+$ in $R(G)$ with respect to which it is an abelian group (verify). Further, define a map $f \cdot g$ from G to R by

$$(f \cdot g)(x) = \Sigma_{y \cdot z = x} f(y) g(z).$$

Observe that the sum in the right-hand side is essentially finite, for f and g both are zero at all but finitely many members of G. Further, if f is zero outside F_1, and g is zero outside F_2, then $f \cdot g$ is zero outside $F_1 \cdot F_2 = \{a \cdot b \mid a \in F_1, b \in F_2\}$. This gives us another binary operation $\cdot$ on $R(G)$. We show that $(R(G), +, \cdot)$ is a ring with identity.

$$((f \cdot g) \cdot h)(x) = \Sigma_{y \cdot z = x}(f \cdot g)(y)h(z) =$$
$$\Sigma_{y \cdot z = x}(\Sigma_{u \cdot v = y}f(u)g(v)) \cdot h(z) = \Sigma_{(u \cdot v) \cdot z = x}f(u)g(v)h(z) =$$
$$\Sigma_{u \cdot t = x}f(u)(\Sigma_{v \cdot z = t}g(v)h(z)) = \Sigma_{u \cdot t = x}f(u)(g \cdot h)(t) = (f \cdot (g \cdot h))(x).$$

This shows that $\cdot$ is associative. Similarly, we can show that $\cdot$ distributes over $+$. The map ℓ from G to R defined by $\ell(e) = 1$ and $\ell(x) = 0$ for all $x \neq e$ is the identity. Thus, $(R(G), +, \cdot)$ is a ring with identity. This ring is called the **semigroup ring** of the ring R over the semigroup G. If G is a group, then we term it as a **group ring**.

Define a map ϕ from R to $R(G)$ by $\phi(a)(e) = a$ and $\phi(a)(x) = 0$ for all $x \neq e$. It is easy to check that ϕ is an embedding of the ring R into the semigroup ring $R(G)$.

Next, define a map ψ from G to $R(G)$ by $\psi(g)(g) = 1$ and $\psi(g)(x) = 0$ for all $x \neq g$. ψ can also be seen to be an embedding of the semigroup G into the semigroup $(R(G), \cdot)$. If we identify R as a subring of $R(G)$ through ϕ and G as sub semigroup of $R(G)$ through ψ (i.e., we identify $\phi(a)$ with a and $\psi(g)$ with g), then every element of $R(G)$ is of the form

$$\Sigma_{g\in G}\alpha_g g,$$

where the sum is essentially a finite sum in the sense that $\alpha_g = 0$ for all but finitely many $g \in G$ (this element is the representation of the map α from G to R given by $\alpha(g) = \alpha_g$). The addition $+$ and the multiplication $\cdot$ are given by

$$\Sigma_{g\in G}\alpha_g g + \Sigma_{g\in G}\beta_g g = \Sigma_{g\in G}(\alpha_g + \beta_g)g,$$

and

$$(\Sigma_{g \in G}\alpha_g g) \cdot (\Sigma_{g \in G}\beta_g g) = \Sigma_{g \in G}(\Sigma_{h \cdot k = g}\alpha_h \beta_k)g.$$

Clearly, under the identification, the semigroup G is a sub-semigroup of the multiplicative semigroup of $R(G)$, and the group $U(G)$ of units of G is a subgroup of the group $U(R(G))$ of units of $R(G)$. In particular, any group G is a subgroup the group $U(R(G))$.

Example 7.6.1 Let us describe the group ring $\mathbb{R}(G)$ of the field $\mathbb{R}$ over the group $G = \{e, g\}$ of order 2. Clearly,

$$\mathbb{R}(G) = \{ae + bg \mid a, b \in \mathbb{R}\}.$$

The addition + is given by

$$(ae + bg) + (ce + dg) = (a+c)e + (b+d)g$$

and the multiplication $\cdot$ is given by

$$(ae + bg) \cdot (ce + dg) = (ac + bd)e + (ad + bc)g.$$

$1e$ is the multiplicative identity. An element $ae + bg$ is a unit if and only if the system

$$aX + bY = 1$$

$$bX + aY = 0$$

of linear equations over $\mathbb{R}$ has a unique solution. Thus, $ae + bg$ is a unit if and only if $a^2 - b^2 \neq 0$, and then, $\frac{a}{a^2-b^2}e - \frac{b}{a^2-b^2}g$ is the inverse of $ae + bg$. The group $U(\mathbb{R}(G))$ of units of $\mathbb{R}(G)$ is given by $U(\mathbb{R}(G)) = \{ae + bg \mid a^2 - b^2 \neq 0\}$. Observe that $\mathbb{R}(G) - U(\mathbb{R}(G))$ is the set of all zero divisors. In particular, $1e + 1g$ is a zero divisor.

Example 7.6.2 Let us describe the group ring $\mathbb{Z}(V_4)$ of the ring $\mathbb{Z}$ over the Klein's four group V_4. Clearly,

$$\mathbb{Z}(V_4) = \{\alpha_0 e + \alpha_1 a + \alpha_2 b + \alpha_3 c \mid \alpha_i \in \mathbb{Z}\},$$

where $V_4 = \{e, a, b, c\}$ is the Klein's four group. The addition + is given by

$$(\alpha_0 e + \alpha_1 a + \alpha_2 b + \alpha_3 c) + (\beta_0 e + \beta_1 a + \beta_2 b + beta_3 c) =$$
$$(\alpha_0 + \beta_0)e + (\alpha_1 + \beta_1)a + (\alpha_2 + \beta_2)b + (\alpha_3 + \beta_3)c.$$

The product $\cdot$ is given by

$$(\alpha_0 e + \alpha_1 a + \alpha_2 b + \alpha_3 c) \cdot (\beta_0 e + \beta_1 a + \beta_2 b + beta_3 c) = \gamma_0 e + \gamma_1 a + \gamma_2 b + \gamma_3 c.$$

where $\gamma_0 = \alpha_0\beta_0 + \alpha_1\beta_1 + \alpha_2\beta_2 + \alpha_3\beta_3$, $\gamma_1 = \alpha_0\beta_1 + \alpha_1\beta_0 + \alpha_2\beta_3 + \alpha_3\beta_2$, $\gamma_2 = \alpha_0\beta_2 + \alpha_2\beta_0 + \alpha_1\beta_3 + \alpha_3\beta_1$, $\gamma_3 = \alpha_0\beta_3 + \alpha_3\beta_0 + \alpha_1\beta_2 + \alpha_2\beta_1$. Find out the group $U(\mathbb{Z}(V_4))$ of units of $\mathbb{Z}(V_4)$.

Now, we study a very particular but universal semigroup ring $R(G)$, where G is an infinite cyclic semigroup with identity which is generated by a symbol X. Thus, $G = <X> = \{X^n \mid n \in \mathbb{N} \cup \{0\}\}$ with the understanding that $X^n = X^m$ if and only if $n = m$. The operation $\cdot$ in G is given by $X^n \cdot X^m = X^{n+m}$. The element X^0 is the identity of G, and it is denoted by e. Note that any two such semigroups are isomorphic. Indeed, G is isomorphic to the semigroup $\mathbb{N} \cup \{0\}$ with usual addition. The symbol X is called an **indeterminate** or **transcendental** element. The semigroup ring $R(G)$ is given by

$$R(G) = \{a_0 + a_1 X + a_2 X^2 + \cdots + a_n X^n \mid a_i \in R\}.$$

Clearly, this ring is generated by $R \cup \{X\}$. We denoted this ring by $R[X]$ and call it the **polynomial ring** over R in one indeterminate. The elements of $R[X]$ are called **polynomials** in one variable. We note the following.

(i) $a_0 + a_1 X + a_2 X^2 + \cdots + a_n X^n = 0$ *if and only if* $a_i = o$ *for all* i.
(ii) $a_0 + a_1 X + \cdots + a_n X^n = b_0 + b_1 X + \cdots + b_n X^n$ *if and only if* $a_i = b_i$ *for all* i.
(iii) $a_0 + a_1 X + \cdots + a_n X^n + b_0 + b_1 X + \cdots + b_m X^m$

$$= (a_0 + b_0) + (a_1 + b_1)X + \cdots + (a_n + b_n)X^n + b_{n+1}X^{n+1} + \cdots + b_m X^m,$$
where $n \leq m$.
(iv) $(a_0 + a_1 X + \cdots + a_n X^n) \cdot (b_0 + b_1 X + \cdots + b_m X^m)$

$$= c_0 + c_1 X + \cdots + c_{n+m} X^{n+m}, \text{ where } c_i = \Sigma_{j+k=i} a_j b_k.$$

If $f(X) = a_0 + a_1 X + \cdots + a_n X^n$, where $a_n \neq 0$, then $a_n X^n$ is called the **leading term** of the polynomial, a_n is called the **leading coefficient**, and n is called the **degree** of the polynomial. Thus, degree of every nonzero polynomial is defined. It can be considered as a map from $R[X]^\star = R[X] - \{0\}$ to $\mathbb{N} \cup \{0\}$. The degree of a nonzero polynomial $f(X)$ is denoted by $deg(f(X))$. Degree of zero polynomial is not defined.

The proof of the following proposition is straightforward verification.

Proposition 7.6.3 *The polynomial ring $R[X]$ is commutative if and only if R is commutative.*   ♯

The proof of the following proposition is immediate from the definition of addition and multiplication of polynomials.

Proposition 7.6.4 *Let $f(X)$ and $g(X)$ be nonzero polynomials such that $f(X) + g(X) \neq 0$. Then, $deg(f(X)+g(X)) \leq max(deg(f(X)), deg(g(X)))$. If $(f(X) \cdot g(X)) \neq 0$, then $deg(f(X) \cdot g(X)) \leq deg(f(X)) + deg(g(X))$.*   ♯

In general, strict inequality may hold. For example, take $f(X) = 1+X$ and $g(X) = 2-X$ in $\mathbb{Z}[X]$. Then, $deg(fX)) = 1 = deg(g(X))$, whereas $deg(f(X)+g(X)) = 0$. Also, consider the nonzero polynomials $f(X) = \bar{1} + \bar{3}X$ and $g(X) = \bar{1} + \bar{2}X$ in $\mathbb{Z}_6[X]$. Then, $f(X) \cdot g(X) = \bar{1} + \bar{5}X$ and so $deg(f(X) \cdot g(X)) = 1$ whereas $deg(f(X)) + deg(g(X)) = 2$. However, we have the following:

Proposition 7.6.5 *If R is an integral domain, then $deg(f(X) \cdot g(X)) = deg(f(X)) + deg(g(X))$.*

Proof If the leading coefficient of $f(X)$ is $a_n \neq 0$ and that of $g(X) = b_m \neq 0$, then the leading term of $f(X) \cdot g(X)$ is $a_n b_m X^{n+m}$, and since R is an integral domain $a_n \cdot b_m \neq 0$. Thus, $deg(f(X) \cdot g(X)) = deg(f(X)) + deg(g(X))$.   ♯

Proposition 7.6.6 *$R[X]$ is an integral domain if and only if R is an integral domain.*

Proof If R is an integral domain, then the proof of the above proposition says that $R[X]$ is also an integral domain. Conversely, if $R[X]$ is an integral domain, then R being a subring of $R[X]$ is an integral domain.   ♯

Proposition 7.6.7 *Let R be an integral domain with identity. Then, units of $R[X]$ are those of R. Thus, $U(R[X]) = U(R)$.*

Proof Since R is a subring of $R[X]$ and the identity of R is that of $R[X]$, units of R are also units of $R[X]$. Let $f(X)$ be a unit of $R[X]$. Then, *there exists $g(X) \in R[X]$* such that $f(X) \cdot g(X) = 1$. Comparing the degrees of both the sides, we see that $deg(f(X)) = 0 = deg(g(X))$. Hence $f(X)$ and $g(X)$ both belong to R. This shows that units of $R[X]$ are also units of R.   ♯

Corollary 7.6.8 *$R[X]$ can never be a field.*

Proof $X \neq 0$, and it cannot be a unit of $R[X]$.   ♯

Remark 7.6.9 If R is not an integral domain, then there may be units of $R[X]$ which are not in R. For example in $\mathbb{Z}_4[X]$, $(\bar{1}+\bar{2}X) \cdot (\bar{1} + \bar{2}X) = \bar{1}$. Thus, $\bar{1} + \bar{2}X$ is a unit in $\mathbb{Z}_4[X]$ which is not in $\mathbb{Z}_4$.

Theorem 7.6.10 (Division Algorithm) *Let R be a commutative integral domain with identity. Let $f(X), g(X) \in R[X]$. Suppose that $g(X) \neq 0$, and the leading coefficient of $g(X)$ is a unit. Then, there exists unique pair $(q(X), r(X))$ in $R[X] \times R[X]$ such that*

$$f(X) = q(X)g(X) + r(X),$$

where $r(X) = 0$ or else $deg(r(X)) < deg(g(X))$.

Proof If $deg(g(X)) = 0$, then $g(X) = a \in R$. Since the leading coefficient of $g(X)$ is assumed to be a unit, $g(X) = a$ is a unit of R. But then

$$f(X) = a^{-1}f(X) \cdot g(X) + 0,$$

and there is nothing to do. Next, suppose that $deg(g(X)) > 0$. If $f(X) = 0$, then

$$f(X) = 0 \cdot g(X) + 0,$$

and again, there is nothing to do. Suppose that $f(X) \neq 0$. The proof in this case is by the induction on the degree of $f(X)$. If $deg(f(X)) = 0 < deg(g(X))$, then

$$f(X) = 0 \cdot g(X) + f(X),$$

where $deg(f(X)) < deg(g(X))$. Thus, the result is true whenever $deg(f(X)) = 0$. Assume that the result is true for all those polynomials whose degree is less than the degree of $f(X)$. Then, we have to prove it for $f(X)$. If $deg(f(X)) < deg(g(X))$, then again

$$f(X) = 0 \cdot g(X) + f(X),$$

and so there is nothing to do. Suppose that $deg(f(X)) \geq deg(g(X))$. Let

$$f(X) = a_0 + a_1 X + \cdots a_r X^r,$$

where $a_r \neq 0$, and

$$g(X) = b_0 + b_1 X + \cdots b_s X^s,$$

where b_s is a unit and $r \geq s$. Consider the polynomial

$$f_1(X) = f(X) - a_r b_s^{-1} X^{r-s} \cdot g(X).$$

If $f_1(X) = 0$, then there is nothing to do. If not, $deg(f_1(X)) < deg(f(X))$. By the induction assumption *there exist* $q_1(X)$ and $r(X)$ such that

$$f_1(X) = q_1(X)g(X) + r(X),$$

where $r(X) = 0$ or else $deg(r(X)) < deg(g(X))$. But then

$$f(X) = f_1(X) + a_r b_s^{-1} X^{r-s} g(X) =$$
$$(q_1(X) + a_r b_s^{-1} X^{r-s})g(X) + r(X) = q(X)g(X) + r(X),$$

where $r(X) = 0$ or else $deg(r(X)) < deg(g(X))$.

Finally, we prove the uniqueness of the pair $(q(X), r(X))$. Suppose that

$$f(X) = q_1(X)g(X) + r_1(X)$$

and

$$f(X) = q_2(X)g(X) + r_2(X),$$

where $r_1(X) = 0$ or else $deg(r_1(X)) < deg(g(X))$ and $r_2(X) = 0$ or else $deg(r_2(X)) < deg(g(X))$. Then,

$$(q_1(X) - q_2(X))g(X) = (r_2(X) - r_1(X)).$$

Suppose that $r_1(X) \neq r_2(X)$. Then, $(r_2(X) - r_1(X)) \neq 0$. Since R is an integral domain, comparing the degrees,

$$deg(q_1(X) - q_2(X)) + deg(g(X)) = deg(r_2(X) - r_1(X)).$$

Since

$$deg(r_2(X) - r_1(X)) \leq max(deg(r_2(X)), deg(-r_1(X))) < deg(g(X)),$$

we arrive at a contradiction. Thus, $r_1(X) = r_2(X)$, and since $g(X) \neq 0$, $q_1(X) = q_2(X)$.   ♯

The polynomial $q(X)$ in the above theorem is called the **quotient** and $r(X)$ is called the **remainder** when $f(X)$ is divided by $g(X)$.

Since every nonzero element of a field is a unit, we have the following.

Corollary 7.6.11 *Let F be a field. Let $f(X)$ and $g(X)$ be polynomials in $F[X]$ such that $g(X) \neq 0$. Then, there exists a unique pair $(q(X), g(X))$ in $F[X] \times F[X]$ such that*

$$f(X) = q(X)g(X) + r(X),$$

where $r(X) = 0$ or else $deg(f(X)) < deg(g(X))$.   ♯

The proof of the above theorem gives an algorithm to find the quotient and the remainder. We illustrate it by means of the following example.

Example 7.6.12 Let $f(X) = 2 + 5X + 8X^2 + 4X^3$ and $g(X) = 2 + 3X + X^2$ be two polynomials in $\mathbb{Z}[X]$. The leading coefficient of $g(X)$ is 1 which is a unit. Now, $f_1(X) = f(X) - 4Xg(X) = 2 - 3X - 4X^2$. Further, $f_1(X) + 4g(X) = 10 + 9X$. Hence

$$f(X) = (4 - 4X)g(X) + (10 + 9X).$$

Thus, the quotient is $4 - 4X$ and the remainder is $10 + 9X$.

Remark 7.6.13 The assumption that the leading coefficient of $g(X)$ is a unit is essential. For example, consider $f(X) = 1 + 2X + 3X^2$ and $g(X) = 3 + 6X$ in $\mathbb{Z}[X]$. Suppose we have pairs $(q(X), r(X))$ in $\mathbb{Z}[X] \times \mathbb{Z}[X]$ such that

$$f(X) = q(X)g(X) + r(X),$$

where $r(X) = 0$ or else $deg(r(X)) < deg(g(X))$. But then $q(X) = a + bX$ for some $a, b \in \mathbb{Z}$ and $r(X) = c \in \mathbb{Z}$. This gives $6b = 3$ which is impossible.

Let $f(X) = a_0 + a_1X + \cdots a_rX^r$ be a polynomial in $R[X]$ and $\alpha \in R$. Then,

$$f(\alpha) = a_0 + a_1\alpha + a_2\alpha^2 + \cdots a_r\alpha^r$$

is called the **specialization or evaluation** of $f(X)$ at α. For each $\alpha \in R$, we have the evaluation map e_α from $R[X]$ to R defined by $e_\alpha(f(X)) = f(\alpha)$. The map e_α is a ring homomorphism from $R[X]$ to R (verify) which is surjective.

Corollary 7.6.14 (Remainder Theorem) *Let R be a commutative integral domain with identity. Let $f(X) \in R[X]$ and $a \in R$. Then, there exists unique $q(X) \in R[X]$ such that*

$$f(X) = q(X) \cdot (X - a) + f(a)$$

Proof By the division algorithm, *there exist* $q(X), r(X) \in R[X]$ such that

$$f(X) = q(X)(X - a) + r(X),$$

where $r(X) = 0$ or else $deg(r(X)) < 1$. But then $r(X) = r \in R$. Specializing the equation at a, we get $f(a) = q(a) \cdot (a - a) + r = r$. Thus, $f(X) = q(X)(X - a) + f(a)$. ♯

Corollary 7.6.15 *$ker\, e_\alpha$ of the evaluation map e_α is $R[X] \cdot (X-\alpha) = \{f(X) \cdot (X-\alpha) \mid f(X) \in R[X]\}$.* ♯

We say that a polynomial $q(X) \in R[X]$ divides a polynomial $f(X)$ if *there exists* $q(X) \in R[X]$ such that $f(X) = q(X)g(X)$. An element $a \in R$ is said to be a **root or a zero** of $f(X)$ if $f(a) = 0$. The following corollary is immediate from the remainder theorem.

Corollary 7.6.16 (Factor Theorem) *Let R be a commutative integral domain with identity and $f(X) \in R[X]$. Then, $X - a$ divides $f(X)$ if and only if a is a root of $f(X)$.* ♯

Corollary 7.6.17 *let R be a commutative integral domain with identity and $f(X) \in R[X]$. Let $a_1, a_2, \ldots, a_r$ be distinct roots of $f(X)$. The $(X - a_1)(X - a_2) \cdots (X - a_r)$ divides $f(X)$.*

Proof The proof is by induction on r. If $r = 1$, then it reduces to the above corollary. Suppose that the result is true for r. Let $a_1, a_2, \ldots, a_r, a_{r+1}$ be distinct roots of $f(X)$. Then, by the induction hypothesis, $(X - a_1)(X - a_2) \cdots (X - a_r)$ divides $f(X)$. Suppose that

$$f(X) = q(X)(X - a_1)(X - a_2) \cdots (X - a_r).$$

Evaluating at a_{r+1}, we get $0 = f(a_{r+1}) = q(a_{r+1})(a_{r+1} - a_1) \cdots (a_{r+1} - a_r)$. Since $a_{r+1} \neq a_i$ for all $i \leq r$ and R is an integral domain, $q(a_{r+1}) = 0$. Again by the above corollary $X - a_{r+1}$ divides $q(X)$. Hence, $(X - a_1)(X - a_2) \cdots (X - a_{r+1})$ divides $f(X)$. ♯

Corollary 7.6.18 *Let R be as above and $f(X)$ a polynomial of degree n. Then, $f(X)$ can have at most n distinct roots.*

Proof If $f(X)$ has r distinct roots $a_1, a_2, \ldots, a_r$, then from the above corollary

$$f(X) = q(X)(X - a_1)(X - a_2) \cdots (X - a_r).$$

Since R is an integral domain, comparing the degrees, we get that $deg(f(X)) \geq r$. ♯

Remark 7.6.19 If R is not an integral domain, then a polynomial of degree n may have more than n distinct roots: For example, $\bar{2} + \bar{2}X$ in $\mathbb{Z}_6[X]$ has $\bar{2}$ and $\bar{5}$ both as roots.

Theorem 7.6.20 *Every finite subgroup of the multiplicative group F^* of a field F is cyclic.*

Proof Let G be a subgroup of F^* of order n. Then, for each divisor d of n, the equation $X^d - 1 = 0$ has at most d solutions(follows from the above corollary). From the illustration 1.4 of Chap. 5, G is cyclic.   ♯

Remark 7.6.21 In fact, every finitely generated subgroup of the multiplicative group of a field is cyclic.

Corollary 7.6.22 *The multiplicative group of a finite field is cyclic.*   ♯

In particular,

Corollary 7.6.23 *The group $U_p = Z_p^*$ of prime residue classes modulo a prime p is cyclic of order $p - 1$.*   ♯

How to find a generator of U_p? This problem will be addressed later. One may ask another natural question: What are m for which U_m is cyclic? Indeed, U_m is cyclic if and only if $m = 2, 4, p^n$, or $2p^n$, where p is an odd prime. The proof of this fact will also follow in Algebra 2.

Remark 7.6.24 Further arithmetical properties of polynomial rings will be discussed in Chap. 11.

Exercises

7.6.1 Let R be a commutative ring with identity and G a semigroup with identity e. ϕ and ψ the embedding of R and G, respectively(observe that $\phi(1) = \psi(e)$). Let R' be a ring with identity together with a ring homomorphism η from R to R' and an identity preserving semigroup homomorphism ρ from $(G, \cdot)$ to $(R', \cdot)$ such that $\eta(1) = \rho(e) = 1'$ the identity of R'. Show that there exists a unique ring homomorphism μ from $R(G)$ to R' such that $\mu o \phi = \eta$ and $\mu o \psi = \rho$.

7.6.2 Let G be a cyclic group. Show that $R(G)$ is homomorphic image of $R(\mathbb{Z})$.

7.6.3 Find the number of elements in $\mathbb{Z}_p(G)$, where $\mid G \mid = n$. Is it an integral domain?

7.6.4 Describe the group ring $\mathbb{Z}(G)$, where G is a cyclic group generated by g. Describe its group of units.

7.6.4 Find the group of units of $\mathbb{Z}_p(G)$, where p is a prime and G is a cyclic group of order p.

7.6.5 Let G be a finite group and R a commutative integral domain with identity. Show that $e + g$ is a zero divisor for all $g \in G$.

7.6.6 Show that the group ring $\mathbb{R}(G)$ described in Example 7.6.1 is isomorphic to the subring of $M_2(\mathbb{R})$ consisting of the matrices of the type

$$\begin{bmatrix} a & b \\ b & a \end{bmatrix}.$$

Hence, describe its group of units as a matrix group.

7.6.7 Describe the group ring $\mathbb{R}(G)$, where G is a cyclic group of order 3. What is the group of units?

7.6.8 Describe the group ring $\mathbb{Z}(G)$, where G is a cyclic group of order 3. What is the group of units?

7.6.9 Describe the group ring $\mathbb{R}(G)$, where G is infinite cyclic group. What is the group of units?

7.6.10 Let G be a group. Show that the map ϵ from $\mathbb{Z}(G)$ to $\mathbb{Z}$ defined by $\epsilon(\Sigma_{g \in G} a_g g) = \Sigma_{g \in G} a_g$ is a ring homomorphism. This map is called the **augmentation** map. The kernel of this map is called the **augmentation ideal**. Show that the augmentation ideal is generated by the set $\{e - g \mid g \in G\}$.

7.6.11 Let R be a commutative ring with identity and G a semigroup with identity. Show that the ring $M_n(R(G))$ of $n \times n$ matrices with entries in the semigroup ring $R(G)$ is tautologically isomorphic to the semigroup ring $M_n(R)(G)$.

7.6.12 Let G be a semigroup with identity such that for any $x \in G$, there are only finitely many pairs $y, z \in G$ such that $yz = x$ (e.g., $\mathbb{N} \bigcup \{0\}$). Let $R((G))$ denote the set of all maps from G to R. Show that $R((G))$ is a ring with respect to the point wise addition and the product defined by

$$(f \cdot g)(x) = \Sigma_{y \cdot z = x} f(y) g(z),$$

and which contains $R(G)$ as a subring.

7.6.13 Describe the centers of the group rings $\mathbb{C}(S_3)$ and $\mathbb{C}(Q_8)$.

7.6.14 Let R be a commutative ring with identity, and R' a ring. Let $\alpha \in R'$, and η a homomorphism from R to R' such that α commutes with each element of $\eta(R)$. Show that η has unique extension ρ to $R[X]$ subject to the condition $\rho(X) = \alpha$.

7.6.15 Let $f(X) = 2 + 5X + 6X^2 + 8X^4$, and $g(X) = 3 + 5X + 5X^2$. Find $f(X) + g(X)$ and $f(X) \cdot g(X)$. Show that $f(X)$ cannot be expressed as $f(X) = q(X)g(X) + r(X)$, where $r(X) = 0$ or else $deg(r(X)) < deg(g(X))$.

7.6.16 Find the remainder when $1 + 6X + 8x^2 + 5X^3$ is divided by $X + 2$.

7.6.17 Show that $X^2 - 2X + 2$ has no root in $\mathbb{Q}$.

7.6.18 Show that $X^2 + X + \bar{1}$ has no root in $\mathbb{Z}_2$.

7.6.19 Show that $X^2 + \bar{1}$ has no root in $\mathbb{Z}_7$.

7.6.20 Show that the polynomial ring $\mathbb{Z}_p[X]$ is an infinite integral domain of characteristic $p \neq 0$. Give an example of an infinite field of characteristic $p \neq 0$.

7.6.21 Show that $f(X) - f(a)$ is divisible by $(X - a)$ for all polynomial $f(X)$. Deduce that $(X - a)$ divides $X^n - a^n$ for all $n \geq 1$.

7.6.22 Suppose that F is an infinite field. Suppose that $f(a) = 0$ for all $a \in F$. Show that $f(X) = 0$.

7.6.23 Show that the ideal of $\mathbb{R}[X]$ generated by $X^2 + 1$, where $\mathbb{R}$ is the field of real numbers, is a maximal ideal.
Hint. If A is an ideal containing $\mathbb{R}[X] \cdot (X^2 + 1)$ properly, then *there exists $f(X) \in A$* such that $X^2 + 1$ does not divide $f(X)$. Apply division algorithm to show that $X + \alpha \in A$ for some $\alpha \in \mathbb{R}$. Again, apply division algorithm to show that $\alpha^2 + 1 \in A$.

7.6.24 Show that $\mathbb{R}[X]/(\mathbb{R}[X](X^2+1))$ is a field isomorphic to the field $\mathbb{C}$ of complex numbers.
Hint. Consider the map $f(X) \mapsto f(i)$ from $\mathbb{R}[X]$ to the field $\mathbb{C}$. Show that it is surjective homomorphism. Find the kernel.

7.6.25 Show that $\mathbb{Z}_7[X]/(\mathbb{Z}_7[X](X^2 + \bar{1}))$ is a field of order 49.

7.6.26 Show that $\mathbb{Z}_2[X]/(\mathbb{Z}_2[X](X^2 + X + \bar{1}))$ is a field of order 4.

7.6.27 Define a map D from $R[X]$ to $R[X]$ as follows: Let $f(X) = a_0 + a_1 X + \cdots + a_n X^n \in R[X]$. Define

$$D(f(X)) = a_1 + 2a_2 X + 3a_3 X^2 + \cdots na_n X^{n-1}.$$

Show that D is a derivation in the sense that

(i) $D(af(X) + bg(X)) = aD(f(X)) + bD(g(X))$,
and

(ii) $D(f(X) \cdot g(X)) = D(f(X)) \cdot g(X) + f(X) \cdot D(g(X))$.

7.6.28 Define D^r inductively as follows:
Define $D^0(f(X)) = f(X)$, $D^1(f(X)) = D(f(X))$. Supposing that $D^r(f(X))$ has already been defined, define $D^{r+1}(f(X)) = D(D^r(f(X)))$. Let $f(X)$ be a polynomial in $R[X]$ of degree n. Show that for each $r \leq n$, there exists a unique polynomial $g(X)$ in $R[X]$ such that $r!g(X) = D^r(f(X))$. We denote the polynomial $g(X)$ by $\frac{D^r(f(X))}{r!}$.

7.6.29 (Taylor's Formula). Let $f(X) \in R[X]$ be a polynomial of degree n. Show that

$$f(a+h) = f(a) + hD(f(a)) + h^2\frac{D^2(f(a))}{2!} + \cdots h^n\frac{D^n(f(a))}{n!}$$

for all $a, h \in R$.

7.6.30 Call a polynomial $f(X)$ of positive degree irreducible if it cannot be expressed as product of polynomials of lower and positive degrees. Let $f(X) \in F[X]$ be an irreducible polynomial of positive degree over a field F. Let F' be a field containing F as a subfield. Let $\alpha \in F'$, and $f(X) \in F[X]$ be a polynomial of least degree of which α is a root. Show that $f(X)$ is an irreducible polynomial in $F[X]$.

7.6.31 Let F' be a field containing F as a subfield. Let $\alpha \in F'$ and $f(X) \in F[X]$. An element $\alpha \in F'$ is said to be a multiple root of $f(X)$ in F' if $(X - \alpha)^2$ divides $f(X)$ in $F'[X]$. Show that α is a multiple root of $f(X)$ if and only if α is also a root of the derivative $D(f(X))$ of $f(X)$.

7.6.32 Suppose that F is a finite field, or it is a field of characteristic 0. Let F' be a field containing F as a subfield which contains all roots of a nonzero irreducible polynomial $f(X)$ in $F[X]$(we shall see in Algebra 2 in the chapter on Galois theory that such a field always exists). Use the above exercise to show that there are exactly n distinct roots of $f(X)$ in F', where n is the degree of $f(X)$.

7.6.33 Is the above result true for infinite fields of characteristic $p \neq 0$? Support.

7.7 Polynomial Ring in Several Variable

Let W denote the set of all nonnegative integers, and $G = W^r$. Then, G is a semigroup with respect to coordinate-wise addition. The identity being $(0.0. \cdots , 0)$. Let R be a ring with identity. Consider the semigroup ring $R(G)$ and the embeddings ϕ of R and ψ of G as described in the beginning of the previous section. Let us denote $\psi((1, 0, 0, \ldots, 0))$ by X_1, $\psi((0, 1, 0, \ldots, 0))$ by X_2, $\ldots, \psi((0, 0, \ldots, 0, 1))$ by X_r. Identify $\phi(a)$ by a. Then, every element of $R(G)$ is uniquely expressible as

$$\Sigma_{(i_1,i_2,\ldots,i_r)\in W^r}\,\alpha_{i_1 i_2\cdots i_r}X_1^{i_1}X_2^{i_2}\cdots X_r^{i_r}.$$

The ring $R(G)$ is generated by $R\bigcup\{X_1,X_2,\ldots,X_r\}$, and it is denoted by $R[X_1,X_2,\ldots,X_r]$. This ring is called the **polynomial ring in r variables**. $\{X_1,X_2,\ldots,X_r\}$ are called the indeterminates of the polynomial ring. An element of $R[X_1,X_2,\ldots,X_r]$ is usually denoted by $f(X_1,X_2,\ldots,X_r)$. The nonzero expressions $\alpha_{i_1 i_2\cdots i_r}X_1^{i_1}X_2^{i_2}\cdots X_r^{i_r}$ appearing in the polynomial $f(X_1,X_2,\ldots,X_r)$ are called the monomials in the polynomial $f(X_1,X_2,\ldots,X_r)$. The sum $i_1+i_2+\cdots+i_r$ is called the degree of the monomial. The degree of a maximum degree monomial appearing in a nonzero polynomial is called the degree of the polynomial. As in case of one variable polynomial, R is an integral domain if and only if $R[X_1,X_2,\ldots,X_r]$ is an integral domain, and then, the degree of the product of two polynomial is sum of their degrees.

Every nonzero element $f(X_1,X_2)\in R[X_1,X_2]$ is uniquely expressible as a member

$$a_0(X_1)\,+\,a_1(X_1)X_2\,+\,a_2(X_1)X_2^2\,+\,\cdots\,+\,a_r(X_1)X_2^r$$

of $R[X_1][X_2]$. This identification of elements of $R[X_1,X_2]$ as elements of $R[X_1][X_2]$ also respects the operations in the corresponding rings. As such, the ring $R[X_1,X_2]$ is same as $R[X_1][X_2]$. More generally, by induction, $R[X_1,X_2,\ldots,X_{r+1}]=R[X_1,X_2,\ldots,X_r][X_{r+1}]$.

Let R be a commutative ring with identity. Let $Map(R^r,R)$ denote the set of all functions from R^r to R. Clearly, $Map(R^r,R)$ is a ring with respect to pointwise addition and multiplication. Define a map $\wp$ from $R[X_1,X_2,\ldots,X_r]$ to $Map(R^r,R)$ by

$$\wp\,(f(X_1,X_2,\ldots,X_r))((\alpha_1,\alpha_2,\ldots,\alpha_r))\,=\,f(\alpha_1,\alpha_2,\ldots\alpha_r).$$

Since R is commutative, $\wp$ is a ring homomorphism. The elements of the image of $\wp$ are called the **polynomial functions**. They are also called the **regular functions** on R^r. In general, a function need not be a polynomial function. For example, the function f from $\mathbb{R}$ to $\mathbb{R}$ defined by $f(t)=sint$ is not a polynomial function.

Fix an element $\overline{\alpha}=(\alpha_1,\alpha_2,\ldots,\alpha_r)\in R^r$. Define $e_{\overline{\alpha}}$ from $R[X_1,X_2,\ldots,X_r]$ to R by

$$e_{\overline{\alpha}}(f(X_1,X_2,\ldots,X_r))\,=\,f(\alpha_1,\alpha_2,\ldots,\alpha_r).$$

Again, since R is commutative, $e_{\overline{\alpha}}$ is a surjective ring homomorphism. This homomorphism is called the **evaluation** map. It is also called the **specialization** map at $\overline{\alpha}$.

Let F be a field. The set F^m is called an **affine m-space**. This set is denoted by $A^m(F)$. Let $f(X_1,X_2,\ldots,X_m)$ be a polynomial in m variable. An element $\overline{\alpha}$ of $A^m(F)$ is called a **zero** of $f(X_1,X_2,\ldots,X_m)$ if $f(\alpha_1,\alpha_2,\ldots,\alpha_m)=0$, or equivalently, $f(X_1,X_2,\ldots,X_m)\in ker\,e_{\overline{\alpha}}$. The set of all zeros of $f(X_1,X_2,\ldots,X_m)$ is denoted by $V(f)$, and it is called a **hypersurface** of the affine m- space $A^m(F)=F^m$.

Example 7.7.1 The circle $S^1 = \{(x, y) \in \mathbb{R}^2 \mid x^2 + y^2 = 1\}$ is a hypersurface $V(f)$ of $A^2(\mathbb{R})$, where $f(X, Y) = X^2 + Y^2 - 1$. If we treat $f(X, Y)$ as polynomial in $\mathbb{R}[X, Y, Z]$, then $V(f)$ is a right circular cylinder which is a hypersurface in $\mathbb{R}^3$.

If F is a field, then $F[X_1, X_2, \ldots, X_m]$ is an integral domain. Its field of quotients is denoted by $F(X_1, X_2, \ldots, X_m)$. The members of $F(X_1, X_2, \ldots, X_m)$ are of the form $\frac{f(X_1, X_2, \ldots, X_m)}{g(X_1, X_2, \ldots, X_m)})$, where $g(X_1, X_2, \ldots, X_m) \neq 0$. In short, this element is denoted by $\frac{f}{g}$.

Proposition 7.7.2 *Let F be an infinite field, and $f(X_1, X_2, \ldots, X_m)$ be a polynomial in m variable. Then, $V(f) = A^m(F)$ if and only if $f(X_1, X_2, \ldots X_m) = 0$.*

Proof If $f(X_1, X_2, \ldots X_m) = 0$, then clearly $V(f) = A^m(F)$. We prove the converse by the induction on m. If $f(X)$ is a nonzero polynomial in $F[X]$, then the set $V(f)$ of zeros of $f(X)$ is a finite. Since F is infinite, $V(f) \neq A^1(F)$. This proves the result for $m = 1$. Assume the result for m. Let $f(X_1, X_2, \ldots, X_{m+1})$ be a nonzero polynomial in $m + 1$ indeterminates. Then,

$$f(X_1, X_2, \ldots, X_{m+1}) =$$
$$a_0(X_1, \ldots, X_m) + a_1(X_1, \ldots, X_m)X_{m+1} + \cdots + a_r(X_1, \ldots, X_m)X_{m+1}^r,$$

where $a_r(X_1, \ldots, X_m)$ is a nonzero polynomial in m indeterminates and $r \geq 1$. Thus, this polynomial is a nonzero polynomial in one variable over the field K, where $K = F(X_1, X_2, \ldots, X_m)$. Since it has only finitely many zeros in K, there is an element $\alpha_{m+1} \in F$ such that $f(X_1, X_2, \ldots, X_m, \alpha_{m+1}) \neq 0$. The polynomial $g(X_1, X_2, \ldots, X_m) = f(X_1, X_2, \ldots, X_m, \alpha_{m+1})$ is a nonzero polynomial in m indeterminates. By the induction hypothesis, there are elements $\alpha_1, \alpha_2, \ldots, \alpha_m$ in F such that $g(\alpha_1, \alpha_2, \ldots, \alpha_m) \neq 0$. This means that $(\alpha_1, \alpha_2, \ldots, \alpha_{m+1}) \notin V(f)$. ♯

Corollary 7.7.3 *Let F be an infinite field. Then, the map $\wp$ from $F[X_1, X_2, \ldots, X_r]$ to $Map(R^r, R)$ defined by*

$$\wp(f(X_1, X_2, \ldots, X_r))((\alpha_1, \alpha_2, \ldots, \alpha_r)) = f(\alpha_1, \alpha_2, \ldots \alpha_r)$$

is an injective map.

Proof From the above proposition, it follows that the $ker\wp = \{0\}$. ♯

Proposition 7.7.4 *Let F be an algebraically closed field in the sense that every polynomial over F has a root in F (e.g., $\mathbb{C}$). Then, $V(f) \neq \emptyset$ for every polynomial f in $F[X_1, X_2, \ldots, X_m]$.*

Proof We prove the result by the induction on m. If $m = 1$, then the result follows from the assumption that F is algebraically closed field. Assume the result for m. Let $f(X_1, X_2, \ldots, X_{m+1})$ be a polynomial in $F[X_1, X_2, \ldots, X_{m+1}]$. If it is a zero polynomial, then there is nothing to do. Suppose that it is a nonzero polynomial. Then,

$$f(X_1, X_2, \ldots, X_{m+1}) =$$
$$a_0(X_1, \ldots, X_m) + a_1(X_1, \ldots, X_m)X_{m+1} + \cdots + a_r(X_1, \ldots, X_m)X_{m+1}^r,$$

where $a_r(X_1, \ldots, X_m)$ is a nonzero polynomial in m indeterminates and $r \geq 1$. By the induction hypothesis, we have $\alpha_1, \alpha_2, \ldots, \alpha_m$ such that $a_r(\alpha_1, \ldots, \alpha_m) = 0$. Consider the polynomial $g(X_{m+1}) = f(\alpha_1, \alpha_2, \ldots, \alpha_m, X_{m+1})$ in $F[X_{m+1}]$. Since F is algebraically closed field, there is $\alpha_{m+1} \in F$ such that $(\alpha_1, \alpha_2, \ldots, \alpha_m, \alpha_{m+1}) \in V(f)$. ♯

Proposition 7.7.5 *Let F_q be a finite field of order $q = p^n$, $n \geq 1$ (in fact, it will follow (see Algebra 2, Chap. 9) that there is one and only one such field up to isomorphism). Let $f(X_1, X_2, \ldots, X_m)$ be a nonzero polynomial over F_q such that the degree of f in each X_i is less than q. Then, $V(f) \neq F_q^m$, i.e, there are elements $\alpha_1, \alpha_2, \ldots, \alpha_m$ in F_q such that $f(\alpha_1, \alpha_2, \ldots, \alpha_m) \neq 0$.*

Proof The proof is by the induction on m. If $f(X_1)$ is a polynomial over F in one indeterminate, then since $deg f(X_1) < q$, it has at the most q roots. Hence, there is an $\alpha \in F_q$ such that $f(\alpha) \neq 0$. Assume that the result is true for m. Let $f(X_1, X_2, \ldots, X_{m+1})$ be a nonzero polynomial over F_q such that the degree of f in each X_i is less than q. Then,

$$f(X_1, X_2, \ldots, X_{m+1}) =$$
$$a_0(X_1, \ldots, X_m) + a_1(X_1, \ldots, X_m)X_{m+1} + \cdots + a_r(X_1, \ldots, X_m)X_{m+1}^r,$$

where $a_r(X_1, \ldots, X_m)$ is a nonzero polynomial in m indeterminates, $1 \leq r < q$, and the degree of $a_r(X_1, \ldots, X_m)$ in each x_i is less than q. By the induction hypothesis, there are elements $\alpha_1, \alpha_2, \ldots, \alpha_m$ in F_q such that $a_r(\alpha_1, \alpha_2, \ldots, \alpha_m) \neq 0$. In turn, $g(X_{m+1}) = f(\alpha_1, \alpha_2, \ldots, \alpha_m, X_{m+1})$ is a polynomial in one indeterminate, and its degree is less than q. Hence, there is an element $\alpha_{m+1} \in F$ such that $g(\alpha_{m+1}) = f(\alpha_1, \alpha_2, \ldots, \alpha_m, \alpha_{m+1}) \neq 0$. ♯

Corollary 7.7.6 *The map $\wp$ from $F_q[X_1, X_2, \ldots X_m]$ to $Map(F_q^m, F_q)$ is a surjective map, i.e., every map from F_q^m to F_q is determined by a polynomial. Indeed, it is determined by a unique polynomial $f(X_1, X_2, \ldots, X_m)$ having the property that the degree of $f(X_1, X_2, \ldots, X_m)$ in each X_i is less than q.*

Proof Consider the subset

$$A = \{f(X_1, X_2, \ldots, X_m) \mid degree\ of\ f\ in\ X_i\ is\ less\ than\ q\ for\ each\ i\}.$$

It follows from the above proposition that $\wp$ restricted to A is an injective map. Clearly, the number of elements in A is q^{q^m} which is the same as the number of elements in the set $Map(F_q^m, F_q)$. Thus, $\wp$ is a bijective map from A to $Map(F_q^m, F_q)$. In turn, it is a surjective map from $F_q[X_1, X_2, \ldots X_m]$ to $Map(F_q^m, F_q)$. ♯

Proposition 7.7.7 *Let $f(X_1, X_2, \ldots, X_m) \in F_q[X_1, X_2, \ldots, X_m]$. Then, there exist polynomials $q_i(X_1, X_2, \ldots, X_m) \in F_q[X_1, X_2, \ldots, X_m]$, $i = 1, 2, \ldots, m$, and a polynomial $r(X_1, X_2, \ldots, X_m)$ such that*

$$f(X_1, X_2, \ldots, X_m) =$$
$$\Sigma_{i=1}^{m} q_i(X_1, X_2, \ldots, X_m)(X_i^q - X_i) + r(X_1, X_2, \ldots, X_m),$$

where $r(X_1, X_2, \ldots, X_m)$ *is zero, or else degree of* $r(X_1, X_2, \ldots, X_m)$ *in each* X_i *is less than* q.

Proof For $m = 1$, this is just the division algorithm theorem. Suppose that $m > 1$. Let $f(X_1, X_2, \ldots, X_m) \in F_q[X_1, X_2, \ldots, X_m]$. Since $X_m^q - X_m$ is a nonzero polynomial in $F_q[X_1, X_2, \ldots, X_{m-1}][X_m]$, by the division algorithm theorem, there are polynomials $q_m(X_1, X_2, \ldots, X_m)$ and $r_m(X_1, X_2, \ldots, X_m)$ in $F_q[X_1, X_2, \ldots, X_{m-1}][X_m]$ such that

$$f(X_1, X_2, \ldots, X_m) =$$
$$q_m(X_1, X_2, \ldots, X_m)(X_m^q - X_m) + r_m(X_1, X_2, \ldots, X_m),$$

where $r_m(X_1, X_2, \ldots, X_m)$ is 0, or else degree of $r_m(X_1, X_2, \ldots, X_m)$ in X_m is less than q. If $r_m(X_1, X_2, \ldots, X_m)$ is 0, then there is nothing to do. Suppose that $r_m(X_1, X_2, \ldots, X_m) \neq 0$. Then, as before, treating it as an element of $F_q[X_1, X_2, \ldots, X_{m-2}, X_m][X_{m-1}]$, we get polynomials $q_{m-1}(X_1, X_2, \ldots, X_m)$ and $r_{m-1}(X_1, X_2, \ldots, X_m)$ such that

$$r_m(X_1, X_2, \ldots, X_m) =$$
$$q_{m-1}(X_1, X_2, \ldots, X_m)(X_{m-1}^q - X_{m-1}) + r_{m-1}(X_1, X_2, \ldots, X_m),$$

where $r_{m-1}(X_1, X_2, \ldots, X_m)$ is 0, or else degree of $r_{m-1}(X_1, X_2, \ldots, X_m)$ in X_{m-1} is less than q. Suppose that $r_{m-1}(X_1, X_2, \ldots, X_m)$ in $X_{m-1} \neq 0$. Since the degree of $r_m(X_1, X_2, \ldots, X_m)$ in X_m is less than q, the degree of $r_{m-1}(X_1, X_2, \ldots, X_m)$ in X_m is less than q. Proceeding inductively, we arrive at the desired result. ♯

The following corollary is immediate from the above proposition.

Corollary 7.7.8 *The kernel A of the map $\wp$ is the ideal generated by the set*

$$\{X_1^q - X_1, X_2^q - X_2, \ldots, X_m^q - X_m\}.$$ ♯

Corollary 7.7.9 $F_q[X_1, X_2, \ldots, X_m]/A$ *is isomorphic to the ring* $Map(F_q^m, F_q)$.   ♯

Theorem 7.7.10 (Artin–Chevalley) *Suppose that* $f(X_1, X_2, \ldots, X_m) \in F_q[X_1, X_2, \ldots, X_m]$ *is a polynomial of degree less than m such that* $f(0, 0, \ldots, 0) = 0$. *Then,* $V(f) \neq \{(0, 0, \ldots, 0)\}$, *i.e., there exist* $\alpha_1, \alpha_2, \ldots, \alpha_m$ *not all zero such that* $f(\alpha_1, \alpha_2, \ldots, \alpha_m) = 0$.

Proof Suppose that $V(f) = \{(0, 0, \ldots, 0)\}$. Consider the polynomial $\phi(X_1, X_2, \ldots, X_m) = 1 - f(X_1, X_2, \ldots, X_m)^{q-1}$. Since $f(0, 0, \ldots, 0) = 0$, $(0, 0, \ldots, 0) \notin V(\phi)$. If $(\alpha_1, \alpha_2, \ldots, \alpha_m) \neq (0, 0, \ldots, 0)$, then by the supposition, $f(\alpha_1, \alpha_2, \ldots, \alpha_m) \neq 0$. But then $(\alpha_1, \alpha_2, \ldots, \alpha_m) \in V(\phi)$. This shows that $V(\phi) = F_q^m - \{(0, 0, \ldots, 0)\}$. Consider the polynomial

$$P(X_1, X_2, \ldots, X_m) = (1 - X_1^{q-1})(1 - X_2^{q-1}) \cdots (1 - X_m^{q-1}).$$

Clearly, $V(\phi) = V(P)$. Consequently, $\phi - P \in ker\wp$. Since the $ker\wp$ is the ideal generated by the set

$$\{X_1^q - X_1, X_2^q - X_2, \ldots, X_m^q - X_m\},$$

and the degree of P is $m(q - 1)$, it follows that the $deg(\phi) \geq m(q - 1)$. Since the degree of f is assumed to be less than m, we arrive at a contradiction. Hence, $V(f) \neq \{(0, 0, \ldots, 0)\}$. ♯

Theorem 7.7.11 (Warning's Theorem) *Let* $f(X_1, X_2, \ldots, X_m)$ *be a member of* $F_q[X_1, X_2, \ldots, X_m]$, *and the degree of* f *is less than* m. *Then,* $| V(f) |$ *is divisible by the characteristic* p *of* F_q.

Before proving the Warning's theorem, we establish the following lemma.

Lemma 7.7.12 *Let* f *be a map from* $\mathbb{N} \bigcup \{0\}$ *to* F_q *defined by* $f(r) = \Sigma_{x \in F_q} x^r$ (*by convention* $x^0 = 1$). *Then,* $f(r) = -1$ *if* $r > 0$ *and* $q - 1$ *divides* r. *Further,* $f(r) = 0$, *otherwise*.

Proof Clearly, $f(0) = q1 = 0$. Suppose that $r \in \mathbb{N}$. If $q - 1$ divides r, then $x^r = 1$ for all $x \neq 0$, and, of course, $0^r = 0$. Hence, $f(r) = (q - 1)1 = -1$. Next, suppose that $q - 1$ does not divide r. Then, there is a nonzero element $a \in F_q$ such that $a^r \neq 1$. Now, $a^r f(r) = a^r \Sigma_{x \in F_q} x^r = \Sigma_{x \in F_q} (ax)^r = f(r)$. Since $a^r \neq 1$, $f(r) = 0$. ♯

Proof of the Warning's Theorem.

Let $f(X_1, X_2, \ldots, X_m) \in F_q[X_1, X_2, \ldots, X_m]$ be a polynomial whose degree is less than m. Consider the polynomial $g(X_1, X_2, \ldots, X_m) = 1 - f(X_1, X_2, \ldots, X_m)^{q-1}$. Clearly, $g(\alpha_1, \alpha_2, \ldots, \alpha_m) = 1$ if and only if $(\alpha_1, \alpha_2, \ldots, \alpha_m) \in V(f)$. Further, $g(\alpha_1, \alpha_2, \ldots, \alpha_m) = 0$ if $(\alpha_1, \alpha_2, \ldots, \alpha_m) \notin V(f)$. Therefore, it is sufficient to show that

$$\Sigma_{(\alpha_1, \alpha_2, \ldots, \alpha_m) \in F_q^m} g(\alpha_1, \alpha_2, \ldots, \alpha_m) = 0.$$

Next, observe that $deg(g) < m(q-1)$. Hence, the degree of each monomial appearing in g is of degree less than $m(q - 1)$, and so the degree of some X_i appearing in each monomial of g is less than $q - 1$. Let $a_{i_1 i_2 \cdots i_m} X_1^{i_1} X_2^{i_2} \cdots X_m^{i_m}$ be a monomial appearing in $g(X_1, X_2, \ldots, X_m)$. We may assume that $i_m < q - 1$. From the above lemma, for fixed members $\alpha_1, \alpha_2, \ldots, \alpha_{m-1}$ of F_q,

$$\Sigma_{\alpha_m \in F_q} a_{i_1 i_2 \cdots i_m} \alpha_1^{i_1} \alpha_2^{i_2} \cdots \alpha_{m-1}^{i_{m-1}} \alpha_m^{i_m} = 0.$$

Hence,

$$\Sigma_{(\alpha_1, \alpha_2, \ldots, \alpha_m) \in F_q^m} a_{i_1 i_2 \cdots i_m} \alpha_1^{i_1} \alpha_2^{i_2} \cdots \alpha_{m-1}^{i_{m-1}} \alpha_m^{i_m} = 0.$$

In turn, it follows that

$$\Sigma_{(\alpha_1,\alpha_2,\ldots,\alpha_m)\in F_q^m} g(\alpha_1, \alpha_2, \ldots, \alpha_m) = 0.$$

♯

Corollary 7.7.13 *Let* $f_1, f_2, \ldots f_r$ *be polynomials over* F_q *in* m *variables. Suppose that* $m > \Sigma_{i=1}^r deg(f_i)$, *and each* f_i *is without constant term (i.e.,* $f_i(\overline{0}) = 0$ *for all i).* *Then,*

$$V(f_1) \bigcap V(f_2) \bigcap \cdots \bigcap V(f_r) \neq \{(0, 0, \ldots, 0)\}.$$

Proof Take $f = f_1 f_2 \cdots f_r$, and apply the Warning's theorem.   ♯

Exercises

7.7.1 Let F be a field, $A^n(F)$ the affine n- set, and $F[X_1, X_2, \ldots X_n]$ the polynomial ring in n variables. Let A be a subset of $F[X_1, X_2, \ldots X_n]$, and $V(A)$ denotes the subset $\bigcap_{f \in A} V(f)$. This defines a map V from the power set of $F[X_1, X_2, \ldots X_n]$ to the power set of $A^n(F) = F^n$. The sets of the form $V(A)$ are called **affine algebraic subset** of the affine n-set. Show that

(i) $V(\{0\}) = A^n(F)$.
(ii) $V(\{1\}) = \emptyset$.
(iii) $V(A) = V(<A>) = V(\sqrt{<A>})$. Thus, V is not injective.
(iv) Show that arbitrary intersection of affine algebraic sets are affine algebraic.
(v) Finite union of affine algebraic sets are affine algebraic.
(vi) The family T of compliments of affine algebraic sets form a topology on $A^n(F)$. This topology is called the Zariski topology on $A^n(F)$.

7.7.2 Show that $\{(cos\ t, sin\ t,\ t) \mid t \in \mathbb{R}\}$, where $\mathbb{R}$ is the field of real numbers, is not an affine algebraic set. Thus, V need not be a surjective map.

7.7.3 Let Y be a subset of the affine n-set $A^n(F)$. Let $I(Y) = \{f \in F[X_1, X_2, \ldots X_n] \mid Y \subseteq V(f)\}$. Show that $I(Y)$ is a radical ideal such that $Y \subseteq V(I(Y))$.

7.7.4 Show that $V(I(V(A))) = V(A)$.

7.7.5 Show that V and I are inclusion reversing maps.

7.7.6 Let Y be a subset of the affine m-set $A^m(F)$. The ring $\Gamma(Y) = F[X_1, X_2, \ldots, X_m]/I(Y)$ is called the **coordinate ring** of Y. Show that the coordinate ring $\Gamma(Y)$ is a reduced ring.

7.7.7 Consider the line $L = V(f)$ in $\mathbb{R}^2$, where $f(X, Y) = Y - mX + c, m \neq 0$. Show that $\Gamma(L)$ is isomorphic to the polynomial ring $\mathbb{R}[t]$ in one indeterminate.

7.7.8 Consider the parabola $P = V(Y - X^2)$ in $\mathbb{R}^2$. Show that $\Gamma(P)$ is isomorphic to $\Gamma(L)$. Observe that P and L are not congruent.

7.7.9 Consider the circle $S^1 = V(X^2 + Y^2 - 1)$ in $\mathbb{R}^2$. Show that $\Gamma(S^1)$ is not isomorphic to $\Gamma(L)$. What is $\Gamma(S^1)$?

7.7.10 Identify the coordinate ring $\Gamma V(Y)$, where $Y = \{X^2 + Y^2 + Z^2 - 1, X - Y\}$.

7.7.11 Can the ring $M_2(\mathbb{R})$ be coordinate ring of algebraic subset? Support.

7.7.12 Is $\Gamma(Y)$ always an integral domain? If not, under what condition it can be an integral domain?

Chapter 8
Number Theory 2

Chap. 3 was devoted to some elementary results in arithmetic such as division algorithm, Euclidean algorithm, fundamental theorem of arithmetic, and solutions of linear Diophantine equations and of linear congruences. Further, in Chaps. 4, 5, and 7, we proved some fundamental results such as Wilson theorem, Euler-Fermat's theorem as applications of group theory and ring theory. In this chapter, we study arithmetic functions, Quadratic residues, Quadratic reciprocity law, nth power residues, and nonlinear Diophantine equations.

8.1 Arithmetic Functions

Definition 8.1.1 A map f from the set $\mathbb{N}$ of natural numbers to the field $\mathbb{C}$ of complex numbers is called an **arithmetic function**.

Let $A(\mathbb{N}, \mathbb{C})$ denote the set of all arithmetic functions on $\mathbb{N}$. Then, $A(\mathbb{N}, \mathbb{C})$ is an abelian group with respect to the pointwise addition. Define a multiplication $\star$ in $A(\mathbb{N}, \mathbb{C})$ by

$$(f \star g)(n) \; = \; \Sigma_{d_1 d_2 \,=\, n} f(d_1) g(d_2).$$

This multiplication is called the **Dirichlet** multiplication. Now,

$$((f \star g) \star h)(n) \; = \; \Sigma_{dd_3 \,=\, n}(f \star g)(d)h(d_3) \; = \; \Sigma_{dd_3 \,=\, n}(\Sigma_{d_1 d_2 \,=\, d})h(d_3) \; =$$
$$\Sigma_{d_1 d_2 d_3 \,=\, n} f(d_1)g(d_2)h(d_3) \; = \; ((f \star (g \star h)(n) \, for \, all \, n.$$

Thus, $\star$ is associative, and so $A(\mathbb{N}, \mathbb{C})$ is a semigroup with respect to $\star$. It is also easy to see that $\star$ distributes over $+$, and so $A(\mathbb{N}, \mathbb{C}), +, \star)$ is a ring. Further,

© Springer Nature Singapore Pte Ltd. 2017
R. Lal, *Algebra 1*, Infosys Science Foundation Series in Mathematical Sciences,
DOI 10.1007/978-981-10-4253-9_8

$$(f \star g)(n) = \Sigma_{d_1 d_2 = n} f(d_1) g(d_2) = (g \star f)(n).$$

for all $f, g \in A(\mathbb{N}, \mathbb{C})$. Thus, $A(\mathbb{N}, \mathbb{C})$ is a commutative ring.

Let e be a map from $\mathbb{N}$ to $\mathbb{C}$ defined by

$$e(n) = \begin{cases} 1 \text{ if n} = 1 \\ 0 \text{ otherwise.} \end{cases}$$

Then, $(f \star e)(n) = \Sigma_{d_1 d_2 = n} f(d_1) e(d_2) = f(n)$, and similarly, $(e \star f)(n) = f(n)$ *for all* $n \in \mathbb{N}$. Thus, e is the identity of the ring $A(\mathbb{N}, \mathbb{C})$ of arithmetic functions.

The exponential function $n \longrightarrow e^n$, the *log* function $n \longrightarrow log(n)$, the factorial function $n \longrightarrow n!$, the function $n \longrightarrow \sqrt{n}$, and the function $n \longrightarrow 1 + \frac{1}{2} + \frac{1}{3} + \cdots + \frac{1}{n} - logn$ are some examples of arithmetic functions. We shall be more interested in functions f such that $f(n)$ reflects some arithmetic properties of n, and not only the size of n. Following are some important such examples.

Example 8.1.2 The Euler's phi function Φ from $\mathbb{N}$ to $\mathbb{N}$ already defined in Chap. 4 ($\Phi(1) = 1$, *and for* $n > 1$, $\Phi(n) = $ the number of positive integers less than n and co-prime to n) is an arithmetic function.

Example 8.1.3 The **divisor function** τ from $\mathbb{N}$ to $\mathbb{N}$ ($\tau(n) = $ the number of divisors of n) is an arithmetic function. **divisor function**.

Example 8.1.4 The function σ from $\mathbb{N}$ to $\mathbb{N}$ defined by

$$\sigma(n) = \Sigma_{d/n} d,$$

is an arithmetic function. This is called the
sum of divisor function.

Example 8.1.5 The function σ_k, $k \geq 1$ defined by

$$\sigma_k(n) = \Sigma_{d/n} d^k$$

is another arithmetic function.

Example 8.1.6 (*Möbius function*) The arithmetic function μ from $\mathbb{N}$ to $\mathbb{Z}$ defined by

  (i) $\mu(1) = 1$,
 (ii) $\mu(n) = 0$, *if* a^2 *divides* n, *for some* $a > 1$, and
(iii) $\mu(n) = (-1)^r$, *if n is product of r distinct primes*
      is called the **Mobius function**.

Example 8.1.7 The identity map $I_{\mathbb{N}}$ from $\mathbb{N}$ to $\mathbb{N}$ is another example of an arithmetic functions.

Example 8.1.8 For $k \geq 1$, we define an arithmetic function Φ_k from $\mathbb{N}$ to $\mathbb{N}$ as follows: $\Phi_k(1) = 1$, and for $n \geq 1$, $\Phi_k(n)$ is the number of k-tuples $(\alpha_1, \alpha_2, \ldots \alpha_k)$ of positive integers such that $\alpha_i \leq n$ *for all i* and the *g.c.d* $(\alpha_1, \alpha_2, \ldots, \alpha_k, n) = 1$. What is Φ_1?

Example 8.1.9 We have another arithmetic function λ from $\mathbb{N}$ to the field $\mathbb{R}$ of real numbers defined by $\lambda(n) = log\ p$, if n is a power of some prime p, and zero, otherwise.

Example 8.1.10 The constant function $\P$ from $\mathbb{N}$ to $\mathbb{N}$ defined by $\P(n) = 1$ is an arithmetic function.

Definition 8.1.11 An arithmetic function f from $\mathbb{N}$ to $\mathbb{C}$ is called a **multiplicative** function if f is not identically zero, and

$$(m, n) = 1\ implies\ that\ f(mn) = f(m)f(n).$$

It is said to be an **absolutely multiplicative** function if

$$f(mn) = f(m)f(n)\ for\ all\ m, n \in \mathbb{N}.$$

Remark 8.1.12 If f is multiplicative nonzero function, then $f(1) = 1$, for $(m, 1) = 1$, and so $f(m) = f(m)f(1)$ *for all* $m \in \mathbb{N}$.

Example 8.1.13 The identity element e of $A(\mathbb{N}, \mathbb{C})$ defined earlier is multiplicative. In fact, it is absolutely multiplicative(verify).

Example 8.1.14 The function $\P$ defined above is multiplicative. In fact, this is also absolutely multiplicative.

Example 8.1.15 The identity function I_N is also absolutely multiplicative.

Proposition 8.1.16 *The divisor function τ is multiplicative, and*

$$\tau(p_1^{\alpha_1} p_2^{\alpha_2} \ldots p_r^{\alpha_r}) = (\alpha_1 + 1)(\alpha_2 + 1) \ldots (\alpha_r + 1),$$

where $p_1, p_2, \ldots, p_r$ are distinct primes.

Proof Suppose that $(m, n) = 1$. It is evident that every divisor d of $m \cdot n$ can be uniquely expressed as $d = d_1 \cdot d_2$, where d_1 is a divisor of m and d_2 is a divisor of n. Thus, the number of divisors of $m \cdot n$ is the product of the number of divisors of m and that of n. This shows that $\tau(m \cdot n) = \tau(m) \cdot \tau(n)$. Further, if p is a prime number, then $1, p, p^2, \ldots p^\alpha$ are precisely the divisors of p^α. Hence, $\tau(p^\alpha) = \alpha + 1$. By the multiplicative properties of τ, the result follows. $\sharp$

Proposition 8.1.17 *Let f and g be multiplicative functions. Then $f \star g$ is also multiplicative.*

Proof Suppose that $(m, n) = 1$ and $uv = mn$. Then, $u = d_1 d_2$ and $v = d_3 d_4$, where d_1/m, d_2/n, d_3/m, and d_4/n. Clearly, $(d_1, d_2) = 1 = (d_3, d_4)$. Also, $d_1 d_3 = m$ and $d_2 d_4 = n$. Since f and g are multiplicative, $f(u) = f(d_1)f(d_2)$, and $g(v) = g(d_3)g(d_4)$. Thus, $f(u)g(v) = f(d_1)g(d_3)f(d_2)g(d_4)$. Hence,

$$(f \star g)(mn) = \Sigma_{u \cdot v = m \cdot n} f(u)g(v) =$$
$$\Sigma_{d_1 d_3 = m, d_2 d_4 = n}(f(d_1)g(d_3))(f(d_2)g(d_4)) =$$
$$(\Sigma_{d_1 d_3 = m} f(d_1)g(d_3)) \cdot (\Sigma_{d_2 d_4 = n} f(d_2)g(d_4)) = (f \star g)(m) \cdot (f \star g)(n).$$

Thus, $f \star g$ is multiplicative.   ♯

Let $M(\mathbb{N}, \mathbb{C})$ denote the set of all multiplicative functions. Following result is the restatement of the above proposition.

Corollary 8.1.18 $M(\mathbb{N}, \mathbb{C})$ *is a sub-semigroup of* $(A(\mathbb{N}, \mathbb{C}), \star)$ *which also contains the identity e.*   ♯

Corollary 8.1.19 *Let f be a multiplicative function. Define a function F by*

$$F(n) = \Sigma_{d/n} f(d).$$

Then F is also multiplicative.

Proof The result follows from the above proposition if we observe that $F = f \star \P$.   ♯

Proposition 8.1.20 *The function σ_k(in particular σ) is multiplicative function, and*

$$\sigma_k(p_1^{\alpha_1} p_2^{\alpha_2} \ldots p_r^{\alpha_r}) = \frac{p_1^{(\alpha_1+1)k}-1}{p_1^k-1} \cdot \frac{p_2^{(\alpha_2+1)k}-1}{p_2^k-1} \ldots \frac{p_r^{(\alpha_r+1)k}-1}{p_r^k-1},$$

where $p_1, p_2, \ldots, p_r$ are distinct primes and $k \geq 1$.

Proof By the definition, $\sigma_k(n) = \Sigma_{d/n} d^k = \Sigma_{d/n} t_k(d)$, where t_k is defined by $t_k(n) = n^k$. Since t_k is multiplicative(in fact, absolutely multiplicative), the multiplicativity of σ_k follows from Corollary 7.1.19. Next,

$$\sigma_k(p^\alpha) = \Sigma_{d/p^\alpha} d^k = 1^k + p^k + (p^2)^k + \cdots + (p^\alpha)^k = \frac{p^{(\alpha+1)k}-1}{p^k-1}.$$

The result follows from the multiplicativity of σ_k.   ♯

Example 8.1.21 The number of divisors of 200 is $\tau(200) = \tau(2^3 \cdot 5^2) = (3+1) \cdot (2+1) = 12$, and the sum of divisors of 200 is $\sigma(200) = \sigma(2^3 \cdot 5^2) = \frac{2^4-1}{2-1} \cdot \frac{5^3-1}{5-1} = 465$.

Proposition 8.1.22 *The Möbius function μ is multiplicative.*

Proof Suppose that $(m, n) = 1$. If m or n is 1, then since $\mu(1) = 1$, $\mu(mn) = \mu(m)\mu(n)$. Suppose that $m \neq 1 \neq n$. If a square greater than 1 divides m or n, then it also divides $m \cdot n$, and so in this case, $\mu(m \cdot n) = 0 = \mu(m)\mu(n)$. Suppose that m is product of r distinct primes and n is product of s distinct primes. Then, since m and n are co-prime, $m \cdot n$ will be product of $r + s$ distinct primes. By the definition, $\mu(m \cdot n) = (-1)^{r+s} = (-1)^r \cdot (-1)^s = \mu(m)\mu(n)$. ♯

Proposition 8.1.23

$$\Sigma_{d/n}\mu(d) = \begin{cases} 1 \ if \ n = 1 \\ 0 \ otherwise \end{cases}$$

Thus, $e(n) = \Sigma_{d/n}\mu(d)$.

Proof Let η denote the function given by $\eta(n) = \Sigma_{d/n}\mu(d)$. We show that $\eta = e$. Since μ is multiplicative, η is multiplicative. If $n = 1$, then $\Sigma_{d/1}\mu(d) = \mu(1) = 1$. Next, for $\alpha \geq 1$, $\eta(p^\alpha) = \Sigma_{d/p^\alpha}\mu(d) = \Sigma_{\beta=0}^{\alpha}\mu(p^\beta) = 1 - 1 + 0 + 0 + \cdots + 0 = 0$. Since η is multiplicative, the result follows. ♯

Corollary 8.1.24 *The function* $\mathbb{1}$ *is the inverse of the Möbius function* μ *in the semigroup* $A(N, C)$ *of arithmetic functions with respect to the Dirichlet product. More precisely,* $\mathbb{1} \star \mu = e = \mu \star \mathbb{1}$.

Proof From the above proposition, we have

$$e(n) = \Sigma_{d/n}\mu(d) = \Sigma_{dd'=n}\mu(d)\mathbb{1}(d') = \mu \star \mathbb{1}(n).$$

Thus, $\mu \star \mathbb{1} = e$. Similarly, $\mathbb{1} \star \mu = e$. ♯

Theorem 8.1.25 (Möbius inversion theorem) *Let* f *be an arithmetic function, and* F *the function defined by*

$$F(n) = \Sigma_{d/n}f(d).$$

Then

$$f(n) = \Sigma_{d/n}\mu(d)F(\tfrac{n}{d}).$$

Proof

$$F(n) = \Sigma_{d/n}f(d) = \Sigma_{dd'=n}f(d)\mathbb{1}(d') = (f \star \mathbb{1})(n).$$

Thus,

$$F = f \star \mathbb{1} = \mathbb{1} \star f.$$

Now,

$$\mu \star F = \mu \star (\mathbb{1} \star f) = (\mu \star \mathbb{1}) \star f = e \star f = f.$$

Hence,

$$f(n) = \Sigma_{dd'=n}\mu(d)F(d') = \Sigma_{d/n}\mu(d)F(\tfrac{n}{d}).$$

♯

Corollary 8.1.26 *An arithmetic function f is multiplicative if and only if the function F defined by*

$$F(n) = \Sigma_{d/n} f(d)$$

is multiplicative.

Proof We have seen that $F = f \star \mathbb{1}$ and $f = \mu \star F$. Since μ and $\mathbb{1}$ are both multiplicative, and the Dirichlet product of multiplicative functions is multiplicative, the result follows.   ♯

Corollary 8.1.27 *The Euler's phi function Φ is multiplicative, and*

$$\Phi(n) = n \cdot \Sigma_{d/n} \frac{\mu(d)}{d}.$$

Proof By Theorem 4.5.37,

$$n = \Sigma_{d/n} \Phi(d)$$

or

$$I_N(n) = \Sigma_{d/n} \Phi(d).$$

Since I_N is multiplicative, by the above corollary, Φ is multiplicative. Further using Möbius inversion theorem

$$\Phi(n) = \Sigma_{d/n} \mu(d) I_N(\tfrac{n}{d}) = n \cdot \Sigma_{d/n} \frac{\mu(d)}{d}.$$ ♯

Proposition 8.1.28 *Let f be a multiplicative function. Then*

$$\Sigma_{d/n} \mu(d) f(d) = \prod_{p/n, p \text{ a prime}} (1 - f(p)).$$

Proof Since μ and f are both multiplicative and product of multiplicative functions is multiplicative, $\mu \cdot f$ is multiplicative. Hence, the function η defined by $\eta(n) = \Sigma_{d/n} \mu(d) f(d)$ is multiplicative. Thus, it is sufficient to show that $\eta(p^\alpha) = 1 - f(p)$, where p is a prime and $\alpha \geq 1$. Now,

$$\eta(p^\alpha) = \Sigma_{\beta=0}^\alpha \mu(p^\beta) f(p^\beta) = \mu(1) f(1) + \mu(p) f(p) = 1 - f(p).$$ ♯

Corollary 8.1.29 $\Sigma_{d/n} \frac{\mu(d)}{d} = \prod_{p/n, p \text{ a prime}} (1 - \frac{1}{p}).$

Proof The map $n \rightsquigarrow \frac{1}{n}$ is a multiplicative function. Apply the above proposition. ♯

Corollary 8.1.30 $\Phi(n) = n \cdot \prod_{p/n, p \text{ a prime}} (1 - \frac{1}{p}).$

Proof $\Phi(n) = n \cdot \Sigma_{d/n} \frac{\mu(d)}{d} = n \cdot \prod_{p/n, p \text{ a prime}} (1 - \frac{1}{p}).$ ♯

Example 8.1.31 The number of positive integers less than 200 and co-prime to 200 is

$$\Phi(200) \;=\; \Phi(2^3 \cdot 5^2) \;=\; 200 \cdot (1 - \tfrac{1}{2}) \cdot (1 - \tfrac{1}{5}) \;=\; 80.$$

The number of positive integers less than 200 and not co-prime to 200 is $200 - 80 = 120$.

Example 8.1.32 $\mu \star \tau = \P$. Equivalently, $\Sigma_{d/n}\mu(d)\tau(\tfrac{n}{d}) = 1$ for all n.

Proof $\tau(n)$, by definition, is the number of divisors of n. Thus, $\tau(n) = \Sigma_{d/n}\P(d)$. Using Mőbius inversion theorem,

$$1 \;=\; \P(n) \;=\; \Sigma_{d/n}\mu(d)\tau(\tfrac{n}{d}) \; \forall n. \qquad\qquad ♯$$

Example 8.1.33 $\mu \star \sigma = I_N$ or equivalently, $n = \Sigma_{d/n}\mu(d)\sigma(\tfrac{n}{d})$ *for all* n.

Proof Since $\sigma(n)$ is the sum of divisors of n,

$$\sigma(n) \;=\; \Sigma_{d/n}d \;=\; \Sigma_{d/n}I_N(d).$$

Using Mőbius inversion theorem,

$$n \;=\; I_N(n) \;=\; \Sigma_{d/n}\mu(d)\sigma(\tfrac{n}{d}). \qquad\qquad ♯$$

Example 8.1.34 $\Sigma_{d/n} \mid \mu(d) \mid = 2^r$, where r is the number of distinct primes dividing n.

Proof The map $n \rightsquigarrow \mid \mu(n) \mid$ is clearly a multiplicative function. Thus, the map χ defined by

$$\chi(n) \;=\; \Sigma_{d/n} \mid \mu(d) \mid$$

is multiplicative. Further,

$$\chi(p^\alpha) \;=\; \Sigma_{d/p^\alpha} \mid \mu(d) \mid = \mu(1) + \mu(p) + 0 + \cdots + 0 = 2.$$

The result follows from the fact that χ is multiplicative. $\qquad\qquad ♯$

Example 8.1.35 $\Sigma_{d/n}(\mu(d))^2(\Phi(d))^2 = \prod_{p/n, p\ a\ prime}(p^2 - 2p + 2)$.

Proof Since μ and Φ are multiplicative and the products of multiplicative functions are multiplicative,
$n \rightsquigarrow (\mu(n))^2(\Phi(n))^2$ is multiplicative. Hence, the function η defined by

$$\eta(n) \;=\; \Sigma_{d/n}(\mu(d))^2(\Phi(d))^2$$

is multiplicative. Thus, it is sufficient to show that $\eta(p^\alpha) = p^2 - 2p + 2$ for all $\alpha \geq 1$. Now, for $\alpha \geq 1$,

$$\eta(p^\alpha) = (\mu(1))^2(\Phi(1))^2 + (\mu(p))^2(\Phi(p))^2 + 0 \cdots + 0 =$$
$$1 + (p-1)^2 = p^2 - 2p + 2. \qquad \qquad \qquad \sharp$$

Exercises

8.1.1 Show that $f \in A(\mathbb{N}, \mathbb{C})$ is a unit if and only if $f(1) \neq 0$. Find the inverse of f assuming that $f(1) \neq 0$.

8.1.2 Show that if f is a nonzero multiplicative function, then $f(1) = 1$.

8.1.3 Show that inverse of a multiplicative function is also multiplicative.

8.1.4 Show that inverse of μ in $M(\mathbb{N}, \mathbb{C})$ is $\P$.

8.1.5 Find the inverse of τ in $M(\mathbb{N}, \mathbb{C})$.

8.1.6 Find the inverse of σ in $M(\mathbb{N}, \mathbb{C})$.

8.1.7 Determine the inverse of Φ.

8.1.8 Find the number of divisors of 1000.

8.1.9 Find the sum of divisors of 1538.

8.1.10 Find the sum of squares of divisors of 1000.

8.1.11 Let f be a multiplicative function. Show that the function h defined by

$$h(n) = \Sigma_{d^2/n} f(d)$$

is a multiplicative function.

8.1.12 Show that $\Sigma_{d^2/n} \mu(d) = |\mu(n)|$.
Hint. Observe that both sides are multiplicative and so verify it for p^α.

8.1.13 Prove the following generalizations of the results about the Euler's phi function to the function Φ_k.

  (i) $n^k = \Sigma_{d/n} \Phi_k(d)$.
  (ii) Φ_k is multiplicative.
  (iii) $\Phi_k(n) = n^k \cdot \prod_{p/n, p \text{ a prime}} (1 - \frac{1}{p^k})$.

8.1.14 Find the number of positive integers less than 1000 and co-prime to 1000. Find the number of positive integers less than 1000 and not co-prime to 1000.

8.1.15 Find the number of pairs $(\alpha, \beta) \in \mathbb{N} \times \mathbb{N}$ such that $\alpha \leq 100$, $\beta \leq 100$, and the g.c.d $(\alpha, \beta, 100) = 1$. Find the number of pairs (α, β) such that $\alpha \leq 100$, $\beta \leq 100$, and $(\alpha, \beta, 100) \neq 1$.

8.1.16 Find $\Sigma_{n=5}^{20} \mu(n)$.

8.1.17 Find $\Sigma_{d/1000} \mu(d)$.

8.1.18 Show that $\Sigma_{d/n} \mu(d) \Phi(d) = \prod_{p/n, p \ a \ prime} (2 - p)$.

8.1.19 Show that $\Sigma_{d/n} \frac{\mu(d)}{\Phi(d)} = \prod_{p/n, p \ a \ prime} \frac{p-2}{p-1}$.

8.1.20 Find the expression for

$$\Sigma_{d/n} \frac{(\mu(d))^2}{(\Phi(d))^2}$$

in terms of the primes dividing n.

8.1.21 Show that $\Sigma_{d/n} \frac{(\mu(d))^2}{\Phi(d)} = \frac{n}{\Phi(n)}$.

8.1.22 Show that $\Sigma_{d/n} (\tau(d))^3 = (\Sigma_{d/n} \tau(d))^2$.

8.1.23 Find $\Sigma_{d/100} \frac{(\mu(d))^2}{(\Phi(d))^2}$.

8.1.24 Show that $\tau(n)$ is even if and only if n is not a square.

8.1.25 Show that $\sigma(n)$ is even if and only if $\sqrt{n}$ and $\sqrt{\frac{n}{2}}$ are irrational numbers.

8.1.26 Show that $\Phi(nm) = \Phi((n,m)) \cdot \Phi([n,m])$.

8.1.27 Show that $\tau(n^2)$ is the number of pairs of positive integers (α, β) such that $[\alpha, \beta] = n$.

8.1.28 Show that $\lambda(n) = -\Sigma_{d/n} \mu(d) \log d$.

8.1.29 Let $\sigma_{odd}(n)$ denote the sum of positive odd divisors of n. Show that

$$\sigma_{odd}(n) = \Sigma_{d/n} (-1)^{\frac{n+d}{d}} \cdot d.$$

8.1.30 Call a number n a **perfect** number if $\sigma(n) = 2n$. Show that if $2^n - 1$ is a prime number, then $2^{n-1}(2^n - 1)$ is a perfect number. Conversely, Euler proved that every even perfect number is of this form. Thus, finding even perfect number is equivalent to finding primes of the form $2^n - 1$. Such primes are called the **Mersenne** primes. It is not known whether there are infinitely many Mersenne primes, or equivalently infinitely many even perfect numbers. It is also not known whether there are odd perfect numbers.

8.1.31 Let $f : \mathbb{N} \times \mathbb{N} \longrightarrow \mathbb{C}$ be a nonzero function. Define a function F from $\mathbb{N} \times \mathbb{N}$ to $\mathbb{C}$ by

$$F(m, n) = \Sigma_{d_1/m, d_2/n} f(d_1, d_2).$$

Prove the following inversion formula:

$$f(m, n) = \Sigma_{d_1/m, d_2/n} \mu(d_1) \mu(d_2) F(\frac{m}{d_1}, \frac{n}{d_2}).$$

8.1.32 The function [ ] from the field $\mathbb{R}$ of real numbers to $\mathbb{Z}$ defined by $[\,](x) =$ *the largest integer* $\leq x$ is called the **bracket** function. The image of x is denoted by $[x]$ and is called the **integral part** of x. Prove the following properties of the bracket function:

   (i) $0 \leq x - [x] < 1.$ $x - [x]$ is called the **fractional** part of x.
  (ii) $[x + 1] = [x] + 1.$
 (iii) The function f defined by $f(x) = [x] + [-x]$ is zero at integers and -1 at nonintegers.
  (iv) $[x + y] \leq [x] + [y].$
   (v) If $a, m \in \mathbb{Z}$ and $m > 0$, then $a - [\frac{a}{m}] \cdot m$ is the remainder when a is divided by m.
  (vi) $[x] + [y] + [x + y] \leq [2x] + [2y].$

8.1.33 Let n be a positive integer and p a prime. Let $ord_p n$ denote the exponent of the highest power of p dividing n. Show that

$$ord_p n! = \Sigma_{i=1}^{\infty}[\frac{n}{p^i}].$$

8.1.34 Let $\Phi(x, n)$ denote the number of positive integers less than or equal to x and co-prime to $n(x \geq 2)$. Show that

$$\Sigma_{d/n}\Phi(\frac{x}{d}, \frac{n}{d}) = [x].$$

Deduce that

$$\Phi(x, n) = \Sigma_{d/n}\mu(d)[\frac{x}{d}].$$

8.1.35 Let f be an arithmetic function and the function F is given by

$$F(n) = \Sigma_{d/n}f(d).$$

Show that

$$\Sigma_{m=1}^{n} F(m) = \Sigma_{m=1}^{n}[\frac{n}{m}]f(m).$$

Deduce that

$$\Sigma_{m=1}^{n}\tau(m) = \Sigma_{m=1}^{n}[\frac{n}{m}].$$

8.1.36 Let f be a function from $\mathbb{N}$ to an abelian group $(A, +)$. Let F be a function from $\mathbb{N}$ to A defined by

$$F(n) = \Sigma_{d/n}f(d).$$

Show that

$$f(n) = \Sigma_{d/n}\mu(d)F(\frac{n}{d}).$$

If the operation of the group is written multiplicatively, then if

$$F(n) = \prod_{d/n} f(d),$$

$$f(n) = \prod_{d/n} (F(\tfrac{n}{d}))^{\mu(d)}.$$

8.1.37 Show that

$$\prod_{d/n} d = n^{\frac{\tau(n)}{2}}.$$

Deduce that

$$n = \prod_{d/n} d^{\mu(\frac{n}{d}) \cdot \frac{\tau(d)}{2}}.$$

8.1.38 Let $f(X) \in \mathbb{Z}[X]$. Define an arithmetic function ψ_f by $\psi_f(n) = $ the number of integers $f(m), 1 \le m \le n$ such that $(f(m), n) = 1$. Show that ψ_f is multiplicative and

$$\psi_f(n) = n \cdot \prod_{p/n, p \ a \ prime} \frac{\psi_f(p)}{p}.$$

8.1.39 Let a be an integer co-prime to n. Show that

$$\prod_{d/n} (a^d)^{\mu(\frac{n}{d})} \equiv 1 \ (mod \ n).$$

8.1.40 Let $\eta(n)$ denote the number of positive integers $m \le n$ such that $\overline{m^2}$ is a generator of the cyclic group $\mathbb{Z}_n$. Find a formula for $\eta(n)$. Calculate $\eta(100)$.

8.1.41 Show that

$$Lim_{n \rightsquigarrow \infty}(\Sigma_{m=1}^{n} \tfrac{1}{m^s})^2 = Lim_{n \rightsquigarrow \infty}(\Sigma_{m=1}^{n} \tfrac{\tau(m)}{m^s})$$

for $s > 1$.

8.2 Higher Degree Congruences

Let m be a positive integer greater than 1. We have already discussed linear Diophantine equations, linear congruences modulo m, or equivalently, linear equations in $\mathbb{Z}_m$. Now, we consider higher degree congruences

$$f(X) = a_0 X^n + a_1 X^{n-1} + \cdots + a_n \equiv 0 (mod \ m),$$

where $a_i \in \mathbb{Z}$ and m does not divide a_0, or equivalently, the polynomial equations

$$\overline{f}(X) = \overline{a_0}X^n + \overline{a_1}X^{n-1} + \cdots + \overline{a_n} = \overline{0},$$

where $\overline{a_0} \neq \overline{0}$ in $\mathbb{Z}_m$. Let

$$m = p_1^{\alpha_1} p_2^{\alpha_2} \cdots p_r^{\alpha_r},$$

where $p_1, p_2, \ldots p_r$ are distinct primes and $\alpha_i \geq 0$ for all i. The solutions of the congruence

$$f(X) \equiv 0 (mod\ m)$$

are precisely the solutions of the simultaneous system

$$f(X) \equiv 0 (mod\ p_1^{\alpha_1}),\ f(X) \equiv 0 (mod\ p_2^{\alpha_2}), \ldots, f(X) \equiv 0 (mod\ p_r^{\alpha_r})$$

of congruences. Further, if u_i is a solution of

$$f(X) \equiv 0 (mod\ p_i^{\alpha_i});\ i = 1, 2, \ldots r,$$

then, since

$$X \equiv u_i (mod\ p_i^{\alpha_i}) \Longrightarrow f(X) \equiv f(u_i)(mod\ p_i^{\alpha_i})$$

(for $(X - u_i)$ divides $f(X) - f(u_i)$), we need to find solutions of the simultaneous system

$$X \equiv u_1 (mod\ p_1^{\alpha_1}), \ldots, X \equiv u_r (mod\ p_r^{\alpha_r}). \tag{8.2.1}$$

of congruences. Thus, the problem of solving higher degree congruences reduces to the following two problems.

Problem 8.2.1 (i) Does there exists a solution of the simultaneous system

$$X \equiv u_1 (mod\ p_1^{\alpha_1}), \ldots, X \equiv u_r (mod\ p_r^{\alpha_r})$$

of congruences, where $p_1, p_2, \ldots, p_r$ are distinct primes?
(ii) If a solution exists, is it unique modulo $m = p_1^{\alpha_1} p_2^{\alpha_2} \cdots p_r^{\alpha_r}$?
(iii) How to find solutions if they exist?

Problem 8.2.2 (i) Does there always exist solution of

$$f(X) \equiv 0 (mod\ p^\alpha),$$

where p is prime? If not, under what conditions a solution exists?
(ii) If a solution exists, then how many distinct modulo p^α solutions are there?
(iii) How to determine the solutions?

Solution to Problem 8.2.1 is in affirmative, and this is precisely the Chinese remainder theorem.

Theorem 8.2.1 (Chinese Remainder Theorem) *Let $\{m_1, m_2, \ldots, m_r\}$ be a set of pairwise co-prime integers. Let $a_1, a_2, \ldots, a_r$ be integers. Then the system of congruences*

$$X \equiv a_1 (mod\ m_1),\ X \equiv a_2 (mod\ m_2), \ldots, X \equiv a_r (mod\ m_r)$$

has a common solution. If a and b are any two common solutions of the above system of congruences, then $a \equiv b (mod\ m)$, where $m = m_1 m_2 \ldots m_r$.

Proof Put $n_i = \frac{m}{m_i}$. Since $\{m_1, m_2, \ldots, m_r\}$ is a set of pairwise co-prime integers

$$(m_i, n_i) = 1\ for\ all\ i = 1, 2, \ldots, r.$$

By the Euclidean algorithm, we can find integers u_i and v_i such that

$$u_i m_i + v_i n_i = 1\ for\ all\ i = 1, 2, \ldots, r.$$

Put $b_i = v_i n_i$, $i = 1, 2, \ldots, r$. Then, $b_i \equiv 1 (mod\ m_i)$, and $b_i \equiv 0 (mod\ n_i)$. Thus,

$$b_i \equiv 1 (mod\ m_i),\ and\ b_i \equiv 0 (mod\ m_j)\ for\ all\ j \neq i,\ i = 1, 2, \ldots, r.$$

Take $a = a_1 b_1 + a_2 b_2 + \cdots + a_r b_r$. Then,

$$a - a_i = a_1 b_1 + a_2 b_2 + \cdots a_{i-1} b_{i-1} + (b_i - 1) a_i + \cdots + a_r b_r$$

is clearly divisible by m_i for all i. Thus,

$$a \equiv a_i (mod\ m_i)\ for\ all\ i,$$

and so a is a common solution to the given system of congruences.

Further, since $\{m_1, m_2, \ldots, m_r\}$ is a set of pairwise co-prime integers,

$$a \equiv b (mod\ m)\ if\ and\ only\ if\ a \equiv b (mod\ m_i)\ for\ all\ i.$$

Thus, there is a unique solution modulo m of the above system of congruences. ♯

Corollary 8.2.2 *Let $\{m_1, m_2, \ldots, m_r\}$ be a set of pairwise co-prime positive integers. Let $\{a_1, a_2, \ldots, a_r\}$ be a set of non negative integers such that $a_i < m_i$ for all i. Then, there is unique smallest non negative integer t such that if we divide t by m_i the remainder is a_i for all i.*

Proof From the above theorem, we can find an integer a such that $a \equiv a_i(mod\ m_i)$ for all i. Divide a by m, and let t be the remainder. Then, $0 \leq t < m$ and

$$t \equiv a_i(mod\ m_i)\ for\ all\ i.$$

This means that t is such that if we divide t by m_i the remainder is a_i for all i. Clearly, t is the smallest such nonnegative integer.   ♯

Example 8.2.3 In this example, we illustrate the algorithm of chinese remainder theorem by means of an example. We find the smallest positive integer a such that if we divide a by 3 the remainder is 2, if we divide a by 5 the remainder is 1, and if we divide a by 7 the remainder is 3. We first find a common solution of the system

$$X \equiv 2(mod\ 3),\ \ X \equiv 1(mod\ 5),\ \ X \equiv 3(mod\ 7)$$

of congruences. Here, $m_1 = 3$, $m_2 = 5$, $m_3 = 7$, $a_1 = 2$, $a_2 = 1$, and $a_3 = 3$. Thus, $n_1 = 5 \times 7 = 35$, $n_2 = 3 \times 7 = 21$, and $n_3 = 3 \times 5 = 15$. Now, $(3, 35) = 1$. Using the Euclidean algorithm, we find that

$$1 = 12 \times 3 + (-1) \times 35.$$

Thus, $b_1 = -35$. Similarly, $b_2 = 21$ and $b_3 = 15$. Therefore, a common solution is given by

$$a_1 b_1 + a_2 b_2 + a_3 b_3 = 2b_1 + 1b_2 + 3b_3 = -4.$$

To find the smallest nonnegative solution, we divide -4 by $m = m_1 m_2 m_3 = 105$ and take the remainder. Since

$$-4 = (-1) \times 105 + 101,$$

101 is the required solution.

Corollary 8.2.4 *Let $\{m_1, m_2, \ldots, m_r\}$ be a set of pairwise co-prime integers, and $m = m_1 m_2 \ldots m_r$. Then*

$$\mathbb{Z}_m \approx \mathbb{Z}_{m_1} \times \mathbb{Z}_{m_2} \times \cdots \times \mathbb{Z}_{m_r}$$

as rings.

Proof Define a map f from the ring $\mathbb{Z}$ of integers to $\mathbb{Z}_{m_1} \times \mathbb{Z}_{m_2} \times \cdots \times \mathbb{Z}_{m_r}$ by

$$f(a) = (\overline{a}, \overline{a}, \ldots, \overline{a}),$$

where $\overline{a}$ at the ith place is the residue class of a in $\mathbb{Z}_{m_i}$. It is straightforward to see that f is a homomorphism of rings. Let $(\overline{a_1}, \overline{a_2}, \ldots, \overline{a_r})$ be an arbitrary element of $\mathbb{Z}_{m_1} \times \mathbb{Z}_{m_2} \times \cdots \times \mathbb{Z}_{m_r}$. By the chinese remainder theorem, there is an integer a

such that $a \equiv a_i \pmod{m_i}$ for all i. Thus, $\overline{a} = \overline{a_i}$ in $\mathbb{Z}_{m_i}$ for all i. This shows that f is surjective. Suppose that $f(a) = (\overline{0}, \overline{0}, \ldots, \overline{0})$. Then, $\overline{a} = \overline{0}$ in $\mathbb{Z}_{m_i}$ for all i. Thus, m_i/a for all i. Since $m_1, m_2, \ldots, m_r$ are pairwise co-prime, m/a, and so $a \in m\mathbb{Z}$. This shows that $ker\ f = m\mathbb{Z}$. The result follows from the fundamental theorem of homomorphism for rings. ♯

Corollary 8.2.5 *Under the hypothesis of the above corollary,*

$$U_m \approx U_{m_1} \times U_{m_2} \times \cdots \times U_{m_r}.$$

Proof We know that the group of units of product of rings is the product of group of units of each of them. The result follows from the above corollary, if we observe that $U(\mathbb{Z}_m) = U_m$. ♯

Another proof of multiplicativity of Φ. Let $(m_1, m_2) = 1$. Then, from the above corollary, $U_m \approx U_{m_1} \times U_{m_2}$. Comparing the orders, we get that $\Phi(m_1 m_2) = \Phi(m_1)\Phi(m_2)$. ♯

In general, a system of congruences need not have any common solution. For example, congruences $X \equiv 0 \pmod 4$ and $X \equiv 1 \pmod 6$ have no common solution (prove it). The following theorem gives us a necessary and sufficient conditions for a system of congruences to have a common solution.

Theorem 8.2.6 *Let $\{m_1, m_2, \ldots, m_r\}$ be a set of nonzero integers. Let $a_1, a_2, \ldots, a_r$ be integers. Then the system*

$$X = u_1 \pmod{m_1}, \ X \equiv a_2 \pmod{m_2}, \ldots, X \equiv a_r \pmod{m_r}$$

of congruences has a common solution if and only if $(m_i, m_j)/(a_i - a_j)$ for all pair (i, j). Further, if a solution exists, it is unique modulo $[m_1, m_2, \ldots, m_r]$.

Proof Suppose that we have a common solution a of the given system of congruences. Then, for every pair (i, j), $a \equiv a_i \pmod{m_i}$ and $a \equiv a_j \pmod{m_j}$. Let $a = a_i + m_i k_i = a_j + m_j k_j$. Then, $a_i - a_j = m_j k_j - m_i k_i$. This shows that $(m_i, m_j)/(a_i - a_j)$.

Conversely, suppose that $(m_i, m_j)/(a_i - a_j)$ for all i, j. We show that there is a common solution a to the system of congruences which is unique modulo $[m_1, m_2, \ldots, m_r]$. The proof is by induction on $r \geq 2$. We are given that $(m_1, m_2)/(a_1 - a_2)$. Any solution of $X \equiv a_1 \pmod{m_1}$ is of the form $a_1 + km_1$. Since (m_1, m_2) divides $(a_2 - a_1)$, *there exists a* k such that $a_1 + km_1 \equiv a_2 \pmod{m_2}$. Further, we know from the theory of linear congruences that k is unique modulo $\frac{m_2}{(m_1, m_2)}$, and so a solution $a_1 + km_1$ is unique modulo $\frac{m_1 m_2}{(m_1, m_2)} = [m_1, m_2]$.

Assume that the result is true for r. Then, we prove it for $r + 1$. Let $m_1, m_2, \ldots, m_{r+1}$ be nonzero integers together with integers $a_1, a_2, \ldots, a_{r+1}$ such that

$(m_i, m_j)/(a_i - a_j)$ for all i, j. Then, we have to show the existence of a common solution to the system of congruences

$$X \equiv a_1(mod\ m_1),\ X \equiv a_2(mod\ m_2), \ldots, X \equiv a_{r+1}(mod\ m_{r+1}).$$

From what we have shown above, it follows that there is a common solution a of the pair of congruences

$$X \equiv a_1(mod\ m_1)\ \text{and}\ X \equiv a_2(mod\ m_2)$$

which is unique modulo $[m_1, m_2]$. Consider the system

$$X \equiv a(mod\ [m_1, m_2]),\ X \equiv a_3(mod\ m_3), \ldots, X \equiv a_{r+1}(mod\ m_{r+1}).$$

Already $(m_i, m_j)/(a_i - a_j)$, $3 \leq i \leq r + 1$, $3 \leq j \leq r + 1$. Since $m_1/(a - a_1)$, $(m_1, m_i)/(a - a_1)$. Also, since $(m_1, m_i)/(a_1 - a_i)$, it follows that $(m_1, m_i)/(a - a_i)$. Similarly, $(m_2, m_i)/(a - a_i)$. But, then it follows that $([m_1, m_2], m_i)$ divides $a - a_i$. By the induction hypothesis, there is a common solution of the system

$$X \equiv a(mod\ [m_1, m_2]),\ X \equiv a_3(mod\ m_3), \ldots, X \equiv a_{r+1}(mod\ m_{r+1})$$

which is unique modulo $[[m_1, m_2], m_3, \ldots, m_{r+1}] = [m_1, m_2, \ldots, m_{r+1}]$. Since solutions of $X \equiv a(mod\ [m_1, m_2])$ are also solutions of the system $X \equiv a_1(mod\ m_1)$, $X \equiv a_2(mod\ m_2)$ of congruences, the result follows.   ♯

Example 8.2.7 Here, again, in this example, we illustrate the algorithm of the above theorem by solving for a common solution of the system

$$X \equiv 1(mod\ 9),\ X \equiv 7(mod\ 15),\ X \equiv 2(mod\ 10)$$

of congruences and getting the smallest positive solution. Observe that the hypothesis of the above theorem is satisfied.

First consider the pair

$$X \equiv 1(mod\ 9)\ \text{and}\ X \equiv 7(mod\ 15)$$

of congruences. Any solution of the congruence equation $X \equiv 1(mod\ 9)$ is of the form $1 + 9\alpha$ for some $\alpha \in \mathbb{Z}$. If $1 + 9\alpha \equiv 7(mod\ 15)$, then $9\alpha \equiv 6(mod\ 15)$ or $3\alpha \equiv 2(mod\ 5)$. Thus, $\alpha = 4$, and so 37 is a common solution of the pair

$$X \equiv 1(mod\ 9)\ \text{and}\ X \equiv 7(mod\ 15)$$

of congruences. Now, $[9, 15] = 45$, and so we try to get a common solution to pair

$$X \equiv 37(mod\ 45)\ \text{and}\ X \equiv 2(mod\ 10)$$

of congruences. Any solution of $X \equiv 37 (mod\ 45)$ is of the form $37 + 45\beta$, and if it is also a solution of $X \equiv 2 (mod\ 10)$, then $37 + 45\beta \equiv 2 (mod\ 10)$, or equivalently, $45\beta \equiv -35 (mod\ 10) \equiv 5 (mod\ 10)$. This means that $9\beta \equiv 1 (mod\ 2)$. Thus, $\beta = 1$, and $37 + 45 \times 1 = 82$ is a common solution to the given system of congruences. Since $[9, 15, 10] = 90$, 82 is the smallest nonnegative solution.

Because of the Chinese remainder theorem, the problem of solving higher degree congruence modulo n reduces to the problem of solving higher degree congruence modulo p^α, where p is a prime. We further reduce this problem to the problem of solving higher degree congruence modulo p.

Theorem 8.2.8 *Let $m \geq 1$ and p a prime. Let*

$$f(X) = a_0 + a_1 X + a_2 X^2 + \cdots + a_n X^n$$

be a polynomial in $\mathbb{Z}[X]$, where p^{m+1} does not divide a_n. If α is a solution to the congruence

$$f(X) \equiv 0 (mod\ p^{m+1}),$$

then it is also a solution of

$$f(X) \equiv 0 (mod\ p^m).$$

If β is a solution of

$$f(X) \equiv 0 (mod\ p^m),$$

then $\alpha = \beta + \gamma \cdot p^m$ is a solution of the congruence

$$f(X) \equiv 0 (mod\ p^{m+1})$$

if and only if γ is a solution of the congruence

$$Df(\beta) \cdot X \equiv -\frac{f(\beta)}{p^m} (mod\ p),$$

where $Df(X)$ denotes the derivative of the polynomial $f(X)$ (see Exercise 7.6.27).

Proof Clearly, a solution of

$$f(X) \equiv 0 (mod\ p^{m+1})$$

is also a solution of

$$f(X) \equiv 0 (mod\ p^m).$$

Let β be a solution of

$$f(X) \equiv 0 (mod\ p^m).$$

Then, $p^m / f(\beta)$. By the Taylor's theorem (see Exercise 7.6.29),

$$f(\beta + \gamma p^m) = f(\beta) + \gamma p^m Df(\beta) + (\gamma p^m)^2 \frac{D^2 f(\beta)}{2!} + \cdots + (\gamma p^m)^n \frac{D^n f(\beta)}{n!}.$$

$\frac{D^k f(\beta)}{k!}$ are integers for $k \geq 1$ (Exercise 7.6.28). Thus,

$$f(\beta + \gamma p^m) \equiv (f(\beta) + \gamma p^m Df(\beta))(mod\ p^{m+1}).$$

Hence,

$$f(\beta + \gamma p^m) \equiv 0(mod\ p^{m+1})$$

if and only if

$$f(\beta) + \gamma p^m Df(\beta) \equiv 0(mod\ p^{m+1}),$$

or equivalently,

$$(Df(\beta)) \cdot \gamma \equiv -\frac{f(\beta)}{p^m}(mod\ p).$$

♯

Corollary 8.2.9 *The number of mod p^{m+1} distinct solutions of*

$$f(X) \equiv 0(mod\ p^{m+1})$$

which are congruent to β modulo p^m is

(i) 0, if p divides $Df(\beta)$ but it does not divide $\frac{f(\beta)}{p^m}$,
(ii) p, if p divides $Df(\beta)$, and it also divides $\frac{f(\beta)}{p^m}$, and
(iii) 1, otherwise.

Proof We know that $aX \equiv b(mod\ m)$ has no solution if (a, m) does not divide b, it has (a, m) distinct modulo m solutions if (a, m) divides b. The result follows from the above theorem.   ♯

Remark 8.2.10 Theorem 8.2.8 gives us a method by which, knowing the solutions of $f(X) \equiv 0(mod\ p)$, we can determine (by induction) all solutions of $f(X) \equiv 0(mod\ p^m)$, $m \in \mathbb{N}$. Thus, the problem of solving higher degree congruences reduces to the problem of solving them modulo different primes.

Example 8.2.11 We illustrate the algorithm of the above theorem and corollary by solving the congruence

$$X^3 + X^2 + X + 1 \equiv 0(mod\ 27).$$

First, let us look at the solutions of

$$X^3 + X^2 + X + 1 \equiv 0(mod\ 3).$$

Note that we have not yet described an algorithm to solve a higher degree congruence modulo p. However, for small p, we can try all integers modulo p.

We see that out of $0, 1, 2$, only 2 is the solution of the above congruence. Thus, by the above theorem, $2 + \gamma \cdot 3$ is a solution of

$$X^3 + X^2 + X + 1 \equiv 0(mod \ 3^2)$$

if and only if γ is a solution of

$$D(X^3 + X^2 + X + 1)(2) \cdot X \equiv -\frac{2^3 + 2^2 + 2 + 1}{3}(mod \ 3).$$

Thus, γ should be a solution of

$$17X \equiv -5(mod \ 3)$$

or equivalently

$$2X \equiv 1(mod \ 3).$$

Thus, $\gamma = 2$ is the unique solution. Hence, $2 + 2 \times 3 = 8$ is the unique solution modulo 9 of congruence

$$X^3 + X^2 + X + 1 \equiv 0(mod \ 3^2).$$

Further, let $8 + \delta \cdot 3^2$ be a solution of

$$X^3 + X^2 + X + 1 \equiv 0(mod \ 3^3).$$

Then, again from the above theorem, δ should be a solution of

$$D(X^3 + X^2 + X + 1)(8) \equiv -\frac{8^3 + 8^2 + 8 + 1}{3^2}(mod \ 3)$$

or equivalently, δ should be a solution of

$$205X \equiv -45(mod \ 3).$$

This, in turn, means that δ should be a solution of

$$2X \equiv 1(mod \ 3).$$

This gives the unique $\delta = 2(mod \ 3)$. Hence, $8 + 2 \times 3^2 = 26$ is the unique solution modulo 27 of the given congruence.

Example 8.2.12 We use the algorithms in this section to solve the congruence

$$f(X) \equiv 0(mod \ 135),$$

where $f(X) = X^3 + X^2 + X + 1$. Now, $135 = 3^3 \times 5$. Thus, we first solve the congruences

$$f(X) \equiv 0 (mod \; 3^3) \qquad\qquad (8.2.2)$$

and

$$f(X) \equiv 0 (mod \; 5) \qquad\qquad (8.2.3)$$

We have already seen in the above example that 26 is the unique solution of 1. Further, we find that 2, 3, and 4 are three distinct solutions modulo 5 of the congruence equation 8.2.3. Thus, there are three distinct (modulo 135) solutions of the congruence

$$f(X) \equiv 0 (mod \; 135)$$

which are common solutions of the following three pairs of congruences:

1. $X \equiv 26 (mod \; 27)$, $X \equiv 2 (mod \; 5)$.
2. $X \equiv 26 (mod \; 27)$, $X \equiv 3 (mod \; 5)$.
3. $X \equiv 26 (mod \; 27)$, $X \equiv 4 (mod \; 5)$.

Using the Chinese remainder theorem, we find that the solutions are 107, 53, and 134.

Exercises

8.2.1 Give an example of a system of congruences which has no common solution.

8.2.2 Find the smallest positive integer x such that if we divide x by 7 the remainder is 2, if we divide x by 10 the remainder is 3, and if we divide x by 9 the remainder is 4.

8.2.3 Find the smallest positive integer x such that $x \equiv 3 (mod \; 10)$, $x \equiv 7 (mod \; 22)$, and $x \equiv -4 (mod \; 11)$.

8.2.4 Find all distinct solutions modulo 343 of the congruence

$$X^3 + 2X^2 + 3X + 35 \equiv 0 (mod \; 343).$$

8.2.5 Find all distinct modulo $49 \times 17 = 833$ solutions of the congruence equation

$$X^3 + 2X^2 + 3X + 35 \equiv 0 (mod \; 833).$$

8.2.6 Let $f(X) \in \mathbb{Z}[X]$ be such that $f(n)$ is eventually prime in the sense that *there exists a* $n_0 \in \mathbb{N}$ such that $f(n)$ is prime *for all* $n \geq n_0$. Show that $f(X)$ is a constant polynomial.

8.2.7 Let $m = p_1^{\alpha_1} p_2^{\alpha_2} \ldots p_r^{\alpha_r}$, where $p_1, p_2, \ldots, p_r$ are all distinct primes. Suppose that

$$f(X) \equiv 0(mod \ p_i^{\alpha_i})$$

has n_i distinct (modulo $p_i^{\alpha_i}$) solutions. Show that

$$f(X) \equiv 0(mod \ m)$$

has $n_1 \cdot n_2 \cdots n_r$ distinct (modulo m) solutions.

8.2.8 Suppose that p is a prime which does not divide a, and also it does not divide n. Show that $X^n \equiv a(mod \ p^m)$ has a solution if and only if $X^n \equiv a(mod \ p)$ has a solution.

8.2.9 Let $m \in \mathbb{N}$ be such that U_m is cyclic group. Let $a \in \mathbb{Z}$ such that $(a, m) = 1$. Show that $X^n \equiv a(mod \ m)$ is solvable if and only if

$$a^{\frac{\Phi(m)}{(n, \Phi(m))}} \equiv 1(mod \ m).$$

In particular, if p is a prime, then

$$X^n \equiv a(mod \ p)$$

has a solution if and only if

$$a^{\frac{p-1}{(n, p-1)}} \equiv 1(mod \ p).$$

8.3   Quadratic Residues and Quadratic Reciprocity

In Sect. 8.2, we observed that to solve a higher degree congruence modulo m, we need to evolve a method to solve higher degree congruence modulo p, where p is a prime. In this section, we study quadratic congruences modulo p.

There are two cases:

(i) $p = 2$.
(ii) p is an odd prime.

Consider the case when $p = 2$. Let $f(X) \in \mathbb{Z}[X]$ be a quadratic polynomial. Then, there are three possibilities

(a) $f(X) \equiv (X^2 + X + 1)(mod \ 2)$.
(b) $f(X) \equiv (X^2 + X)(mod \ 2)$.
(c) $f(X) \equiv (X^2 + 1)(mod \ 2)$.

If $f(X) \equiv (X^2 + X + 1)(mod \ 2)$, then $f(X) \equiv 0(mod \ 2)$ has no solution, for 2 divides $x^2 + x$ for all $x \in \mathbb{Z}$. If $f(X) \equiv (X^2 + X)(mod \ 2)$, then 0 and 1 both are solutions. Finally, if $f(X) \equiv (X^2 + 1)(mod\ 2)$, then 1 is the unique solution of $f(X) \equiv 0(mod\ 2)$.

Now, we assume that p is an odd prime. Let

$$f(X) = a_0 + a_1 X + a_2 X^2$$

be a polynomial in $\mathbb{Z}[X]$ of degree 2 such that p does not divide a_2. Since $(p, a_2) = 1$, *there exists a b_2* such that $(b_2, p) = 1$, and $b_2 a_2 \equiv 1(mod\ p)$. Further, since $(b_2, p) = 1$,

$$f(X) \equiv b_2 f(X)(mod\ p) \equiv (X^2 + b_2 a_1 X + b_2 a_0)(mod\ p).$$

Thus, any quadratic polynomial

$$f(X) = a_0 + a_1 X + a_2 X^2,$$

where p does not divide a_2 is congruent to a polynomial of the type

$$X^2 + bX + c$$

modulo p. Since p is an odd prime, $(p, 2) = 1 = (p, 4)$. Thus, there exists an integer d such that $d \cdot 2 \equiv 1(mod\ p)$, $d^2 \cdot 4 \equiv 1(mod\ p)$, and $(d, p) = 1$. Now,

$$X^2 + bX + c \equiv (X^2 + 2dbX + b^2 d^2 - b^2 d^2 + c)(mod\ p) \equiv$$
$$((X + bd)^2 + c - b^2 d^2)(mod\ p).$$

Thus, the solutions of
$$X^2 + bX + c \equiv 0(mod\ p)$$

are same as the solutions of

$$(X + bd)^2 \equiv (b^2 d^2 - c)(mod\ p).$$

This has solution if and only if

$$Y^2 \equiv (b^2 d^2 - c)(mod\ p)$$

has a solution, and if u is a solution, then $u - bd$ is a solution of

$$X^2 + bX + c \equiv 0(mod\ p).$$

Observe that there are at most two distinct solutions mod p. This motivates to have the following definition.

Definition 8.3.1 Let $m \in \mathbb{N}$, $m \geq 2$. Let $a \in \mathbb{Z}$ be such that $(a, m) = 1$. We say that a is a **quadratic residue mod m** if

$$X^2 \equiv a(mod\ m)$$

is solvable.

The following proposition follows as particular case of the discussions in the beginning of Sect. 8.2 of this chapter.

Proposition 8.3.2 *Let* $m = p_1^{\alpha_1} p_2^{\alpha_2} \dots p_r^{\alpha_r}$, *where* $p_1, p_2, \dots p_r$ *are distinct primes, and* $\alpha_i \geq 1$ *for all i. Then a is a quadratic residue mod m if and only if a is quadratic residue mod* $p_i^{\alpha_i}$ *for all i.*   ♯

Proposition 8.3.3 *Let p be an odd prime, and a an integer co-prime to p. Then a is a quadratic residue modulo* p^m, $m \geq 1$ *if and only if a is quadratic residue modulo p. Further, then there are only two distinct solutions modulo* p^m *of*

$$X^2 \equiv a(mod\ p^m).$$

Proof The proof is by induction on m. If $m = 1$, the conclusion is the same as the hypothesis. Assume the result for m. We have to prove it for $m + 1$. Thus, $X^2 \equiv a(mod\ p^m)$ has a solution if and only if $X^2 \equiv a(mod\ p)$ has a solution, and then, there are exactly two solutions modulo p^m. Note that $X^2 \equiv a(mod\ p^m)$ means that $(X^2 - a) \equiv 0(mod\ p^m)$, and also $X^2 \equiv a(mod\ p^{m+1})$ means that $(X^2 - a) \equiv 0(mod\ p^{m+1})$. Applying Theorem 8.2.8 to the polynomial $X^2 - a$, we observe that corresponding to every solution β of $(X^2 - a) \equiv 0(mod\ p^m)$, there is a unique solution of $(X^2 - a) \equiv 0(mod\ p^{m+1})$ if and only if p does not divide $Df(\beta) = 2\beta$. Since β is supposed to be a solution of $(X^2 - a) \equiv 0(mod\ p^m)$, and p does not divide a, it follows that p does not divide β. Since p is an odd prime, it does not divide 2β. The result follows by induction.   ♯

The following proposition takes care of the case $p = 2$.

Proposition 8.3.4 *Let a be an odd integer. Then*

(i) $X^2 \equiv a(mod\ 2)$ *has a unique solution mod 2.*
(ii) $X^2 \equiv a(mod\ 4)$ *has a solution if and only if* $a \equiv 1(mod\ 4)$, *and then it has exactly 2 distinct solutions mod 4.*
(iii) $X^2 \equiv a(mod\ 2^m)$, $m \geq 3$ *has a solution if and only if* $a \equiv 1(mod\ 8)$, *and then it has exactly 4 distinct solutions mod 8.*

Proof (i) Clearly, 1 is the unique solution of $X^2 \equiv a(mod\ 2)$.
(ii) If b is a solution of $X^2 \equiv a(mod\ 4)$, then since a is odd, b is also odd. Suppose that $b = 1 + 2l$. Then,

$$b^2 = 1 + 4l + 4l^2 \equiv a(mod\ 4),$$

and hence,

$$a \equiv 1(mod\ 4).$$

Further, then out of 4 residue classes modulo 4, only $1(mod\ 4)$ and $3(mod\ 4)$ are the solutions of

$$X^2 \equiv a(mod\ 4) \equiv 1(mod\ 4).$$

(iii) Suppose that $b = 1 + 2l$ is a solution of

$$X^2 \equiv a(mod\ 2^m),\ m \geq 3.$$

Then,

$$1 + 4l^2 + 4l \equiv a(mod\ 2^m),\ m \geq 3$$

or

$$1 + 4l(l+1) \equiv a(mod\ 2^m),\ m \geq 3.$$

This implies that

$$1 + 4l(l+1) \equiv a(mod\ 8).$$

Since $2/l(l+1)$,

$$a \equiv 1(mod\ 8).$$

Further, suppose that $a \equiv 1(mod\ 8)$, then out of 8 residue classes modulo 8, $1(mod\ 8), 3(mod\ 8), 5(mod\ 8)$, and $7(mod\ 8)$ are 4 solutions of

$$X^2 \equiv a(mod\ 8) \equiv 1(mod\ 8).$$

Also, each solution of

$$X^2 \equiv a(mod\ 2^m),\ m \geq 3$$

is also a solution of

$$X^2 \equiv a(mod\ 8).$$

Conversely, we show that each solution of

$$X^2 \equiv a(mod\ 8)$$

determines a unique solution of

$$X^2 \equiv a(mod\ 2^m),\ m \geq 3.$$

This we prove by induction on m. If $m = 3$, there is nothing to do. Assume the result for m. Let b be a solution of

$$X^2 \equiv a(mod\ 2^m).$$

Then, $b + \lambda 2^{m-1}$ is a solution of

$$X^2 \equiv a (mod\ 2^{m+1})$$

if and only if

$$(b + \lambda 2^{m-1})^2 \equiv a (mod\ 2^m)$$

or

$$b^2 + \lambda^2 2^{2m-2} + 2^m \lambda b \equiv a (mod\ 2^{m+1})$$

or

$$2^m \lambda b \equiv (a - b^2)(mod\ 2^{m+1}).$$

Since $2^m / (a - b^2)$,

$$\lambda b \equiv \frac{a-b^2}{2^m} (mod\ 2).$$

Since b is odd (for a is odd), there is a unique $\lambda (mod\ 2)$ satisfying the above congruence. Thus, corresponding to each solution of

$$X^2 \equiv a (mod\ 2^m),$$

we have a unique solution of

$$X^2 \equiv a (mod\ 2^{m+1}).$$

The result follows from the principle of induction. ⊓⊔

Theorem 8.3.5 (Euler's Criterion) *Let p be an odd prime, and a an integer such that p does not divide a. Then a is quadratic residue modulo p if and only if*

$$a^{\frac{p-1}{2}} \equiv 1 (mod\ p).$$

Proof Suppose that a is a quadratic residue mod p. Then, there is an integer b such that

$$b^2 \equiv a (mod\ p).$$

Since $(a, p) = 1$, $(b, p) = 1$. By the Fermat's theorem,

$$b^{p-1} \equiv 1 (mod\ p).$$

Hence,

$$a^{\frac{p-1}{2}} \equiv (b^2)^{\frac{p-1}{2}} (mod\ p) \equiv b^{p-1}(mod\ p) \equiv 1 (mod\ p).$$

Next, suppose that a is not a quadratic residue mod p. Let b be a positive integer less than p and so co-prime to p also. Then,

$$bX \equiv a(mod\ p).$$

has a unique solution modulo p. Since a is not a quadratic residue modulo p, this solution can not be $b(mod\ p)$. Thus, there is an element $c \neq b$, $1 \leq c \leq p - 1$ such that

$$bc \equiv a(mod\ p).$$

This shows that we can pair the set $\{1, 2, \ldots, p - 1\}$ such that product of any pair is congruent to $a(mod\ p)$. Taking the product of $1, 2, \ldots, p - 1$, we obtain that

$$(p - 1)! \equiv a^{\frac{p-1}{2}} (mod\ p).$$

By the Wilson theorem, $(p - 1)! \equiv (-1)(mod\ p)$. Hence, if a is not a quadratic residue, then

$$a^{\frac{p-1}{2}} \equiv (-1)(mod\ p). \qquad \qquad \sharp$$

The following corollary follows from the previous 4 results.

Corollary 8.3.6 *Let* $m = 2^{\alpha_1} p_2^{\alpha_2} p_3^{\alpha_3} \ldots p_r^{\alpha_r}$, *where* $p_2, p_3, \ldots, p_r$ *are distinct odd primes,* $\alpha_1 \geq 0$, *and* $\alpha_i \geq 1$ *for all* $i \geq 2$. *Let* a *be an integer such that* $(a, m) = 1$. *Then* a *is a quadratic residue modulo* m *if and only if*

(i) $a^{\frac{p_i-1}{2}} \equiv (mod\ p_i)$ *for all* $i \geq 2$.
(ii) $a \equiv 1(mod\ 4)$ *if* $\alpha_1 = 2$, *and* $a \equiv 1(mod\ 8)$ *if* $\alpha_1 \geq 3$.

Further,

$$X^2 \equiv a(mod\ m)$$

has 2^{r-1} *solutions, if* $\alpha_1 \leq 1$, 2^r *solutions, if* $\alpha_1 = 2$, *and* 2^{r+1} *solutions, if* $\alpha_1 \geq 3$. $\qquad \qquad \sharp$

The rest of the section is devoted to develop a method by which we can determine whether a given integer a is a quadratic residue mod p, p an odd prime.

Definition 8.3.7 Let a be an integer and p a positive prime. Define the Symbol $(\frac{a}{p})$ as follows:

(i) $(\frac{a}{p}) = 1$ if p does not divide a, and a is quadratic residue modulo p.
(ii) $(\frac{a}{p}) = -1$ if p does not divide a, and a is not a quadratic residue modulo p.
(iii) $(\frac{a}{p}) = 0$ if p divides a.

Let P denote the set of all positive primes. The map $(-)$ from $\mathbb{Z} \times P$ to $\{1, 0, -1\}$ defined by $(-)((a, p)) = (\frac{a}{p})$ is called the **Legendre symbol** map.

Proposition 8.3.8 *Let* p *be an odd prime. Then*

(i) $a^{\frac{p-1}{2}} \equiv (\frac{a}{p})(mod\ p)$ for all $a \in \mathbb{Z}$.

(ii) $a \equiv b(mod\ p)$ implies that $(\frac{a}{p}) = (\frac{b}{p})$ for all $a, b \in \mathbb{Z}$.

(iii) $(\frac{ab}{p}) = (\frac{a}{p}) \cdot (\frac{b}{p})$ for all $a, b \in \mathbb{Z}$.

Proof If p/a, then $a^{\frac{p-1}{2}} \equiv 0(mod\ p)$, and by the definition, $(\frac{a}{p}) = 0$, and so in this case, $a^{\frac{p-1}{2}} \equiv (\frac{a}{p})(mod\ p)$. Suppose that p does not divide a. Then, by the Euler's theorem, we have the following:

(a) If a is a quadratic residue mod p, then $a^{\frac{p-1}{2}} \equiv 1(mod\ p)$.

(b) If a is not a quadratic residue mod p, then $a^{\frac{p-1}{2}} \equiv -1(mod\ p)$.

By the definition of the Legendre symbol, it follows that

$$a^{\frac{p-1}{2}} \equiv (\frac{a}{p})(mod\ p).$$

(i) Since $a \equiv b(mod\ p)$ if and only if $a^{\frac{p-1}{2}} \equiv b^{\frac{p-1}{2}} (mod\ p)$, the result follows from part (i)

(ii) $(ab)^{\frac{p-1}{2}} = a^{\frac{p-1}{2}}b^{\frac{p-1}{2}} \equiv (\frac{a}{p})(\frac{b}{p})(mod\ p)$. Also, $(ab)^{\frac{p-1}{2}} \equiv (\frac{ab}{p})(mod\ p)$. Since the absolute value of any Legendre symbol is less than or equal to 1, the result follows.   ♯

The following corollary follows from the part (ii) of the above proposition.

Corollary 8.3.9 $(\frac{a}{p}) = (\frac{r}{p})$, where r is the remainder obtained when a is divided by p.   ♯

The following corollary is immediate from part (iii) of the proposition.

Corollary 8.3.10 *Product of any two quadratic residue mod p is a quadratic residue mod p. Product of any two nonquadratic residue is a quadratic residue. Product of a quadratic residue and a non quadratic residue is a nonquadratic residue.*   ♯

Corollary 8.3.11 *We have a surjective homomorphism from $\mathbb{Z}_p^* = U_p$ to the group $\{1, -1\}$ given by $\bar{a} \rightsquigarrow (\frac{a}{p})$.*

Proof Suppose that $\bar{a} = \bar{b}$. Then, $a \equiv b(mod\ p)$, and so $(\frac{a}{p}) = (\frac{b}{p})$. Hence, we have a map given by $a \rightsquigarrow (\frac{a}{p})$. Also since

$$(\frac{ab}{p}) = (\frac{a}{p})(\frac{b}{p}),$$

the map is a group homomorphism. Clearly, $(\frac{1}{p}) = 1$. Further, there are $\frac{p-1}{2}$ distinct solutions of $X^{\frac{p-1}{2}} = \bar{1}$ in U_p. Hence, there are only $\frac{p-1}{2}$ distinct members $\bar{a}$ of U_p for which $(\frac{a}{p}) = 1$. For the rest of $\frac{p-1}{2}$ elements $\bar{a}$ of U_p, $(\frac{a}{p}) = -1$. Hence, the given map is surjective.   ♯

Corollary 8.3.12 $(\frac{-1}{p}) = (-1)^{\frac{p-1}{2}}$. *In particular,* -1 *is a quadratic residue mod p if and only if p is of the form* $4n + 1$.

Proof From the Euler's theorem, $(\frac{-1}{p}) = (-1)^{\frac{p-1}{2}}$. *Further,* $(-1)^{\frac{p-1}{2}} = 1$ *if and only if* $p = 4n + 1$ *for some n.*   ♯

Proposition 8.3.13 *There are infinitely many primes of the form* $4n + 1$.

Proof 5 is a prime of the form $4n + 1$. Suppose that there are only finitely many primes $p_1, p_2, \ldots, p_r$ of the form $4n + 1$. Consider

$$a = (2p_1 p_2 \ldots p_r)^2 + 1.$$

By the fundamental theorem of arithmetic, there is a prime p dividing a. Clearly, p is odd, and

$$(2p_1 p_2 \ldots p_r)^2 \equiv -1(mod\ p).$$

This shows that -1 is a quadratic residue modulo p. By the above corollary, p is of the form $4n + 1$. Clearly, $p \neq p_i$ for all i. This is a contradiction to the supposition.   ♯

Lemma 8.3.14 (Gauss Lemma) *Let p be an odd prime, and a be an integer such that p does not divide a. Consider the group* U_p *of prime residue classes modulo p. Let*

$$X = \{\bar{1}, \bar{2}, \ldots, \overline{\tfrac{p-1}{2}}\}$$

and

$$Y = \{\overline{\tfrac{p+1}{2}}, \overline{\tfrac{p+3}{2}}, \ldots, \overline{p-1}\}.$$

Then

$$(\tfrac{a}{p}) = (-1)^{|\bar{a}X \cap Y|}.$$

Proof We have

$$X \bigcup Y = U_p. \tag{8.3.1}$$

and

$$-Y = X. \tag{8.3.2}$$

Also,

$$\bar{a}X = \bar{a}X \bigcap U_p = (\bar{a}X \bigcap X) \bigcup (\bar{a}X \bigcap Y). \tag{8.3.3}$$

By (8.3.2), $-(\bar{a}X \bigcap Y) \subseteq X$. Let $\bar{b} \in (\bar{a}X \bigcap X) \bigcap -(\bar{a}X \bigcap Y)$. Then, $\bar{b} = \overline{am_1} = -\overline{am_2}$, where $1 \leq m_1, m_2 \leq \frac{p-1}{2}$. But, then $\overline{a(m_1 + m_2)} = \bar{0}$. Since $(a, p) = 1$, $\overline{m_1 + m_2} = \bar{0}$. Also, $2 \leq m_1 + m_2 \leq p - 1$. This is a contradiction. Thus,

$$(\bar{a}X \bigcap X) \bigcap -(\bar{a}X \bigcap Y) = \emptyset. \tag{8.3.4}$$

Counting the number of elements, we get that

$$X = (\bar{a}X \cap X) \bigcup -(\bar{a}X \cap Y).$$

Taking the product of the elements of both sides

$$\bar{1} \cdot \bar{2} \cdot \ldots \cdot \overline{\tfrac{p-1}{2}} = \bar{1} \cdot \bar{2} \cdot \ldots \cdot \overline{\tfrac{p-1}{2}} \cdot \bar{a}^{\frac{p-1}{2}} \cdot \overline{(-1)^{|\bar{a}X \cap Y|}}.$$

Hence,

$$\bar{1} = \bar{a}^{\frac{p-1}{2}} \cdot \overline{(-1)^{|\bar{a}X \cap Y|}}$$

or equivalently,

$$\bar{a}^{\frac{p-1}{2}} = \overline{(-1)^{|\bar{a}X \cap Y|}}.$$

Thus,

$$a^{\frac{p-1}{2}} \equiv (-1)^{|\bar{a}X \cap Y|} (mod\ p).$$

By the Euler's theorem, the result follows. ♯

Corollary 8.3.15 $(\tfrac{2}{p}) = (-1)^{\frac{p^2-1}{8}}$.

Proof Let X and Y be as in the above lemma. Then, by the Gauss lemma,

$$(\tfrac{2}{p}) = (-1)^{\lambda},$$

where λ is the number of elements in the set $\bar{2}X \cap Y$. Now,

$$\bar{2}X \cap Y = \{\overline{2m} \mid \tfrac{p-1}{4} < m \le \tfrac{p-1}{2}\}.$$

Thus, λ is the number of integers m such that $\tfrac{p-1}{4} < m \le \tfrac{p-1}{2}$. Since p is an odd prime, one and only one of the following four cases hold.

(i) $p = 8k + 1,\ k \ge 1$.
(ii) $p = 8k + 3,\ k \ge 0$.
(iii) $p = 8k + 5,\ k \ge 0$.
(iv) $p = 8k + 7,\ k \ge 0$.

Consider the case (i). In this case, λ is the number of integers m such that $\frac{8k+1-1}{4} < m \le \frac{8k+1-1}{2}$. This is clearly $2k$. Thus, in this case $(\tfrac{2}{p}) = (-1)^{2k} = 1$. Also in this case, $(-1)^{\frac{p^2-1}{8}} = 1$. Therefore, in case (i), the result is true.

In case (ii), $p = 8k + 3$, and then, λ is the number of integers m such that $2k + \tfrac{1}{2} < m \le 4k + 1$. Thus, λ is the number of integers m such that $2k + 1 \le$

$m \leq 4k + 1$. In this case, $\lambda = 2k + 1$, and so $(\frac{2}{p}) = (-1)^{2k+1} = -1$. Further, $(-1)^{\frac{(8k+3)^2-1}{8}} = -1$. Therefore, in case (ii) also, the result is verified.

In case (iii), $p = 8k + 5$, and λ is the number of integers m such that $\frac{8k+4}{4} < m \leq 4k + 2$. Thus, λ is the number of integers m such that $2k + 1 < m \leq 4k + 2$. Thus, in this case, $\lambda = 2k + 1$, and $(\frac{2}{p}) = -1 = (-1)^{\frac{(8k+5)^2-1}{8}}$. Hence, in this case also, the result is true.

Finally, consider the case (iv) when $p = 8k + 7$. In this case, λ is the number of integers m such that $\frac{8k+6}{4} < m \leq 4k + 3$ which is the same as the number of integers m such that $2k + 2 \leq m \leq 4k + 3$. Clearly, $\lambda = 2k + 2$, and so $(\frac{2}{p}) = 1 = (-1)^{\frac{(8k+7)^2-1}{8}} = (-1)^{\frac{p^2-1}{8}}$. This completes the proof.   ♯

As an application of the above result, we have the following:

Corollary 8.3.16 *There are infinitely many primes of the form* $8k + 7$.

Proof Clearly, 7 is a prime of this form. Suppose that there are only finitely many primes $p_1, p_2, \ldots, p_r$ of the form $8k + 7$. Consider $a = (4p_1 p_2 \ldots p_r)^2 - 2$. Clearly, 2 is a quadratic residue modulo every prime divisor of a. Obviously, there are odd prime divisors of a. It is clear from the proof of the above corollary that 2 is a quadratic residue modulo an odd prime p if and only if p is of the form $8k + 1$ or it is of the form $8k + 7$. Since product of numbers of forms $8k + 1$ is also numbers of same form, if all odd prime divisors of a are of the form $8k + 1$, then

$$a = 16p_1^2 p_2^2 \ldots p_r^2 - 2 = 2(8l + 1)$$

for some l. This implies that $8t^2 - 1 = 8l + 1$, where $t = p_1 p_2 \ldots p_r$. This is a contradiction, for it would mean that 8 divides 2. Hence, not all odd prime divisors are of the form $8k + 1$. Thus, there is a prime divisor of a of the form $8k + 7$. Clearly, this prime is different from $p_1, p_2, \ldots p_r$. This again is contradiction to the supposition that there are only finitely many primes of the form $8k + 7$.   ♯

Remark 8.3.17 The above result is a particular case of a more general theorem due to Dirichlet which says that if a and b are co-prime, then there are infinitely many primes of the form $ax + b$.

Corollary 8.3.18 *Let p be an odd prime. Then, $\bar{2}$ is a generator of U_p in each of the following two cases:*

(i) p is of the form $4q + 1$, where q is an odd prime.
(ii) p is of the form $2q + 1$, where q is a prime of the form $4k + 1$.

Also $-\bar{2}$ is a generator of U_p, if p is of the form $2q + 1$, where q is a prime of the form $4k + 3$.

Proof $p = 4q + 1$. Let m be the order of $\bar{2}$ in U_p. Then, $m/(p - 1)$. Since q is an odd prime, $m = 2, 4, q, 2q$ or $4q$. If $m = 2$, then $2^2 \equiv 1 \pmod{p}$. But, then $p = 3$ is not of the form $4q + 1$. If $m = 4$, then $2^4 \equiv 1 \pmod{p}$. But, then

$p = 5$ is not of the form $4q + 1$ with q is a prime. If $m = q$ or $m = 2q$, then $2^{2q} = 2^{\frac{p-1}{2}} \equiv 1(mod\ p)$. But $(\frac{2}{p}) = (-1)^{\frac{p^2-1}{8}} = -1$. This is a contradiction to the Euler's theorem. This shows that $m = p - 1$. The rest of the corollary can be proved similarly and is left as an exercise.   ♯

The following theorem known as *Gauss Quadratic Reciprocity Law* is extremely useful to calculate the Legendre symbol and, in turn, to determine whether an integer is quadratic residue modulo m. This is the Golden theorem of Gauss. There are several proofs of the theorem. We give a combinatorial proof.

Theorem 8.3.19 (Gauss Quadratic Reciprocity Law) *Let p and q be odd primes. Then*

$$(\tfrac{p}{q}) \cdot (\tfrac{q}{p}) = (-1)^{\frac{p-1}{2} \cdot \frac{q-1}{2}}.$$

Proof By the Gauss lemma,

$$(\tfrac{q}{p}) = (-1)^{|\bar{q}X \cap Y|},$$

and

$$(\tfrac{p}{q}) = (-1)^{|\bar{p}Z \cap T|},$$

where

$$X = \{\bar{1}, \bar{2}, \ldots, \overline{\tfrac{p-1}{2}}\},$$

$$Y = \{\overline{\tfrac{p+1}{2}}, \overline{\tfrac{p+3}{2}}, \ldots, \overline{p-1}\}$$

are subsets of U_p,

$$Z = \{\bar{1}, \bar{2}, \ldots, \overline{\tfrac{q-1}{2}}\},$$

and

$$T = \{\overline{\tfrac{q+1}{2}}, \overline{\tfrac{q+3}{2}}, \ldots, \overline{q-1}\}$$

are subsets of U_q. Put $\lambda = |\bar{q}X \cap Y|$ and $\mu = |\bar{p}Z \cap T|$. It is sufficient to show that

$$\lambda + \mu \equiv \tfrac{p-1}{2} \cdot \tfrac{q-1}{2}(mod\ 2).$$

Let

$$A = \{1, 2, \ldots, \tfrac{p-1}{2}\},$$

and

$$B = \{1, 2, \ldots, \tfrac{q-1}{2}\}.$$

Let

$$P = \{(x, y) \in A \times B \mid -\tfrac{p}{2} < qx - py < 0\},$$

$$Q = \{(x, y) \in A \times B \mid -\tfrac{q}{2} < py - qx < 0\},$$

$$R = \{(x, y) \in A \times B \mid qx - py < -\tfrac{p}{2}\},$$

and

$$S = \{(x, y) \in A \times B \mid py - qx < -\tfrac{q}{2}\}.$$

Clearly, the sets P, Q, R, and S are pairwise disjoint. We show that

$$A \times B = P \bigcup Q \bigcup R \bigcup S$$

It is easily observed that

$$P \bigcup Q \bigcup R \bigcup S \subseteq A \times B.$$

Let $(x, y) \in A \times B$. Then, $qx - py \neq 0$ for otherwise x has to be at least p and y has to be at least q. Also, since $qx - py$ is an integer and p and q are odd, $qx - py \neq \pm\tfrac{p}{2}$ and $qx - py \neq \pm\tfrac{q}{2}$. Thus,

$$(qx - py) \in (-\infty, -\tfrac{p}{2}) \bigcup (-\tfrac{p}{2}, 0) \bigcup (0, \tfrac{q}{2}) \bigcup (\tfrac{q}{2}, \infty).$$

Now,

$$(qx - py) \in (-\infty, -\tfrac{p}{2}) \Longrightarrow (x, y) \in R,$$

$$(qx - py) \in (-\tfrac{p}{2}, 0) \Longrightarrow (x, y) \in P,$$

$$(qx - py) \in (0.\tfrac{q}{2}) \Longrightarrow (x, y) \in Q$$

and

$$(qx - py) \in (\tfrac{q}{2}, \infty) \Longrightarrow (x, y) \in S.$$

This shows that

$$A \times B = P \bigcup Q \bigcup R \bigcup S.$$

In turn,

$$\tfrac{p-1}{2} \cdot \tfrac{q-1}{2} = \mid P \mid + \mid Q \mid + \mid R \mid + \mid S \mid. \tag{8.3.5}$$

Now, suppose that $(x, y) \in P$. Then, $-\tfrac{p}{2} < qx - py < 0$ or equivalently $\tfrac{p+1}{2} \leq qx - py + p \leq p - 1$. Hence, $\overline{qx} = qx - py + p$ belongs to Y. Thus, we have a map ϕ from P to $\overline{q}X \bigcap Y$ defined by

$$\phi((x, y)) = \overline{qx}.$$

Suppose that $\phi((x, y)) = \phi((x', y'))$. Then, $\overline{qx} = \overline{qx'}$. This means that p divides $qx - qx'$. Since q is co-prime to p and $1 \leq x \leq p - 1$, $1 \leq x' \leq p - 1$, $x = x'$. Hence, $(x, y), (x, y') \in P$. But, then $-\frac{p}{2} < (qx - py) < 0$ and $-\frac{p}{2} < (qx - py') < 0$ or equivalently

$$\frac{q}{p}x < y < \frac{q}{p}x + \frac{1}{2} \text{ and } \frac{q}{p}x < y' < \frac{q}{p}x + \frac{1}{2}.$$

Hence, $y = y'$. This shows that ϕ is injective.

Let $\overline{qx} \in \overline{q}X \cap Y$, where $1 \leq x \leq \frac{p-1}{2}$. Then, there exists r such that $\overline{qx} = \overline{r}$, where $\frac{p+1}{2} \leq r \leq p - 1$. Since $\overline{qx} = \overline{r} = \overline{r - p}$, there exists a $y \in B$ such that $qx - py = r - p$. Clearly, then

$$-\frac{p}{2} < (qx - py) < 0,$$

and so

$$\frac{q}{p}x < y < \frac{q}{p}x + \frac{1}{2}.$$

Already, $1 \leq x \leq \frac{p-1}{2}$, $1 \leq y \leq \frac{q-1}{2}$. This shows that $(x, y) \in P$, and $\phi((x, y)) = \overline{qx}$. Thus, ϕ is also surjective. Hence,

$$\lambda = |P|. \tag{8.3.6}$$

Similarly,

$$\mu = |Q|. \tag{8.3.7}$$

Consider the map η from $A \times B$ to itself defined by

$$\eta((x, y)) = (\tfrac{p+1}{2} - x, \tfrac{q+1}{2} - y).$$

This is clearly a bijective map. Also,

$$(x, y) \in R \text{ if and only if } (py - qx) > \tfrac{p}{2},$$

and

$$(py - qx) > \tfrac{p}{2} \text{ if and only if } p(\tfrac{q+1}{2} - y) - q(\tfrac{p+1}{2} - x) = \tfrac{p-q}{2} - (py - qx) < -\tfrac{q}{2}.$$

Thus,

$$(x, y) \in R \text{ if and only if } \eta((x, y)) \in S.$$

This shows that

$$|R| = |S|. \tag{8.3.8}$$

From (8.3.5) to (8.3.8), we obtain that

$$\tfrac{p-1}{2} \cdot \tfrac{q-1}{2} = (\lambda + \mu)(mod\ 2).$$

The result follows.   ♯

The following example illustrates an algorithm to determine the Legendre symbol $(\tfrac{a}{p})$ and thus to determine whether a is a quadratic residue mod p or not.

Example 8.3.20 We determine the Legendre symbol $(\tfrac{2125}{641})$ and thereby determine whether 2125 is a quadratic residue modulo 641 (observe that 641 is a prime number which is the smallest prime dividing $2^{2^5} + 1$). Also, we find the remainder when 2125^{320} is divided by 641.

If we divide 2125 by 641, the remainder is 202. Hence, by Corollary 8.3.9, $(\tfrac{2125}{641}) = (\tfrac{202}{641})$. By Proposition 8.3.8(iii), we have $(\tfrac{202}{641}) = (\tfrac{2}{641})(\tfrac{101}{641})$. By Corollary 8.3.15,

$$(\tfrac{2}{641}) = (-1)^{\tfrac{641^2-1}{8}} = 1.$$

Hence,

$$(\tfrac{2125}{641}) = (\tfrac{101}{641}). \tag{8.3.9}$$

By the Gauss quadratic reciprocity law,

$$(\tfrac{101}{641}) \cdot (\tfrac{641}{101}) = (-1)^{\tfrac{101-1}{2} \cdot \tfrac{641-1}{2}} = 1.$$

Hence,

$$(\tfrac{101}{641}) = (\tfrac{641}{101}) = (\tfrac{35}{101}) = (\tfrac{5}{101}) \cdot (\tfrac{7}{101}). \tag{8.3.10}$$

Again by the quadratic reciprocity,

$$(\tfrac{5}{101}) \cdot (\tfrac{101}{5}) = (-1)^{\tfrac{5-1}{2} \cdot \tfrac{101-1}{2}} = 1.$$

Hence

$$(\tfrac{5}{101}) = (\tfrac{101}{5}) = (\tfrac{1}{5}) = 1. \tag{8.3.11}$$

By the quadratic reciprocity,

$$(\tfrac{7}{101}) \cdot (\tfrac{101}{7}) = (-1)^{\tfrac{7-1}{2} \cdot \tfrac{101-1}{2}} = 1.$$

Hence,

$$(\tfrac{7}{101}) = (\tfrac{101}{7}) = (\tfrac{3}{7}). \tag{8.3.12}$$

Also

$$(\tfrac{3}{7}) \cdot (\tfrac{7}{3}) = (-1)^{\tfrac{7-1}{2} \cdot \tfrac{3-1}{2}} = -1.$$

Hence,

$$(\tfrac{3}{7}) = -(\tfrac{7}{3}) = -(\tfrac{1}{3}) = -1.$$

Thus,

$$(\tfrac{7}{101}) = -1. \tag{8.3.13}$$

Substituting the value of $(\tfrac{5}{101})$ and $(\tfrac{7}{101})$ in Eq. 8.3.10, we get

$$(\tfrac{101}{641}) = -1.$$

From Eq. 8.3.9, we get

$$(\tfrac{2125}{641}) = -1.$$

This implies that 2125 is not a quadratic residue modulo 641. By Proposition 8.3.8 (i),

$$(2125)^{\frac{641-1}{2}} \equiv (\tfrac{2125}{641})(mod\ 641) \equiv -1(mod\ 641) \equiv 640(mod\ 641).$$

Thus, if we divide $(2125)^{320}$ by 641, the remainder is 640.

Proposition 8.3.21 *Let a be an integer and p an odd prime such that p does not divide a. Then*

$$aX^2 + bX + c \equiv 0(mod\ p). \tag{8.3.14}$$

has a solution if and only if $b^2 \equiv 4ac(mod\ p)$, or else

$$(\tfrac{b^2-4ac}{p}) = 1.$$

Further, in the second case it has two distinct solutions.

Proof Since p does not divide a, there is an integer u co-prime to p such that

$$ua \equiv 1(mod\ p). \tag{8.3.15}$$

and so the solutions of the given Eq. 8.3.14 are same as those of

$$uaX^2 + ubX + uc \equiv 0(mod\ p).$$

The given Eq. 8.3.14 has a solution if and only if the above equation has a solution. Since $ua \equiv 1(mod\ p)$,

$$uaX^2 + ubX + uc \equiv (X^2 + ubX + uc)(mod\ p).$$

Thus, the given Eq. 8.3.14 has a solution if and only if

$$X^2 + ubX + uc \equiv 0(mod\ p). \tag{8.3.16}$$

has a solution, and then, they have same solutions. Since p is an odd integer, there is an integer v such that

$$2v \equiv 1(mod\ p). \tag{8.3.17}$$

But, then

$$X^2 + ubX + uc \equiv (X^2 + 2vubX + uc)(mod\ p). \tag{8.3.18}$$

Now,

$$X^2 + 2vubX + uc = (X + vub)^2 - (v^2u^2b^2 - uc).$$

Thus,

$$X^2 + 2vubX + uc \equiv 0(mod\ p)$$

has a solution if and only if

$$Y^2 \equiv (v^2u^2b^2 - uc)(mod\ p) \tag{8.3.19}$$

has a solution, and then, they have same number of solutions. Now, Eq. 8.3.19 has unique trivial solution if

$$v^2u^2b^2 - uc \equiv 0(mod\ p),$$

and has two distinct solutions if and only if

$$(\tfrac{v^2u^2b^2 - uc}{p}) = 1. \tag{8.3.20}$$

From (8.3.15), $a^2u^2 \equiv 1(mod\ p)$, and from (8.3.17), $4v^2 \equiv 1(mod\ p)$. Hence,

$$4v^2a^2u^2b^2 \equiv b^2(mod\ p), \tag{8.3.21}$$

and

$$4a^2uc \equiv 4ac(mod\ p). \tag{8.3.22}$$

Also, since $4a^2$ is co-prime to p, from (8.3.21) and (8.3.22), we see that

$$(v^2u^2b^2 - uc) \equiv (b^2 - 4ac)(mod\ p).$$

It follows that (8.3.19) has a unique solution if and only if $b^2 \equiv 4ac(mod\ p)$, and it has two distinct solution if and only if

$$(\tfrac{b^2 - 4ac}{p}) = 1. \qquad\qquad ♯$$

Since (8.3.14) and (8.3.19) have same set of solutions, the result follows.

Example 8.3.22 To determine the number of solutions of

$$X^2 \equiv 13(mod\ m) \qquad\qquad (8.3.23)$$

where $m = 2^2 \times 3^2 \times (17)^3$. Since $13 \equiv 1(mod\ 4)$, from Corollary 8.3.6, it follows that (8.3.23) has a solution if and only if

$$(13)^{\frac{3-1}{2}} \equiv 1(mod\ 3),$$

and

$$(13)^{\frac{17-1}{2}} \equiv 1(mod\ 17).$$

Equivalently,

$$(\tfrac{13}{3}) = 1 = (\tfrac{13}{17}).$$

Now,

$$(\tfrac{13}{3}) = (\tfrac{1}{3}) = 1.$$

By the Gauss law of quadratic reciprocity,

$$(\tfrac{13}{17}) \cdot (\tfrac{17}{13}) = (-1)^{\frac{13-1}{2} \cdot \frac{17-1}{2}} = 1.$$

Hence,

$$(\tfrac{13}{17}) = (\tfrac{17}{13}) = (\tfrac{4}{13}) = (\tfrac{2}{13})^2 = 1.$$

Thus, $X^2 \equiv 13(mod\ m)$ has a solution, and again by Corollary 8.3.6, there are $2^3 = 8$ distinct solutions modulo m. The reader is asked to find them.

Example 8.3.23 Let p be a prime dividing $X^2 - X + 1$. Then, $p \equiv 1(mod\ 12)$ or $p \equiv 7(mod\ 12)$.

Proof Since $X^2 - X = X(X - 1)$ is even, $X^2 - X + 1$ is always an odd integer. Hence, any prime p dividing $X^2 - X + 1$ is an odd prime. Let p be a prime dividing $X^2 - X + 1$, then since p is odd, $(2, p) = 1$. Hence, there is an integer l such that

$$2l \equiv 1(mod\ p). \qquad\qquad (8.3.24)$$

In turn,

$$(X^2 - X + 1) \equiv (X^2 - 2lX + 1)(mod\ p).$$

Thus, p divides $X^2 - 2lX + 1$ also. Since

$$X^2 - 2lX + 1 = (X - l)^2 - (l^2 - 1),$$

$$(X - l)^2 \equiv (l^2 - 1)(mod\ p).$$

This shows that $l^2 - 1$ is a quadratic residue modulo p, and hence,

$$\left(\tfrac{l^2-1}{p}\right) = 1.$$

Since $\left(\tfrac{4}{p}\right) = \left(\tfrac{2}{p}\right)^2 = 1$, we have

$$\left(\tfrac{4l^2-4}{p}\right) = 1.$$

Since $2l \equiv 1(mod\ p)$, $(4l^2 - 4) \equiv -3(mod\ p)$. Hence $1 = \left(\tfrac{-3}{p}\right) = \left(\tfrac{-1}{p}\right) \cdot \left(\tfrac{3}{p}\right)$. Thus,

$$\left(\tfrac{3}{p}\right) = \left(\tfrac{-1}{p}\right) = (-1)^{\frac{p-1}{2}}. \tag{8.3.25}$$

By the Gauss quadratic reciprocity law,

$$\left(\tfrac{3}{p}\right) \cdot \left(\tfrac{p}{3}\right) = (-1)^{\frac{p-1}{2}\cdot\frac{3-1}{2}} = (-1)^{\frac{p-1}{2}} = \left(\tfrac{3}{p}\right).$$

Hence

$$\left(\tfrac{p}{3}\right) = 1. \tag{8.3.26}$$

Since p is an odd prime, $p \equiv 1(mod\ 3)$ or $p \equiv 2(mod\ 3)$. Since $\left(\tfrac{2}{3}\right) = (-1)^{\frac{3^2-1}{8}} = -1$, we have

$$p \equiv 1(mod\ 3). \tag{8.3.27}$$

Again, since p is an odd prime, $p \equiv 1(mod\ 4)$ or $p \equiv 3(mod\ 4)$. In case $p \equiv 1(mod\ 4)$, $4/(p - 1)$, and also by (8.3.27), $3/(p - 1)$. But, then 12 divides $p - 1$. This means that $p \equiv 1(mod\ 12)$. In case $p \equiv 3(mod\ 4)$, since $p \equiv 1(mod\ 3)$, by the chinese remainder theorem, $p \equiv 7(mod\ 12)$.   ♯

Example 8.3.24 Let $\mathbb{Z} \times \mathbb{Z}$ denote the set of points on the plane with integral coordinates. Then, the parabola

$$X^2 + 19Y = 2$$

does not intersect $\mathbb{Z} \times \mathbb{Z}$.

Proof To say that the given parabola intersects the lattice $\mathbb{Z} \times \mathbb{Z}$ is to say that $X^2 + 19Y = 2$ has an integral solution. Equivalently, $X^2 \equiv 2(mod\ 19)$ will have solutions. But

$$\left(\tfrac{2}{19}\right) = (-1)^{\frac{(19)^2-1}{8}} = -1.   ♯$$

Example 8.3.25 To determine primes p for which 10 is a quadratic residue modulo p. To say that 10 is a quadratic residue modulo p is to say that

$$1 = \left(\tfrac{10}{p}\right) = \left(\tfrac{2}{p}\right) \cdot \left(\tfrac{5}{p}\right).$$

Thus, 10 is a quadratic residue modulo p if and only if one of the following holds.

(i) $\left(\tfrac{2}{p}\right) = 1 = \left(\tfrac{5}{p}\right)$.

(ii) $\left(\tfrac{2}{p}\right) = -1 = \left(\tfrac{5}{p}\right)$.

Consider the case (1):

$$\left(\tfrac{2}{p}\right) = 1 \;\; \textit{if and only if } p \equiv 1 (mod\ 8) \textit{ or } p \equiv 7 (mod\ 8). \qquad (8.3.28)$$

By the Gauss quadratic reciprocity law,

$$\left(\tfrac{p}{5}\right) \cdot \left(\tfrac{5}{p}\right) = (-1)^{\frac{p-1}{2} \cdot \frac{5-1}{2}} = 1.$$

Hence,

$$\left(\tfrac{5}{p}\right) = \left(\tfrac{p}{5}\right). \qquad (8.3.29)$$

Thus,

$$\left(\tfrac{5}{p}\right) = 1 \;\; \textit{if and only if } \left(\tfrac{p}{5}\right) = 1.$$

Now,

$$\left(\tfrac{p}{5}\right) = 1 \;\; \textit{if and only if } p \equiv 1 (mod\ 5) \textit{ or } p \equiv 4 (mod\ 5). \qquad (8.3.30)$$

By the Chinese remainder theorem,

$$[p \equiv 1 (mod\ 8) \textit{ and } p \equiv 1 (mod\ 5)] \textit{ if and only if } p \equiv 1 (mod\ 40),$$

$$[p \equiv 1 (mod\ 8) \textit{ and } p \equiv 4 (mod\ 5)] \textit{ if and only if } p \equiv 9 (mod\ 40),$$

$$[p \equiv 7 (mod\ 8) \textit{ and } p \equiv 1 (mod\ 5)] \textit{ if and only if } p \equiv 31 (mod\ 40),$$

and

$$[p \equiv 7 (mod\ 8) \textit{ and } p \equiv 4 (mod\ 5)] \textit{ if and only if } p \equiv 39 (mod\ 40).$$

Thus, in case (i), p is congruent to 1, 9, 31, or 39 modulo 40.

Consider the case (ii):

$$\left(\tfrac{2}{p}\right) = -1 \;\; \textit{if and only if } [p \equiv 3 (mod\ 8) \textit{ or } p \equiv 5 (mod\ 8)].$$

Also, $(\frac{5}{p}) = -1$ *if and only if* $(\frac{p}{5}) = -1$ and

$$(\frac{p}{5}) = -1 \ \textit{if and only if} \ [p \equiv 2(mod\ 5) \ \textit{or} \ p \equiv 3(mod\ 5)].$$

Again using Chinese remainder theorem, we find that in this case p is congruent to 3, 13, 27, or 37 modulo 40. This shows that 10 is a quadratic residue modulo p if and only if when p is divided by 40 the remainder is one of the 1, 3, 9, 13, 27, 31, 37, or 39.

Exercises

8.3.1 Determine whether 150 is quadratic residue modulo 131.

8.3.2 Compute $(\frac{69}{59})$.

8.3.3 Find the remainder when $(60)^{35}$ is divided by 71.

8.3.4 Determine whether the congruence equation

$$40X^2 + 12X + 6 \equiv 0(mod\ 23)$$

has a solution. Determine the number of distinct solutions modulo 23, if it has any.

8.3.5 Determine whether 221 is a quadratic residue modulo $4 \times 7^2 \times (13)^3$. Find the number of distinct solutions of

$$X^2 \equiv 221(mod\ 4 \times 7^2 \times (13)^3),$$

if it has any.

8.3.6 Determine whether

$$12X^2 + 7X + 3 \equiv 0(mod\ 511225)$$

has a solution. How many distinct modulo 511225 solutions are there?

8.3.7 Show that the number of distinct solutions of $X^2 \equiv a(mod\ p)$ is $1 + (\frac{a}{p})$, where p does not divide a.

8.3.8 Show that $\Sigma_{i=1}^{p-1}(\frac{i}{p}) = 0$.

8.3.9 Show that if $(p, a) = 1$, then

$$\Sigma_{i=1}^{p-1}(\frac{ai+b}{p}) = 0.$$

8.3.10 Let p be a prime dividing $X^4 - X^2 + 1$. Show that $p \equiv 1(mod\ 12)$.

8.3.11 Determine whether the parabola $Y^2 = 641X + 3$ intersects the lattice $\mathbb{Z} \times \mathbb{Z}$.

8.3.12 Determine primes p for which 15 is a quadratic residue mod p.

8.3.13 Suppose that a is a quadratic residue modulo all but finitely many primes. Show that a is a perfect square.

8.3.14 Show that the ellipse $X^2 + 5Y^2 = p$ where p is a prime intersects the lattice $Z \times Z$ if $p \equiv 1 (mod\ 20)$ or $p \equiv 9 (mod\ 20)$.

8.3.15 Suppose that $\left(\frac{-10}{p}\right) = 1$. Show that the ellipse $X^2 + 10Y^2 = p^2$ intersects the lattice $\mathbb{Z} \times \mathbb{Z}$.

Chapter 9
Structure Theory of Groups

9.1 Group Actions, Permutation Representations

Let G be a group with identity e and X a set. A map $\star : G \times X \longrightarrow X$ is called an **action** of G on X if

(i) $\star((e, x)) = x$ for all x in X, and
(ii) $\star((g_1 g_2, x)) = \star((g_1, \star((g_2, x))))$ for all g_1, g_2 in G and x in X.

If we denote the image $\star((g, x))$ by $g \star x$, then the conditions (i) and (ii) read as

(i) $e \star x = x$ for all x in X, and
(ii) $(g_1 g_2) \star x = g_1 \star (g_2 \star x)$ for all g_1, g_2 in G and x in X.

We say that G acts on X through the action $\star$. We also say that X is a G-set under the action $\star$.

Before having some interesting examples and applications of group actions, we shall develop the theory of group actions to some extent.

Suppose that G acts on X through an action $\star$. Let $g \in G$. Define a map f_g from X to X by $f_g(x) = g \star x$. Suppose that $f_g(x_1) = f_g(x_2)$. Then, $g \star x_1 = g \star x_2$. This implies that $x_1 = e \star x_1 = (g^{-1} g) \star x_1 = g^{-1} \star (g \star x_1) = g^{-1} \star (g \star x_2) = (g^{-1} g) \star x_2 = e \star x_2 = x_2$. Thus, f_g is injective. Next, if $y \in X$, then $f_g(g^{-1} \star y) = g \star (g^{-1} \star y) = (g g^{-1}) \star y = e \star y = y$. Hence, f_g is also surjective. This shows that $f_g \in Sym(X)$.

Define a map ρ from G to $Sym(X)$ by $\rho(g) = f_g$. Now, $\rho(g_1 g_2)(x) = f_{g_1 g_2}(x) = (g_1 g_2) \star x = g_1 \star (g_2 \star x) = f_{g_1}(f_{g_2}(x)) = (\rho(g_1) o \rho(g_2))(x)$ for all $g_1, g_2 \in G$ and $x \in X$. Hence, $\rho(g_1 g_2) = (\rho(g_1) o \rho(g_2))$ for all $g_1, g_2 \in G$. This shows that ρ is a homomorphism from G to $Sym(X)$. Such a homomorphism is called a **permutation representation** of G on X. Thus, given an action $\star$ of G on X, we have a representation of G on X. Conversely, given any representation ρ of G on X, we have an action $\star$ of G on X defined by $g \star x = \rho(g)(x)$ such that the corresponding representation is the same as ρ (verify).

© Springer Nature Singapore Pte Ltd. 2017
R. Lal, *Algebra 1*, Infosys Science Foundation Series in Mathematical Sciences,
DOI 10.1007/978-981-10-4253-9_9

Thus, there is a faithful way to view an action as a representation, and a representation as an action, and they are related by

$$g \star x \ = \ \rho(g)(x).$$

Let $\star$ be an action of G on X with the corresponding representation ρ. Then, the kernel of ρ is $\{g \in G \mid \rho(g) \ = \ I_G\} \ = \ \{g \in G \mid \rho(g)(x) \ = \ x \text{ for all } x \in X\} \ = \ \{g \in G \mid g \star x \ = \ x \text{ for all } x \in X\}$. Therefore, $\{g \in G \mid g \star x \ = \ x \text{ for all } x \in X\}$ is a normal subgroup of G. This subgroup is called **isotropy group or the stabilizer** the of the action $\star$ of G on X, and it is denoted by $Stab(G, X)$.

By the fundamental theorem of homomorphism, we have the following proposition.

Proposition 9.1.1 $G/Stab(G, X)$ *is isomorphic to the subgroup $\rho(G)$ of $Sym(X)$, where ρ is the corresponding representation.*   ♯

Proposition 9.1.2 *Let $\star$ be an action of G on X and $x \in X$. Then, $G_x \ = \ \{g \in G \mid g \star x \ = \ x\}$ is a subgroup of G.*

Proof Since $e \star x \ = \ x, e \in G_x$. Let $a, b \in G_x$. Then, $a \star x \ = \ x \ = \ b \star x$. But, then $x \ = \ e \star x \ = \ a^{-1} \star (a \star x) \ = \ a^{-1} \star (b \star x) \ = \ (a^{-1}b) \star x$. This shows that $a^{-1}b \in G_x$.   ♯

The subgroup $G_x \ = \ \{g \in G \mid g \star x \ = \ x\}$ of G is called the **isotropy group at x** or the **stabilizer of x**. This is also called the **local isotropy group at x**. Clearly, $Stab(G, X) \ = \ \bigcap_{x \in X} G_x$.

Let G be a group which acts on X. Define a relation $\sim$ on X as follows:

$$x \sim y \text{ if and only if } g \star x \ = \ y \text{ for some } g \in G.$$

The relation $\sim$ is reflexive, for $e \star x \ = \ x$. It is symmetric, for $g \star x \ = \ y$ implies that $g^{-1} \star y \ = \ x$. It is transitive, for $g \star x \ = \ y$ and $h \star y \ = \ z$ implies that $hg \star x \ = \ z$. Thus, $\sim$ is an equivalence relation.

The equivalence class of X modulo $\sim$ determined by x is $\{y \in X \mid x \sim y\} \ = \ \{g \star x \mid g \in G\}$. This set is denoted by $G \star x$, and it is called the **orbit** of the action through x.

From the properties of equivalence relations and equivalence classes (see Proposition 2.4.4), we have the following:

(i) $X \ = \ \bigcup_{x \in X} G \star x$.
(ii) Distinct orbits are disjoint.
(iii) $G \star x$ and $G \star y$ are same if and only if $g \star x \ = \ y$ for some $g \in G$.

The following proposition relates the orbit $G \star x$ through x and the isotropy group G_x at x.

Proposition 9.1.3 *Let $\star$ be an action of G on X. Then, there is a bijective map from the set $G/^l G_x$ of left cosets of G modulo G_x to the orbit $G \star x$ through x.*

Proof Suppose that $gG_x = hG_x$. Then, $g^{-1}h \in G_x$. In turn, $g^{-1}h \star x = x$. It follows that $g \star x = h \star x$. Thus, we have a map f from $G/^lG_x$ to $G \star x$ defined by $f(gG_x) = g \star x$. Clearly, this mapping is also surjective. Next, suppose that $f(gG_x) = f(hG_x)$. Then, $g \star x = h \star x$, or equivalently, $g^{-1}h \in G_x$. Consequently, $gG_x = hG_x$. Thus, f is also an injective map. ♯

Corollary 9.1.4 G_x *is of finite index in G if and only if $G \star x$ is finite. Further, then* $[G : G_x] = | G \star x |$. *In particular, if G is finite, then $[G : G_x] = | G \star x |$ for all* $x \in X$, *and so $| G \star x |$ divides $| G |$ for all $x \in X$. Also, if X is finite, then isotropy subgroups at all points are of finite indexes, and $[G : G_x] = | G \star x |$ for all $x \in X$.* ♯

Let G be a group which acts on X through an action $\star$. An element $x \in X$ is called a **fixed point** if $g \star x = x$ for all $g \in G$. Thus, x is a fixed point if and only if $G_x = G$, or equivalently, $G \star x = \{x\}$. The set of all fixed points of the action $\star$ is denoted by X^G, and it is called the **fixed point set** of the action. Thus, $X^G = \{x \in X \mid g \star x = x \text{ for all } g \in G\}$.

An element $x \in X$ is a fixed point if and only if the orbit $G \star x$ is singleton. Thus, X^G is the union of all those orbits which are singletons.

Let A be a set obtained by choosing one and only one member from each orbit different from singleton. Thus, if $x \in A$, then $G \star x \neq \{x\}$. Further, $G \star x \cap G \star y = \emptyset$ for all $x, y \in A$, $x \neq y$. Since X is union of all orbits,

$$X = X^G \bigcup \left(\bigcup_{x \in A} G \star x \right). \tag{9.1.1}$$

Suppose that X is finite. Then, since distinct orbits are disjoint, we have

$$| X | = | X^G | + \Sigma_{x \in A} | G \star x | \tag{9.1.2}$$

Observe that each term under summation in the above equation is greater than 1. By Corollary 9.1.4, we have

$$| X | = | X^G | + \Sigma_{x \in A} [G : G_x] \tag{9.1.3}$$

Here again, each term under summation in R.H.S. is greater than 1 and divides $| G |$ (Lagrange theorem).

Equation (9.1.3) is called the **class equation or class formula** for the action $\star$.

Proposition 9.1.5 *Let G be a finite group of order p^n, $n \geq 1$, where p is a prime. Let X be a finite set on which G acts. Then,*

$$| X | \equiv | X^G | \ (mod \ p).$$

Proof Consider the class formula

$$| X | = | X^G | + \Sigma_{x \in A} [G : G_x]$$

of the action, where each term under summation in the right hand side is greater than 1 and also divides $\mid G \mid = p^n$ (by the Lagrange theorem). This shows that the second term of the right hand side is a multiple of p. The result follows.   ♯

Corollary 9.1.6 *Let G be a finite group of order p^n, $n \geq 1$, where p is prime. Let X be a finite set containing m elements on which G acts. Suppose that p does not divide m. Then, $X^G \neq \emptyset$.*

Proof From Proposition 9.1.5, $m = \mid X \mid \equiv \mid X^G \mid \pmod{p}$. Since p does not divide m, $\mid X^G \mid \neq 0$. This means that $X^G \neq \emptyset$.   ♯

So far, we developed elementary theory of group actions. Now, we give some examples and applications to the structure theory of finite groups.

Example 9.1.7 Let G be a group and X the set part of G. Take $\star$ to be the binary operation of G. Then, G acts on G. This action is called the **left multiplecation of G on G or left regular action of G on G**. The isotropy group G_x at x is given by $G_x = \{g \in G \mid gx = x\} = \{e\}$ (left cancellation law). The stablizer $Stab(G, G)$ of the left regular action of G on G is also $\{e\}$. The representation determined by left regular action of G on G is the homomorphism ρ of the Cayley's theorem (see Theorem 6.2.21). Clearly, ρ is injective. The representation ρ is called the **regular permutation representation** of G.

Example 9.1.8 Let G be a group and $X = G$. Define $g \star x = gxg^{-1}$, $g, x \in G$. It is easy to verify that $\star$ is an action of G on G. This action is called the **inner conjugation** of G on G. The isotropy group of the inner conjugation of G on G at $x \in G$ is $G_x = \{g \in G \mid gxg^{-1} = x\} = \{g \in G \mid gx = xg\}$. Recall (Definition 4.4.15) that this subgroup of G is called the **centralizer** of x in G, and it is denoted by $C_G(x)$. An element $a \in G$ is called a **conjugate** of an element $b \in G$ if there is an element $g \in G$ such that $gag^{-1} = b$. Thus, the orbit through $x \in G$ of the inner conjugation of G on G is the set $G \star x = \{gxg^{-1} \mid g \in G\}$ of all conjugates of x in G. This is called the **conjugacy class of G determined by x**, and it is denoted by $\bar{x}$.

Corollary 9.1.4 applied to the inner conjugation of G on G gives the following:

Corollary 9.1.9 *Let G be a finite group. Then, the number of conjugates to an element $x \in G$ is equal to the index of the centralizer $C_G(x)$ of x in G. In particular, the number of conjugates to an element of G is a divisor of the order of G.*   ♯

The stablizer $Stab(G, G)$ of the inner conjugation of G on G is $\{g \in G \mid gxg^{-1} = x \text{ for all } x \in G\} = \{g \in G \mid gx = xg \text{ for all } x \in G\}$. Recall (Definition 4.4.15) that this is the **center** $Z(G)$ of G. The representation ρ determined by the inner conjugation of G on G is given by $\rho(g) = f_g$, where f_g is the inner automorphism of G determined by g. Thus, $\rho(G)$ is the group $Inn(G)$ of all inner automorphisms of G. By the fundamental theorem of homomorphism, we have the following:

Theorem 9.1.10 $G/Z(G) \approx Inn(G)$.   ♯

An element $x \in G$ is a fixed point of the inner conjugation of G on G if and only if $gxg^{-1} = x$ for all $g \in G$, or equivalently, $xg = gx$ for all $g \in G$. Thus, $x \in G$ is a fixed point of the inner conjugation of G on G if and only if $x \in Z(G)$. This shows that the fixed point set G^G of the inner conjugation of G on G is the center $Z(G)$ of G. Applying the Class equation (9.1.3) for the inner conjugation, we get the following:

Theorem 9.1.11 *If G is finite, then*

$$| G | = | Z(G) | + \Sigma_{x \in A}[G : C_G(x)] \qquad (9.1.4)$$

where A is a set obtained by choosing one and only one member from each conjugacy class which is not singleton. Each term under summation in the R.H.S. of (9.1.4) is greater than 1, and also divides $| G |$. ♯

Equation (9.1.4) is called the **classical class equation** of G.

Example 9.1.12 Let G be a group and X the set $S(G)$ of all subgroups of G. Define an action $\star$ of G on X by $g \star H = gHg^{-1}$ (note that gHg^{-1} is a subgroup of G and $\star$ is indeed an action). This action is called the **inner conjugation of G on subgroups of G**.

Let $H \in S(G)$. Then, the Isotropy group G_H of the inner conjugation action at H is given by $G_H = \{g \in G \mid gHg^{-1} = H\} = \{g \in G \mid gH = Hg\}$. Recall (Definition 4.4.15) that this subgroup of G is called the **Normalizer of H in G**, and it is denoted by $N_G(H)$. Since $hH = H = Hh$ for all $h \in H$, $H \subseteq N_G(H)$. Further, if K is a subgroup of G such that $H \trianglelefteq K$, then $gH = Hg$ for all $g \in K$, and so $K \subseteq N_G(H)$. Thus, we have the following:

Proposition 9.1.13 *The normalizer $N_G(H)$ of H in G is the largest subgroup of G in which H is normal.* ♯

A subgroup H of a group G is called a **conjugate** to a subgroup K of G if there is an element $g \in G$ such that $K = gHg^{-1}$. Thus, the orbit $G \star H$ at H of the inner conjugation of G on subgroups of G is the set $G \star H = \{gHg^{-1} \mid g \in G\}$ of all conjugates of H in G. Proposition 9.1.3 applied to the inner conjugation of G on the set $S(G)$ of all subgroups of G gives the following corollary.

Corollary 9.1.14 *There is a bijection from the set $G/^l N_G(H)$ of left cosets of G modulo $N_G(H)$ to the set of all conjugates of H. If G is finite, then the number of conjugates to H is equal to $[G : N_G(H)]$. The number of conjugates to H is a divisor of the order of G.* ♯

An element $H \in S(G)$ will be a fixed point of the inner conjugation of G on $S(G)$ if and only if $gHg^{-1} = H$ for all $g \in G$, or equivalently, $gH = Hg$ for all $g \in G$. Thus, H is a fixed point if and only if $H \trianglelefteq G$, and so the fixed point set $(S(G))^G$ is the set of all normal subgroups of G. The class formula for the inner conjugation of G on $S(G)$ becomes

$$| S(G) | = | (S(G))^G | + \Sigma_{H \in A}[G : N_G(H)] \qquad (9.1.5)$$

where each term under summation in the R.H.S. is greater than 1 and A is the set obtained by choosing one and only one member from each conjugacy class of subgroups which is not singleton.

Example 9.1.15 Let G be a group and H a subgroup of G. Suppose that $x_1 H = x_2 H$. Then, $(gx_1)^{-1}(gx_2) = x_1^{-1}x_2 \in H$. Hence, $gx_1 H = gx_2 H$. Thus, we have an action $\star$ of G on $G/^l H$ defined by $g \star xH = gxH$. This action is called the **left multiplication** of G on $G/^l H$. The isotropy group G_{xH} of the action of G on $G/^l H$ at xH is given by $G_{xH} = \{g \in G \mid gxH = xH\} = \{g \in G \mid x^{-1}gx \in H\} = \{g \in G \mid g \in xHx^{-1}\} = xHx^{-1}$. The orbit $G \star xH = \{gxH \mid g \in G\} = G/^l H$. The stablizer $Stab(G, G/^l H) = \bigcap_{x \in G} G_{xH} = \bigcap_{x \in G} xHx^{-1}$. This subgroup of G is called the **core of H in G**, and it is denoted by $Core_G(H)$.

Proposition 9.1.16 *$Core_G(H)$ of H in G is the largest normal subgroup of G contained in H.*

Proof Clearly, $Core_G(H) = \bigcap_{x \in G} xHx^{-1}$ is a normal subgroup of G. If K is a normal subgroup of G contained in H, then $K = xKx^{-1} \subseteq xHx^{-1}$ for all $x \in G$. This shows that $K \subseteq Core_G(H)$.   ♯

Since the representation ρ determined by the left multiplication action of G on $G/^l H$ is a homomorphism whose kernel is $Core_G(H)$, we have the following:

Proposition 9.1.17 *$G/Core_G(H)$ is isomorphic to a subgroup of $Sym(G/^l H)$.*   ♯

Example 9.1.18 This example is from the dynamics of projectile. (The acceleration g due to gravity is assumed to be constant). A projectile is completely determined by a point on the path of the projectile at a particular instant and the velocity of the particle at that point. Consider $\mathbb{R}^6$ whose first three coordinates determine the position of a particle in the space, and the last three components give the components of the velocity of the particle along x, y, and z axes (z axis is vertical), respectively, at that instant. Consider the additive group $\mathbb{R}$ of reals. We have the following action $\star$ of $\mathbb{R}$ on $\mathbb{R}^6$ defined by

$$t \star (\alpha, \beta, \gamma, u_1, u_2, u_3) = (u_1 t + \alpha, u_2 t + \beta, u_3 t - \tfrac{1}{2}gt^2 + \gamma, u_1, u_2, u_3 - gt),$$

where t represents time parameter. It is easy to check that $\star$ is an action.

Example 9.1.19 Consider the general linear group $GL(n, \mathbb{R})$. If we treat the members of $\mathbb{R}^n$ as column vectors, then the matrix multiplication from left defines an action of $GL(n, \mathbb{R})$ on $\mathbb{R}^n$. There are only two orbits of this action, viz., $\{\overline{0}\}$ and $\mathbb{R}^n - \{\overline{0}\}$. Further, all subgroups of $GL(n, \mathbb{R})$ act on $\mathbb{R}^n$. In particular, Consider the action of the special orthogonal group $SO(3)$ on the Euclidean 3-space $\mathbb{R}^3$. Let $\overline{e_1}$ denote the unit vector $[1, 0, 0]$. Then, $SO(3) \cdot \overline{e_1}^t$ is the unit sphere S^2 with center origin. Indeed, for any $A \in SO(3)$,

$$(A \cdot \overline{e_1}^t)^t \cdot A \cdot \overline{e_1}^t \ = \ \overline{e_1} \cdot \overline{e_1}^t \ = \ 1.$$

This shows that $SO(3) \cdot \overline{e_1}^t \subseteq S^2$. Also given any unit column vector $\overline{X}^t$ in S^2, let $\overline{Y}$ and $\overline{Z}$ be a pair of unit vectors such that $\{\overline{X}, \overline{Y}, \overline{Z}\}$ is a set of pairwise orthogonal unit vectors. Let $A \ = \ [\overline{X}^t, \overline{Y}^t, \overline{Z}^t]$. Then, $A \in O(3)$. Interchanging the second and the third column, if necessary, we may assume that $A \in SO(3)$. Clearly $A \cdot \overline{e_1}^t \ = \ \overline{X}^t$. This shows that $S^2 \subseteq SO(3) \cdot \overline{e_1}^t$. Thus, S^2 is an orbit of the action passing through the unit vector $\overline{e_1}^t$.

Similarly, the sphere $S^2(r)$ with center origin and radius r, $r > 0$ is also an orbit. Indeed, $S^2(r) \ = \ SO(3) \cdot r\overline{e_1}^t$. $\{\overline{0}^t\}$ is the trivial orbit. Evidently, these are all the orbits of the action.

Now, the isotropy subgroup $SO(3)_{\overline{e_1}^t} \ = \ \{A \in SO(3) \mid A \cdot \overline{e_1}^t \ = \ \overline{e_1}^t\}$ can easily be seen to be the subgroup of $SO(3)$ consisting of the matrices of the type

$$\begin{bmatrix} 1 & 0 & 0 \\ 0 & cos\theta & \pm sin\theta \\ 0 & \mp sin\theta & cos\theta \end{bmatrix},$$

where $0 \leq \theta \leq \pi$. This subgroup is clearly isomorphic to $SO(2)$. Without any loss, we may denote it by $SO(2)$. Thus, we have the bijective map η from $SO(3)/^r SO(2)$ to S^2 given by $\eta(A \cdot SO(2)) \ = \ A \cdot \overline{e_1}^t$. This map is also a topological homeomorphism.

Now, we apply the theory of group actions to the structure theory of finite groups.

Proposition 9.1.20 *Let G be a group of order p^n, $n \geq 1$. Let H be a nontrivial normal subgroup of G. Then, $H \cap Z(G) \neq \{e\}$. In particular, $Z(G) \neq \{e\}$.*

Proof Since $H \trianglelefteq G$, G acts on H through inner conjugation. The fixed point set $H^G \ = \ \{h \in H \mid ghg^{-1} \ = \ h \text{ for all } g \in G\} \ = \ \{h \in H \mid h \in Z(G)\} \ = \ H \cap Z(G)$. Consider the class formula

$$p^r \ = \mid H \mid = \mid H^G \mid + \Sigma_{x \in A}[G : G_x]$$

Since $H \neq \{e\}$, $r \geq 1$. Thus, L.H.S. is divisible by p and the second term in the R.H.S. is also a multiple of p. Hence, p divides $\mid H^G \mid$. Since $e \in H^G \ = \ H \cap Z(G)$, $H \cap Z(G)$ contains at least p elements. Taking $H \ = \ G$, we observe that the $Z(G) \neq \{e\}$.   ♯

Corollary 9.1.21 *Let G is a group of order p^2, where p is a prime. Then, G is abelian.*

Proof Consider $Z(G)$. By the above proposition, $\mid Z(G) \mid = \ p \text{ or } p^2$. If $\mid Z(G) \mid \ = \ p^2$, then $Z(G) \ = \ G$, and so G is abelian. We show that $\mid Z(G) \mid \neq p$. Suppose that $\mid Z(G) \mid \ = \ p$, then $\mid G/Z(G) \mid \ = \ p$, and so $G/Z(G)$ is cyclic. By Theorem

5.2.32, it follows that G is abelian. But, then $Z(G) = G$, and so $| Z(G) | = p^2$. This is a contradiction to the supposition.   ♯

The following theorem classifies all groups of order p^2, where p is a prime.

Theorem 9.1.22 *There are only two isomorphism classes of groups of order p^2, where p is a prime. One is the cyclic group of order p^2, and the other is direct product of two cyclic groups of order p.*

Proof If G contains an element of order p^2, then it is cyclic. Suppose that it contains no element of order p^2. Then, order of each nontrivial element of G is p. Let a be an element of G of order p. Then, $|< a >| = p$, and so there exists an element $b \notin < a >$. Clearly, b is also of order p, and $< a > \neq < b >$. Since any nonidentity element of a cyclic group of order p is a generator of the group, $< a > \bigcap < b > = \{e\}$. Hence, $G = < a > < b >$. Also, since G is abelian (by the above corollary), by Proposition 5.2.4, it is internal direct product of $< a >$ and $< b >$. By Proposition 5.2.3, G is isomorphic to the external direct product $< a > \times < b >$.   ♯

Proposition 9.1.23 *Let G be a group of order p^n, $n \geq 1$, where p is a prime. Then, to every divisor of p^n, there is a subgroup, in fact, a normal subgroup of that order.*

Proof The proof is by induction on n. If $n = 1$, then G is prime cyclic, and there is nothing to do. Assume that the result is true for a group of order p^n. Let G be a group of order p^{n+1}. By Proposition 9.1.20, $Z(G) \neq \{e\}$. Let $a \in Z(G)$, $a \neq e$. Suppose that $| a | = p^s$, $s \geq 1$. Clearly, a is of order p or else $a^{p^{s-1}}$ is of order p. Thus, there is an element $b \in Z(G)$ of order p. Take $H = < b >$. Then, H is a subgroup of G of order p. Since H is contained in $Z(G)$, it is normal in G. Consider the group $K = G/H$. Clearly, $| G/H | = p^n$. By the induction hypothesis, corresponding to every divisor p^t, $0 \leq t \leq n$, there is a normal subgroup L of G/H of order p^t. By Proposition 5.2.29, there is a unique normal subgroup K of G containing H such that $L = K/H$. Since $| L | = p^t$, by the Lagrange theorem, $| K | = p^{t+1}$.   ♯

The following proposition is a generalization of the result: 'Every subgroup of index 2 is normal'.

Proposition 9.1.24 *Let G be a finite group, and p the smallest prime dividing the order of G. Then, every subgroup of G of index p is normal in G.*

Proof Let H be a subgroup of G of index p, where p is the smallest prime dividing order of G. Then, $G/^l H$ contains p elements. Consider the action $\star$ of G on $G/^l H$ defined by $g \star x H = gx H$. The stablizer of this action is the $Core_G(H)$ which is the largest normal subgroup of G contained in H. Also, $G/Core_G(H)$ is isomorphic to a subgroup of $Sym(G/^l H) \approx S_p$. Since p is the smallest prime dividing the order of G, $(| G/Core_G(H) |, | S_p |) = 1$ or p. If $| G/Core_G(H) | = 1$, then $G = Core_G(H) \subseteq H$, a contradiction to the supposition that $[G : H] = p$. Hence, $[G : Core_G(H)] = [G : H]$. Since $Core_G(H) \subseteq H$, $H = Core_G(H) \triangleleft G$.   ♯

Exercises

9.1.1 Let G be a group which acts on X through the action $\star$. Show that $G_{g\star x} = gG_x g^{-1}$ for all $g \in G$ and $x \in X$. Deduce that $\{G_{g\star x} \mid g \in G\}$ is a complete conjugacy class of subgroups.

9.1.2 Let G be a group which acts on X. We also say that X is a G-set. A subset Y of X is called a G-subset if the action of G on X induces an action on Y, i.e., $g \star y \in Y$ for all $g \in G$ and $y \in Y$. Show that a subset of X is a G-subset if and only if it is a union of some of the orbits.

9.1.3 Let G be a group which acts on X and also on Y. A map f from X to Y is called a G-equivariant map if $f(g \star x) = g \star f(x)$ for all $g \in G$ and $x \in X$. Let f from X to Y be a G-equivariant map. Show that $G_x \subseteq G_{f(x)}$ for all $x \in X$.

9.1.4 Let G be a group and H, K be subgroups. Then, G acts on $G/^l H$, and it also on $G/^l K$ through left multiplication. Show that there is a G-equivariant map from $G/^l H$ to $G/^l K$ if and only if there exists an element $a \in G$ such that $aHa^{-1} \subseteq K$, and then $f_a : G/^l H \longrightarrow G/^l K$ defined by $f_a(xH) = xa^{-1}K$ is an equivariant map.
Hint. Suppose that $f(H) = a^{-1}K$.

9.1.5 Call a bijective G equivariant map to be a **G-equivalence**. Show that $G/^l H$ is G equivalent to $G/^l K$ if and only if H and K are conjugates.

9.1.6 Call an action $\star$ of G on X to be a **transitive** action if there is only one orbit, i.e., given any $x, y \in X$, there is an element $g \in G$ such that $g \star x = y$. Suppose that G acts transitively on X. Let $x \in X$. Show that the map f from $G/^l G_x$ to X defined by $f(gG_x) = g \star x$ is a G equivalence.

9.1.7 Find the number of S_4 equivalence classes of transitive S_4 actions.

9.1.8 Show that every G-set is disjoint union of transitive G-sets.

9.1.9 Describe the action of the additive group $\mathbb{R}$ of real numbers on a suitable Euclidean space which describes a motion under a central force of attraction following the inverse square law.

9.1.10 Describe all the equivalence classes of actions of Q_8 on a set containing 3 elements. How many of them are there?

9.1.11 (Cauchy–Frobenius). Suppose that G acts on a finite set X. Show that the number of orbits is $\frac{1}{|G|}\Sigma_{g\in G} \mid \{x \in X \mid g \star x = x\} \mid$. Deduce that if G acts transitively, then $\Sigma_{g\in G} \mid \{x \in X \mid g \star x = x\} \mid = \mid G \mid$.
Hint: Consider the set $\Omega = \{(g, x) \in G \times X \mid g \star x = x\}$. Let $\{G \star x_1, G \star x_2, \ldots, G \star x_r\}$ be the set of distinct orbits of the action. Then, $\mid \Omega \mid = \Sigma_{g\in G} \mid \{x \in X \mid g \star x = x\} \mid$. On the other hand, $\mid \Omega \mid = \Sigma_{i=1}^{r} \Sigma_{x \in G \star x_i} \mid G_x \mid = \Sigma_{i=1}^{r} \mid G \mid$. Equate the two.

9.1.12 Let G be a finite group and $\{a_1, a_2, \ldots, a_r\}$ be a set obtained by choosing one and only one member from each conjugacy class of G. Then, show that

$$\frac{1}{|C_G(a_1)|} + \frac{1}{|C_G(a_2)|} + \cdots + \frac{1}{|C_G(a_r)|} = 1.$$

Deduce that there are only finitely many finite groups (upto isomorphism) which has a fixed number of conjugacy classes (the number of conjugacy classes of a group G is called the **class number**). Describe groups having class numbers 1, 2 and 3.

9.1.13 Let H be a maximal subgroup of G. Suppose that H is not normal in G. Show that there are $[G : H]$ conjugates to H.

9.1.14 Let H be a proper subgroup of G. Show that $G \neq \bigcup_{g \in G} gHg^{-1}$.

9.1.15 Let G be a group which has a proper subgroup of finite index. Show that G has a proper normal subgroup of finite index. Deduce that an infinite group having a subgroup of finite index cannot be simple.

9.1.16 Show that $C_G(H) \trianglelefteq N_G(H)$.

9.1.17 Let H be a subgroup of finite index of a finitely generated group G. Show that H is also finitely generated.
Hint. Let S be a right transversal to H in G. Then, S is finite. Let Y be a finite set of generators of G. For each $y \in Y$, and for each $x \in S$, let $\sigma(x, y)$ be the element of H determined by the equation $xy = \sigma(x, y)z$, where $z \in S$. Show that $\{\sigma(x, y) \mid x \in S \text{ and } y \in Y\}$ generates H.

9.1.18 Let H be a subgroup of finite index in G. Show that $Core_G(H)$ is a normal subgroup of finite index in G.

9.1.19 A transitive action of a group G on X is said to be regular if $G_x = \{e\}$ for all $x \in X$. Show that if G is regular, then $|X| = |G|$, and it is equivalent to the left multiplication on G.

9.1.20 Show that every transitive faithful action of an abelian group is regular.

9.1.21 Show that $\bigcap_{H \in S(G)} N_G(H) \trianglelefteq G$, and the corresponding quotient group is isomorphic to a subgroup of $Sym(S(G))$.

9.1.22 Let G be a group of order $p^n, n \geq 1$, where p is a prime. Let r denote the number of subgroups of G and s the number of normal subgroups of G. Show that p divides $r - s$.

9.1.23 Call an action of G on X to be doubly transitive if given $x_1 \neq x_2$ and $y_1 \neq y_2$ in X, there is a $g \in G$ such that $g \star x_1 = y_1$ and $g \star x_2 = y_2$. Suppose that a finite group G acts doubly transitively on X. Show that $\Sigma_{g \in G} | \{x \in X \mid g \star x = x\} |^2 = 2 |G|$.

9.1.24 Let G be a group which acts transitively on X. Show that G acts doubly transitively if and only if for each $x \in X$, the isotropy subgroup G_x at x acts transitively on $X - \{x\}$.

9.1.25 Let G be a finite group which acts doubly transitively on a set X containing n elements. Show that $n(n-1)$ divides the order of G.

9.1.26 Call an action of G on X to be sharply doubly transitive if given $x_1 \neq x_2$ and $y_1 \neq y_2$ in X, there is a unique $g \in G$ such that $g \star x_1 = y_1$ and $g \star x_2 = y_2$. Suppose that a finite group G acts doubly transitively on a set X containing n elements. Show that G acts sharply doubly transitively if and only if G is of order $n(n-1)$.

9.1.27* Let G be a group which acts sharply doubly transitively on a set X. Show that G contains several elements of order 2 (called involutions), and all elements of order 2 form a complete single conjugacy class. Further, show that either all isotropy subgroups G_x contain a unique element of order 2, or no isotropy subgroup contain an element of order 2.

9.1.28 Let F be any field. Define multiplication $\cdot$ on $G = F \times F^*$ by

$$(x, a) \cdot (y, b) = (x + ay, ab)$$

Show that G is a group with respect to this multiplication. Define a map $\star$ from $G \times F$ to F by

$$(x, a) \star y = x + ay$$

Show that $\star$ is an action which is sharply doubly transitive action on F.

9.2 Sylow Theorems

Structure of a finite abelian group is well understood. Indeed, finite abelian groups are completely classfied (see Sect. 9.3). However, understanding the structure of a finite nonabelian group is extremely difficult problem. Perhaps, it is beyond dream to classify nonabelian finite groups. Mathematicians always roam around this problem. In the last section, we obtained some structural information about finite groups of prime power orders. In this section, we study finite groups by analyzing prime power order subgroups of the group. Basic results in this direction are Sylow theorems. The following is the Sylow 1st theorem.

Theorem 9.2.1 (Sylow) *Let G be a finite group and p^r divides $|G|$, where p is prime. Then, G has a subgroup of order p^r.*

Proof The proof is by induction on $|G|$. If $|G| = 1$, then there is nothing to do. Assume that the result is true for all those groups whose orders are less than $|G|$. Then, we have to prove the result for G. Suppose that $p^r, r \geq 1$ divides the order of G. Consider the class formula

$$| G | = | Z(G) | + \Sigma_{x \in A}[G : C_G(x)]$$

for the inner conjugation action of G on itself. The terms under summation in the R.H.S. are greater than 1, and they also divide $| G |$. There are two cases:

(i) p divides $| Z(G) |$.

(ii) p does not divide $| Z(G) |$.

Consider the case (i). Since $Z(G)$ is abelian, and p divides $| Z(G) |$, by the Cauchy theorem for abelian groups (Theorem 5.2.31), there exists a subgroup H of $Z(G)$ of order p. Since every subgroup of $Z(G)$ is normal in G, $H \unlhd G$. Now, consider G/H. Clearly, $| G/H | = \frac{|G|}{|H|} = \frac{|G|}{p} < | G |$. Also p^{r-1} divides $| G/H |$. By the induction hypothesis, G/H has a subgroup L of order p^{r-1}. By Proposition 5.2.29, there is a unique subgroup K of G containing H such that $L = K/H$. Since $| L | = p^{r-1}$, by the Lagrange theorem, $| K | = p^r$. But, then K is a subgroup of G of order p^r.

Now, consider the case (ii). Since p does not divide $| Z(G) |$, and p divides $| G |$, at least one term under summation in the R.H.S. of the classical class formula is not divisible by p. Suppose that p does not divide $[G : C_G(x)]$, where $[G : C_G(x)]$ is greater than 1. Thus, p does not divide $\frac{|G|}{|C_G(x)|}$. Since p^r divides $| G |$, it follows that p^r divides $| C_G(x) |$. Further, since $[G : C_G(x)] > 1$, $| C_G(x) | < | G |$. By the induction hypothesis, $C_G(x)$ has a subgroup of order p^r, and so G has a subgroup of order p^r.   ♯

Corollary 9.2.2 *Let G be a finite group, and a prime p divides the order of G. Then, G contains an element of order p.*

Proof From the above theorem, G has a subgroup H of order p. Any nonidentity element of H is of order p.   ♯

Definition 9.2.3 A group G is said to be a **p-group** if order of each element of G is a power of p.

Corollary 9.2.4 *A finite group G is a p-group if and only if $| G | = p^n$ for some n.*

Proof If order of G is p^n, then since order of each element of G divides order of G, order of each element of G is a power of p. Conversely, if order of each element of G is a power of p, then no other prime q can divide the order of G, for otherwise, by the above corollary, G will have an element of order q.   ♯

Definition 9.2.5 A maximal p-subgroup of G is called a **Sylow p-subgroup** of G.

Remark 9.2.6 Every group has a Sylow p-subgroup (may be $\{e\}$) for every prime p. This is an easy consequence of Zorn's Lemma if we observe that union of a chain of p-subgroups is a p-subgroup. For finite groups, we have the following corollary:

Corollary 9.2.7 *Let G be a finite group. Then, G has a Sylow p-subgroup which is of order p^m, where p^m divides $| G |$ but p^{m+1} does not divide $| G |$.*

Proof Since G is finite, there exists m such that p^m divides $\mid G \mid$ but p^{m+1} does not divide $\mid G \mid$. From the Sylow 1st theorem, there is a subgroup P (say) of G of order p^m. Since there is no higher power of p dividing $\mid G \mid$, P is a Sylow p-subgroup of G.   ♯

Example 9.2.8 Consider S_3. The subgroup A_3 is the only Sylow 3-subgroup of S_3. The subgroups $\{I, (12)\}, \{I, (23)\}$ and $\{I, (13)\}$ are all Sylow 2-subgroups. For the rest of the primes, $\{I\}$ is the only Sylow subgroup.

Example 9.2.9 Consider the group A_4. V_4 is the only maximal 2-subgroup of A_4, and hence, this is the only Sylow 2-subgroup of A_4. Further, $\{I, (123), (132)\}, \{I, (124), (142)\}, \{I, (134), (143)\}$, and $\{I, (234), (243)\}$ are all Sylow 3-subgroups of A_4. There is no other prime dividing the order of A_4.

Example 9.2.10 Consider the group S_4 which is of order $24 = 2^3 \times 3$. There are 4 subgroups of order 3 which are all Sylow 3-subgroups. They are $\{I, (123), (132)\}$, $\{I, (124), (142)\}, \{I, (134), (143)\}$, and $\{I, (234), (243)\}$. For the Sylow 2-subgroups, consider the Klein's four subgroup V_4 which is a normal subgroup of S_4. Let $H = \{I, (12)\}$. Then, HV_4 is a subgroup of order 8. Thus, it is a Sylow 2-subgroup of S_4. If we take $K = \{I, (13)\}$, and $L = \{I, (23)\}$, then KV_4 and LV_4 are also Sylow 2-subgroups of S_4. They are all.

Proposition 9.2.11 *Let P be a Sylow p-subgroup of G. Then, any conjugate gPg^{-1} of P is also a Sylow p-subgroup of G.*

Proof Let P be a Sylow p-subgroup of G. Since the subgroup gPg^{-1} is isomorphic to P, it is p-subgroup. Further, if P' is also a p-subgroup of G such that $gPg^{-1} \subseteq P'$, then $P \subseteq g^{-1}P'g$. Since $g^{-1}P'g$ is also a p-subgroup, and P is a Sylow (and so maximal) p-subgroup, it follows that $P = g^{-1}P'g$. In turn, $gPg^{-1} = P'$. This shows that gPg^{-1} is a maximal p-subgroup, and so it is a Sylow p-subgroup.   ♯

Corollary 9.2.12 *If P is a unique Sylow p-subgroup of G, then it is normal.*   ♯

Corollary 9.2.13 *The intersection of all Sylow p-subgroups of G is normal in G.*

Proof Let $\wp$ denote the set of all Sylow p-subgroups of G. Then, $g(\bigcap_{P\in\wp})g^{-1} = \bigcap_{P\in\wp} gPg^{-1} = \bigcap_{P\in\wp} P$, for $\{gPg^{-1} \mid P \in \wp\} = \wp$. This shows that $\bigcap_{P\in\wp}$ is a normal subgroup of G.   ♯

Proposition 9.2.14 *Every p-subgroup of a finite group is contained in a Sylow p-subgroup.*

Proof Let H be a p-subgroup of a finite group G. If H is a maximal p-subgroup, then it is a Sylow p-subgroup. If not, then there is a p-subgroup H_1 containing H properly. H_1 may be a Sylow p-subgroup. If not, proceed. Since G is finite, the process stops after finitely many steps giving us a Sylow p-subgroup of G.   ♯

Remark 9.2.15 The result of the above proposition is true even for infinite groups. This follows from an easy application of Zorn's Lemma.

Proposition 9.2.16 *Let P be a Sylow p-subgroup of a finite group G, and H a p-subgroup such that $HP = PH$. Then, $H \subseteq P$.*

Proof Since $HP = PH$, HP is a subgroup. Also $| HP | = \frac{|H| \cdot |P|}{|H \cap P|}$ is a power of p. Thus, HP is a p-subgroup of G. Clearly, $P \subseteq HP$. Since P is a Sylow p-subgroup, $P = HP$. Hence, $H \subseteq HP = P$.   ♯

Corollary 9.2.17 *Let P_1 and P_2 be Sylow p-subgroups of a finite group G such that $P_1 P_2 = P_2 P_1$. Then, $P_1 = P_2$.*

Proof This is immediate from the above proposition.   ♯

Corollary 9.2.18 *Let P_1 and P_2 be Sylow p-subgroups of a finite group G such that $g P_1 g^{-1} \subseteq P_1$ for all $g \in P_2$. Then, $P_1 = P_2$.*

Proof If $g P_1 g^{-1} \subseteq P_1$, then $g P_1 g^{-1} = P_1$, for $g P_1 g^{-1}$ is also a Sylow p-subgroup. Thus, $g P_1 g^{-1} = P_1$ for all $g \in P_2$, and so $g P_1 = P_1 g$ for all $g \in P_2$. This means that $P_1 P_2 = P_2 P_1$, and hence, from the above corollary, $P_1 = P_2$.   ♯

Corollary 9.2.19 *Let P be a Sylow p-subgroup of a finite group G. Then, $N_G(P)$ contains a unique Sylow p-subgroup of G.*   ♯

Following is the Sylow 2nd theorem.

Theorem 9.2.20 (Sylow 2) *Let G be a finite group and a prime p divides the order of G. Then, the set of all Sylow p-subgroups of G form a single complete conjugacy class of subgroups (i.e., conjugate of a Sylow p-subgroup is a Sylow p-subgroup, and any two Sylow p-subgroups of G are conjugate). Further, the number m of Sylow p-subgroups of G is of the form $1 + kp$, $k \geq 0$ (i.e., $m \equiv 1(mod\ p)$).*

Proof Let P_1 be a Sylow p-subgroup of G, and $X = \{P_1, P_2, \ldots, P_m\}$ be the set of all conjugates of P_1. Then, we need to show that all Sylow p-subgroups of G are in X, and $m \equiv 1(mod\ p)$. Since X is complete conjugacy class of subgroups of G, the group G, and so also P_1 act on X through inner conjugation. Consider the class formula

$$m = | X | = | X^{P_1} | + \Sigma_{P_j \in A}[P_1 : (P_1)_{P_j}] \qquad (9.2.1)$$

for the inner conjugation of P_1 on X, where X^{P_1} denotes the fixed point set of the action, and A is a set obtained by choosing one and only one member from each nonsingleton orbit. Suppose that $P_l \in X^{P_1}$. Then, $g P_l g^{-1} = P_l$ for all $g \in P_1$. But, then $P_1 P_l = P_l P_1$, and hence, $P_1 = P_l$ (Corollary 9.2.17). Thus, $X^{P_1} = \{P_1\}$, and so $| X^{P_1} | = 1$. Since each term under summation in the R.H.S. of the class equation is greater than 1, and also divides $| P_1 |$, it follows that the second term in the R.H.S. of (9.2.1) is a multiple of p. Thus, $m \equiv 1(mod\ p)$. It remains to show that each Sylow p-subgroup of G is a member of X. Suppose, if possible, that there is a Sylow p-subgroup P' of G which is not in X. Then, P' also acts on X through inner conjugation. Suppose $P_l \in X^{P'}$. Then, $g P_l g^{-1} = P_l$ for all $g \in P'$, and so

$P'P_l = P_l P'$. Again from Corollary 9.2.17, $P_l = P'$. Since $P' \notin X$, $X^{P'} = \emptyset$, and so $| X^{P'} | = 0$. Looking at the class formula for this action, we observe that $m \equiv 0 (mod\ p)$. This is a contradiction to the already observed fact that $m \equiv 1 (mod\ p)$. Thus, all Sylow p-subgroups of G are in X. ♯

Corollary 9.2.21 *If p^m divides $| G |$, and p^{m+1} does not divide $| G |$, then all Sylow p-subgroups of G are of order p^m.*

Proof From the above theorem, all Sylow p-subgroups are conjugate. Also all conjugate subgroups have same orders, and there is a Sylow p-subgroup of order p^m (Sylow 1st theorem). The result follows. ♯

Since the number of conjugates to a subgroup is equal to the index of the normalizer of that subgroup in the group, the following corollary follows from the Lagrange theorem.

Corollary 9.2.22 *The number $m = 1 + kp$ of Sylow p-subgroups of G is a divisor of $| G |$.* ♯

Following corollary gives a sufficient condition for a Sylow subgroup to be normal.

Corollary 9.2.23 *Let p be a prime divisor of $| G |$. Suppose that $1 + kp$ divides $| G |$ only if $k = 0$. Then, there is a unique Sylow p-subgroup of G which is normal.*

Proof From the above corollary, there is a unique Sylow p-subgroup of G, and since conjugate to a Sylow p-subgroup is again a Sylow p-subgroup, it follows that the Sylow p-subgroup is normal. ♯

Corollary 9.2.24 *Let H be a normal p-subgroup of a finite group G. Then, H is contained in each Sylow p-subgroup of G (i.e., $\bigcap_{P \in \wp} P$ is the largest p-subgroup of G which is normal in G).*

Proof Let H be a normal p-subgroup of G. Then, by Proposition 9.2.14, H is contained in a Sylow p-subgroup P (say) of G. Let P' be any Sylow p-subgroup of G. Then, by Theorem 9.2.20, $P' = gPg^{-1}$ for some $g \in G$. Thus, $gHg^{-1} \subseteq gPg^{-1} = P'$. Since H is supposed to be normal in G, $gHg^{-1} = H$, and hence, $H \subseteq P'$. ♯

Proposition 9.2.25 *A finite group G is direct product of its Sylow subgroups if and only if all Sylow subgroups are normal.*

Proof If G is direct product of all its Sylow subgroups, then by Corollary 5.2.25, all Sylow subgroups of G are normal. Conversely, suppose that all Sylow subgroups of G are normal. Suppose that $| G | = p_1^{\alpha_1} p_2^{\alpha_2} \cdots p_r^{\alpha_r}$, where $p_1, p_2, \ldots, p_r$ are distinct primes and $\alpha_i > 0$. Let P_i be the Sylow p_i subgroup of G (since each Sylow p_i subgroup is normal, they are unique). Then, by the hypothesis, each $P_i \trianglelefteq G$. Since products of normal subgroups are normal $P_1 P_2 \ldots P_{i-1} P_{i+1} P_{i+2} \ldots P_r$

is normal in G for each i. Next, since $P_1 \cap P_2 = \{e\}$, $| P_1 P_2 | = | P_1 || P_2 | = p_1^{\alpha_1} \cdot p_2^{\alpha_2}$. Proceeding inductively, we find that $| P_1 || P_2 | \ldots | P_t | = p_1^{\alpha_1} p_2^{\alpha_2} \ldots p_t^{\alpha_t}$ for all t. Thus, $| G | = | P_1 || P_2 | \ldots | P_r |$, and hence, $G = P_1 P_2 \ldots P_r$. Also $P_i \cap (P_1 P_2 \ldots P_{i-1} P_{i+1} P_{i+2} \ldots P_r) = \{e\}$. It follows from Corollary 5.2.25 that G is direct product of all its Sylow subgroups.   ♯

Corollary 9.2.26 *Let G be a group of order $p_1^{n_1} p_2^{n_2} \ldots p_r^{n_r}$, where $p_1, p_2, \ldots p_r$ are distinct primes, and $n_i \in \{1, 2\}$. Then, G is abelian if and only if all its Sylow subgroups are normal.*

Proof If G is abelian, then all its subgroups, and in particular all its Sylow subgroups are normal. Further, if all Sylow subgroups of G are normal, then G is direct product its Sylow subgroups $P_1, P_2, \ldots, P_r$. Each P_i, being of order $p_i^{n_i}$, $n_i \leq 2$, is abelian. Since direct product of abelian groups are abelian, G is abelian.   ♯

Since direct product of cyclic groups of co-prime orders are cyclic, we have the following corollary.

Corollary 9.2.27 *Let G be a group of order $p_1 p_2 \ldots p_r$, where $p_1, p_2, \ldots, p_r$ are distinct primes. Then, G is cyclic if and only if all its Sylow subgroups are normal.* ♯

Corollary 9.2.28 *Let $p_1, p_2, \ldots, p_r$ be distinct primes. Then, there are exactly 2^t nonisomorphic abelian groups of order $p_1^{n_1} p_2^{n_2} \ldots p_r^{n_r}$, $n_i \in \{1, 2\}$, where $t = \Sigma_{i=1}^r n_i - r$. In particular, there is only one (upto isomorphism) abelian group of order $p_1 p_2 \ldots p_r$ which is cyclic.*

Proof Let G be an abelian group of order $p_1^{n_1} p_2^{n_2} \ldots p_r^{n_r}$. Then, G is isomorphic to the external direct product of its Sylow subgroups $P_1, P_2, \ldots, P_r$ which are of orders $p_1^{n_1}, p_2^{n_2}, \ldots, p_r^{n_r}$, respectively. If $n_i = 1$, then P_i is necessarily cyclic of order p_i. If $n_i = 2$, then P_i has 2 possibilities, viz. $\mathbb{Z}_{p_i^2}$, and $\mathbb{Z}_{p_i} \times \mathbb{Z}_{p_i}$. In turn, there are exactly 2^t possibilities for G, where $t = \Sigma_{i=1}^r n_i - r$. In particular, if G is abelian group of order $p_1 p_2 \ldots p_r$, then it is isomorphic to direct product of distinct prime cyclic groups, and so it is cyclic.   ♯

Let G be a finite group, and H a subgroup of G. Let P be a Sylow p-subgroup of G. Then, $H \cap P$ need not be a Sylow p-subgroup of H. For example, $P = \{I, (234), (243)\}$ is a Sylow 3-subgroup of S_4, and $S_3 = \{I, (12), (23), (13), (123), (132)\}$ is a subgroup of S_4 whereas $S_3 \cap P = \{I\}$ is not a Sylow 3-subgroup of S_3. However, we have the following:

Proposition 9.2.29 *Let H be a subgroup of a finite group G, and P a Sylow p-subgroup of G such that HP is a subgroup of G (in particular H may be normal or P may be normal in G). Then, $H \cap P$ is a Sylow p-subgroup of H. Further, if $H \trianglelefteq G$, then HP/H is a Sylow p-subgroup of G/H.*

Proof Since $H \cap P$ is a subgroup of P, it is a p-subgroup of H. Also $\frac{|H|}{|H \cap P|} = \frac{|HP|}{|P|}$. Since HP is a subgroup of G, $[G : P] = [G : HP] \cdot [HP : P]$. Further, since P is a Sylow p-subgroup of G, $[G : P]$ is co-prime to p, and so $[HP : P]$ is also co-prime to p. This means that p does not divide $\frac{|H|}{|H \cap P|}$. Hence, $H \cap P$ is a Sylow p-subgroup of H.

Next, if H is a normal subgroup of G, then $HP/H \approx P/(H \cap P)$ (Noether isomorphism theorem) is a p-subgroup of G/H. Also $\frac{|G/H|}{|HP/H|} = \frac{\frac{|G|}{|H|}}{\frac{|HP|}{|H|}} = \frac{|G|}{|HP|}$ is co-prime to p. This shows that HP/H is a Sylow p-subgroup of G/H. ♯

Under a homomorphism, image of a Sylow p-subgroup need not be a Sylow p-subgroup, though it is a p-subgroup. For example, the trivial homomorphism from S_4 to A_4 does not take any Sylow 3-subgroup to a Sylow 3-subgroup. However, from the fundamental theorem of homomorphism, and the above proposition, we get

Proposition 9.2.30 *Under a surjective homomorphism, image of a Sylow p-subgroup is a Sylow p-subgroup.* ♯

Inverse image of a Sylow p-subgroup under a homomorphism need not even be a p-subgroup. For example, the first projection from $A_4 \times \mathbb{Z}_2$ to A_4 is a surjective homomorphism, and the inverse image of the Sylow 3-subgroup $A_3 = \{I, (123), (132)\}$ is $A_3 \times \mathbb{Z}_2$ which is not a 3-subgroup. However, we have the following:

Proposition 9.2.31 *Let G_1 be a finite group, and f a surjective homomorphism from G_1 to G_2. Let P be a Sylow p-subgroup of G_2. Then, $f^{-1}(P)$ is a Sylow p-subgroup of G_1 if and only if the kernel of f is a p-subgroup of G_1.*

Proof By the fundamental theorem of homomorphism, $f^{-1}(P)/\ker f \approx P$. Thus, $f^{-1}(P)$ is a p-subgroup if and only if $\ker f$ is a p-subgroup. Further, then, $[G_1 : f^{-1}(P)] = [G_1/\ker f : f^{-1}(P)/\ker f] = [G_2 : P]$ is co-prime to p. This shows that $f^{-1}(P)$ is a Sylow p-subgroup of G_1. ♯

Applications of Sylow Theorems

9.2.1. Let G be group of order pq, where p and q are primes. Suppose that $p = q$. Then, $|G| = p^2$, and so G is abelian. There are only two possibilities for G. It is isomorphic to $\mathbb{Z}_{p^2}$ or to $\mathbb{Z}_p \times \mathbb{Z}_p$. Next, suppose that $p > q$. Then, $1 + kp$ will divide pq only when $k = 0$. Thus, there is a unique Sylow p-subgroup P of order p, which, therefore, is normal in G. In particular, a group of order pq cannot be simple. Further, suppose that q does not divide $p - 1$. Then, $1 + kq$ also cannot divide pq, unless it is 1. But, then, Sylow q-subgroup Q is also normal in G. From Corollary 9.2.27, it follows that G is cyclic. Thus, for example, every group of order 15 is cyclic. Let G be a group of order pqr, where p, q and r are distinct primes such that $1 + k_1 p$ divides pqr only if $k_1 = 0$, $1 + k_2 q$ divides pqr only if $k_2 = 0$ and $1 + k_3 r$ divides pqr only if $k_3 = 0$. Then, all Sylow subgroups of G are normal, and again by Corollary 9.2.27, G is cyclic (for example, every group of order 1001 is cyclic). A complete classification of groups of order pq will follow in the last illustration of this section.

9.2.2. Every group of order $5^2 \cdot 7^2 = 1225$ is abelian, and there are four isomorphism classes of groups of order 1225: Indeed, $1 + 5k$ divides $5^2 \cdot 7^2$ only when $k = 0$, $1 + 7l$ divides $5^2 \cdot 7^2$ only when $l = 0$. Thus, there is a unique Sylow 5-subgroup P (say) which is of order 5^2, and also a unique Sylow 7-subgroup Q

(say). In turn, P and Q are normal subgroups of G. But, then, G is direct product of P and Q. Since direct product of abelian groups are abelian, G is abelian. Further, $P \approx Z_{5^2}$ or $P \approx Z_5 \times Z_5$. Similarly, $Q \approx Z_{7^2}$ or $Q \approx Z_7 \times Z_7$. Thus, G is isomorphic to one and only one of the following groups: (i) $Z_{1225} \approx Z_{5^2} \times Z_{7^2}$, (ii) $Z_5 \times Z_5 \times Z_{7^2}$, (iii) $Z_5 \times Z_5 \times Z_7 \times Z_7$, or (iv) $Z_{5^2} \times Z_7 \times Z_7$.

9.2.3. Let G be a group of order $7^3 5^2$. We describe the center of G in case it is nonabelian: By the Sylow theorem, there is a unique Sylow 7-subgroup, and also a unique Sylow 5-subgroup. Let H denote the Sylow 7-subgroup, and K the Sylow 5-subgroup. Then, G is the direct product $H \times K$. Since K (being of order 5^2) is abelian, and G is assumed to be nonabelian, H is nonabelian. The center $Z(H)$ of H cannot be of order 7^2, for otherwise $H/Z(H)$ will be cyclic, a contradiction (Theorem 5.2.32). Hence, the center of H is a cyclic group of order 7. By Corollary 9.1.21, and Theorem 9.1.22, $Z(K) = K$ is either Z_{25} or $Z_5 \times Z_5$. Thus, there are exactly two possibilities for the center $Z(G)$ of G, viz. $Z_7 \times Z_{25}$ and $Z_7 \times Z_5 \times Z_5$.

9.2.4. Let G be a group of order $p^n \cdot t$, where p is a prime, $n \geq 2$ and $t < 1 + 2p$. Then, G has a normal subgroup of order p^n, or if not, $t = 1 + p$ and it has a normal subgroup of order p^{n-1}.

Proof. If $t = p$, then by Proposition 9.1.23, G has a normal subgroup of order p^n. Suppose $t \neq p$. The number m of Sylow p-subgroups of G is of the form $1 + kp$ which divides $p^n \cdot t$. Since it has no common factor with p^n, it divides t. Thus, it is 1, unless $t = 1 + p$. Hence, the Sylow p-subgroup is normal, unless $t = 1 + p$. Suppose that $t = p + 1$, i.e., $|G| = p^n \cdot (p+1)$, and Sylow p-subgroup is not normal. In this case, the number of Sylow p-subgroups is $1 + p$. Let $X = \{P_1, P_2, \ldots, P_{1+p}\}$ be the set of all Sylow p-subgroups of G. Let $H = P_1 \cap P_2$ and $|H| = p^r$. Now, $|P_1 P_2| = \frac{|P_1| \cdot |P_2|}{|H|} = p^{2n-r} \leq p^n \cdot (p+1)$. Hence, $p^{n-r} \leq p + 1$. This implies that $n - r \leq 1$, and so $r \geq n - 1$. Since P_1 and P_2 are distict, $|H| = p^{n-1}$. Consider the action of H on X through the inner conjugation. The class formula for this action is

$$1 + p = |X| = |X^H| + \Sigma_{P_j \in A}[H : H_{P_j}],$$

where A is as usual. Since $H = P_1 \cap P_2$, $h \star P_1 = hP_1h^{-1} = P_1$ for all $h \in H$, and also $h \star P_2 = hP_2h^{-1} = P_2$ for all $h \in H$. Thus, P_1 and P_2 belong to X^H, and so $|X^H| \geq 2$. Since the second term of the class formula is a multiple of p, it follows that $X^H = X$. Thus, $hP_ih^{-1} = P_i$ for all i and for all $h \in H$. This shows that $HP_i = P_iH$ for all i. Since each P_i is a Sylow p-subgroup, $H = P_1 \cap P_2 \subseteq P_1 \cap P_2 \cap \cdots \cap P_{1+p} \subseteq P_1 \cap P_2$. Therefore, H, being the intersection of all Sylow p-subgroups, is a normal subgroup.

9.2.5. Let G be a group of order $p \cdot q \cdot r$, where p, q, r are distinct primes with $p < q < r$. Then, sylow r-subgroup of G is normal.

Proof. Suppose that neither Sylow q-subgroup nor Sylow r-subgroup is normal in G. Let $m_1 = 1 + k_1 r$ be the number of Sylow r-subgroups, and $m_2 = 1 + k_2 q$ the number of Sylow q-subgroups of G. Since these Sylow subgroups are not normal, $m_1 > 1, m_2 > 1$, and they also divide $|G| = pqr$. Clearly, $m_1 \neq p$ and $m_1 \neq q$.

Since $p < q < r, m_1 = 1 + k_1 r = p \cdot q$. Thus, there are $p \cdot q$ distinct groups of order r. Similarly, $m_2 = 1 + k_2 q$ is at least r, and so there are at least r distinct subgroups of order q. Since every nonidentity element of a cyclic group of prime order is a generator of the cyclic group, we see that distinct Sylow r-subgroups have only identity element in common. Thus, there are $p \cdot q \cdot (r - 1)$ elements in G of order r. Similarly, there are at least $r \cdot (q - 1)$ distinct elements of order q. Further, there should be at least $p - 1$ elements of order p. Therefore, G should contain at least $p \cdot q \cdot (r - 1) + r \cdot (q - 1) + p = p \cdot q \cdot r + (q - 1) \cdot (r - p)$ elements. Since $p < r$ and $q > 1$, the second term is positive. But, this would mean that G contains more than $p \cdot q \cdot r$ elements. This is a contradiction. It follows that Sylow r-subgroup is normal, or Sylow q-subgroup is normal. If Sylow r-subgroup is normal, we are done. Suppose that the Sylow q-subgroup Q is normal. Let R be a Sylow r-subgroup of G. Then, QR is also a subgroup of G of order $q \cdot r$, and the index of QR is p. Since p is the smallest prime dividing the order of G, QR is normal in G. Further, $| QR | = q \cdot r$, and $r > q$. It follows that R is normal in QR, and there is only one Sylow r-subgroup in QR. Let $g \in G$. Then, $gRg^{-1} \subseteq gQRg^{-1} = QR$, for $(QR \lhd G)$. Thus, R and gRg^{-1} are both Sylow r-subgroups of QR. Hence, $gRg^{-1} = R$. Since g is arbitrary element of G, $R \lhd G$.   ♯

9.2.6. Let G be a finite group, and p a prime which divides the order of G. Let P be a Sylow p-subgroup of G, and H a subgroup of G such that $N_G(P) \subseteq H$. Then, $N_G(H) = H$. In particular, $N_G(N_G(P)) = N_G(P)$.

Proof. Clearly, $H \subseteq N_G(H)$. Let $g \in N_G(H)$. Then, $gHg^{-1} = H$. Since $N_G(P) \subseteq H$, $P \subseteq H$. Hence, $gPg^{-1} \subseteq gHg^{-1} = H$. Thus, P and gPg^{-1} are both Sylow p-subgroups of H. By the Sylow theorem, they are conjugate in H. Hence, there exists an element $h \in H$ such that $hPh^{-1} = gPg^{-1}$, i.e., $h^{-1}gP = Ph^{-1}g$. This shows that $h^{-1}g \in N_G(P) \subseteq H$. Since $h \in H$, $g \in H$.   ♯

Structure of Groups of Order pq

Let G be a group of order $p \cdot q$, where p and q are primes. If $p = q$, then $| G | = p^2$. By Theorem 8.1.22, G is isomorphic to one of the following:

(i) Z_{p^2}.
(ii) $Z_p \times Z_p$.

Assume that $p \neq q$. Without any loss, we may assume that $p > q$. Then, by the Sylow 2nd theorem, there is unique Sylow p-subgroup P (say) which is normal in G. It is of order p, and so cyclic. Suppose that $P = < a >$, where a is an element of order p. Now there are two cases:

(i) q does not divide $p - 1$.
(ii) q divides $p - 1$.

Consider the case (i). In this case, Sylow q-subgroup is also normal (Sylow 2nd theorem). Hence, by Corollary 9.2.27, G is necessarily cyclic group of order pq, which, therefore, is isomorphic to $Z_{pq} = Z_p \times Z_q$.

Now, consider the case (ii). In this case p is of the form $1 + kq$. The number m of Sylow q-subgroups is also of the form $1 + lq$ (by Sylow 2nd theorem), and it divides $p \cdot q$. Thus, there are two possibilities:

(a) $m = 1$.
(b) $m = 1 + kq = p$.

In case (a), there is unique Sylow q-subgroup Q, which is normal and also cyclic of order q. It follows that G is direct product of a cyclic group P and a cyclic group Q which are of co-prime orders. Therefore, in this case G is isomorphic to Z_{pq}.

Next, suppose that G has $p = 1 + kq$ Sylow q-subgroups. Let $Q = <b>$ be a Sylow q-subgroup. Then, $|b| = q$. Since $P \cap Q = \{e\}$, $|PQ| = p \cdot q$, and so $G = PQ$. In other words

$$G = \{a^i b^j, 0 \le i < p, \ 0 \le j < q\}. \tag{9.2.2}$$

To determine the group G, we need to determine a rule by which we can multiply $a^i b^j$ and $a^k b^l$. In other words given i, j, k and l, to determine $u, v, \ 0 \le u < p$ and $0 \le v < q$ such that $(a^i b^j) \cdot (a^k b^l) = a^u b^v$. Since $P \lhd G$, $bab^{-1} \in P = <a>$. Suppose that

$$bab^{-1} = a^r, \ 0 \le r < p.$$

Obviously $r \ne 0$. If $r = 1$, then $ba = ab$, and then G will become abelian, a contradiction to the supposition that Sylow q-subgroup is not normal. Hence,

$$bab^{-1} = a^r, \ 1 < r < p. \tag{9.2.3}$$

Since $|b| = q$, $b^q = e$, and so $a = b^q a b^{-q} = a^{r^q}$. Thus, $a^{r^q - 1} = e$. Since $|a| = p$, p divides $r^q - 1$. This means that

$$r^q \equiv 1 \ (mod \ p), \tag{9.2.4}$$

and of course, $r \not\equiv 1 \ (mod \ p)$.

Clearly, $(r, p) = 1$, and so $\bar{r} \in U_p$, $\bar{r} \ne \bar{1}$. Since U_p is a cyclic group of order $p - 1$ (Corollary 7.6.23), and q divides $p - 1$, $H = \{\bar{a} \in U_p \mid \bar{a}^q = \bar{1}\}$ is a cyclic subgroup of U_p of order q. Thus, there are $q - 1$ solutions of the equation

$$x^q \equiv 1 \ (mod \ p), \ 1 < x < p \tag{9.2.5}$$

Now, given any solution r of this equation, we can determine the product in G as follows:

$$a^i b^j \cdot a^k b^l = a^i b^j a^k b^{-j} b^{j+l} = a^i (b^j a b^{-j})^k b^{j+l} = a^i (a^{r^j})^k b^{j+l} =$$
$$a^{i + r^j k} b^{j+l} = a^u b^v,$$

where u is the remainder obtained when $i + r^j k$ is divided by p, and v is th remainder obtained when $j + l$ is divided by q. Thus, G is determined once we know r with the required property.

Now, it appears that there may be $q - 1$ such groups corresponding to each solution of Eq. (9.2.5). But they are all isomorphic. The proof of this fact is as follows: Let r and s be two numbers such that $r^q \equiv 1 \ (mod \ p)$, $1 < r < p$ and $s^q \equiv 1 \ (mod \ p)$, $1 < s < p$. Then, $\bar{r}$ and $\bar{s}$ are nontrivial elements of H which is a cyclic group of prime order q. Since any nontrivial element of a cyclic group of prime order is a generator of the group, $\bar{r}$ and $\bar{s}$ both generate H. Thus, we have m, n, $1 \leq m < q$ and $1 \leq n < q$ such that

$$r^m \equiv s \ (mod \ p). \tag{9.2.6}$$

and

$$s^n \equiv r \ (mod \ p). \tag{9.2.7}$$

Let $G_1 = \{a^i b^j \mid 0 \leq i \leq p - 1, 0 \leq j \leq q - 1, a^p = e = b^q \text{ and } bab^{-1} = a^r\}$. and $G_2 = \{a^i c^j \mid 0 \leq i \leq p - 1, 0 \leq j \leq q - 1, a^p = e = c^q \text{ and } cac^{-1} = a^s\}$. Define a map ϕ from G_1 to G_2 by

$$\phi(a^i b^j) = a^i c^{nj},$$

and a map ψ from G_2 to G_1 by

$$\psi(a^i c^j) = a^i b^{mj}.$$

Since $r^m \equiv s \ (mod \ p)$ and $s^n \equiv r \ (mod \ p)$, $r^{mn} = (r^m)^n \equiv r \ (mod \ p)$. Since $|\bar{r}| = q$ in U_p, q divides $mn - 1$. Thus,

$$mn \equiv 1 (mod \ q). \tag{9.2.8}$$

Now $\psi o \phi(a^i b^j) = \psi(a^i c^{nj}) = a^i b^{mnj} = a^i b^j$, for by Eq. (9.2.8), $mnj \equiv j (mod \ q)$. Thus, $\psi o \phi = I_{G_1}$. Similarly, $\phi o \psi = I_{G_2}$. This shows that ϕ is bijective. Further,

$$\phi(a^i b^j) = \phi(a^{i+r^j k} b^{j+l}) = a^{i+r^j k} c^{n(j+l)}. \tag{9.2.9}$$

and

$$\phi(a^i b^j) \cdot \phi(a^k b^l) = a^i c^{nj} \cdot a^k c^{nl} = a^{i+s^{nj} k} c^{n(j+l)}. \tag{9.2.10}$$

Since $s^n \equiv r \ (mod \ p)$, $s^{nj} \equiv r^j (mod \ p)$, or $s^{nj} k \equiv r^j k (mod \ p)$. Since $|a| = p$, $a^{i+s^{nj} k} = a^{i+r^j k}$. Thus, from (9.2.9) and (9.2.10), it follows that ϕ is an isomorphism.

To summarize, we have proved the following theorem.

Theorem 9.2.32 *Let p and q be primes, and G a group of order $p \cdot q$.*

(a) If $p = q$, then G is isomorphic to one and only one of the following two groups:

(i) Z_{p^2}.

(ii) $Z_p \times Z_p$.

(b) If $p \neq q$, $p > q$ and q does not divide $p - 1$, then G is isomorphic to Z_{pq}.

(c) If q divides $p - 1$, then G is isomorphic to one and only one of the following two groups:

(i) Z_{pq}.

(ii) $G = \{a^i b^j \mid 0 \leq i \leq p - 1, 0 \leq j \leq q - 1\}$, where the multiplication in G is determined by the following rule: Take any nontrivial solution r of the equation $x^q \equiv 1 \pmod{p}$, $2 \leq x \leq p - 1$, and then $(a^i b^j)(a^k b^l) = a^u b^v$, where u is the remainder obtained when $i + r^j k$ is divided by p, and v is the remainder obtained when $j + l$ is divided by q.

Example 9.2.33 Any group of order 15 is cyclic, for $15 = 5 \cdot 3$ and 3 does not divide $5 - 1 = 4$.

Example 9.2.34 Let G be a nonabelian group of order $2p$, where p is an odd prime. Then, $r = p - 1$ is the only solution $x^2 \equiv 1 \pmod{p}$, $2 \leq x \leq p - 1$. Thus, $G = \{a^i b^j \mid 0 \leq i \leq p - 1, 0 \leq j \leq 1\}$, and the multiplication is given by $(a^i b^j) \cdot (a^k b^l) = a^u b^v$, where u is the remainder obtained when $i + (p - 1)^j k$ is divided by p, and v is the remainder when $j + l$ is divided by 2 (observe that $bab^{-1} = a^{p-1} = a^{-1}$). Since there is only one nonabelian group of order $2p$, $p \geq 3$, and the dihedral group D_p of the group of isometries (see Example 4.1.28) of a regular polygon of p sides is nonabelian of order $2p$, G is isomorphic to D_p. Indeed, the isomorphism is given by the map $a^i b^j \rightsquigarrow \rho^i \sigma^j$, where ρ is the rotation about the center of the polygon through an angle $\frac{2\pi i}{p}$, and σ is a reflexion about a line joining a vertex of the polygon with the middle point of the opposite edge.

Example 9.2.35 Let G be a nonabelian group of order 21. We find that 2 and 4 are the solutions of the equation $x^3 \equiv 1 \pmod{7}$, $1 < x < 7$. Taking $r = 2$ the product $(a^2 b^2)(a^6 b)$, for example, is $a^{2 + 2^2 \cdot 6} b^3 = a^5 b^0 = a^5$.

Exercises

9.2.1 Find all Sylow subgroups of A_4, S_4, A_5, and S_5.

9.2.2 Find Sylow p-subgroups of the group $GL(2, \mathbb{Z}_p)$ of invertible 2×2 matrices with entries in the field $\mathbb{Z}_p$.

9.2.3 Show that every group of order 35 is cyclic.

9.2.4 Show that every group of order 4199 is cyclic.

9.2.5 Let G be a group of order pq, where p and q are prime. Show that G cannot be simple.

9.2.6 Let G be a group of order pq, where p and q are primes such that $q < p < 2q + 1$. Show that G is cyclic.

9.2.7 Show that every noncyclic group of order 6 is isomorphic to S_3.

9.2.8 Give a presentation of a nonabelian group of order 33.

9.2.9 Show that every group of order 48,841 is abelian. Find the number of nonisomorphic groups of this order.

9.2.10 Show that every group of order 99 is abelian.

9.2.11 Show that a group of order 12 cannot be simple.

9.2.12 Show that a group of order $17 \cdot 23 \cdot 29$ cannot be simple. Is it cyclic?

9.2.13 Let G be a group of order $p^3 \cdot (p+1) \cdot m$ such that $(p+1) \cdot m < p^2$, where p is prime. Show that G has a normal subgroup of order p^3 or a normal subgroup of order p^2.

9.2.14 Show that every group of order 216 contains a normal subgroup of order 27 or a normal subgroup of order 9.

9.2.15 Show that a group whose order is less than 60 is simple if and only if it is cyclic group of prime order.

9.2.16 Show that a group of order 28 having a normal subgroup of order 4 is abelian.

9.2.17 How many nonisomorphic groups of order 55 are there? How many elements of order 5 in a nonabelian group of order 55 are there?

9.2.18 Show that a group of order $p^4 \cdot (p+1)$ has a normal subgroup of order p^4 or a normal subgroup of order p^3.

9.2.19 Let H be a proper subgroup of a group G such that $|G| = p^n, n \geq 2$. Show that there exists $x \in G - H$ such that $xHx^{-1} = H$.

9.2.20 Find the number of Sylow p-subgroups of S_p.

9.2.21 Show that a subgroup of order p^{n-1} of a group of order p^n is normal.

9.2.22 Let G be a group of order $p^n \cdot q$, where p and q are primes. Let P be a Sylow p-subgroup of G. Show that either P is normal or $N_G(P) = P$.

9.2.23 Let P be a normal Sylow p-subgroup of a finite group G and f an endomorphism of G. Show that $f(P) \subseteq P$.

9.2.24 Let H be a normal subgroup of G such that p does not divide $|G/H|$. Show that H contains all Sylow p-subgroups of G.

9.2.25 Let H be a normal subgroup of a finite group G, and P a Sylow p-subgroup of G which is contained in H as a normal subgroup. Show that $P \unlhd G$.

9.2.26 Let G be a group of order $p^n \cdot t$, where p is a prime and $n \geq 2$. Assume that $t \leq 2p$. Show that G cannot be simple.

9.2.27 Let G be a group and $a \neq e$. Suppose that a is of finite order, and also the conjugacy class determined by a is finite. Show that G has a nontrivial finite normal subgroup.

9.2.28 Find a nontrivial solution r of $x^3 \equiv 1(mod\ 13)$, and then find $(a^5b^2) \cdot (a^7b)$, where $a^{13} = e = b^3$ and $bab^{-1} = a^r$.

9.2.29 Show that if a is conjugate to b, then $C_G(a)$ is conjugate to $C_G(b)$.

9.2.29* Show that every simple group of order 60 is isomorphic to A_5.
Hint. Show the existence of a subgroup H of order 12, and then consider the action of G on $G/^l H$.

9.2.30 Let G be a group of order p^2q. Show that G has a normal Sylow subgroup.

9.2.31* Let G be group of order p^3q. Suppose that G has no normal Sylow subgroups. Show that G is isomorphic to S_4. Deduce that no group of order p^3q can be simple.

9.2.32* Let G be a group of order p^mq, where p and q are distinct primes. Suppose that Sylow p-subgroup of G is not normal in G. Show the following:

   (i) There are q Sylow p-subgroups.
  (ii) Let K be a maximal member among the intersections of distinct pairs of Sylow p-subgroups, and $H = N_G(K)$. Show that H has at least two distinct Sylow p-subgroups.
 (iii) Show further that H contains exactly q Sylow p-subgroups.
 (iv) Deduce that all Sylow p-subgroups of H are contained in K.
  (v) Show that every Sylow p-subgroup of a finite subgroup is obtained by taking the intersection of a Sylow p-subgroup of the group with that subgroup.
 (vi) Use (v) to deduce that $K = \{e\}$.
 (vii) Deduce that the intersection of any two distinct Sylow p-subgroup is trivial.
(viii) Counting the elements of G, show that there is unique Sylow q-subgroup of G.
 (ix) Deduce that G cannot be simple.

Remark 9.2.36 More generally, no group of order p^mq^n can be simple, where p and q are primes. This was proved by Burnside in the beginning of the 20th century using the representation theory. The proof will be given in the chapter on representation theory, Algebra 2. A nonrepresentation theoretic proof of this fact was obtained quite late with the works of Thompson, Goldschmidt, Bendor and Matsuyama.

9.2.33 Describe all nonabelian groups of order 8.

9.2.34 A finite group G is called a **CLT** group, if corresponding to every divisor m of the order of the group G, there is a subgroup of order m. Show the following:

(i) Every finite p-group is a CLT group.
(ii) Suppose that every Sylow subgroup of G is normal. Show that it is a CLT group.
(iii) Every finite abelian group is a CLT group.
(iv) Give an example to show that subgroup of a CLT group need not be a CLT group.
(v) Give an example to show that quotient of a CLT group need not be a CLT group.
(vi) Show that S_4 a CLT group.

9.2.35* Describe all nonabelian groups of order p^3, where p is a prime.

9.2.36* Describe all groups of order pqr, where p, q, r are primes.

9.2.37* Let G be a finite p-group, where p is an odd prime. Show that all subgroups of G are normal if and only if G is abelian. Deduce that all subgroups of an odd ordered group G is normal if and only if G is abelian. Give an example to show that the result is not true for $p = 2$.
Hint. Observe that every nontrivial normal subgroup of a p-group contains a nontrivial element of the center. Use Induction.

9.3 Finite Abelian Groups

In this section, we classify all finite abelian groups. Since all Sylow subgroups will be normal, a finite abelian group is direct product of its Sylow subgroups. As a consequence, it follows that two finite abelian groups are isomorphic if and only if there corresponding Sylow subgroups are isomorphic. Thus, it is sufficient to classify all finite abelian groups of prime power orders.

Theorem 9.3.1 *Let G be a finite abelian p-group of order p^n. Then, there exist positive integers $n_1 \geq n_2 \geq \cdots \geq n_r$, $n_1 + n_2 + \cdots + n_r = n$ together with elements $x_1, x_2, \ldots, x_r$ in G such that $| x_i | = p^{n_i}$ for all i, and G is the direct product*

$$< x_1 > \times < x_2 > \times \cdots \times < x_r > .$$

Proof The proof is by induction on n. If $n = 1$, then G itself is cyclic generated by any nonidentity element of G. Assume that the result is true for all those abelian groups of orders p^m, where $m < n$. Let G be an abelian group of order p^n. We have to prove the result for G. Let x_1 be an element of maximum order p^{n_1} in G. If $n_1 = n$, then G is cyclic and there is nothing to prove. Suppose that $n_1 < n$. Consider the group $H = G/ < x_1 >$. Clearly, $| H | = \frac{|G|}{|<x_1>|} = p^{n-n_1}$, $n - n_1 < n$. By the induction hypothesis,

$$H \; = \; < \nu(y_2) > \times < \nu(y_3) > \times \cdots \times < \nu(y_r) >,$$

where ν is the quotient map from G to $H \; = \; G/ < X_1 >$, $y_2, y_3, \ldots, y_r \in G$, $|\nu(y_i)| = \; p^{n_i}$, and $n_2 \geq n_3 \geq \cdots \geq n_r$. Since p^{n_1} is maximum among the orders of elements of G, and $|\nu(y)|$ divides $|y|$ for all $y \in G$, it follows that $n_1 \geq n_2 \geq \cdots \geq n_r$.

Next, we show the existence of $z_i \in G$ such that $\nu(z_i) \; = \; \nu(y_i)$, and $|z_i| = |\nu(z_i)| = |\nu(y_i)| = \; p^{n_i}$ for all $i \geq 2$. Since $|\nu(y_i)| = |(y_i + < x_1 >)| = p^{n_i}$, $p^{n_i} y_i + < x_1 > \; = \; < x_1 >$ for all $i \geq 2$. Hence, $p^{n_i} y_i \in < x_1 >$ for all $i \geq 2$. Suppose that $p^{n_i} y_i \; = \; p^{t_i} a_i x_1$, $i \geq 2$, where $(p, a_i) \; = \; 1$ and $t_i \leq n_1$. If $t_i = n_1$, then $p^{n_i} y_i \; = \; 0$ and $|y_i|$ divides p^{n_i}. Since $p^{n_i} \; = \; |\nu(y_i)|$ divides $|y_i|$, it follows that $|y_i| = |\nu(y_i)|$, and we can take $z_i \; = \; y_i$. Suppose that $t_i < n_1$. Since $(a_i, p^{n_1}) \; = \; 1$, by the Euclidean algorithm, there are integers u and v such that $u a_i + v p^{n_1} \; = \; 1$. But, then $x_1 \; = \; u a_i x_1$, and so $|a_i x_1| = |x_1| = \; p^{n_1}$. Thus, $|p^{n_i} y_i| = |p^{t_i} a_i x_1| = \; p^{n_1 - t_i}$. Hence, $|y_i| = \; p^{n_1 - t_i + n_i}$. Since p^{n_1} is largest among the orders of elements of G, $n_1 - t_i + n_i \leq n_1$. This shows that $n_i \leq t_i$. Take $z_i \; = \; y_i - p^{t_i - n_i} a_i x_1$. Then, $\nu(z_i) \; = \; \nu(y_i)$, and $|z_i| = \; p^{n_i} = |\nu(y_i)|$.

Finally, we shall show that

$$G \; = \; < x_1 > \times < z_2 > \times \cdots \times < z_r > .$$

Let $x \in G$. Then, $\nu(x) \in H$. Since $H = \; < \nu(y_2) > \times < \nu(y_3) > \times \cdots \times < \nu(y_r) > \; = \; < \nu(z_2) > \times < \nu(z_3) > \times \cdots \times < \nu(z_r) >$, it follows that

$$\nu(x) \; = \; \alpha_2 \nu(z_2) + \alpha_3 \nu(z_3) + \cdots + \alpha_r \nu(z_r).$$

for some integers $\alpha_2, \alpha_3, \ldots, \alpha_r$. This means that

$$x + < x_1 > \; = \; (\alpha_2 z_2 + \alpha_3 z_3 + \cdots + \alpha_r z_r) + < x_1 > .$$

Hence, $x - \alpha_2 z_2 - \alpha_3 z_3 - \cdots - \alpha_r z_r$ belongs to $< x_1 >$. In other words, $x = \alpha_1 x_1 + \alpha_2 z_2 + \cdots + \alpha_r z_r$ for some integers $\alpha_1, \alpha_2, \ldots, \alpha_r$.

Next, suppose that $\alpha_1 x_1 + \alpha_2 z_2 + \cdots + \alpha_r z_r \; = \; 0$. Then, $\alpha_2 \nu(z_2) + \cdots \alpha_r \nu(z_r) \; = \; < x_1 >$ the zero of H. Since $H \; = \; < \nu(z_2) > \times \cdots \times < \nu(z_r) >$, $\alpha_i \nu(z_i) \; = \; < x_1 >$ for all $i \geq 2$. Since $|z_i| = |\nu(z_i) >|$, it follows that $\alpha_i z_i = \; 0$ for all $i \geq 2$. In turn, $\alpha_1 x_1$ is also 0. This shows that $G \; = \; < x_1 > \times < z_2 > \times \cdots \times < z_r >$. ♯

Corollary 9.3.2 *If G is a finite abelian group, then G is a direct product of cyclic groups of prime power orders.*

Proof Every finite abelian group is direct product of its Sylow subgroups which are of prime power orders. The result follows from the above theorem. ♯

Theorem 9.3.3 *Let G and G' be finite abelian p-groups. Suppose that*

$$G = \langle x_1 \rangle \times \langle x_2 \rangle \times \cdots \times \langle x_r \rangle$$

and

$$G' = \langle y_1 \rangle \times \langle y_2 \rangle \times \cdots \times \langle y_s \rangle .$$

*are direct decompositions of G and G', respectively, as direct product of cyclic groups
with* $| x_i | = p^{n_i}$, $| y_j | = p^{m_j}$, $n_1 \geq n_2 \geq \cdots \geq n_r$, *and* $m_1 \geq m_2 \geq \cdots \geq m_s$.
Then, G is isomorphic to G' if and only if

(i) $r = s$.
(ii) $n_i = m_i$ *for all i.*

Proof If part of the proof is evident because any two cyclic groups of same orders are
isomorphic, and if $H \approx H'$ and $K \approx K'$, then $H \times K$ is isomorphic to $H' \times K'$. The
proof of the converse is by induction on $max(r, s)$. If $max(r, s) = 1$, then G and G'
are cyclic groups of same prime power orders and so they are isomorphic. Assume
that the result is true for all cases for which $max(r, s) < m$, $m > 1$. Suppose that
$max(r, s) = m$ in the representation of G and G' and σ is an isomorphism from G
to G'. Clearly, p^{n_1} is the maximum among the orders of G and p^{m_1} is the maximum
among the orders of G'. Since under an isomorphism orders of elements remain the
same, $p^{n_1} = p^{m_1}$, and so $n_1 = m_1$. Let

$$\sigma(x_1) = \beta_1 y_1 + \beta_2 y_2 + \cdots + \beta_s y_s . \tag{9.3.1}$$

$| x_1 | = | \sigma(x_1) | = p^{n_1}$ and $n_1 = m_1 > m_j$ for all j. Thus, $| \beta_j y_j | = p^{n_1}$ for some
j. After rearranging, we can assume that $| \beta_1 y_1 | = p^{n_1}$ (note that if $| \beta_j y_j | = p^{n_1}$,
then, since $| \beta_j y_j |$ divides $| y_j |$, $| y_j | = p^{n_1}$). We show that

$$G' = \langle \sigma(x_1) \rangle \times \langle y_2 \rangle \times \cdots \times \langle y_s \rangle .$$

Since $| y_1 | = p^{n_1} = | \beta_1 y_1 |$, $(\beta_1, p^{n_1}) = 1$. By the Euclidean algorithm, there
exist $u, v \in \mathbb{Z}$ such that $u\beta_1 + vp^{n_1} = 1$. Hence, $y_1 = u\beta_1 y_1 = u(\sigma(x_1) - \beta_2 y_2 - \cdots - \beta_s y_s)$. Since $\{y_1, y_2, \ldots, y_s\}$ generates G', $\{\sigma(x_1), y_2, \ldots, y_s\}$ also
generates G'. Further, suppose that

$$\delta_1 \sigma(x_1) + \delta_2 y_2 + \cdots + \delta_s y_s = 0.$$

Substituting the value of $\sigma(x_1)$ from 1, we get

$$\delta_1 \beta_1 y_1 + (\delta_1 \beta_2 + \delta_2) y_2 + \cdots + (\delta_1 \beta_s + \delta_s) y_s = 0.$$

Since $G' = \langle y_1 \rangle \times \langle y_2 \rangle \times \cdots \times \langle y_s \rangle$, $\delta_1 \beta_1 y_1 = (\delta_1 \beta_2 + \delta_2) y_2 = \cdots = (\delta_1 \beta_s + \delta_s) y_s = 0$. Since $| \beta_1 y_1 | = p^{n_1}$, p^{n_1} divides δ_1. Hence, $\delta_1 y_i = 0$ for all
$i \geq 2$. This shows that $\delta_i y_i = 0$ for all $i \geq 2$, and so $\delta_1 \sigma(x_1) = 0$. Thus,

$$G' = \langle \sigma(x_1) \rangle \times \langle y_2 \rangle \times \cdots \times \langle y_s \rangle .$$

Since σ is an isomorphism from G to G' such that $\sigma(<x_1>) = <\sigma(x_1)>$, it induces an isomorphism from $G/<x_1>$ to $G'/<\sigma(x_1)>$. Also

$$G/<x_1> \approx <x_2> \times \cdots \times <x_r>$$

and

$$G'/<\sigma(x_1)> \approx <y_2> \times \cdots \times <y_s>.$$

By the induction assumption $r - 1 = s - 1$ and $n_i = m_i$ for all $i \geq 1$.   ♯

Corollary 9.3.4 *The number of isomorphism classes of abelian groups of order p^n is $p(n)$, where $p(n)$ is the number of partitions of n.*

Proof From the above theorem, it is clear that there are as many isomorphism classes of abelian groups of order p^n as many elements in the set $\{(n_1, n_2, \ldots, n_r) \mid n_1 \geq n_2 \geq \cdots \geq n_r$ and $n_1 + n_2 + \cdots + n_r = n\}$. The number of elements in this set, by definition, is $p(n)$ the partition of n.   ♯

Since any two finite abelian groups are isomorphic if and only if their Sylow subgroups are isomorphic, we have the following corollary:

Corollary 9.3.5 *Let $n = p_1^{n_1} p_2^{n_2} \ldots p_r^{n_r}$, where $p_1, p_2, \ldots, p_r$ are distinct primes. Then, the number of isomorphism classes of abelian groups of order n is $p(n_1) \cdot p(n_2) \ldots p(n_r)$, where p is the partition function.*   ♯

Exercises

9.3.1 Find the number of nonisomorphic abelian groups of order 144. Also list a member from each isomorphism class.

9.3.2 Show that there are as many isomorphism classes of abelian groups of order $p_1^{n_1} p_2^{n_2} \ldots p_r^{n_r}$ as many conjugacy classes in $S_{n_1} \times \cdots \times S_{n_r}$.

9.3.3 Let G and G' be finite abelian groups such that for each n, G and G' have same number of elements of order n. Show that G is isomorphic to G'.

9.4 Normal Series and Composition Series

Let G be a group with operator set Ω. A Ω-**subnormal series** of G is a descending chain

$$G = G_1 \trianglerighteq G_2 \trianglerighteq \cdots \trianglerighteq G_n \trianglerighteq G_{n+1} = \{e\} \tag{9.4.1}$$

of Ω-subgroups of G, where $G_{i+1} \trianglelefteq G_i$ for all i. (Note that a term in a Ω-subnormal series need not be Ω-normal in G. Indeed, they are called Ω-subnormal subgroups). If $\Omega = \emptyset$, then a Ω-subnormal series is simply said to be a **subnormal series** of G.

If $\Omega = Inn(G)$, then all the terms of Ω-subnormal series are normal subgroups of G, and in this case, we say that it is a **normal** series of G. If $\Omega = Aut(G)$, then all the terms of Ω-subnormal series are characteristic subgroups, and in this case, we term it as a **characteristic** series of G. If $\Omega = End(G)$, then all the terms of Ω-subnormal series are fully invariant subgroups, and in this case, we term it as a **fully invariant** series of G.

We say that the series 1 is **without repetitions**, if $G_i \neq G_{i+1}$ for all i. The number n is called the **length** of the series 1.

The Ω-quotient groups $G_1/G_2, G_2/G_3, \ldots, G_n/G_{n+1}$ are called the **factors** of the series 1.

Let

$$G = H_1 \trianglerighteq H_2 \trianglerighteq \cdots \trianglerighteq H_m \trianglerighteq H_{m+1} = \{e\}. \tag{9.4.2}$$

be also a Ω-subnormal series of G. We say that (9.4.1) is **refinement** of (9.4.2) if there exists an injective map σ from $\{1, 2, \ldots, m\}$ to $\{1, 2, \ldots, n\}$ such that $H_i = G_{\sigma(i)}$ for all i. Thus, if (9.4.1) is refinement of (9.4.2), then $m \leq n$. We say that (9.4.1) is a **proper** refinement of (9.4.2) if (i) (9.4.1) is refinement of (9.4.2), and (ii) $m < n$.

A Ω-subnormal series of a Ω group G which is without repetitions is called a Ω-**composition series** of G, if it has no proper refinement which is a Ω-subnormal series, and which is without repetitions. If $\Omega = \emptyset$, then a Ω-composition series is simply said to be a **composition series** of G. If $\Omega = Inn(G)$, then Ω-composition series are called a **principal** or **chief** series of G. If $\Omega = Aut(G)$, Ω-composition series are called a **principal characteristic** series of G. If $\Omega = End(G)$, then Ω-composition series are called a **principal fully invariant** series of G.

Ω-subnormal series (9.4.1) and (9.4.2) are said to be **equivalent**, if

(i) $m = n$, and

(ii) there is a bijective map between the set of factors of (9.4.1) and that of (9.4.2) such that the corresponding factors are Ω-isomorphic. We shall also express this by saying that the factors of (9.4.1) are Ω-isomorphic to the factors of (9.4.2) after some rearrangement.

Example 9.4.1 $\mathbb{Z} \triangleright 6\mathbb{Z} \triangleright 24\mathbb{Z} \triangleright \{0\}$ is a normal series of the additive group $\mathbb{Z}$ of integers which is without repetitions. The length of this normal series is 3, the factors are $\mathbb{Z}/6\mathbb{Z} \approx \mathbb{Z}_6, 6\mathbb{Z}/24\mathbb{Z} \approx \mathbb{Z}_4$, and $24\mathbb{Z}/\{0\} \approx \mathbb{Z}$.

Example 9.4.2 The series $S_4 \triangleright A_4 \triangleright V_4 \triangleright \{I\}$ is a normal series of S_4 which is a proper refinement of the normal series $S_4 \triangleright V_4 \triangleright \{I\}$, and both are without repetitions.

Example 9.4.3 The series $S_4 \triangleright A_4 \triangleright V_4 \triangleright \{I, (12)(34)\} \triangleright \{I\}$ is a subnormal series of S_4. Indeed, this is a composition series of S_4, for it is without repetitions, and it cannot be refined further without admitting repetitions (note that there is no subgroup of S_4 in between S_4 and A_4 except S_4 and A_4, and similarly, there is no subgroup between A_4 and V_4 except A_4 and V_4, etc.). The series $S_4 \triangleright A_4 \triangleright V_4 \triangleright \{I\}$ is a principal series of S_4. Indeed, it is a principal Characteristic series of G. Is it a fully invariant series?

Example 9.4.4 The normal series $\mathbb{Z} \rhd 8\mathbb{Z} \rhd 24\mathbb{Z} \rhd 48\mathbb{Z} \rhd \{0\}$ of $\mathbb{Z}$ is equivalent to the normal series $\mathbb{Z} \rhd 3\mathbb{Z} \rhd 6\mathbb{Z} \rhd 48\mathbb{Z} \rhd \{0\}$ (verify). The subnormal series $S_4 \rhd A_4 \rhd V_4 \rhd \{I, (12)(34)\} \rhd \{I\}$ of S_4 is equivalent to the subnormal series $S_4 \rhd A_4 \rhd V_4 \rhd \{I, (13)(24)\} \rhd \{I\}$ (verify and write another subnormal series of S_4 which is equivalent to the above two subnormal series). The normal series $C_6 \rhd C_3 \rhd \{e\}$ and the normal series $C_6 \rhd C_2 \rhd \{e\}$ of a cyclic group C_6 of order 6 (C_3 is the cyclic subgroup of order 3 of C_6, and C_2 is that of order 2) are equivalent. Note that they are also principal fully invariant series.

Two subnormal series of a group may not have a common refinement. For example, $S_4 \rhd V_4 \rhd \{I, (12)(34)\} \rhd \{I\}$ and $S_4 \rhd V_4 \rhd \{I, (13)(24)\} \rhd \{I\}$ have no common refinements. However, we have the following theorem of Schreier.

Theorem 9.4.5 (Schreier) *Let G be a group with operator set Ω. Then, any two Ω-subnormal series (Ω-normal series) of G have equivalent refinements.*

We need the following theorem known as **Zassenhauss Lemma** or **third isomorphism theorem** for the proof of the Schreier's theorem.

Theorem 9.4.6 (Zassenhauss Lemma) *Let G be a group with operator set Ω. Let G_1, G_2 be Ω-subgroups of G. Suppose that G_1' is Ω-normal subgroup of G_1, and G_2' a Ω-normal subgroup of G_2. Then, $(G_1 \cap G_2')G_1'$ is Ω-normal subgroup of $(G_1 \cap G_2)G_1'$, and $(G_1' \cap G_2)G_2'$ is Ω-normal subgroup of $(G_1 \cap G_2)G_2'$. Further, the corresponding factors $(G_1 \cap G_2)G_1'/(G_1 \cap G_2')G_1'$ and $(G_1 \cap G_2)G_2'/(G_1' \cap G_2)G_2'$ are Ω-isomorphic.*

Proof Since G_2' is Ω-normal subgroup of G_2, $(G_1 \cap G_2')$ is Ω-normal in G_1. Again, since G_1' is Ω-normal in G_1, the product $(G_1 \cap G_2')G_1'$ is a Ω-subgroup, in fact, a Ω-normal subgroup (product of Ω-normal subgroups are Ω-normal subgroups) of G_1. Similarly, it can be observed that all subgroups in the lemma are Ω-subgroups. Let $h = uv \in (G_1 \cap G_2')G_1'$, where $u \in G_1 \cap G_2'$, $v \in G_1'$, and $g = xy \in (G_1 \cap G_2)G_1'$, where $x \in G_1 \cap G_2$ and $y \in G_1'$. Then,

$$ghg^{-1} = xyuv(xy)^{-1} = xux^{-1}(ux^{-1})^{-1}y(ux^{-1})x(vy^{-1})x^{-1}.$$

Observe that $ux^{-1} \in G_1$, and $y \in G_1'$. Since $G_1' \unlhd G_1$, $(ux^{-1})^{-1}y(ux^{-1}) \in G_1'$. Also, since $vy^{-1} \in G_1'$, and $x \in G_1$, $xvy^{-1}x^{-1} \in G_1'$. Thus, $ghg^{-1} = xux^{-1}w$, where $w \in G_1'$. Further, since $x, u \in G_1$, $xux^{-1} \in G_1$. Again, $u \in G_2', x \in G_2$, and since $G_2' \unlhd G_2$, $xux^{-1} \in G_2'$. Hence, $xux^{-1} \in G_1 \cap G_2'$. This shows that $ghg^{-1} \in (G_1 \cap G_2')G_1'$. Thus, $(G_1 \cap G_2')G_1' \unlhd (G_1 \cap G_2)G_1'$. Similarly, interchanging the role of indexes 1 and 2, we can show that $(G_1' \cap G_2)G_2' \unlhd (G_1 \cap G_2)G_2'$.

Now, we show that $(G_1 \cap G_2)/(G_1 \cap G_2')(G_1' \cap G_2)$ is Ω-isomorphic to $(G_1 \cap G_2)G_1'/(G_1 \cap G_2')G_1'$. Take $H = (G_1 \cap G_2)$, and $K = (G_1 \cap G_2')G_1'$. Then, H is a Ω-subgroup of $(G_1 \cap G_2)G_1'$, and K a Ω-normal subgroup of $(G_1 \cap G_2)G_1'$. By the Noether second isomorphism theorem, $H/(H \cap K)$ is Ω-isomorphic to HK/K (note that all isomorphism theorems hold for group with operators). Now, $HK = (G_1 \cap G_2)(G_1 \cap G_2')G_1' = (G_1 \cap G_2)G_1'$, for $(G_1 \cap G_2') \subseteq$

$(G_1 \cap G_2)$. Thus, $HK/K = (G_1 \cap G_2)G_1'/(G_1 \cap G_2')G_1'$. Further, $H \cap K = ((G_1 \cap G_2')G_1') \cap (G_1 \cap G_2) = ((G_1 \cap G_2')G_1') \cap G_2$. Clearly, $(G_1 \cap G_2') (G_1' \cap G_2) \subseteq ((G_1 \cap G_2')G_1') \cap G_2$. Let $g \in ((G_1 \cap G_2')G_1') \cap G_2$. Then, $g \in G_2$ and $g = xy$, where $x \in (G_1 \cap G_2')$ and $y \in G_1'$. Since $g, x \in G_2$, $y \in G_2$. Thus, $g = xy \in (G_1 \cap G_2')(G_1' \cap G_2)$. This shows that $H \cap K = (G_1 \cap G_2')(G_1' \cap G_2)$, and hence, $(G_1 \cap G_2)/(G_1 \cap G_2')(G_1' \cap G_2)$ is Ω-isomorphic to $(G_1 \cap G_2)G_1'/ (G_1 \cap G_2')G_1'$. Interchanging the role of the indexes, we see that $(G_1 \cap G_2)/ (G_1 \cap G_2')(G_1' \cap G_2)$ is Ω-isomorphic to $(G_1 \cap G_2)G_2'/(G_1' \cap G_2)G_2'$. The result follows from the fact that the relation of Ω-isomorphism is symmetric as well as transitive. ♯

Proof of the Schreier's theorem. Let

$$G = G_1 \trianglerighteq G_2 \trianglerighteq \cdots \trianglerighteq G_n \trianglerighteq G_{n+1} = \{e\} \tag{9.4.3}$$

and

$$G = H_1 \trianglerighteq H_2 \trianglerighteq \cdots \trianglerighteq H_m \trianglerighteq H_{m+1} = \{e\} \tag{9.4.4}$$

be two Ω-subnormal series of G. For $1 \leq i \leq n+1$ and $1 \leq j \leq m+1$, put

$$G_{ij} = (G_i \cap H_j)G_{i+1}, G_{n+2} = \{e\}, H_{ji} = (G_i \cap H_j)H_{j+1}, H_{m+2} = \{e\}.$$

By the Zassenhauss lemma, G_{ij+1} is Ω-normal in G_{ij}, H_{ji+1} is Ω-normal in H_{ji}, and

$$G_{ij}/G_{ij+1} \approx H_{ji}/H_{ji+1} \tag{9.4.5}$$

for all i, j. Now, $G_{i1} = (G_i \cap H_1)G_{i+1} = (G_i \cap G)G_{i+1} = G_iG_{i+1} = G_i$, and $G_{im+1} = (G_i \cap H_{m+1})G_{i+1} = \{e\}G_{i+1} = G_{i+1}$. Thus, for each i, we have the segment

$$G_i = G_{i1} \trianglerighteq G_{i2} \trianglerighteq \cdots \trianglerighteq G_{im+1} = G_{i+1}, \tag{9.4.6}$$

where each term is Ω-normal in the preceding term. Similarly, for each j, we have the segment

$$H_j = H_{j1} \trianglerighteq H_{j2} \trianglerighteq \cdots \trianglerighteq H_{jm+1} = H_{j+1}, \tag{9.4.7}$$

where each term is Ω-normal in the preceding term. Insert the segment (9.4.6) between G_i and G_{i+1} for all i in the Ω-subnormal series (9.4.3), and the segment (9.4.7) between H_j and H_{j+1} for all j in the Ω-subnormal series (9.4.4). We get refinements of (9.4.3) and (9.4.4), respectively, which are equivalent by (9.4.5). ♯

Remark 9.4.7 The proof of the Schreier's theorem gives an algorithm to determine equivalent refinements of any two subnormal series provided we have an algorithm to determine HK and $H \cap K$ for subgroups H and K appearing in the subnormal series. For example, in additive group $\mathbb{Z}$ of integers, $m\mathbb{Z} \cap n\mathbb{Z} = [m.n]\mathbb{Z}$, and $m\mathbb{Z} + n\mathbb{Z} = (m, n)\mathbb{Z}$, where $[m, n]$ is l.c.m. and (m, n) is the g.c.d. of m and n.

Example 9.4.8 Consider the normal series

$$\mathbb{Z} \triangleright 12\mathbb{Z} \triangleright 120\mathbb{Z} \triangleright 240\mathbb{Z} \triangleright \{0\}$$

and

$$\mathbb{Z} \triangleright 25\mathbb{Z} \triangleright 100\mathbb{Z} \triangleright \{0\}$$

of $\mathbb{Z}$. Using the algorithm of the Schreier's theorem, we get the refinement

$$\mathbb{Z} \triangleright 4\mathbb{Z} \triangleright 12\mathbb{Z} \triangleright 60\mathbb{Z} \triangleright 120\mathbb{Z} \triangleright 240\mathbb{Z} \triangleright 1200\mathbb{Z} \triangleright \{0\}$$

of the first normal series, and the refinement

$$\mathbb{Z} \triangleright 5\mathbb{Z} \triangleright 25\mathbb{Z} \triangleright 100\mathbb{Z} \triangleright 300\mathbb{Z} \triangleright 600\mathbb{Z} \triangleright 1200\mathbb{Z} \triangleright \{0\}$$

of the second normal series which are equivalent.

Example 9.4.9 We determine equivalent refinements of

$$\mathbb{R} \triangleright \mathbb{Q} \triangleright \mathbb{Z} \triangleright \{0\}$$

and

$$\mathbb{R} \triangleright \mathbb{Q} + 2\pi\mathbb{Z} \triangleright 2\pi\mathbb{Z} \triangleright \{0\}.$$

Here, $G_{12} = (\mathbb{R} \bigcap (\mathbb{Q} + 2\pi\mathbb{Z})) + \mathbb{Q} = \mathbb{Q} + 2\pi\mathbb{Z} + \mathbb{Q} = \mathbb{Q} + 2\pi\mathbb{Z}$, $G_{13} = (\mathbb{R} \bigcap 2\pi\mathbb{Z}) + \mathbb{Q} = \mathbb{Q} + 2\pi\mathbb{Z}$, $G_{14} = \mathbb{Q} = G_{22}$. Further, $G_{23} = (\mathbb{Q} \bigcap 2\pi\mathbb{Z}) + \mathbb{Z} = \{0\} + \mathbb{Z} = \mathbb{Z}$. Observe that $\mathbb{Q} \bigcap 2\pi\mathbb{Z} = \{0\}$, for π is irrational. Similarly, $G_{24} = \mathbb{Z} = G_{32}$ and $G_{33} = \{0\}$. Thus, we get a refinement

$$\mathbb{R} \triangleright \mathbb{Q} + 2\pi\mathbb{Z} \triangleright \mathbb{Q} \triangleright \mathbb{Z} \triangleright \{0\}$$

of the first normal series, and similarly, we get a refinement

$$\mathbb{R} \triangleright \mathbb{Q} + 2\pi\mathbb{Z} \triangleright \mathbb{Z} + 2\pi\mathbb{Z} \triangleright 2\pi\mathbb{Z} \triangleright \{0\}$$

of the second normal series which are equivalent.

Proposition 9.4.10 *We can remove all repetitions from any two equivalent Ω-subnormal series of a group without affecting their equivalence.*

Proof The proof is by the induction on the length of normal series. If both the Ω-subnormal series have length of 1, then there is nothing to do. Suppose that the result is true for all Ω equivalent subnormal series which have length n. Let

$$G = G_1 \trianglerighteq G_2 \trianglerighteq \cdots \trianglerighteq G_n \trianglerighteq G_{n+1} \trianglerighteq G_{n+2} = \{e\} \qquad (9.4.8)$$

and

$$G = H_1 \trianglerighteq H_2 \trianglerighteq \cdots \trianglerighteq H_n \trianglerighteq H_{n+1} \trianglerighteq H_{n+2} = \{e\} \qquad (9.4.9)$$

be Ω-equivalent subnormal series. If there are no repetitions, then there is nothing to do. Suppose that $G_i = G_{i+1}$. Since (9.4.8) is equivalent to (9.4.9), there is a bijection σ from the set $\{G_1/G_2, G_2/G_3, \ldots, G_{n+1}/G_{n+2}\}$ to $\{H_1/H_2, H_2/H_3, \ldots, H_{n+1}/H_{n+2}\}$ such that the corresponding factors are Ω-isomorphic. Let $\sigma(G_i/G_{i+1}) = H_j/H_{j+1} \approx G_i/G_{i+1}$. Since $G_i = G_{i+1}, H_j = H_{j+1}$. Removing G_{i+1} from 1 and H_{j+1} from 2, we get two Ω-equivalent subnormal series of length n, and so by induction assumption, we can remove all repetitions from these Ω-subnormal series without affecting their equivalence. ♯

Ω-Composition Series

A group need not have any composition series. For example, the additive group $\mathbb{Z}$ of integers have no composition series: given any normal series $\mathbb{Z} \triangleright m_1\mathbb{Z} \triangleright m_2\mathbb{Z} \triangleright \cdots m_t\mathbb{Z} \triangleright \{0\}$, which is without repetitions, we have a proper refinement $\mathbb{Z} \triangleright m_1\mathbb{Z} \triangleright m_2\mathbb{Z} \triangleright \cdots \triangleright m_t\mathbb{Z} \triangleright 2m_t\mathbb{Z} \triangleright \{0\}$ which is without repetitions. However, if a group has a Ω-composition series, then it is essentially unique.

Theorem 9.4.11 (Jordan Holder Theorem) *Any two Ω-composition series of a group are equivalent.*

Proof By the Schreier's theorem, any two Ω-composition series of a group will have equivalent refinements. A Ω-composition series, by definition, is a Ω-subnormal series without repetitions, and which cannot be refined further without admitting repetitions. Thus, if we remove all repetitions from the equivalent refinements of the two composition series, then we arrive at the original composition series. From the above proposition, removing all repetitions from equivalent subnormal series does not affect their equivalence. It follows that any two composition series of a group are equivalent. ♯

Taking $\Omega = \emptyset \, (Inn(G), \, Aut(G), \, End(G))$, we get the following corollary:

Corollary 9.4.12 *Any two Composition series (Principal series, Principal Characteristic series, Principal Fully invariant series) are equivalent.* ♯

Let G be a group which has a composition series. Then, the length of any two composition series are same. This common number is called the **composition length** or **Jordan Holder length** of G, and it is denoted by $l(G)$. Thus, $l(S_4) = 4$, for $S_4 \triangleright A_4 \triangleright V_4 \triangleright \{I, (12)(34)\} \triangleright \{I\}$ is a composition series of S_4. For $n \geq 5$, $S_n \triangleright A_n \triangleright \{I\}$ is the only composition series of S_n. Thus, $l(S_n) = 2$ for $n \geq 5$. Similarly, $l(Q_8), = 3, l(A_4) = 3, l(A_n) = 1$ for $n \geq 5$. $l(G) = 1$ if and only if G is simple. An isomorphism takes a composition series to a composition series, and hence, isomorphic groups have same composition length. Thus, composition length is an invariant of the group structure($G \approx G'$ *implies that* $l(G) = l(G')$). But $l(G) = l(G')$ does not imply that $G \approx G'$ ($l(S_3) = 2 = l(S_5)$).

Let

$$G = G_1 \rhd G_2 \rhd \cdots \rhd G_n \rhd G_{n+1} = \{e\}$$

be a Ω-composition series of a Ω-group G. Then, it cannot be refined further without admitting repetitions. Thus, we cannot put anything in between G_i and G_{i+1} for any i. This means that G_{i+1} is maximal Ω-normal subgroup of G_i for all i. Conversely, if each G_{i+1} is a maximal Ω-normal subgroup of G_i for all i, then we cannot refine the given Ω-subnormal series further without admitting repetitions. Thus, we have the following proposition.

Proposition 9.4.13 *A Ω-subnormal series*

$$G = G_1 \rhd G_2 \rhd \cdots \rhd G_n \rhd G_{n+1} = \{e\},$$

is a Ω-composition series if and only if each G_{i+1} is maximal Ω-normal in G_i.   ♮

Clearly, H is maximal Ω-normal in G if and only if G/H is nontrivial Ω-simple in the sense that it has no proper Ω-normal subgroup. Thus, we have the following:

Corollary 9.4.14 *A Ω-subnormal series is a Ω-composition series if and only if all its factors are nontrivial Ω-simple groups.*   ♮

Since factors of equivalent Ω-subnormal series are Ω-isomorphic, and a group which is Ω-isomorphic to a Ω-simple group is Ω-simple, from the previous corollary, we get the following corollary.

Corollary 9.4.15 *A Ω-subnormal series, which is equivalent to a Ω-composition series, is a Ω-composition series.*   ♮

Proposition 9.4.16 *Let G be a group with operator Ω which has a Ω-composition series. Then, any Ω-subnormal series of G which is without repetitions can be refined to a Ω-composition series.*

Proof Let

$$G = G_1 \rhd G_2 \rhd \cdots \rhd G_n \rhd G_{n+1} = \{e\} \qquad (9.4.10)$$

be a Ω-composition series and

$$G = H_1 \rhd H_2 \rhd \cdots \rhd H_m \rhd H_{m+1} = \{e\} \qquad (9.4.11)$$

a Ω-subnormal series without repetitions. By the Schreier's theorem, (9.4.10) and (9.4.11) have equivalent refinements. After removing the repetitions from the refinement of (9.4.10), it reduces to (9.4.10), and after removing the repetitions from the refinement of (9.4.11), it will remain to be a refinement of (9.4.11). Thus, there is a refinement of (9.4.11) which is equivalent to the composition series (9.4.10). Since any Ω-subnormal series which is equivalent to a Ω-composition series is a Ω composition series, (9.4.11) has a refinement which is a Ω-composition series.   ♮

Proposition 9.4.17 *Let G be a group with operator set Ω which has a Ω-composition series. Let H be a Ω-normal subgroup of G. Then, H and G/H both will have Ω-composition series, and*

$$l(G) = l(H) + l(G/H)$$

Proof If $H = \{e\}$ or $H = G$, then there is nothing to do. Suppose that $H \neq \{e\}$ and $H \neq G$. Then, $G \triangleright H \triangleright \{e\}$ is a Ω-subnormal series of G which is without repetitions. From the above proposition, it can be refined to a Ω-composition series

$$G = G_1 \triangleright G_2 \triangleright \cdots \triangleright G_{i-1} \triangleright H \triangleright \cdots \triangleright G_{n+1} = \{e\}.$$

Then,

$$H \triangleright G_{i+1} \triangleright \cdots \triangleright G_{n+1} = \{e\}$$

is a Ω composition series of H. Further, by the first isomorphism theorem, $(G_j/H)/(G_{j+1}/H)$ is Ω-isomorphic to G_j/G_{j+1} for all $j \leq i - 1$. Hence, $(G_j/H)/(G_{j+1}/H)$ is Ω-simple for all $j \leq i - 1$. Thus,

$$G/H = G_1/H \triangleright G_2/H \triangleright \cdots \triangleright G_{i-1}/H \triangleright \{H\}$$

is a Ω-composition series of G/H. Clearly, $l(G) = l(H) + l(G/H)$. ♯

The following corollary is a consequence of the fundamental theorem of homomorphism, and the above corollary.

Corollary 9.4.18 *Let G be a Ω-group which has a Ω-composition series. Then, any Ω-homomorphic image of G has a Ω-composition series. If f from G to G' is a surjective Ω-homomorphism, then $l(G) = l(G') + l(\ker f)$.* ♯

Proposition 9.4.19 *Let H be a Ω-normal subgroup of G such that H and G/H both have Ω-composition series. Then, G has a Ω composition series, and $l(G) = l(H) + l(G/H)$.*

Proof Let

$$H = H_1 \triangleright H_2 \triangleright \cdots H_{n+1} = \{e\}$$

be a Ω-composition series of H, and

$$G/H = G_1/H \triangleright G_2/H \triangleright \cdots \triangleright G_{m+1}/H = \{H\}$$

a Ω-composition series of G/H. Then, H_j/H_{j+1} is Ω-simple. Further, G_i/G_{i+1}, being Ω-isomorphic to $(G_i/H)/(G_{i+1}/H)$, is Ω-simple. Hence,

$$G = G_1 \triangleright G_2 \triangleright \cdots \triangleright G_{m+1} \triangleright H_2 \triangleright \cdots \triangleright H_{n+1} = \{e\}$$

is a Ω-composition series of G(note that $G_{m+1} = H$). Clearly, $l(G) = l(H) + l(G/H)$. ♯

Proposition 9.4.20 *Let H_1 be a Ω-subgroup of G, and H_2 a Ω-normal subgroup of G. Suppose that H_1 and H_2 have Ω-composition series. Then, $H_1 H_2$ has a Ω-composition series, and $l(H_1 H_2) = l(H_1) + l(H_2) - l(H_1 \bigcap H_2)$.*

Proof By the Noether isomorphism theorem, $H_1 H_2 / H_2 \approx H_1 / H_1 \bigcap H_2$. Since H_1 has a Ω-composition series, $H_1 / H_1 \bigcap H_2$ also has a Ω-composition series. Hence, $H_1 H_2 / H_2$ has a Ω-composition series. Since H_2 is supposed to have a Ω-composition series, by Proposition 9.4.19, $H_1 H_2$ has a Ω-composition series. Also $l(H_1 H_2 / H_2) = l(H_1 / H_1 \bigcap H_2)$. Hence, $l(H_1 H_2) - l(H_2) = l(H_1) - l(H_1 \bigcap H_2)$.   ♯

Proposition 9.4.21 *Every finite Ω-group has a Ω-composition series.*

Proof The proof is by induction on order of G. If $| G | = 1$, there is nothing to do. Assume that the result is true for all those groups whose orders are less than n. Let G be a Ω-group of order n. If G is Ω-simple, then $G \triangleright \{e\}$ is the Ω-composition series of G. Suppose that G is not Ω-simple. Let H be a nontrivial proper normal Ω-subgroup of G. Then, $| H | < n$, and also $| G/H | < n$. By the induction hypothesis H and G/H both have Ω-composition series. By Proposition 9.4.19, G has a Ω-composition series.   ♯

Remark 9.4.22 The main guiding problem in the theory of finite groups is the classification of finite groups, or in particular, classification of finite groups of a given order. Every finite group has a composition series. As such, this classification problem reduces to the following two problems:

Problem 1 Classify all finite simple groups.

Problem 2 Given finite groups H and K, classify the class of groups G having H as a normal subgroup with G/H isomorphic to K.

The classification of finite simple groups was achieved in 1980. There are four types of finite simple groups to be described in another volume of the book. However, the solution to the problem 2 is beyond dream to mathematicians, and usually, partial solutions to this problem are addressed in the theory of extensions and cohomology of groups (see Chap. 10, Algebra 2, or any book on cohomology of groups).

Proposition 9.4.23 *An abelian group has a composition series if and only if it is finite.*

Proof If G is finite then, by the Proposition 9.4.21, it has a composition series. Let G be an abelian group, and

$$G = G_1 \triangleright G_2 \triangleright \cdots \triangleright G_n \triangleright G_{n+1} = \{e\}$$

a composition series of G. Then, G_i / G_{i+1} is abelian as well as simple for all i. Since an abelian group is simple if and only if it is a cyclic group of prime order, G_i / G_{i+1} is finite cyclic group of order p_i, where p_i is some prime. Thus, $G_n \approx G_n / \{e\}$ is finite of order p_n. Since G_{n-1} / G_n is finite of order p_{n-1}, G_{n-1} is also finite of order $p_{n-1} p_n$ (observe that if H is finite and G/H is also finite, then G is also finite and $| G | = | H | \cdot | G/H |$). Similarly, G_{n-2} is also finite of order $p_{n-2} p_{n-1} p_n$. Proceeding inductively, we find that G is finite of order $p_1 p_2 \ldots p_n$.   ♯

Recall that a subgroup H of a group G is called a **subnormal subgroup** of G if it appears in a subnormal series of G. We write $H \unlhd \unlhd G$ to say that H is a subnormal subgroup of G. A subnormal subgroup need not be normal. For example $\{I, (12)(34)\}$ is a subnormal subgroup of S_4 ($S_4 \rhd V_4 \rhd \{I, (12)(34)\} \rhd \{I\}$ is a normal series of S_4). An arbitrary subgroup need not be a subnormal subgroup for example $\{I, (12)\}$ is a subgroup of S_3 which is not subnormal ($S_3 \rhd A_3 \rhd \{I\}$ is the only nontrivial normal series of S_3).

The following proposition is an immediate consequence of the fact that the inverse image of a normal subgroup under a homomorphism is a normal subgroup, and the image of a normal subgroup under a surjective homomorphism is a normal subgroup.

Proposition 9.4.24 *Inverse image of a subnormal subgroup under a homomorphism is a subnormal subgroup, and under a surjective homomorphism, image of a subnormal subgroup is a subnormal subgroup.*   ♯

Corollary 9.4.25 *In the correspondence theorem, subnormal subgroups correspond.*   ♯

Corollary 9.4.26 *A subgroup K/H of G/H is subnormal if and only if K is subnormal in G.*   ♯

Proposition 9.4.27 *A group G is simple if and only if it has no nontrivial proper subnormal subgroups.*

Proof If a group G has no nontrivial proper subnormal subgroups, then, in particular, it has no nontrivial proper normal subgroups, and so, it is simple. Conversely, if G is simple, then $G \rhd \{e\}$ is the only normal series of G, and so, it has no nontrivial proper subnormal subgroups of G.   ♯

Remark 9.4.28 It may be an interesting problem to study groups all of whose subgroups are subnormal. Finite groups all of whose subgroups are subnormal are known as nilpotent groups. We shall discuss such groups in Chap. 10.

Proposition 9.4.29 *Let G be a group which has a composition series, and $H \unlhd \unlhd G$. Then, H has a composition series, and $l(H) < l(G)$.*

Proof Since $H \unlhd \unlhd G$, there is a normal series, which we can take to be without repetitions, and in which H is a term. Since G has a composition series, it can be refined to a composition series of G. The subchain of this composition series starting from H onward is a composition series of H. Clearly, $l(H) < l(G)$.   ♯

Example 9.4.30 An arbitrary subgroup of a group need not have a composition series even if the group has a composition series. Consider, for example, the group A_∞ which is simple (Proposition 6.2.43), and so, it has composition series. Let H be the subgroup of A_∞ generated by the set $X = \{(4n \, 4n+1)(4n+2 \, 4n+3) \mid n \in \mathbb{N}\}$. Since X is infinite and elements of X commute, H is infinite abelian group. It follows that H has no composition series.

Now, we find a necessary and sufficient condition for a Ω-group to have a Ω-composition series.

A Ω-group G is said to satisfy **descending chain condition** (D.C.C.) for Ω-subnormal subgroups if given any chain

$$G = G_1 \unrhd G_2 \unrhd \cdots \unrhd G_n \unrhd G_{n+1} \unrhd \cdots,$$

where each G_{n+1} is normal Ω-subgroup of G_n, there exists $n_0 \in \mathbb{N}$ such that $G_n = G_{n_0}$ for all $n \geq n_0$.

A Ω-group G is said to satisfy **ascending chain condition** (A.C.C.) for Ω-subnormal subgroups if given any ascending chain

$$H_1 \unlhd H_2 \unlhd \cdots \unlhd H_n \unlhd H_{n+1} \unlhd \cdots$$

of Ω-subnormal subgroups of G, there exists $n_0 \in \mathbb{N}$ such that $H_n = H_{n_0}$ for all $n \geq n_0$.

One notices that if G satisfies D.C.C. (A.C.C.), then any Ω-subnormal subgroup of G and any Ω-quotient group of G also satisfies D.C.C. (A.C.C.).

Example 9.4.31 A finite Ω-group satisfies D.C.C. as well as A.C.C., for we cannot have an infinite properly ascending or properly descending chain of Ω-subgroups.

Example 9.4.32 The additive group $\mathbb{Z}$ of integers satisfies A.C.C., for given any subgroup $m\mathbb{Z}$ of $\mathbb{Z}$, $m\mathbb{Z} \subset n\mathbb{Z}$ if and only if n is a proper divisor of m. Since there are only finitely many proper divisors of m, only finitely many subgroups of $\mathbb{Z}$ can contain $m\mathbb{Z}$. It does not satisfy D.C.C., for we have an infinite properly descending chain $\mathbb{Z} \rhd 2\mathbb{Z} \rhd 2^2\mathbb{Z} \rhd \cdots 2^n\mathbb{Z} \rhd 2^{n+1}\mathbb{Z} \rhd \cdots$ of $\mathbb{Z}$.

Example 9.4.33 A group may satisfy D.C.C. but not A.C.C.: Let p be a prime integer, and $\mathbb{Q}_p = \{\frac{m}{p^r} \mid m \in \mathbb{Z}, r \in \mathbb{N} \bigcup \{0\}\}$. Then, $\mathbb{Q}_p$ is a subgroup of the additive group of rationals, and since it is abelian, all its subgroups are subnormal. Let $G = \mathbb{Q}_p/\mathbb{Z}$. We have an infinite properly ascending chain

$$\{\mathbb{Z}\} \lhd \tfrac{1}{p}\mathbb{Z}/\mathbb{Z} \lhd \tfrac{1}{p^2}\mathbb{Z}/\mathbb{Z} \lhd \cdots \lhd \tfrac{1}{p^n}\mathbb{Z}/\mathbb{Z} \lhd \cdots$$

of subgroups of G. Thus, G does not satisfy A.C.C. We show that it satisfies D.C.C. Let H be a subgroup of $\mathbb{Q}_p$ containing $\mathbb{Z}$ properly. Then, for some natural number m co-prime to p and $\alpha \in \mathbb{N}$, $\frac{m}{p^\alpha} \in H - \mathbb{Z}$. By the Euclidean algorithm, there exist $u, v \in \mathbb{Z}$ such that $um + vp^\alpha = 1$. Hence, $\frac{1}{p^\alpha} = v + u\frac{m}{p^\alpha} \in H$. Thus, $\frac{m}{p^\alpha} \in H$ if and only if $\frac{1}{p^\alpha} \in H$. Let α_0 be the least positive integer such that $\frac{1}{p^{\alpha_0}} \in H$. Then,

$$H = \{\tfrac{m}{p^{\alpha_0}} \mid m \in \mathbb{Z}\} = \tfrac{1}{p^{\alpha_0}}\mathbb{Z}.$$

Clearly, there are only finitely many subgroups of $\mathbb{Q}_p$ contained in H. Thus, there are only finitely many subgroups of G contained in the subgroup $H/\mathbb{Z}$ of G. This means that G satisfies D.C.C.

Remark 9.4.34 Since $\mathbb{Q}_p$ does not satisfy (as G does not satisfy) A.C.C., and since it is a subgroup of the additive group $\mathbb{Q}$, the additive group $\mathbb{Q}$ of rationals also does not satisfy A.C.C. Observe that $\mathbb{Q}$ does not satisfy D.C.C. also (for otherwise $\mathbb{Z}$ would satisfy D.C.C). Clearly, the additive group of reals, the additive group of complex numbers, the multiplicative group of nonzero reals, the multiplicative group complex numbers, and the circle group do not satisfy any of the chain conditions.

Theorem 9.4.35 *A Ω-group has a Ω-composition series if and only if it satisfies both the chain conditions for Ω-subnormal subgroups.*

Proof Suppose that G has a Ω-composition series, and $l(G) = n_0$. Since every Ω-subnormal series without repetitions can be refined to a Ω-composition series, and any two Ω-composition series are equivalent, we cannot have a Ω-subnormal series without repetitions which is of length greater than n_0. Thus, G satisfies D.C.C., for otherwise, we can extract a Ω-subnormal series without repetitions of arbitrary length. G also satisfies A.C.C., for otherwise, we can extract a properly ascending chain

$$H_1 \lhd H_2 \lhd \cdots \lhd H_{n_0+1}$$

of Ω-subnormal subgroups. Since H_{n_0+1} is a Ω-subnormal subgroup of G, the above segment can be enlarged to a Ω-subnormal series of G which is without repetitions, and whose length, therefore, is at least $n_0 + 1$. This is a contradiction to the fact that there is no Ω-subnormal series without repetitions of length greater than $n_0 + 1$.

Conversely, suppose that G satisfies A.C.C. as well as D.C.C. We first show that G has a maximal normal Ω-subgroup. The trivial subgroup $\{e\}$ may be maximal normal Ω-subgroup. If not, there is a nontrivial proper normal Ω-subgroup H_1 of G. $\{e\} \lhd H_1 \lhd G$. H_1 may be maximal normal Ω-subgroup. If not, there is a normal Ω-subgroup H_2 of G such that $\{e\} \lhd H_1 \lhd H_2 \lhd G$. H_2 may be maximal normal Ω-subgroup. If not, proceed in the same way. This process stops after finitely many steps giving us a maximal normal Ω-subgroup, because of A.C.C. Let G_2 be a maximal normal Ω-subgroup of G. If $G_2 = \{e\}$, then $G \rhd \{e\}$ is a Ω-composition series of G. Suppose that $G_2 \neq \{e\}$. Then, since every Ω-subnormal subgroup of G also satisfies A.C.C., from the previous argument, G_2 will have a maximal normal Ω-subgroup G_3 (say). If $G_3 = \{e\}$, then $G \rhd G_2 \rhd \{e\}$ is a composition series of G. If not, proceed similarly. This process stops after finitely many steps giving us a Ω-composition series, because of D.C.C. ♯

Taking $\Omega = Inn(G)$, we get the following corollary:

Corollary 9.4.36 *A group has a principal series if and only if it satisfies both the chain conditions for normal subgroups.* ♯

Exercises

9.4.1 Find refinements of

$$\mathbb{Z} \rhd 8\mathbb{Z} \rhd 48\mathbb{Z} \rhd \{0\}$$

and
$$\mathbb{Z} \rhd 7\mathbb{Z} \rhd 42\mathbb{Z} \rhd \{0\}$$

which are equivalent.

9.4.2 Find refinements of

$$\mathbb{R} \rhd \mathbb{Q} + \sqrt{2}\mathbb{Z} \rhd \mathbb{Z} \rhd \{0\}$$

and

$$\mathbb{R} \rhd \mathbb{Q} + \sqrt{3}\mathbb{Z} \rhd \mathbb{Z} \rhd \{0\}$$

which are equivalent.

9.4.3 List all composition series of $\mathbb{Z}_{12}$, and also find their length.

9.4.4 Let $< a >$ be a cyclic group of order $p_1^{\alpha_1} p_2^{\alpha_2} \ldots p_r^{\alpha_r}$, where $p_1, p_2, \ldots, p_r$ are distinct primes. Find a composition series of G and show that its length is $\alpha_1 + \alpha_2 + \cdots + \alpha_r$. Find the number of composition series of G.

9.4.5 Prove the fundamental theorem of arithmetic using the Jordan Holder Theorem.

9.4.6 Suppose that H and G/H both satisfy A.C.C. (D.C.C.). Show that G also satisfies A.C.C. (D.C.C.).

9.4.7 Let $H \lhd G$. Show that G/H has a composition series if and only if following two conditions hold.
  (i) Given any chain
$$H_1 \lhd H_2 \rhd \cdots \rhd H_n \rhd \cdots$$

of subnormal subgroups of G containing H, there exists $n_0 \in \mathbb{N}$ such that $H_n = H_{n_0}$ for all $n \geq n_0$.
  (ii) Given any chain

$$G = G_1 \rhd G_2 \rhd \cdots \rhd G_n \rhd G_{n+1} \rhd \cdots$$

such that each G_{n+1} is normal in G_n, there exists $n_0 \in \mathbb{N}$ such that $G_n H = G_{n_0} H$ for all $n \geq n_0$.

9.4.8 Show that none of the groups (i) $(S^1, \cdot)$, (ii) $(\mathbb{R}^+, \cdot)$ and (iii) $(\mathbb{C}^*, \cdot)$ satisfy any of D.C.C. or A.C.C.

9.4.9 Let G be a group all of whose subgroups are subnormal subgroups. Then, show that G has a composition series if and only if it is finite.

9.4.10 Let G and G' be finite abelian groups such that $l(G) = l(G')$. Show that $| G | = | G' |$.

9.4.11 Let $H \lhd \lhd G$ and K a subgroup of G. Show that $H \cap K \lhd \lhd K$.

9.4.12 Show that the intersection of two subnormal subgroups is a subnormal subgroup.

9.4.13 Let H and K be subgroups of G which have composition series. Suppose that $HK = KH$. Can we conclude that HK has a composition series?

9.4.14 Show that G satisfies A.C.C. for subgroups if and only if every subgroup of G is finitely generated.

9.4.15 Let G be a group which satisfies A.C.C. for subgroups. Show that G has a maximal subgroup.

9.4.16 Show that the additive group $\mathbb{Q}$ of rationals, the additive group $\mathbb{R}$ of reals, the multiplicative group $\mathbb{R}^\star$ of nonzero reals, the multiplicative group $\mathbb{R}^+$ of positive reals, the additive group $\mathbb{C}$ of complex numbers, the multiplicative group $\mathbb{C}^\star$ of nonzero complex numbers, the circle group S^1, and the group P of roots of unity have no maximal subgroups and none of them are finitely generated.

9.4.17 Does the group of nonsingular $n \times n$ matrices over reals satisfy A.C.C.? Does it have a maximal subgroup?

9.4.18 Does the group of $n \times n$ matrices of determinant 1 over reals satisfy A.C.C.? Does it have a maximal subgroup?

9.4.19 Find all principal series, principal characteristic series, and also principal fully invariant series of S_4.

9.4.20 Show that a group which has a composition series has also a principal series. In turn, show that it also has a principal characteristic, and a principal fully invariant series. Is the reverse implication true? Support.

9.4.21 Give an example of an indecomposable group which has no principal series.

9.4.22* What can we say about union of a chain of subnormal subgroups? Is it always subnormal?

9.4.23* Let H and K be subnormal subgroups of a finite group. Show that the subgroup generated by $H \cup K$ is subnormal. Can we conclude the result in infinite groups also?

9.4.24* Let H and K be subnormal subgroups of G such that $HK = KH$. Show that HK is subnormal.

Chapter 10
Structure Theory Continued

This chapter deals with the Remak-Krull-Schmidt Theorem on direct decomposition, structure theory of solvable and nilpotent groups together with the presentation theory of groups.

10.1 Decompositions of Groups

We refer Sect. 5.2 of Chap. 5 to recall the basic definitions and properties of direct products. In this section, we establish the Remak–Krull–Schmidt Theorem and study completely decomposable (in particular semisimple) groups. We also discuss other type of decompositions of groups.

Theorem 10.1.1 (Remak–Krull–Schmidt) *Let G be a group with operator Ω which has a Ω-principal series (equivalently, it satisfies A.C.C and D.C.C for normal Ω-subgroups). Then, G can be expressed as a direct product of finitely many Ω-indecomposable subgroups. Further, the representation of G as direct product of Ω-indecomposable subgroups is essentially unique in the following sense: If*

$$G \; = \; H_1 \times H_2 \times \cdots \times H_n \; = \; K_1 \times K_2 \times \cdots \times K_m,$$

where H_i and K_j are Ω-indecomposable subgroups of G, then
(i) $m = n$,
(ii) there is a central automorphism σ of G such that after some rearrangement $\sigma(H_i) \; = \; K_i$ for all i, and

$$G \; = \; K_1 \times K_2 \times \cdots \times K_r \times H_{r+1} \times H_{r+2} \times \cdots \times H_n.$$

for all $r \leq n$.

In particular, we have the following corollary:

© Springer Nature Singapore Pte Ltd. 2017

R. Lal, *Algebra 1*, Infosys Science Foundation Series in Mathematical Sciences, DOI 10.1007/978-981-10-4253-9_10

Corollary 10.1.2 *If G is a group which has a principal series (characteristic series, fully invariant series), then G can be expressed as a direct product of finitely many indecomposable (characteristically indecomposable, fully invariantly indecomposable) subgroups. Further, then, the representation of G as direct product of indecomposable (characteristically indecomposable, fully invariantly indecomposable) subgroups is essentially unique in the following sense: If*

$$G \; = \; H_1 \times H_2 \times \cdots \times H_n \; = \; K_1 \times K_2 \times \cdots \times K_m,$$

where H_i and K_j are indecomposable (characteristically indecomposable, fully invariantly indecomposable) subgroups of G, then
 (i) $m = n$,
 (ii) there is a central automorphism σ of G such that after some rearrangement $\sigma(H_i) \; = \; K_i$ for all i, and

$$G \; = \; K_1 \times K_2 \times \cdots \times K_r \times H_{r+1} \times H_{r+2} \times \cdots \times H_n.$$

for all $r \leq n$.   ♯

We need some Lemmas to prove the Remak–Krull–Schmidt Theorem.

Lemma 10.1.3 *Let G be a group with operator Ω. Let σ be a normal Ω-endomorphism of G such that $\sigma^2(G) \; = \; \sigma(G)$ and $ker\sigma^2 \; = \; ker\sigma$. Then, $G = \sigma(G) \times ker\sigma$.*

Proof Since σ is a normal Ω-endomorphism of G, $\sigma(G)$ is a normal Ω-subgroup of G. Let $g \in G$. Since $\sigma^2(G) \; = \; \sigma(G)$, there is an element $h \in G$ such that $\sigma(g) \; = \; \sigma^2(h)$. But, then, $(\sigma(h))^{-1}g \in ker\sigma$. This shows that $g \in \sigma(G)ker\sigma$. Thus, $G = \sigma(G)ker\sigma$. Let $g \in \sigma(G) \bigcap ker\sigma$. Then, $g \; = \; \sigma(h)$ for some $h \in G$. Also, $e = \sigma(g) \; = \; \sigma^2(h)$. Hence, $h \in ker\sigma^2 \; = \; ker\sigma$. This shows that $g \; = \; \sigma(h) \; = \; e$. Thus, $\sigma(G) \bigcap ker\sigma \; = \; \{e\}$. From Corollary 5.2.25, it follows that $G \; = \; \sigma(G) \times ker\sigma$. ♯

Lemma 10.1.4 (Fitting) *Let G be a group with operator Ω which has a Ω-principal series. Let σ be a normal Ω-endomorphism of G. Then, there exists a natural number m such that $\sigma^n(G) \; = \; \sigma^m(G)$ for all $n \geq m$, and $G \; = \; \sigma^m(G) \times ker\sigma^m$.*

Proof Since σ is a normal Ω-endomorphism of G, image of a normal Ω-subgroup of G is again a normal Ω-subgroup of G. In turn, $\sigma^n(G)$ is a normal Ω-subgroup of G for all natural number n. Since G has a Ω-principal series, it satisfies D.C.C as well as A.C.C for normal Ω-subgroup of G. Thus, the descending chain $G \supseteq \sigma(G) \supseteq \sigma^2(G) \supseteq \ldots\ldots \supseteq \sigma^n(G) \supseteq \sigma^{n+1}(G) \supseteq \ldots\ldots$, and the ascending chain $ker\sigma \subseteq ker\sigma^2 \subseteq \ldots\ldots \subseteq ker\sigma^n \subseteq \ldots$. eventually terminate after finitely many terms giving us a natural number m such that $\sigma^n(G) \; = \; \sigma^m(G)$ and $ker\sigma^n \; = \; ker\sigma^m$ for all $n \geq m$. Put $\rho \; = \; \sigma^m$. Then, $\rho^2(G) \; = \; \sigma^{2m}(G) \; = \; \sigma^m(G) \; = \; \rho(G)$ and $ker\rho^2 \; = \; ker\sigma^{2m} \; = \; ker\sigma^m \; = \; ker\rho$. From the previous lemma, $G \; = \; \rho(G) \times ker\rho \; = \; \sigma^m(G) \times ker\sigma^m$. ♯

Corollary 10.1.5 *Let G be a Ω-indecomposable group which has a Ω-principal series. Let σ be a normal Ω-endomorphism of G. Then, either σ is nilpotent (in the sense that for some m $\sigma^m(g) = e$ for all $g \in G$), or else it is a central Ω-automorphism of G.*

Proof From the Fitting Lemma, there is a natural number m such that $G = \sigma^m(G) \times \ker \sigma^m$. Since G is indecomposable, $\sigma^m(G) = G$ and $\ker \sigma^m = \{e\}$, or else $\sigma^m(G) = \{e\}$ and $\ker \sigma^m = G$. The result follows. ♯

Recall that two Ω-endomorphism σ_1 and σ_2 are summable if $\sigma_1 + \sigma_2$ is also an endomorphism (and so also, a Ω-endomorphism) of G. Indeed, it is so if and only if the elements of $\sigma_1(G)$ commute with elements of $\sigma_2(G)$. Further, if σ_1 and σ_2 are summable, and σ is an endomorphisms, then $\sigma o \sigma_1$ and $\sigma o \sigma_2$ ($\sigma_1 o \sigma$ and $\sigma_2 o \sigma$) are also summable, and $\sigma o (\sigma_1 + \sigma_2) = \sigma o \sigma_1 + \sigma o \sigma_2$ $((\sigma_1 + \sigma_2) o \sigma = \sigma_1 o \sigma + \sigma_2 o \sigma)$.

Corollary 10.1.6 *Let G be a nontrivial Ω-indecomposable group which has a Ω-principal series. Let $\{\sigma_1, \sigma_2, \ldots, \sigma_n\}$ be a set of pairwise summable normal Ω-endomorphisms. Further, suppose that $\{\sigma_1 + \sigma_2 + \cdots + \sigma_n\}$ is an automorphism. Then, σ_i is a central automorphism for some i.*

Proof By induction, it is sufficient to prove the result for $n = 2$. Let σ_1 and σ_2 be summable normal Ω-endomorphisms such that $\sigma = \sigma_1 + \sigma_2$ is an automorphism. Suppose also that neither σ_1 nor σ_2 is an automorphism. Then, from the above corollary, σ_1 and σ_2 are both nilpotent. As observed above $\tau_1 = \sigma^{-1} o \sigma_1$ and $\tau_2 = \sigma^{-1} o \sigma_2$ is a pair of summable normal Ω-endomorphisms such that $\tau_1 + \tau_2 = I_G$. Also, since σ_1 and σ_2 are nilpotent, and σ is an automorphism, it follows that none of the τ_1 and τ_2 are automorphisms. As such, they are nilpotent. Let m be a natural numbers such that τ_1^m and τ_2^m are trivial maps. Also, $\tau_2 o \tau_1 + \tau_2^2 = \tau_2 o I_G = I_G o \tau_2 = \tau_1 o \tau_2 + \tau_2^2$. This shows that $\tau_1 o \tau_2 = \tau_2 o \tau_1$. We can use the bionomial theorem to conclude that $(\tau_1 + \tau_2)^{2m}$ is also a trivial map. This is a contradiction, for $\tau_1 + \tau_2$ is the identity map on G. Hence, the supposition that neither σ_1 nor σ_2 is an automorphism is false. ♯

Lemma 10.1.7 *Let G be a Ω-group which has a Ω-principal series. Then, every normal injective (surjective) endomorphism of G is a central automorphism of G.*

Proof Let σ be an injective normal Ω-endomorphism of G. Then, σ^n is injective for each n. Since G has a Ω-principal series, $\sigma^n(G) = \sigma^{n+1}(G)$ for some n. Let $x \in G$. Then, there is a $y \in G$ such that $\sigma^n(x) = \sigma^{n+1}(y)$. Since σ^n is injective, $x = \sigma(y)$. This shows that σ is also surjective. Next, suppose that σ is a surjective normal Ω-endomorphism of G. Then, σ^n is surjective for each n. Since G has a Ω-principal series, $\ker \sigma^n = \ker \sigma^{n+1}$ for some n. Suppose that $\sigma(x) = e$. Since σ^n is surjective, $x = \sigma^n(y)$ for some $y \in G$. But, then $\sigma^{n+1}(y) = e$, and so $y \in \ker \sigma^{n+1} = \ker \sigma^n$. Hence, $x = \sigma^n(y) = e$. This shows that σ is injective also. ♯

Proof of Theorem 10.1.1 Existence of decomposition: Let G be a group with operator set Ω which has a Ω-principal series. The proof of the existence of the decomposition is on the induction of the length of the principal series. If the length of the Ω principal series of G is 1, then it is Ω-simple, and so it is indecomposable. Assume that the result is true for all those Ω-groups which have Ω-principal series of length less than that of G. If G is Ω-indecomposable, then there is nothing to do. Suppose that G is Ω-decomposable. Suppose that $G = H \times K$, where H and K are nontrivial proper Ω-normal subgroups of G. Clearly, H and K have Ω-principal series of length less than that of G. By the induction hypothesis, H and K both are direct products of finitely many Ω-indecomposable groups. In turn, G can be expressed as direct product of Ω-indecomposable groups.

Uniqueness of the decomposition: The proof is by the induction on $max(m, n)$. Suppose that $max(m, n) = 1$. Then, since G is nontrivial $m = 1 = n$, and there is nothing to do. Assume the induction hypothesis. Suppose that

$$G = H_1 \times H_2 \times \cdots \times H_n = K_1 \times K_2 \times \cdots \times K_m,$$

where H_i and K_j are Ω-indecomposable subgroups of G, and $max(m, n) > 1$. Let $p_1, p_2, \ldots, p_n$ be the projections corresponding to the decomposition

$$G = H_1 \times H_2 \times \cdots \times H_n,$$

and $q_1, q_2, \ldots, q_m$ that corresponding to the decomposition

$$G = K_1 \times K_2 \times \cdots \times K_m.$$

Then, $\{p_1, p_2, \ldots, p_n\}$ and $\{q_1, q_2, \ldots, q_m\}$ are sets of pairwise summable normal and idempotent Ω-endomorphisms such that

$$p_1 + p_2 + \cdots + p_n = I_G = q_1 + q_2 + \cdots + q_m,$$

and

$$p_i o p_j = 0 = q_k o q_l \, for \, i \neq j \, and \, k \neq l.$$

Thus,

$$p_1 = p_1 o p_1 = p_1 o I_G o p_1 = \sum_1^m p_1 o q_i o p_1$$

Since p_1 restricted to H_1 is the identity map I_{H_1} on H_1, it follows that

$$I_{H_1} = \sum_1^m \eta_i,$$

where η_i denote the restriction of $p_1 o q_i$ to H_1. Clearly, $\{\eta_i | 1 \leq i \leq m\}$ is the set pairwise summable normal Ω-endomorphisms of H_1. It follows from Corollary 10.1.6 that η_i is a central Ω-automorphism of H_1 for some i. After rearranging, we may assume that η_1 is a central Ω-automorphism of H_1. As such, q_1 restricted to H_1 is an injective Ω-homomorphism from H_1 to K_1, and p_1 restricted to K_1 a surjective Ω-homomorphism from K_1 to H_1 which commute with all induced inner automorphisms of G on H_1 and K_1, respectively. Let ρ_1 denote the restriction of q_1op_1 to K_1. Then, ρ_1 is a normal Ω-endomorphism of K_1. Now,

Image ρ_1^2

$= \rho_1^2(K_1)$

$= (q_1p_1q_1p_1)(K_1)$

$= (q_1p_1q_1)(H_1)$ (for p_1 restricted to K_1 a surjective map from K_1 to H_1)

$= q_1(H_1)$ (for p_1q_1 restricted to H_1 is an automorphism of H_1)

$= (q_1p_1)(K_1)$ (for p_1 restricted to K_1 a surjective map from K_1 to H_1)

$=$ Image ρ_1.

Also,

ker ρ_1^2

$= \{x \in K_1 \mid (q_1p_1q_1p_1)(x) = e\}$

$= \{x \in K_1 \mid (q_1p_1(x) = e\}$ (for p_1q_1 restricted to H_1 is an automorphism)

$=$ ker ρ_1.

From Lemma 10.1.3, it follows that $K_1 = \rho_1(K_1) \times ker\,\rho_1$. Since K_1 is indecomposable, $\rho_1(K_1) = K_1$ and $ker\,\rho_1 = \{e\}$, or else $\rho_1(K_1) = \{e\}$ and $ker\,\rho_1 = K_1$. Since p_1 restricted to K_1 is surjective map from K_1 to H_1 and q_1 restricted to H_1 is injective, $ker\,\rho_1 \neq K_1$. Thus, ρ_1, which is the restriction of q_1op_1 to K_1, is a normal Ω-automorphism of K_1. This ensures that p_1 restricted to K_1 is an injective Ω-homomorphism from K_1 to H_1, and q_1 restricted to H_1 is a surjective Ω-homomorphism from H_1 to K_1. Consequently, p_1 restricted to K_1 is a Ω-isomorphism from K_1 to H_1, and q_1 restricted to H_1 is a Ω-isomorphism from H_1 to K_1 which commute with all inner automorphisms of G.

Now, consider the subgroup $L = K_1H_2H_3..........H_n$ of G. Clearly, K_1 and $H_2H_3..........H_n$ are normal subgroups of L. Let $x \in K_1 \cap (H_2H_3..........H_n)$. Then, $q_1(p_1(x)) = e$. Since q_1p_1 restricted to K_1 is an automorphism of K_1, it follows that $x = e$. Thus, $K_1 \cap (H_2H_3..........H_n) = \{e\}$. This shows that L is the direct product $K_1 \times H_2 \times H_3 \times .... \times H_n$. It also follows that $q_1p_1 + p_2 + p_3 + ... + p_n$ is a normal injective endomorphism of G whose image is L. From Lemma 10.1.7, it follows that $L = G$. Clearly, $q_1p_1 + p_2 + p_3 + ... + p_n$ is a central automorphism of G which takes H_1 isomorphically to K_1. It also induces Ω-isomorphism from $G/H_1 = H_2 \times H_3 \times .... \times H_n$ to $G/K_1 = H_2 \times H_3 \times .... \times H_n = K_2 \times K_3 \times .... \times K_m$. The result follows by the induction. ♯

Exercises

10.1.1 Show that S_n, A_n, Q_8, and D_8 are indecomposable groups.

10.1.2 Show that any extra special p-group is indecomposable.

10.1.3 Show that $Q_8 \times Q_8$ is not isomorphic to $Q_8 \times D_8$.

10.1.4 Express all groups up to order 15 as direct product of indecomposable groups.

10.1.6 Is the additive group $\mathbb{R}$ of real numbers indecomposable? Support.

10.2 Solvable Groups

Let G be a group. Recall that an element of the form $aba^{-1}b^{-1}$ is called a commutator, and it is denoted by (a, b). Notice that $(a, b)ba = ab$ and $(a, b) = e$ if and only if a and b commute. Recall further that the subgroup $[G, G]$ generated by all commutators is called the commutator or the derived subgroup of G. Since $(a, b)^{-1} = (b, a)$ is a commutator, every element of $[G, G]$ is a product of commutators. The commutator subgroup $[G, G]$ is characterized (see Theorem 5.2.34 and Remark 5.2.35) by the property that it is the smallest normal subgroup of G by which if we factor we get an abelian group, or equivalently, G/G' is the largest quotient group of G which is abelian. G/G' is also called the abelianizer of G. Further, if a subgroup H contains G', then it is necessarily normal. Clearly, $G' = \{e\}$ if and only if G is abelian. If G is a nonabelian simple group, then $G' = G$. A group G is said to be a **perfect** if its commutator subgroup G' is G itself. Thus, a nonabelian simple group is always perfect. Observe that the commutator subgroup $(H \times K)'$ of the product $H \times K$ is the product $H' \times K'$ of their commutators. Thus, product of nonabelian simple groups is also perfect, but it is not simple.

Remark 10.2.1 Product of two commutators need not be a commutator (refer to Examples 10.2.6 and 10.2.7).

Let A and B be subsets of a group G. The subgroup generated by $\{(a, b) \mid a \in A \text{ and } b \in B\}$ is denoted by $[A, B]$. Since $(a, b)^{-1} = (b, a)$, $[A, B] = [B, A]$.

Proposition 10.2.2 *Let H and K be normal subgroups of G. Then, $[H, K] = [K, H]$ is also a normal subgroup of G, and $[H, K] \subseteq H \cap K$.*

Proof Since $\{(h, k) \mid h \in H \text{ and } k \in K\}$ generates $[H, K]$, it is sufficient to show that $g(h, k)g^{-1} \in [H, K]$ for all $g \in G$, $h \in H$ and $k \in K$. This is evident, for $g(h, k)g^{-1} = (ghg^{-1}, gkg^{-1})$ (verify). Further, since H is normal in G, $kh^{-1}k^{-1} \in H$, and so $(h, k) \in H$. Thus, $[H, K] \subseteq H$. Similarly, $[H, K] \subseteq K$.   ♯

Corollary 10.2.3 $[G, G] \trianglelefteq G$.   ♯

Example 10.2.4 $Q_8/\{1, -1\}$, being a group of order 4, is abelian (in fact it is isomorphic to V_4). Thus, $Q_8' \subseteq \{1, -1\}$. Further, since Q_8 is nonabelian $Q_8' \neq \{1\}$. Hence, $Q_8' = \{1, -1\}$.

Example 10.2.5 For $n \geq 3$, $S_n' = A_n$: Since $S_n/A_n \approx \{1, -1\}$ is abelian, $S_n' \subseteq A_n$. Further, since product of any two transpositions is expressible as products of cycles of length 3, A_n, $n \geq 3$ is generated by cycles of length 3. Therefore, it is sufficient to observe that any cycle of length 3 is a commutator. Clearly, $(ijk) = (ij)(ik)(ij)^{-1}(ik)^{-1}$.

Example 10.2.6 Let F be a field. Let G denote the set consisting of triples $(a(x), b(y), f(x, y))$, where $a(x) = \Sigma_{i=1}^3 a_i x_i$, $b(y) = \Sigma_{j=1}^3 b_j y_j$ are 3-linear expressions (linear form) with coefficients a_i, $b_j \in F$, and $f(x, y) = \Sigma_{i,j=1}^{3} a_{ij} x_i y_j$ the bilinear expression (form) with coefficients $a_{ij} \in F$. We define the binary operation $\cdot$ in G by

$$(a(x), b(y), f(x, y)) \cdot (c(x), d(y), g(x, y)) =$$
$$(a(x) + c(x), b(y) + d(y), f(x, y) + a(x)d(y) + g(x, y)).$$

It can be easily checked that G is a group with respect to this binary operation. Note that if F is infinite, then G is infinite. Now, assume that F is finite and contains q elements. Then, G contains $q^3 q^3 q^9 = q^{15}$ elements. Let H denote the set of pairs $(a(x), b(y))$, where $a(x)$ and $b(y)$ are the 3-linear expressions as explained above. Then, H is an abelian group with respect to coordinate wise addition. The map η from G to H given by $\eta((a(x), b(y), f(x, y)) = (a(x), b(y))$ is a surjective homomorphism. Since H is abelian, $[G, G] \subseteq ker \, \eta = \{(0, 0, f(x, y)) \mid f(x, y) \; is \; bilinear\}$. Indeed, we show that $[G, G] = ker \, \eta$. By the definition of the product, it is evident that $[(a(x), 0, 0), (0, b(y), 0)] = (0, 0, a(x)b(y))$. It is also clear that $(0, 0, f(x, y)) \cdot (0, 0, g(x, y)) = (0, 0, f(x, y) + g(x, y))$. Another basic fact is that for any bilinear expression $f(x, y)$, there exist linear forms $a(x), b(y), c(x), d(y), u(x), v(y)$ such that $f(x, y) = a(x)b(y) + c(x)d(y) + u(x)v(y)$. This shows that $[G, G] = ker \, \eta$. Next, it is easily observed that $[(a(x), b(y), f(x, y)), (c(x), d(y), g(x, y))] = (0, 0, a(x)d(y) - c(x)b(y))$. It follows that $[G, G]$ contains exactly q^9 elements, whereas the number of commutators is at the most q^6. Thus, there are many products of commutators which are not commutators.

Example 10.2.7 $SL(n, \mathbb{R})' = SL(n, \mathbb{R})$: The group $SL(n, \mathbb{R})$ is generated by the set of all matrices of the types E_{ij}^{λ} (transvections) (see Exercise 6.4.7). It is sufficient, therefore, to show that each E_{ij}^{λ}, $i \neq j$ is a commutator. Suppose that $n \geq 3$. Let $k \neq i$, $k \neq j$. Then, it can be observed that $[E_{ik}^{1}, E_{kj}^{\lambda}] = E_{ij}^{\lambda}$, $i \neq j$. Thus, $SL(n, \mathbb{R})' = SL(n, \mathbb{R})$ for all $n \geq 3$. Now, assume that $n = 2$. Then, it can be checked that E_{12}^{λ} is the commutator $[A, E_{12}^{\lambda}]$, where A is the matrix whose first row first column entry is $\sqrt{2}$, the second row second column entry is $\sqrt{2}^{-1}$, and the other entries are 0. Similarly, E_{21}^{λ} is also a commutator. This shows that $SL(2, \mathbb{R})' = SL(2, \mathbb{R})$. Thus, $-I_2 \in [SL(2, \mathbb{R}), SL(2, \mathbb{R})]$. It can be shown that $-I_2$ is not a commutator. This gives another example where product of commutators is not a commutator.

Example 10.2.8 $GL(n, \mathbb{R})' = SL(n, \mathbb{R})$: Since $GL(n, \mathbb{R})/SL(n, \mathbb{R}) \approx \mathbb{R}^*$ (the determinant map from $GL(n, \mathbb{R})$ to $\mathbb{R}^*$ is a surjective homomorphism whose kernel is $SL(n, \mathbb{R})$) is abelian, $GL(n, \mathbb{R})' \subseteq SL(n, \mathbb{R})$. Also, $SL(n, \mathbb{R}) = SL(n, \mathbb{R})' \subseteq GL(n, \mathbb{R})'$.

Proposition 10.2.9 *Let f be a surjective homomorphism from H to K. Then, $f(H') = K'$.*

Proof Since $f((a, b)) = f(aba^{-1}b^{-1}) = f(a)f(b)f(a)^{-1}f(b)^{-1} = (f(a), f(b))$, it follows that $f(H') \subseteq K'$. Further, let (c, d) be a commutator in K. Since f is surjective, there exist $a, b \in H$ such that $f(a) = c$ and $f(b) = d$. But, then $(c, d) = (f(a), f(b)) = f((a, b)) \in f(H')$. Thus, all commutators of K are in $f(H')$, and so $K' \subseteq f(H')$.

Corollary 10.2.10 *Let $H \trianglelefteq G$. Then, $(G/H)' = G'H/H$.*

Proof The quotient map ν from G to G/H is a surjective homomorphism, and so, $(G/H)' = \nu(G') = G'H/H$.   ♯

Definition Let G be a group. Define subgroups G^n of G inductively as follows: Define $G^1 = G'$. Assuming that G^n has already been defined, define $G^{n+1} = [G^n, G^n]$. Thus, we get a series

$$G = G^0 \trianglerighteq G^1 \trianglerighteq G^2 \trianglerighteq \cdots \trianglerighteq G^n \trianglerighteq G^{n+1} \trianglerighteq \cdots$$

of G. This series is called the **commutator Series** or the **derived Series** of G, and G^n is called the nth term of the derived series.

Remark 10.2.11 It follows by induction and Proposition 10.2.2 that each G^n is a normal subgroup of G. In fact, it follows by induction and the fact that $f((a, b)) = (f(a), f(b))$, that each G^n is fully invariant subgroup in the sense that it is invariant under all endomorphisms of G.

Definition 10.2.12 A group G is said to be a **solvable** group (or **soluble**) if the derived series of G terminates to $\{e\}$ after finitely many steps. The smallest n such that $G^n = \{e\}$ is called the **derived length** of G.

If H is a subgroup of G, then it follows by induction that $H^n \subseteq G^n$ for all n. Thus, we have the following proposition:

Proposition 10.2.13 *Subgroup of a solvable group is solvable.*   ♯

If f is a surjective homomorphism from H to K, then again, by induction, it follows that $f(H^n) = K^n$ for all n. Thus, we have the following proposition:

Proposition 10.2.14 *Homomorphic image of a solvable group is solvable. In particular, quotient group of a solvable group is solvable.*   ♯

Following is a necessary and sufficient condition for a group to be solvable.

Proposition 10.2.15 *A group G is solvable if and only if it has a normal series with abelian factors.*

Proof Suppose that G is solvable and $G^n = \{e\}$. Then,

$$G = G^0 \trianglerighteq G^1 \trianglerighteq \cdots \trianglerighteq G^n = \{e\}$$

is a normal series such that all terms are commutator subgroups of the preceding terms, and so all factors are abelian. Conversely, suppose that G has a normal series

$$G = G_0 \trianglerighteq G_1 \trianglerighteq G_2 \trianglerighteq \cdots G_m = \{e\}$$

all of whose factors are abelian. We show, by induction, that $G^i \subseteq G_i$ for all i. Clearly, $G^0 = G = G_0$. Assume that $G^i \subseteq G_i$. Then, since G_i/G_{i+1} is abelian, $(G_i)' \subseteq G_{i+1}$. But, then $G^{i+1} = (G^i)' \subseteq G_i' \subseteq G_{i+1}$. Thus, $G^m \subseteq G_m = \{e\}$, and so $G^m = \{e\}$. This shows that G is solvable. ♯

Proposition 10.2.16 *Let G be a group and H a solvable normal subgroup of G such that G/H is solvable. Then, G is also solvable.*

Proof Since G/H is solvable, it has a normal series

$$G/H = G_0/H \trianglerighteq G_1/H \trianglerighteq G_2/H \trianglerighteq \cdots \trianglerighteq G_n/H = \{H\}$$

with abelian factors. Further, since H is solvable, it has a normal series

$$H = H_0 \trianglerighteq H_1 \trianglerighteq \cdots \trianglerighteq H_m = \{e\}$$

with abelian factors. Consider the normal series

$$G = G_0 \trianglerighteq G_1 \trianglerighteq G_2 \trianglerighteq \cdots G_n = H = H_0 \trianglerighteq H_1 \trianglerighteq \cdots \trianglerighteq H_m = \{e\}$$

of G. Since $(G_i/H)/(G_{i+1}/H) \approx G_i/G_{i+1}$ (first isomorphism theorem), each factor of the above normal series is abelian, and hence, G is solvable. ♯

Proposition 10.2.17 *Let H and K be solvable subgroups of a group G. Suppose that $K \trianglelefteq G$. Then, HK is a solvable subgroup.*

Proof Since $K \trianglelefteq G$, $HK = KH$, and so HK is a subgroup of G. Since H is solvable, $H/H \cap K$ is solvable. By the Noether second isomorphism theorem, $H/H \cap K$ is isomorphic to HK/K. Hence, HK/K and K are solvable. From the previous proposition, HK is solvable. ♯

Corollary 10.2.18 *A maximal solvable normal subgroup of a group (if exists) is the largest solvable normal subgroup.*

Proof If M is a maximal solvable normal subgroup and H a solvable normal subgroup, then by the above proposition, HM is also a solvable normal subgroup. Since M is supposed to be maximal, $HM = M$, and so $H \subseteq M$. ♯

Corollary 10.2.19 *Every finite group has the largest solvable normal subgroup.* ♯

The largest solvable normal subgroup of a group G (if exists) is called the **radical** of G, and it is denoted by $R(G)$. Observe that radical of $G/R(G)$ is trivial. A group G is called **semisimple** if $R(G) = \{e\}$. Thus, in the theory of groups, it is important to study solvable groups and also semisimple groups. There is a nice structure theory of semisimple groups and also of solvable groups.

Remark 10.2.20 Product of arbitrary solvable subgroups need not be solvable, even if $HK = KH$: Consider the alternating group A_5 of degree 5. The subgroup A_4 of permutations in A_5 fixing the symbol 5 is a solvable subgroup ($A_4' = V_4$, and $V_4' = \{e\}$). Let P be a Sylow 5-subgroup of A_5. Then, P, being cyclic, is solvable ($P' = \{e\}$). Clearly, $A_5 = A_4 P$ is not solvable (since A_5 is nonabelian simple, $A_5' = A_5$, and so the derived series will never terminate to $\{e\}$).

Proposition 10.2.21 *Let $G \neq \{e\}$ be a solvable group. Then, G has a nontrivial abelian normal subgroup.*

Proof Since $G \neq \{e\}$ and it is solvable, there exists $m \in \mathbb{N}$ such that $G^m \neq \{e\}$ and $G^{m+1} = \{e\}$. But, then G^m is nontrivial abelian normal subgroup of G.   ♯

Corollary 10.2.22 *A solvable group is simple if and only if it is a cyclic group of prime order.*

Proof Assume that G is solvable simple group. If G is a solvable nontrivial group, then $G' \neq G$, for otherwise all terms of the derived series would be G, and so it will never terminate to $\{e\}$. Since G is assumed to be simple, $G' = \{e\}$, and so G is abelian. Since an abelian simple group is prime cyclic, the result follows.   ♯

Corollary 10.2.23 *A solvable group has a composition series if and only if it is finite.*

Proof Suppose that G is solvable and has a composition series. Since subgroups and quotient groups of solvable groups are solvable, it follows that all composition factors of G are solvable as well as simple. But, then they are prime cyclic (above proposition). Hence, the group G is finite. The converse is immediate.   ♯

Now, we give some examples of solvable groups. Every abelian group G is solvable, for then, $G^1 = G' = \{e\}$. S_3 is solvable, for $S_3 \triangleright A_3 \triangleright \{I\}$ is a normal series of S_3 with abelian factors. S_4 is also solvable, for $S_4 \triangleright A_4 \triangleright V_4 \triangleright \{I\}$ is a normal series of S_4 with abelian factors. The Quaternion group Q_8 is solvable, for $Q_8 \triangleright \{1, -1, i, -i\} \triangleright \{1\}$ is a normal series of Q_8 with abelian factors.

S_n, $n \geq 5$ is not solvable, for $S_n \triangleright A_n \triangleright \{I\}$, and $S_n \triangleright \{I\}$ are the only normal series, and none of them are with abelian factors.

Example 10.2.24 Every group of order pq is solvable, where p and q are primes: If $p = q$, then $| G | = p^2$, and so it is abelian. Hence, it is solvable. Suppose that $p \neq q$, and $p > q$. Then, the Sylow p-subgroup P of G is normal. Clearly, P and G/P (being prime cyclic) are solvable. Hence, G is solvable.

Example 10.2.25 Every group of order $100 = 5^2 \cdot 2^2$ is solvable: Sylow 5-subgroup of G is normal (Sylow 2nd theorem), and also solvable (being abelian). Also G/P, being abelian, is solvable. Hence, G is solvable.

Example 10.2.26 Every group of prime power order is solvable: Let $|G| = p^n$. The proof is by induction on n. If $n = 1$, then G is prime cyclic, and so solvable. Assume that the result is true for all groups of orders p^m, $m < n$. Since G is prime power order, $Z(G) \neq \{e\}$. Hence, $|G/Z(G)| = p^m$ for some $m < n$. By the induction hypothesis, $G/Z(G)$ is solvable. Since $Z(G)$ (being abelian) is solvable, G is solvable.

Example 10.2.27 Let G be a group of order pqr, where p, q and r are distinct primes with $p < q < r$. Then, G is solvable: We know that the Sylow $r-$ subgroup R of G is normal. R, being prime cyclic, is solvable. Further, $|G/R| = pq$, and so G/R is also solvable. Hence, G is solvable.

Remark 10.2.28 From a result of P. Hall, it follows that all C.L.T. groups are solvable. But, the converse is not true: For example, A_4 is solvable, and it has no subgroup of order 6.

Now, we state some results (without proofs) for the sake of information to the readers.

Theorem 10.2.29 *If the order of a group is product of distinct primes, then it is solvable.*   ♯

Theorem 10.2.30 *Every group of order $p^m q^n$ is solvable.*   ♯

The proof of this theorem can be found in Algebra 2, Chap. 9.

Theorem 10.2.31 *Every group of odd order is solvable.*   ♯

Obviously, groups in the above theorems are nonsimple groups.

Remark 10.2.32 The last theorem was conjectured by Burnside in the beginning of the 20th century, and it was proved by W. Fiet and J.G. Thompson in 1963. Since every nonabelian simple group is nonsolvable, the theorem ensures that every nonabelian finite simple group is of even order. Consequently, every nonabelian finite simple group contains an involution (an element of order 2). One of the main problems in theory of finite group was the classification of finite simple groups. The period from 1950 to 1970 was very crucial for the classification problem. Brauer, Fowler, Janko, and their students showed that the problem of determining finite simple groups with given centralizer of an involution is tractable. A two major steps' strategy for the classification was adopted.
Step1. Determine the possibilities for the centralizers of involutions in a finite simple group.
Step2. Determine all finite simple groups with a given centralizer of an involution in the group.
The development during this period was greatly influenced by the works of Janko, Suzuki, Thompson, Feit, Glauberman, Ree, and others. The classification of finite simple groups was complete in early eighties.

Following is a generalization of Sylow theorems for finite solvable groups: Let π be a set of prime numbers. A number m is said to be π-number, if every prime dividing m belongs to π. It is said to be a π' number, if no prime number dividing m is in π. A finite group G is said to be a π-group, if $|G|$ is a π number. A subgroup H of a finite group G is said to be a **Hall π-subgroup** of G if it is a π-subgroup of G, and $[G, H]$ is a π' number.

Theorem 10.2.33 (P. Hall) *Let G be a finite solvable group, and π a set of prime divisors of G. Then, G has a Hall π-subgroup. Further, any two Hall π-subgroups are conjugates.* ♯

Exercises

10.2.1 A subset A of a group G is said to be a normal (fully invariant) subset, if it is invariant under all inner automorphisms (endomorphisms). Let A and B be normal (fully invariant) subset of G. Show that $[A, B]$ is a normal (fully invariant) subgroup of G.

10.2.2 Let G be a group, and f a group homomorphism from G to an abelian group A. Show that there exists a unique homomorphism $\overline{f}$ from G/G' to A such that $\overline{f}(xG') = f(x)$ for all $x \in G$.

10.2.3 Find the commutator subgroups of the alternating groups.

10.2.4 Show that every finitely generated group has a radical.

10.2.5 Show that every group of order $7^2 \cdot 4$ is solvable.

10.2.6 Show, by means of an example, that a subgroup of a solvable group need not be subnormal.

10.2.7 Find commutator subgroup of a nonabelian group of order pq, where p and q are primes with $p > q$.

10.2.8 Show that every group of order $5^3 \cdot 6$ is solvable.

10.2.9 Show that any group whose order is less than 60 is solvable.

10.2.10 Show, by means of an example, that a product of solvable subgroups need not be solvable.

10.2.11 Show that every group of order $p^n q$ is solvable, where p, q are primes.

10.2.12 Show that if every subgroup of G is normal (such groups are called Dedekind groups), then G is solvable.

10.2.13 Prove the following identities known as Witt-Hall identities.

(i) $(a, b) \cdot (b, a) = e$.

(ii) $(a, bc) = (a, b)(a, c)((c, a), b)$.

(iii) $(ab, c) = (b, c)((c, b), a)(a, b)$.

(iv) $((b, a), c^a) \cdot ((a, c), b^c) \cdot ((c, b), a^b) = e$.

(v) $((b, a), c) \cdot ((c, b), a) \cdot ((a, c), b) = (b, a)(c, a)(c, b)^a(a, b)(a, c)^b(b, c)^a(a, c)$ $(c, a)^b$.

for all $a, b, c \in G$, where a^b means bab^{-1}.

10.2.14 Let $T(n, \mathbb{R})$ denote the set of all $n \times n$ nonsingular upper triangular matrices with real entries. Show that $T(n, \mathbb{R})$ is a group with matrix multiplication. Interpret the members of the commutator subgroup of $T(n, \mathbb{R})$. Show that $T(n, \mathbb{R})$ is solvable. Find its derived length.

10.2.15 Find the radical of $GL(n, \mathbb{R})$ and also of $SL(n, \mathbb{R})$.

10.3 Nilpotent Groups

Let G be a group. Define subgroups $L_n(G)$ inductively as follows: Define $L_0(G) = G$, $L_1(G) = [G, L_0(G)] = [G, G] = G'$. Supposing that $L_n(G)$ has already been defined, define $L_{n+1}(G) = [G, L_n(G)]$. It follows from Proposition 10.2.2 that each $L_n(G)$ is normal (in fact, fully invariant) in G. Thus, we get a descending chain

$$G = L_0(G) \trianglerighteq L_1(G) \trianglerighteq L_2(G) \trianglerighteq \cdots \trianglerighteq L_n(G) \trianglerighteq \cdots$$

of normal subgroups of G. This chain is called the **lower central series** of G.

Since $[S_3, S_3] = A_3 = [S_3, A_3]$,

$$S_3 = S_3 \trianglerighteq A_3 \trianglerighteq A_3 \trianglerighteq \cdots$$

is the lower central series of S_3. Similarly,

$$S_4 = S_4 \trianglerighteq A_4 \trianglerighteq A_4 \trianglerighteq \cdots$$

is the lower central series of S_4. Note that the lower central series, in general, is different from the derived series.

Next, define normal subgroups $Z_n(G)$ of G inductively as follows: Define $Z_0(G) = \{e\}$, $Z_1(G) = Z(G)$. Observe that $Z_1(G)/Z_0(G) = Z(G/Z_0(G))$ is the center of $G/Z_0(G)$. Supposing that $Z_n(G)$ has already been defined, define $Z_{n+1}(G)$ by the equation

$$Z_{n+1}(G)/Z_n(G) = Z(G/Z_n(G)).$$

Thus, we get an ascending chain

$${e} = Z_0(G) \trianglelefteq Z_1(G) \trianglelefteq Z_2(G) \trianglelefteq \cdots \trianglelefteq Z_n(G) \trianglelefteq \cdots$$

of normal subgroups of G. This chain is called the **upper central series** of G.

Since the center of S_n is trivial for all $n \geq 3$, the upper central series of S_n is

$${e} \trianglelefteq {e} \trianglelefteq {e} \cdots$$

for all $n \geq 3$.

Remark 10.3.1 It follows by induction that the terms of the lower central series are fully invariant, whereas the terms of upper central series are characteristic subgroups. Note that center, in general, is not a fully invariant subgroup.

Theorem 10.3.2 *The lower central series of G terminates to ${e}$ at the nth step if and only if the upper central series terminates to G at the nth step (i.e., $L_n(G) = {e}$ if and only if $Z_n(G) = G$).*

Proof Suppose that $L_n(G) = {e}$. We show, by induction on i, that $L_{n-i}(G) \subseteq Z_i(G)$ for all i. For $i = 0$, $L_{n-0}(G) = L_n(G) = {e} = Z_0(G)$, and the result is true for 0. Assume that $L_{n-i}(G) \subseteq Z_i(G)$. We show that $L_{n-(i+1)}(G) \subseteq Z_{i+1}(G)$. Let $a \in L_{n-i-1}(G)$. Since $L_{n-i}(G) = [G, L_{n-i-1}(G)]$, $xax^{-1}a^{-1} \in L_{n-i}(G) \subseteq Z_i(G)$ for all $x \in G$. This shows that $xZ_i(G)aZ_i(G) = aZ_i(G)xZ_i(G)$ for all $x \in G$. Thus, $aZ_i(G)$ belongs to the center of $G/Z_i(G)$. By the definition $a \in Z_{i+1}(G)$. Putting $i = n$, we find that $Z_n(G) = G$.

Conversely, suppose that $Z_n(G) = G$. We show, by induction, that $L_i(G) \subseteq Z_{n-i}(G)$ for all i. For $i = 0$, $L_0(G) = G = Z_n(G)$, and the result is true for $i = 0$. Assume that $L_i(G) \subseteq Z_{n-i}(G)$. We show that $L_{i+1}(G) \subseteq Z_{n-i-1}(G)$. From the definition of $L_{i+1}(G)$, it suffices to show that $xax^{-1}a^{-1} \in Z_{n-i-1}(G)$ for all $x \in G$ and $a \in L_i(G)$. Let $a \in L_i(G)$. Since $L_i(G)$ is assumed to be contained in $Z_{n-i}(G)$, $a \in Z_{n-i}(G)$. But, then by the definition of Z_{n-i}, $aZ_{n-i-1}(G)$ belongs to the center of $G/Z_{n-i-1}(G)$. Thus, $xZ_{n-i-1}(G)aZ_{n-i-1}(G) = aZ_{n-i-1}(G)xZ_{n-i-1}(G)$ for all $x \in G$. This means that $xa(ax)^{-1} \in Z_{n-i-1}(G)$ for all $x \in G$. This completes the proof of the fact that $L_i(G) \subseteq Z_{n-i}(G)$ for all i. Putting $i = n$, we get that $L_n(G) \subseteq Z_0(G) = {e}$.   ♯

Definition 10.3.3 A group G is said to be **nilpotent** if $L_n(G) = {e}$, or equivalently, $Z_n(G) = G$ for some n. A group G is said to be **nilpotent of class n** if $L_n(G) = {e}$ but $L_{n-1}(G) \neq {e}$, or equivalently, $Z_n(G) = G$ but $Z_{n-1}(G) \neq G$.

Proposition 10.3.4 *Every nilpotent group is solvable.*

Proof The result follows from the simple observation (do it by induction) that the nth term G^n of the derived series of G is contained in $L_n(G)$ the nth term of the lower central series of G. Thus, if $L_n(G) = {e}$ for some n, then $G^n = {e}$.   ♯

Proposition 10.3.5 *Subgroup of a nilpotent group is nilpotent. Homomorphic image (and so also quotient group) of a nilpotent group is also nilpotent.*

Proof Let H be a subgroup of a nilpotent group G. It follows, by induction, that $L_i(H) \subseteq L_i(G)$ for all i. Thus, if $L_n(G) = \{e\}$, then $L_n(H) = \{e\}$. Further, if f is a surjective homomorphism from H to K, then it follows, by induction, that $f(L_n(H)) = L_n(K)$ for all n. The result follows. ♯

Proposition 10.3.6 *Direct product of finitely many nilpotent groups is nilpotent.*

Proof It is sufficient to prove that the direct product of two nilpotent groups is nilpotent. Let H and K be groups. It follows, by induction, that $L_n(H \times K) = L_n(H) \times L_n(K)$. Thus, if $L_n(H) = \{e\}$ and $L_m(K) = \{e\}$, then $L_k(H \times K) = \{(e, e)\}$, where $k = max(m.n)$. ♯

Proposition 10.3.7 *Let G be a nontrivial nilpotent group. Then, $Z(G) \neq \{e\}$ (thus, in a nontrivial nilpotent group, there is an element different from identity, which commutes with each element of the group).*

Proof Suppose that $Z(G) = \{e\}$. Then, $Z_1(G) = Z(G) = \{e\}$. Since $Z_2(G)/Z_1(G) = Z_2(G)/\{e\} = Z(G/Z_1(G)) = Z(G/\{e\}) = Z(G)/\{e\} = \{e\}/\{e\}$, $Z_2(G) = \{e\}$. Proceeding inductively, we see that $Z_n(G) = \{e\}$ for all n. Hence, $Z_n(G)$ can never be G, and so G is not nilpotent. ♯

Remark 10.3.8 A solvable group need not be nilpotent: S_3 is solvable. Since $Z(S_3) = \{I\}$, S_3 is not nilpotent. Further, unlike the case of solvable groups, even if a normal subgroup H and the quotient group G/H are nilpotent, G need not be nilpotent: A_3 is normal nilpotent subgroup of S_3 such that $S_3/A_3 \approx \{1, -1\}$ is also nilpotent, but S_3 is not nilpotent.

Proposition 10.3.9 *Let H be a normal subgroup of G contained in the center $Z(G)$ of G such that G/H is nilpotent. Then, G is nilpotent.*

Proof Suppose that G/H is nilpotent. By the first isomorphism theorem, $G/Z(G) \approx (G/H)/(Z(G)/H)$. Since quotient of a nilpotent group is nilpotent, $G/Z(G)$ is nilpotent. We show, by induction, that $Z_i(G/Z(G)) = Z_{i+1}(G)/Z(G)$ for all i. $Z_0(G/Z(G)) = \{Z(G)\} = Z(G)/Z(G) = Z_1(G)/Z(G)$. Assume that $Z_i(G/Z(G)) = Z_{i+1}(G)/Z(G)$. Then,

$$Z_{i+1}(G/Z(G))/Z_i(G/Z(G)) = Z((G/Z(G))/Z_i(G/Z(G))) = Z((G/Z(G))/(Z_{i+1}(G)/Z(G))).$$

Further, we know that $(G/Z(G))/(Z_{i+1}(G)/Z(G)) \approx G/Z_{i+1}(G)$, and the canonical isomorphism ϕ is given by

$$\phi(aZ(G) \cdot (Z_{i+1}(G)/Z(G)) = aZ_{i+1}(G)).$$

The center $Z(G/Z_{i+1}(G))$, by definition, is $Z_{i+2}(G)/Z_{i+1}(G)$. Also,

$$Z((G/Z(G))/(Z_{i+1}(G)/Z(G))) = \phi^{-1}(Z_{i+2}(G)/Z_{i+1}(G)) = (Z_{i+2}(G)/Z(G))/(Z_{i+1}(G)/Z(G)) = (Z_{i+2}(G)/Z(G))/Z_i(G/Z(G)).$$

Hence, $Z_{i+1}(G/Z(G)) = Z_{i+2}(G)/Z(G)$. By the principle of induction, it follows that $Z_i(G/Z(G)) = Z_{i+1}(G)/Z(G)$ for all i. If $G/Z(G)$ is nilpotent, then $Z_{n+1}(G)/Z(G) = Z_n(G/Z(G)) = G/Z(G)$ for some n. But, then $Z_{n+1}(G) = G$, and so G is nilpotent (the class of nilpotency of G is one more than that of $G/Z(G)$). ♯

Corollary 10.3.10 G is nilpotent if and only if $G/Z(G)$ is nilpotent.   ♯

Corollary 10.3.11 Every finite p-group is nilpotent.

Proof Every finite p-group is of order p^n for some n. The proof is by induction on n. If $|G| = p$, then G is prime cyclic, and so it is nilpotent. Assume that every group of order p^m, $m < n$ is nilpotent. Let G be a group of order p^n. Then, $Z(G) \neq \{e\}$. Hence, $|G/Z(G)| = p^m$, $m < n$. By the induction assumption, $G/Z(G)$ is nilpotent, and so from the above proposition, G is nilpotent.   ♯

Remark 10.3.12 An infinite p-group need not be nilpotent. In fact, there are uncountably many groups, known as Tarski groups, which are infinite 2-generator simple groups all of whose proper subgroups are of same prime (sufficiently large) order.

Corollary 10.3.13 If all Sylow subgroups of a finite group are normal, then the group is nilpotent.

Proof If all Sylow subgroups are normal, then it is direct product of its Sylow subgroups. Since prime power order groups are nilpotent (above corollary), and the direct product of nilpotent groups are nilpotent, the result follows.   ♯

Proposition 10.3.14 Let H and K be normal subgroups of G such that G/H and G/K are nilpotent groups of classes α and β, respectively. Then, $G/(H \cap K)$ is a nilpotent group of class at most $max(\alpha, \beta)$.

Proof Define a map $\eta : G \longrightarrow G/H \times G/K$ by $\eta(x) = (xH, xK)$. Then, η is a homomorphism with kernel $H \cap K$. By the fundamental theorem of homomorphism, $G/H \cap K$ is isomorphic to a subgroup of $G/H \times G/K$. Since product of nilpotent groups is nilpotent, and subgroup of a nilpotent group is nilpotent, it follows that $G/H \cap K$ is nilpotent of class at most $max(\alpha, \beta)$.   ♯

Proposition 10.3.15 Let H be a normal nilpotent subgroup of G of class α, and K a normal nilpotent subgroup of class β. Then, HK is nilpotent group of class at most $\alpha + \beta$.

Proof The proof is by induction on $\alpha + \beta$. If $\alpha + \beta = 1$, then $\alpha = 1$, $\beta = 0$ or $\alpha = 0$, $\beta = 1$. If $\alpha = 1$ and $\beta = 0$, then $K = \{e\}$ and $HK = H$ is already nilpotent of class $\alpha = \alpha + 0$. Similarly, if $\alpha = 0$, then $H = \{e\}$ and $HK = K$ is nilpotent of class β. Suppose that the result is true in all cases when the sum of nilpotency classes is less than $\alpha + \beta$. Let H and K be normal nilpotent subgroups of classes α and β, respectively. Since H is a normal subgroup of G, $Z(H)$, being characteristic subgroup of H, is normal in G. Similarly,

$Z(K)$ is also normal in G. Clearly, $H/Z(H)$ and $KZ(H)/Z(H)$ are normal subgroups of $G/Z(H)$. Further, $H/Z(H)$ is a nilpotent group of class at most $\alpha - 1$ and $KZ(H)/Z(H) \approx K/K \cap Z(H)$ is nilpotent of class at most β. Hence, by the induction hypothesis, $H/Z(H) \cdot KZ(H)/Z(H) = HK/Z(H)$ is nilpotent of class at most $\alpha + \beta - 1$. Similarly, $HK/Z(K)$ is nilpotent of class at most $\alpha + \beta - 1$. Hence, $HK/Z(H) \times HK/Z(K)$ is nilpotent of class at most $\alpha + \beta - 1$. From the previous proposition $HK/Z(H) \cap Z(K)$ is nilpotent of class at most $\alpha + \beta - 1$. Since $Z(H) \cap Z(K) \subseteq Z(HK)$, it follows that HK is nilpotent of class at most $\alpha + \beta$. The result follows by the principle of induction. ♯

Remark 10.3.16 Product of a normal nilpotent subgroup with a nilpotent subgroup need not be nilpotent: The group S_3 is not nilpotent, whereas it is product of the normal nilpotent subgroup A_3 and the nilpotent subgroup $\{I, (12)\}$. But, it is true (a result of Wielandt) that product of any two nilpotent subgroups (if it is a subgroup) is solvable. Observe that product of solvable subgroups need not be solvable. For example, $A_5 = A_4 H$, where H is the subgroup of A_5 generated by a cycle of length 5. Clearly, A_4 and H are solvable, whereas A_5 is not solvable.

Corollary 10.3.17 *Maximal normal nilpotent subgroup is the largest nilpotent normal subgroup.*

Proof If H is a maximal normal nilpotent subgroup, and K is a normal nilpotent subgroup, then by the above proposition, HK is also normal and nilpotent. Since H is maximal, $HK = H$, and so $K \subseteq H$. ♯

Definition 10.3.18 The largest normal nilpotent subgroup (if exists) is called the **Fitting subgroup** of G, and it is denoted by $Fit(G)$.

Let H be a subgroup of a group G. Define a sequence $\{N_G^n(H) \mid n \in \mathbb{N} \bigcup \{0\}\}$ of subgroups of G inductively as follows: Define $N_G^0(H) = H$. Assuming that $N_G^n(H)$ has already been defined, define $N_G^{n+1}(H) = N_G(N_G^n(H))$. Thus, we get an ascending chain

$$H = N_G^0(H) \trianglelefteq N_G^1(H) \trianglelefteq \cdots \trianglelefteq N_G^n(H) \trianglelefteq \cdots$$

Proposition 10.3.19 *Let G be a nilpotent group of class n. Then, $N_G^n(H) = G$ for all subgroups H of G.*

Proof Suppose that G is nilpotent of class n. Then, $Z_n(G) = G$. We show, by induction on r, that $Z_r(H) \subseteq N_G^r(H)$ for all r. $Z_0(G) = \{e\} \subseteq H = N_G^0(H)$. Assume that $Z_r(G) \subseteq N_G^r(H)$. Let $a \in Z_{r+1}(G)$. Then, $aZ_r(G) \in Z(G/Z_r(G))$. Hence, $aZ_r(G)xZ_r(G) = xZ_r(G)aZ_r(G)$ for all $x \in G$. But, then $axa^{-1}x^{-1} \in Z_r(G)$, and so $axa^{-1}x^{-1} \in N_G^r(H)$ for all $x \in G$. In particular, $axa^{-1} \in N_G^r(H)$ for all $x \in N_G^r(H)$. This shows that $a \in N_G(N_G^r(H)) = N_G^{r+1}(H)$. By the principle of induction, $Z_r(G) \subseteq N_G^r(H)$ for all r. In particular, $G = Z_n(G) \subseteq N_G^n(H)$, and so $N_G^n(H) = G$. ♯

Definition 10.3.20 A group G is said to satisfy the **normalizer condition** if every proper subgroup is properly contained in its normalizer.

Corollary 10.3.21 *Every nilpotent group satisfies normalizer condition.*

Proof Let G be a nilpotent group of class n. Then, $N_G^n(H) = G$ for all subgroups H. If H is a proper subgroup which is not properly contained in $N_G(H)$, then $N_G(H) = H$, and so $N_G^n(H) = H \neq G$. This is a contradiction. ♯

Proposition 10.3.22 *If a finite group satisfies normalizer condition, then all its Sylow subgroups are normal.*

Proof Suppose that G satisfies normalizer condition. Let P be a Sylow p-subgroup of G. By the illustration 9.2.6, $N_G(N_G(P)) = N_G(P)$. Since the group satisfies the normalizer condition, $N_G(P)$ cannot be a proper subgroup of G. Thus, $N_G(P) = G$, and so P is normal in G. ♯

Combining above results, we obtain the following corollaries.

Corollary 10.3.23 *A finite group is nilpotent if and only if all its Sylow subgroups are normal in G.* ♯

Corollary 10.3.24 *A finite nilpotent group is direct product of its Sylow subgroups.* ♯

Thus, finite nilpotent groups are precisely products of prime power ordered groups.

Corollary 10.3.25 *A finite nilpotent group is a C.L.T. group.*

Proof Since a finite p-group is a C.L.T. group and direct product of C.L.T. groups is C.L.T. groups, the result follows from the above corollary. ♯

Remark 10.3.26 A C.L.T. group need not be nilpotent. For example, S_3 is C.L.T. group but it is not nilpotent. However, a C.L.T. group is always solvable.

Theorem 10.3.27 (Wielandt) *A finite group is nilpotent if and only if all its maximal subgroups are normal.*

Proof Let G be a finite nilpotent group and M a maximal subgroup of G. Then, since it satisfies normalizer condition, M is properly contained in $N_G(M)$. Since M is maximal, $N_G(M) = G$, and so M is normal. Conversely, suppose that all maximal subgroups of G are normal. We show that all Sylow subgroups of G are normal. Let P be a Sylow p-subgroup of G. Suppose that $N_G(P) \neq G$. Since G is finite, there is a maximal subgroup M containing $N_G(P)$. But, then $N_G(M) = M$, a contradiction to the supposition that all maximal subgroups are normal. Thus, all Sylow subgroups are normal, and so G is nilpotent. ♯

Frattini Subgroup

Let G be a group. If G has no maximal subgroups, then we define the **Frattini subgroup** of G to be G itself. If G has maximal subgroups, then the **Frattini subgroup** of G is defined to be the intersection of all maximal subgroups of G. Thus, if we agree to the convention that intersection of an empty family of subsets of a set X is X itself, then the Frattini subgroup is the intersection of all maximal subgroups of G. We use the notation $\Phi(G)$ for the Frattini subgroup of G. Thus, if M denotes the family of all maximal subgroups of G, then $\Phi(G) = \bigcap_{H \in M} H$.

Let G be a group and $\alpha \in Aut(G)$. Let H be a subgroup of G. Then, H is maximal if and only if $\alpha(H)$ is maximal. In fact, α induces a bijective map from M to itself given by $H \rightsquigarrow \alpha(H)$. Further, since α is bijective it preserves intersection, and so $\alpha(\Phi(G)) = \Phi(G)$. Hence, the Frattini subgroup $\Phi(G)$ is a characteristic subgroup. In particular, it is also normal.

Example 10.3.28 Maximal subgroups of $\mathbb{Z}$ are of the form $p\mathbb{Z}$, where p is a prime. Since there is no integer except 0 which is multiple of all primes, it follows that the intersection of all maximal subgroups of $\mathbb{Z}$ is $\{0\}$. Hence, $\Phi(\mathbb{Z}) = \{0\}$.

Example 10.3.29 Maximal subgroups of Q_8 are $\{1, -1, i, -i\}, \{1, -1, j, -j\}, \{1, -1, k, -k\}$, and so $\Phi(Q_8) = \{1, -1\}$.

Example 10.3.30 $\Phi(S_3) = \{I\}$, for A_3 and $\{I, (12)\}$ are maximal subgroups of S_3. $\Phi(S_4) = \{I\}$, for A_4 and V_4 are the only proper nontrivial normal subgroups, and they are not contained in all maximal subgroups. Also $\Phi(S_n) = \{I\}$, $n \geq 5$, for the only nontrivial proper normal subgroup of S_n is A_n, and A_n is not contained in all maximal subgroups (for example, it is not contained in S_{n-1}).

Theorem 10.3.31 (Wielandt) *A finite group G is nilpotent if and only if $G' \subseteq \Phi(G)$.*

Proof Let G be a finite nilpotent group and H a maximal subgroup. Then, from the previous theorem of Wielandt, H is normal subgroup of G. Since H is maximal, G/H is a group without proper subgroups, and so it is prime cyclic. In particular, it is abelian. But, then $G' \subseteq H$. This shows that $G' \subseteq \Phi(G)$. Conversely, suppose that $G' \subseteq \Phi(G)$. Then, every maximal subgroup of G contains G', and so every maximal subgroup is normal. By the previous theorem of Wielandt, G is nilpotent. ♯

Now, we characterize elements of the Frattini subgroup $\Phi(G)$.

Definition 10.3.32 An element $x \in G$ is called a **nongenerator** of G if whenever a set S generates G, $S - \{x\}$ also generates G.

Thus, a nongenerator has essentially no role in generating the group. e is always a nongenerator, for any subgroup will contain e. 0 is the only nongenerator of $\mathbb{Z}$: 1 cannot be a nongenerator, for $< 1 > = \mathbb{Z}$ but $< \emptyset > \neq \mathbb{Z}$. Similarly, -1 cannot be a nongenerator. Let $m \in \mathbb{Z}, m \neq 0, |m| > 1$. Then, there exists $n \neq \pm 1$ such that $(m, n) = 1$. But, then $< \{m, n\} > = \mathbb{Z}$ and $< \{n\} > = n\mathbb{Z} \neq \mathbb{Z}$.

1 and -1 are the only nongenerators of Q_8(verify).

Theorem 10.3.33 *The Frattini subgroup $\Phi(G)$ is the set of all nongenerators of G.*

Proof Let X denote the set of all nongenerators of G. We have to show that $\Phi(G) = X$. Suppose that $x \notin \Phi(G)$. Then, there exists a maximal subgroup H of G such that $x \notin H$. But, then $< H \bigcup \{x\} > = G$ and $< H > = H \neq G$. This shows that $x \notin X$, and so $X \subseteq \Phi(G)$. Next, suppose that $x \notin X$. Then, there exists a subset S of G such that $x \notin S$, $< S \bigcup \{x\} > = G$ and $< S > \neq G$. Let $H = < S >$. Let F denote the set of all proper subgroups of G which contain H. Then, $F \neq \emptyset$, for $H \in F$. Thus, F is a nonempty partially ordered set with respect to inclusion. Let $\{H_\alpha \mid \alpha \in I\}$ be a chain in F. Since union of a chain of subgroups is a subgroup, $\bigcup_{\alpha \in I} H_\alpha$ is a subgroup of G. Since each H_α is a proper subgroup of G containing H, $x \notin H_\alpha$ for all $\alpha \in I$. Hence, $x \notin \bigcup_{\alpha \in I} H_\alpha$, and so $\bigcup_{\alpha \in I} H_\alpha \in F$. This shows that every chain in F has an upper bound. By the Zorn's Lemma, F has a maximal element L (say). Then, (i) $x \notin L$ and (ii) L is a maximal subgroup, for if K is a subgroup of G such that L is properly contained in K, then $K \notin F$, and so $x \in K$. But, then $G = < S \bigcup \{x\} > \subseteq K$. This shows that $x \notin \Phi(G)$. Hence, $\Phi(G) \subseteq X$. ♯

Theorem 10.3.34 *Frattini subgroup of a finite group is always nilpotent.*

Proof Let G be a finite group, and P a Sylow p-subgroup of the Frattini subgroup. It is sufficient to show that P is normal in $\Phi(G)$. In fact, we show that P is normal in G. Let $g \in G$. Since $\Phi(G) \trianglelefteq G$, $gPg^{-1} \subseteq \Phi(G)$. Thus, P and gPg^{-1} are both Sylow p-subgroups of $\Phi(G)$, and so they are conjugate in $\Phi(G)$. Hence, there exists $u \in \Phi(G)$ such that $gPg^{-1} = uPu^{-1}$. But, then $u^{-1}g \in N_G(P)$. Thus, $g \in uN_G(P)$ for some $u \in \Phi(G)$. This shows that $G = \Phi(G)N_G(P) = < \Phi(G) \bigcup N_G(P) >$. Since every element of $\Phi(G)$ is a nongenerator, and since G is finite, $G = < N_G(P) > = N_G(P)$. This proves that $P \trianglelefteq G$. ♯

To summarize, we have the following:

Theorem 10.3.35 *Let G be a finite group. Then, the following conditions are equivalent.*

  (i) *G is nilpotent.*
 (ii) *G satisfies normalizer condition.*
(iii) *Every subgroup of G is subnormal.*
 (iv) *There is n such that $N_G^n(H) = G$ for all subgroups H of G.*
  (v) *Every Sylow subgroup of G is normal.*
 (vi) *G is direct product of prime power ordered groups.*
(vii) *Every maximal subgroup of G is normal.*
(viii) *$G' \subseteq \Phi(G)$.* ♯

Finite nilpotent groups are direct product of prime power ordered groups. The problem of classification of finite nilpotent groups reduces to the classification of finite p-groups. This is far more a complex problem. Here we shall restrict ourselves to some elementary facts about finite p-groups.

Groups of order p^3.

It follows from the structure theorem of finite abelian groups, that there are three nonisomorphic abelian groups of order p^3, and they are (i)$\mathbb{Z}_{p^3}$, (ii) $\mathbb{Z}_{p^2} \oplus \mathbb{Z}_p$, and (iii) $\mathbb{Z}_p \oplus \mathbb{Z}_p \oplus \mathbb{Z}_p$.

Let G be a nonabelian group of order p^3. Then, $Z(G) \neq \{e\}$. Since G is non-abelian, $Z(G) \neq G$. $| Z(G) | \neq p^2$, for otherwise $G/Z(G)$ would be cyclic, and this is impossible. Thus, $Z(G)$ is a cyclic group of order p. Also, since G is nonabelian and $G/Z(G)$ is abelian, $G' = Z(G)$ (check that $\Phi(G)$ is also $Z(G)$). Since $G/Z(G)$ is noncyclic abelian group of order p^2, $G/Z(G)$ is direct product of two cyclic groups of order p. Suppose that

$$G/Z(G) = < aZ(G) > \oplus < bZ(G) >$$

Clearly, $a^p \in Z(G)$, $b^p \in Z(G)$, and also the commutator $(a, b) \in G' = Z(G)$. Suppose that $(a, b) = c$. Further, $(a^i, b) = a^i b a^{-i} b^{-1} = a^i (bab^{-1})^{-i} = a^i (c^{-1}a)^{-i} = a^i a^{-i} c^i = c^i$. Similarly, $(a, b^i) = c^i$. Hence, $c = (a, b)$ generates $Z(G)$. We have the following three cases:

(i) $G = < a, b \mid a^p = e = b^p = (a, b)^p = (a, (a, b)) = (b, (a, b)) >$.

(ii) $G = < a, b \mid a^{p^2} = e = b^{p^2} = (a, b)^p = (a, (a, b)) = (b, (a, b)) >$.

and

(iii) $G = < a, b \mid a^{p^2} = e = b^p, a^p = (a, b) >$.

In case (i) the group is of exponent p, and it cannot occur if $p = 2$. Suppose that p is an odd prime. Then, groups described in (i) and (iii) are nonisomorphic (prove it). We show that the group described in case (ii) is isomorphic to that described in case (iii). The map which takes b in (iii) to ab in (ii), and a to a gives an isomorphism. If $p = 2$, then the group in (ii) is the Quaternion group Q_8, and the group in (iii) is the dihedral group D_8, and they are not isomorphic. Thus, there are two nonabelian groups of order p^3, and three abelian groups of order p^3. Overall there are five nonisomorphic groups of order p^3.

In what follows, we will use some results from linear algebra, and the reader may refer to Algebra 2 or any other book on linear algebra for the purpose. The reader may also skip the proof till they acquire basic knowledge of linear algebra.

A set S of generators of G is said to be a minimal set of generators or an irreducible set of generators, if no proper subset of S generates G. Every finite group has a minimal set of generators (prove it). The additive group $\mathbb{Q}$ of rationals has no minimal set of generators (prove it). It is easy to show (do it as exercise) that if a group has a minimal set of generators, then every set of generators contains a minimal set of generators. Distinct minimal set of generators may contain distinct number of elements. For example, $\{1\}$ and $\{5, 7\}$ both are minimal set of generators of $\mathbb{Z}$. For finite p-groups, we have the following theorem.

Theorem 10.3.36 (Burnside Basis Theorem) *Let G be a finite p-group of order p^n. Suppose that $G/\Phi(G)$ contains p^r elements. Then, any minimal set of generators of G contains r elements (r is called the **rank** of the group).*

Proof Since G is nilpotent, $G' \subseteq \Phi(G)$, and so $G/\Phi(G)$ is abelian. Further, given any maximal subgroup H of G, it is normal, and so G/H is prime cyclic. Hence, $a^p \in H$ for all maximal subgroup H of G. This shows that $G/\Phi(G)$ an abelian group all of whose elements are of order p. This means that $G/\Phi(G)$ is a vector space over $\mathbb{Z}_p$ containing p^r elements. Thus, $Dim_{\mathbb{Z}_p} G/\Phi(G) = r$, and so every minimal set of generators of $G/\Phi(G)$ contains exactly r elements. Further, if S generates G, then $\nu(S) = \{a\Phi(G) \mid a \in S\}$ generates $G/\Phi(G)$. Conversely, suppose that S is such that $\nu(S)$ generates $G/\Phi(G)$. Then, $G = \nu^{-1}(< \nu(S) >) = < S > \Phi(G) = < S \bigcup \Phi(G) >$. Since $\Phi(G)$ is a set of nongenerators, $< S > = G$. The result follows.   ♯

Let G be a finite p-group of order p^n and suppose that $G/\Phi(G)$ contains p^r elements. Consider the group $Aut(G)$ of automorphisms of G. Since every automorphism of G takes $\Phi(G)$ to itself, it induces an automorphism of the vector space $G/\Phi(G)$. Thus, we get a homomorphism η from $Aut(G)$ to the group $GL(r, p)$ of all automorphisms of the vector space $G/\Phi(G)$. It is an elementary fact from linear algebra that

$$| GL(r, p) | = (p^r - 1)(p^r - p)(p^r - p^2) \cdots (p^r - p^{r-1}).$$

The $ker\ \eta = \{\alpha \in Aut(G) \mid \eta(\alpha) = I_{G/\Phi(G)}\}$. Let us denote this normal subgroup of $Aut(G)$ by H. Then, $Aut(G)/H$ is isomorphic to a subgroup of $GL(r, p)$.

Let $\{x_1, x_2, \ldots, x_r\}$ be an ordered irreducible set of generators for G. Then, $\{u_1x_1, u_2x_2, \ldots, u_rx_r\}$ is again an ordered irreducible set of generators of G for all $u_1, u_2, \ldots, u_r \in \Phi(G)$. If $\alpha \in H$, then $\alpha(x_1) = u_1x_1, \alpha(x_2) = u_2x_2, \ldots, \alpha(x_r) = u_rx_r$ for some $u_1, u_2, \ldots, u_r \in \Phi(G)$. Let S be the set of all ordered set of generators of forms $u_1x_1, u_1x_2, \ldots, u_rx_r$, where $u_1, u_2, \ldots u_r \in \Phi(G)$. Clearly, $| S | = (p^{n-r})^r = p^{(n-r)r}$.

The group H acts on S through an action $\star$ given by $\alpha \star ((y_1, y_2, \ldots, y_r)) = (\alpha(y_1), \alpha(y_2), \ldots, \alpha(y_r))$. If $\alpha \in H$ fixes any element of S, then it fixes each member of a generator of G, and so it is identity (every local isotopy group is trivial). Hence, every orbit of the action contains exactly h elements, where $h = | H |$. Thus, h divides $| S | = p^{(n-r)r}$. Since $Aut(G)/H$ is isomorphic to a subgroup of $GL(r, p)$, $| Aut(G)/G |$ divides $| GL(r, p) | = (p^r - 1)(p^r - p) \cdots (p^r - p^{r-1})$. Hence, the order of Aut (G) divides $p^{(n-r)r}(p^r - 1)(p^r - p) \cdots (p^r - p^{r-1})$.

Now, we classify a very special type of p-groups known as extra special p-groups.

A finite nonabelian p-group is called an **extra-special p-group** if $Z(G) = G'$, and it is a cyclic group of order p. Thus, every nonabelian group of order p^3 is an extra special p-group.

A group G is said to be the **central product** of its normal subgroups $G_1, G_2, \ldots, G_r$ if

(i) $G = G_1G_2 \cdots G_r$,
(ii) $Z(G) \subseteq G_i$ for all i,
(iii) $G/Z(G)$ is direct product of $G_1/Z(G), G_2/Z(G), \ldots, G_r/Z(G)$, and
(iv) $[G_i, G_j] = \{e\}$ for all $i \neq j$.

Theorem 10.3.37 *Let G be a finite extra special group. Then, it is central product of nonabelian groups of order p^3. In particular, G is of order p^{2m+1} for some m.*

Proof Since $G' = Z(G)$, $(x, a_1a_2) = (x, a_1)(x, a_2)((a_2, x), a_1) = (x, a_1)(x.a_2)$. In particular, $(x, a^p) = (x, a)^p = e$ for all $x \in G$. Hence, $a^p \in Z(G) = G'$ for all $a \in G$. Thus, $G/Z(G)$ is an elementary abelian p-group, and so it is a vector space over the field Z_p. Let c be a generator of $Z(G)$. Note that $(xu, yv) = (x, y)$ for all $u, v \in Z(G)$. Thus, (x, y) depends only on cosets $xZ(G)$ and $yZ(G)$. This defines a map f from $G/Z(G) \times G/Z(G)$ to Z_p by $c^{f(xZ(G),yZ(G))} = (x, y)$. It is straight forward to verify that f is a skew-symmetric bilinear form on the vector space $G/Z(G)$. Further, if $f(xZ(G), yZ(G)) = o$ for all $yZ(G) \in G/Z(G)$, then $(x, y) = e$ for all $y \in G$. But, then, $x \in Z(G)$, and so $xZ(G) = Z(G)$ the zero of $G/Z(G)$. This shows that f is a nondegenerate skew-symmetric bilinear form. It is a standard result (see Algebra 2) of linear algebra that the vector space $G/Z(G)$ is of even dimension $2m$ for some m, and there is a basis

$$\{x_1Z(G), x_2Z(G), \ldots, x_mZ(G), y_1Z(G), y_2Z(G), \ldots, y_mZ(G)\}$$

of $G/Z(G)$ such that

(i) $f(x_iZ(G), y_iZ(G)) = 1$ for all i,
(ii) $f(x_iZ(G), y_jZ(G)) = 0$ for $i \neq j$, and also
(iii) $f(x_iZ(G), x_jZ(G)) = 0 = f(y_iZ(G), y_jZ(G))$ for all i, j.

Let G_i be the subgroup generated by $\{x_i, y_i\}$. Then, G_i is a nonabelian group of order p^3. Since $Z(G) = \Phi(G)$ and $G/Z(G)$ is generated by

$$\{x_1Z(G), x_2Z(G), \ldots, x_mZ(G), y_1Z(G), y_2Z(G), \ldots, y_mZ(G)\},$$

$\{x_1, y_1, x_2, y_2, \ldots, x_m, y_m\}$ generate G. Hence
(i) $G = G_1G_2 \cdots G_m$.
Further, since $f(x_iZ(G), y_iZ(G)) = 1$, $(x_i, y_i) = c \in G_i$. Thus,
(ii) $Z(G) \subseteq G_i$.
(iii) Clearly, $G/Z(G)$ is direct product of $G_1/Z(G), G_2/Z(G), \ldots G_m/Z(G)$.
Again, for $i \neq j$ $f(x_iZ(G), x_jZ(G)) = 0 = f(x_iZ(G), y_jZ(G))$. Hence
(iii) $[G_i, G_j] = \{e\}$ for all $i \neq j$.
This shows that G is central product of $G_1, G_2, \ldots, G_n$. ♯

Remark 10.3.38 It is easy to observe that the converse of the above proposition is also true.

Exercises

10.3.1 Let G be a group of order p^4. Can $Z(G) = G' = \Phi(G)$? Support.

10.3.2 Find the radical of $GL(n, \mathbb{R})$ and also of $SL(n, \mathbb{R})$.

10.3.3 Show that every group of order $5^2 \cdot 7^3$ is nilpotent. Find the number of nonisomorphic groups of this order and list them.

10.3.4 Find the number of nilpotent groups of order $p_1^2 p_2^2 \cdots p_r^2$.

10.3.5 Let G be a finite nilpotent group, and H a nontrivial normal subgroup of G. Show that $H \cap Z(G) \neq \{e\}$.

10.3.6 Show that a minimal normal subgroup of a nilpotent subgroup is contained in the center.

10.3.7 Let G be a nilpotent group, and H a maximal normal abelian subgroup of G. Show that $H = C_G(H)$.

10.3.8 Show that $T(n, \mathbb{R})$ is not nilpotent.

8.7.9. Show that the group $U(n, \mathbb{R})$ of upper triangular matrices all of whose diagonal entries are 1 is a nilpotent group. Find the index of nilpotency.

10.3.10 Let G be a finite nilpotent group and p divides $| G |$. Show that p divides $| Z(G) |$.

10.3.11 Show that a finite group G is nilpotent if and only if elements of co-prime orders commute.

10.3.12 Can S_3 be the Frattini subgroup of a finite group? Support.

10.3.13 Show that $G_1 \approx G_2$ implies that $\Phi(G_1) \approx \Phi(G_2)$. Show, by means of an example, that the converse is not true.

10.3.14 Show that a finite group G is nilpotent if and only if $G/\Phi(G)$ is direct product of cyclic groups of prime orders.

10.3.15 Let G be a finite group. Show that G is nilpotent if and only if $G/\Phi(G)$ is nilpotent.

10.3.16 Can $G/\Phi(G) \approx Q_8$? Support.

10.3.17 Let G_1 and G_2 be groups and $G = G_1 \times G_2$. Show that $\Phi(G)$ is a subgroup of $\Phi(G_1) \times \Phi(G_2)$. Show that they are equal provided that G_1 and G_2 both are finite (or at least finitely generated). In general, it is not known whether the equality always holds. This problem is related to the problem of existence of maximal subgroups of a simple group.

10.3.18 Characterize finite nilpotent group G for which G/G' is cyclic.

10.4 Free Groups and Presentations of Groups

Consider a group G which is generated by an element x and it is subject to only one condition $x^n = e$. Clearly, this is a cyclic group of order n which is isomorphic to $\mathbb{Z}_n$. If a group G is generated by an element x subject to the conditions $x^n = e = x^m$, then it is a cyclic group of order d, where d is the greatest common divisor of m and n. Let us consider a group G generated by $\{a, b\}$ subject to the conditions $a^n = b^m = e$, $bab^{-1} = a^r$. Then, $a = b^m a b^{-m} = a^{r^m}$, and so, $a^{r^m-1} = e$. In turn, $a^d = e$, where d is the g.c.d. of $r^m - 1$ and n. It turns out that $G = \{a^i b^j \mid 0 \le i \le d - 1, 0 \le j \le m - 1\}$ is a group of order dm. The multiplication in G is given by $a^i b^j a^k b^l = a^u b^v$, where u is the remainder obtained when $r^j k + i$ is divided by d and v is the remainder obtained when $j + l$ is divided by m. Description of a nonabelian group of order pq is an example of this kind.

One of the central problems in group theory with tremendous applications to topology, geometry, and other branches of mathematics and physics is to realize the group, once a set of generators together with relations is given. For example, to every knot there is a group attached, called the knot group, which is invariant of the knot, and whose generators and relations can be obtained. It is, therefore, one of the main problems in knot theory, and in turn, in topology, to have an effective procedure by which one can distinguish groups given by generators and relations. Poincare, Dehn, and Tietze realized the importance and the gravity of the problem, and they made some initial and original contributions in this direction.

In this section, we introduce formal theory of presentations of groups.

We have the following universal problem.

Let X be a set. Does there exist a pair (G, i), where G is a group and i a map from X to G such that given any pair (H, j), where H is a group and j a map from X to H, there is a unique homomorphism η from G to H which makes the following diagram commutative?

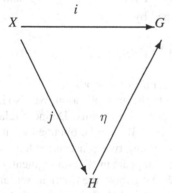

We show that the solution to the above problem exists, and it is unique up to isomorphism.

Uniqueness. Let (G, i) and (H, j) be solutions to the above problem. Then, there exist a homomorphism η from G to H and a homomorphism ρ from H to G such that the diagram

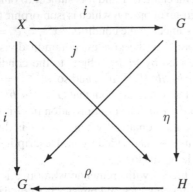

is commutative. Thus, $\rho \circ \eta$ and I_G both will make the left upper half triangle commutative. From the universal property of (G, i), it follows that $\rho \circ \eta = I_G$. Similarly, $\eta \circ \rho = I_H$. Thus, η is an isomorphism.

Existence. Recall that if A and B are sets, then A^B denotes the set of all maps from B to A. For each $n \in \mathbb{N}$, let $\bar{n}$ denote the set $\{1, 2, \ldots, n\}$. Consider the set $\Sigma(X) = \bigcup_{n \in \mathbb{N}} (X \times \{1, -1\})^{\bar{n}} \bigcup \{\emptyset\}$. Elements of $\Sigma(X)$ are called the **words** in X. $\emptyset$ is called the empty word. A map f from $\bar{n}$ to $X \times \{1, -1\}$ is an element of $\Sigma(X)$. This element f is formally and conveniently denoted by

$$x_1^{\epsilon_1} x_2^{\epsilon_2} \cdots x_n^{\epsilon_n},$$

where $f(i) = (x_i, \epsilon_i)$ for all $i \in \bar{n}$. We also denote the empty word by 1. We define product in $\Sigma(X)$ by juxtaposition. Thus, if $W_1 = x_1^{\alpha_1} x_2^{\alpha_2} \cdots x_r^{\alpha_r}$ and $W_2 = y_1^{\beta_1} y_2^{\beta_2} \cdots y_s^{\beta_s}$, then the product $W_1 W_2$ is defined by

$$W_1 W_2 = x_1^{\alpha_1} x_2^{\alpha_2} \cdots x_r^{\alpha_r} y_1^{\beta_1} y_2^{\beta_2} \cdots y_s^{\beta_s}$$

Given a word $W = x_1^{\alpha_1} x_2^{\alpha_2} \cdots x_r^{\alpha_r}$, W^{-1} denotes the word $x_r^{-\alpha_r} x_{r-1}^{-\alpha_{r-1}} \cdots x_1^{-\alpha_1}$. The empty word 1 acts as identity with respect to the multiplication. We also identify 1^{-1} with 1. It is clear that $\Sigma(X)$ is a semigroup with identity. Define a relation R on $\Sigma(X)$ as follows: $(W_1, W_2) \in R$ if and only if W_2 can be obtained from W_1 by insertion of words of the types $x^\alpha x^{-\alpha}$ between any two consecutive letters of the word W_1, or deletion of such words whenever it appears in W_1 as a segment. Clearly, R is an equivalence relation on $\Sigma(X)$. Let $F(X)$ denote the quotient set, and $[W]$ denote the equivalence class determined by the word W. It is evident that if W_1 is related to W_1' and W_2 is related to W_2', then $W_1 W_2$ is related to $W_1' W_2'$. This shows that we have a binary operation on $F(X)$ defined by $[W_1][W_2] = [W_1 W_2]$. It is also clear that $F(X)$ is a group with respect to this operation. We have a map i from X to

$F(X)$ defined by $i(x) = [x^1]$. We show that the pair $(F(X), i)$ is a solution to the universal problem. Let j be a map from X to a group H. Then, we have a map ϕ from $\Sigma(X)$ to H given by $\phi(x_1^{\alpha_1} x_2^{\alpha_2} \cdots x_r^{\alpha_r}) = (j(x_1))^{\alpha_1} (j(x_2))^{\alpha_2} \cdots (j(x_r))^{\alpha_r}$ which is also a homomorphism. It is evident that if W_1 is related to W_2, then $\phi(W_1) = \phi(W_2)$. Thus, ϕ induces a homomorphism η from $F(X)$ to H. Clearly, η makes the required diagram commutative. It is also clear that the only homomorphism which can make the required diagram commutative is η.

The pair $(F(X), i)$ thus obtained is called the **free group** on X. It is clear that i is an injective map, and we identify X as a set of generators of $F(X)$.

Free groups satisfy the following lifting property, also called the projective property, in the category of groups.

Theorem 10.4.1 *Let β be a surjective homomorphism from G_1 to G_2. Let $F(X)$ be the free group on X, and f a homomorphism from $F(X)$ to G_2. Then, there exists a homomorphism η (not necessarily unique), called a lifting of f, such that the following diagram is commutative.*

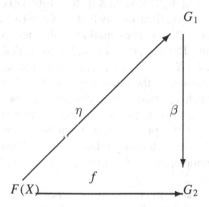

Proof Consider the following diagram, where the maps i, β and f are given.

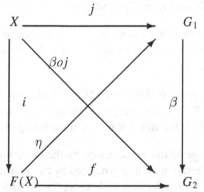

Since β is surjective, $\beta^{-1}(f(i(x))) \neq \emptyset$ for all $x \in X$. By the axiom of choice, we get a map j from X to G_1 such that $\beta \circ j = f \circ i$. Thus, the square and the triangles

X $F(X)$ G_2 and X G_1 G_2 are commutative. From the universal property of the free group, there exists a unique homomorphism η from $F(X)$ to G_1 such that the triangle X $F(X)$ G_1 is commutative. Clearly, $(\beta o \eta)oi = \beta o(\eta oi) = \beta oj = foi$. Thus, $\beta o \eta$ and f both make the triangle X $F(X)$ G_2 commutative. From the universal property of the free group, it follows that $\beta o \eta = f$, and so the triangle $F(X)$ G_1 G_2 is commutative.   ♯

Theorem 10.4.2 *Every group is quotient of a free group.*

Proof Let G be group. Let $F(G)$ denote the free group on the set part of the group G. We have the identity map I_G from the set G to the group G. From the universal property of free group, there exists a unique homomorphism η from $F(G)$ to G such that $\eta oi = I_G$. Thus, η is surjective. The result follows from the fundamental theorem of homomorphism.   ♯

Given any set $\mathfrak{R}(X)$ of words in X, it defines a unique group $F(X)/H$, where H is the normal subgroup of $F(X)$ generated by $\{[W] \mid W \in \mathfrak{R}(X)\}$. If a group G is isomorphic to the group $F(X)/H$ described above, then we say that $< X \; ; \; \mathfrak{R}(X) >$ is a **presentation** of G. In particular, $< X \; ; \; \emptyset >$ is presentation of the free group $F(X)$, also called the **free presentation**. The elements of $\mathfrak{R}(X)$ are called the **defining relators** of the presentation. In case $X = \{x_1, x_2, \ldots, x_r\}$ is a finite set and $\mathfrak{R}(X) = \{R_1, R_2, \ldots R_t\}$ is also finite, we denote the presentation $< X \; ; \; \mathfrak{R}(X) >$ by $< x_1, x_2, \ldots, x_r \; ; \; R_1, R_2, \ldots R_t >$. Such a presentation is called a finite presentation. In general, a group need not have a finite presentation (the additive group $\mathbb{Q}$ of rational numbers does not have any finite presentation, for it is not finitely generated. In fact, it cannot have finitely many defining relators also). Clearly, a group has so many presentations. For example, $< x; x^2 >$ and $< x; x^4, x^6 >$, both are presentations of a cyclic group of order 2.

Theorem 10.4.2 says that every group has a presentation. In general, it is very difficult to recognize a group by its presentation. We give some simple examples of presentations. Presentations for nonabelian groups of order pq and nonabelian groups of order p^3 have already been discussed in earlier sections.

Theorem 10.4.3 (von Dyck's Theorem) *Let $< X \; ; \; \mathfrak{R}(X) >$ be a presentation of a group G and $< X \; ; \; \mathfrak{R}'(X) >$ be a presentation of a group G' such that $\mathfrak{R}(X) \subseteq \mathfrak{R}'(X)$. Then, G' is a homomorphic image of G.*

Proof Since $\mathfrak{R}(X) \subseteq \mathfrak{R}'(X)$, the normal subgroup H generated by $\mathfrak{R}(X)$ is contained in the normal subgroup H' generated by $\mathfrak{R}'(X)$. By the first isomorphism theorem $F(X)/H' \approx (F(X)/H)/(H'/H)$. Thus, G' is isomorphic to a quotient group of G. ♯

Example 10.4.4 Let us describe and recognize the group given by the presentation $< x, y \; ; \; x^3, y^2, xyx^{-2}y >$. We look for a group which is generated by two elements, one of order 3 and the other of order 2. The simplest such group is S_3 which is generated by the cycle $\sigma = (123)$ and the transposition $\tau = (12)$. It is also evident that $\sigma \tau \sigma^{-2} \tau$ is the identity permutation.

Let $F(x, y)$ denote the free group on the set $\{x, y\}$. From the universal property of the free group, we have a unique surjective homomorphism η from $F(x, y)$ to S_3 with $\eta(x) = \sigma$ and $\eta(y) = \tau$. Let H be the normal subgroup of $F(x, y)$ generated by the set $\{x^3, y^2, xyx^{-2}y\}$ of defining relations. Since σ^3, τ^2 and $\sigma\tau\,\sigma^{-2}\tau$ are identity permutations, it follows that $H \subseteq ker\,\eta$. From the fundamental theorem of homomorphism, η induces a surjective homomorphism $\overline{\eta}$ from $F(x, y)/H$ to S_3.

Now, we try to enumerate the cosets of $F(x, y)$ modulo H and find a bound for the order of $F(x, y)/H$, if it happens to be finite. H is the identity coset. Consider the set $S = \{H, [x]H, [x^2]H, [y]H, [y][x]H = [yx]H, [yx^2]H\}$ of cosets of $F(x, y)$ mod H. We show that $[x]S = S = [y]S$. Clearly, $[x]H, [x][x]H = [x^2]H, [x][x^2]H\,[x^3]H = H$ belong to S. Again, $[x][y]H = [xy]H = [yx^2]H$, for $[xyx^{-2}y] \in H$. Further, $[x][yx]H = [xyx]H = [y]H$, for $[xyxy^{-1}] = [xyxy] = xyyx^2 \in H$. Also, $[x][yx^2]H = [xy]H[x^2]H = [yx^2]H[x^2]H\,[yx^4]$ $H\,[yx]H$. This shows that $[x]S = S$. Similarly, $[y]S = S$. Since $\{[x], [y]\}$ generate $F(x, y)$, it follows that S is closed under product of any element of $F(x, y)$ from left. Hence, S is the set of all cosets of $F(x, y)$ modulo H. This shows that the group $F(x, y)/H$ contains at the most six elements. Since η is surjective homomorphism from $F(x, y)/H$ to S_3, it is bijective. Thus, the given presentation describes S_3.

More generally, the following example gives a presentation of S_n for each n.

Example 10.4.5 The symmetric group $S_n, n \geq 2$ has a presentation

$$< x_1, x_2, \ldots, x_{n-1} ;\ x_i^2, (x_j x_{j+1})^3, (x_k x_l)^2 \mid 1 \leq i \leq n - 1, 1 \leq j \leq n - 2, \mid$$
$$k - l \mid > 2, 1 < k \leq n - 1, 1 \leq l \leq n - 1 > .$$

Proof Let X denote the set $\{x_1, x_2, \ldots, x_{n-1}\}$. Define a map α from X to S_n by $\alpha(x_i) = (i, i + 1)$. From the universal property of a free group, we have a unique homomorphism η from $F(X)$ to S_n, which takes x_i to $(i, i + 1)$. Since S_n is generated by the set $\{(i, i + 1) \mid 1 \leq i \leq n - 1\}$, η is a surjective homomorphism. Further, $\eta(x_i^2) = (\eta(x_i))^2 = (i, i + 1)^2 = I$. Similarly, it is easy to observe that η takes all the defining relators of the given presentation to identity. This means that the normal subgroup H of $F(X)$ generated by the set of defining relators is contained in the kernel of η. Thus, η induces a surjective homomorphism (fundamental theorem of homomorphism) from $F(X)/H$ to S_n.

Thus, to show that η is an isomorphism, it is sufficient to show that $F(X)/H$ contains at most $n!$ elements. This we prove by induction on n. If $n = 2$, $F(X)/H$ is cyclic group of order 2. Assume the result for $n - 1, n \geq 3$. Let K be the subgroup of $F(X)/H$ generated by $\{[x_1], [x_2], \ldots, [x_{n-2}]\}$. Clearly, there is a surjective homomorphism from the group with presentation

$$< x_1, x_2, \ldots, x_{n-2};\ x_i^2, (x_j x_{j+1})^3, (x_k x_l)^2 \mid 1 \leq i \leq n - 2, 1 \leq j \leq n - 3, \mid$$
$$k - l \mid \geq 2, 1 \leq k \leq n - 2, 1 \leq l \leq n - 2 >$$

to K. Thus, by the induction hypothesis $\mid K \mid \leq (n - 1)!$. We show that there can be at most n cosets of $F(X)/H$ modulo K. More precisely, we show that every element of

$F(X)/H$ lies in at least one of the left cosets $[x_1 x_2 \cdots x_{n-1}]K$, $[x_2 x_3 \cdots x_{n-1}]K$, $\cdots$
$[x_{n-1}]K$, K. It is sufficient to show that product of an element from left to an element of some of these cosets is again an element of some of these cosets. Since $[x_1], [x_2], \ldots, [x_n]$ generate $F(X)/H$, it is sufficient to show that for each i, product from left of $[x_i]$ to the cosets above is again one of those cosets. Suppose that $i \le j - 2 \le n - 3$. Then, since $(x_i x_j)^2$ is a relator, $[x_i x_j]^2 = [x_i][x_j][x_i][x_j] = e$ in $F(X)/H$. Since x_i^2 and x_j^2 are also relators , it follows that $[x_i]^2 = [x_j]^2 = e$. Hence, in this case $[x_i][x_j] = [x_j][x_i]$, and so
$[x_i] \cdot [x_j x_{j+1} \cdots x_{n-1}]K$
$= [x_i][x_j][x_{j+1}] \cdots [x_{n-1}]K$
$= [x_j][x_{j+1}] \cdots [x_{n-1}][x_i]K$
$= [x_j][x_{j+1}] \cdots [x_{n-1}]K$
$= [x_j x_{j+1} \cdots x_{n-1}]K$,
for $[x_i] \in K$. Next, suppose that $i = j - 1$. Then,

$$[x_i][x_j][x_{j+1}] \cdots [x_{n-1}]K = [x_i x_{i+1} x_{i+2} \cdots x_{n-1}]K.$$

If $i = j$, then also

$$[x_i][x_i x_{i+1} \cdots x_{n-1}]K = [x_{i+1} x_{i+2} \cdots x_{n-1}]K,$$

for x_i^2 is a relator. Suppose that $i \ge j + 1$. Then
$[x_i][x_j x_{j+1} \cdots x_{n-1}]K$
$= [x_j x_{j+1} \cdots x_i x_{i-1} x_i x_{i+1} \cdots x_{n-1}]K$
$= [x_j][x_{j+1}] \cdots [x_i x_{i-1} x_i][x_{i+1} x_{i+2} \cdots x_{n-1}]K$.
Since $(x_{i-1} x_i)^3$, x_{i-1}^2 and x_i^2 are relators, it follows that $[x_i x_{i-1} x_i] = [x_{i-1} x_i x_{i-1}]$, and further, since $[x_{j-1}]$ commutes with all $[x_k]$, $k > i$, we find that
$[x_i][x_j x_{j+1} \cdots x_{n-1}]K$
$= [x_j x_{j+1} \cdots x_{n-1}][x_{i-1}]K$
$= [x_j x_{j+1} \cdots x_{n-1}]K$. The proof is complete.   ♯

The method used in the above two examples to determine an upper bound of the order of a group given by a presentation is called the **coset enumeration** method. Many variants of this method are used to recognize finite groups given by presentations. One can develop a program and take help of computer to describe finite groups given by presentations. For details, see 'Handbook on Computational Group Theory' by D.F. Holt, B. Eick, and O Brien.

Motivated from some classification problems in Topology, Dehn, in 1911, posed the following three problems known as **Dehn's fundamental problems**.

1. **Word problem**. Given a presentation $< X ; \Re(X) >$, to find an algorithm which decides, in finite number of steps, whether a word W defines the identity element in $F(X)/H$.

2. **Conjugacy problem** or **Transformation problem**. Given a presentation $< X ; \Re(X) >$, to find an algorithm which decides, in finite number of steps, whether a pair of words W_1 and W_2 determine conjugate elements in $F(X)/H$.

3. **Isomorphism problem**. Given two presentations $< X ; \mathfrak{R}(X) >$ and $< Y ; \mathfrak{R}'(Y) >$, to find an algorithm which decides, in finite number of steps, if they define isomorphic groups.

Clearly, the solution to the second problem gives a solution to the first problem, for a word W defines the identity element if and only if it defines the element which conjugate to the element defined by the empty word. The third problem is the most difficult.

In general, for a given presentation, there may not exist any algorithm for the word problem (conjugacy problem, isomorphism problem). In fact, Novikov and Boone in 1955, and later many more people gave examples of finite presentations, where such an algorithm, even for word problem, does not exist. Here, we shall describe these algorithms for some simple presentations.

Theorem 10.4.6 *Word and conjugacy problems for a free presentation are solvable.*

Proof Let $< X ; >$ be a free presentation. Let W be a word in the defining symbols. We define free reduction $\rho(W)$ of a word W by the induction on the length of the word W. If $l(W) = 0$, then W is the empty word $\emptyset$ and we define $\rho(\emptyset) = \emptyset$. Assume that $\rho(W)$ has been defined for all words of length at most n with further property that $l(\rho(W)) \leq l(W)$. Consider a word $W x_{n+1}^{\alpha_{n+1}}$, where $\alpha_{n+1} = \pm 1$. Suppose that $\rho(W) = y_1^{\beta_1} y_2^{\beta_2} \cdots y_s^{\beta_s}$. Then define $\rho(W x_{n+1}^{\alpha_{n+1}}) = \rho(W) x_{n+1}^{\alpha_{n+1}}$ provided that $y_s \neq x_{n+1}$ or $\beta_s \neq -\alpha_{n+1}$. If $x_{n+1} = y_s$ and also $\beta_s = -\alpha_{n+1}$, then define $\rho(W x_{n+1}^{\alpha_{n+1}}) = \rho(\rho(W) x_{n+1}^{\alpha_{n+1}})$. Clearly, $l(\rho(W x_{n+1}^{\alpha_{n+1}})) \leq l(W x_{n+1}^{\alpha_{n+1}})$. The definition of the map ρ is complete. The following properties of ρ can be proved easily by the induction on the length of the word.

  (i) $\rho(\emptyset) = \emptyset$.
 (ii) $[W] = [\rho(W)]$ in the free group for all words W.
(iii) $\rho(W)$ is the word of smallest length in the class $[W]$.
 (iv) $\rho(\rho(W)) = \rho(W)$.
  (v) $[W_1] = [W_2]$ if and only if $\rho(W_1) = \rho(W_2)$.
 (vi) $\rho(W_1 W_2) = \rho(\rho(W_1) W_2) = \rho(W_1 \rho(W_2)) = \rho(\rho(W_1) \rho(W_2))$.

It is clear that W defines identity in the free presentation if and only $\rho(W) = \emptyset$. This gives the solution of word problem for free presentation. ρ is called the **free reduction**.

For the conjugacy problem, we define $\sigma(W)$ for each word W by induction on $l(\rho(W))$. If $\rho(W) = \emptyset$, then we define $\sigma(W) = \emptyset$. Assume that $\sigma(W)$ has already been defined for all those words for which $l(\rho(W)) \leq n$. Suppose that $\rho(W) = y_1^{\beta_1} y_2^{\beta_2} \cdots y_{n+1}^{\beta_{n+1}}$. Then, $\sigma(W) = \rho(W)$ provided that $y_1 \neq y_{n+1}$ or $\beta_1 \neq -\beta_{n+1}$. If $y_1 = y_{n+1}$ and $\beta_1 = -\beta_{n+1}$, then define $\sigma(W) = y_2^{\beta_2} y_3^{\beta_3} \cdots y_n^{\beta_n}$. It can be checked by the induction that words W_1 and W_2 define conjugate elements in the free group if and only if $\sigma(W_1) = \sigma(W_2)$. This solves the conjugacy problem for free presentation. The map σ is called the **cyclic reduction**. ♯

Corollary 10.4.7 *Every free group is torsion-free in the sense that every nonidentity element is of infinite order.*

Proof Let $[W] = [\rho(W)]$ be a nonidentity element of the free group. Then, $\rho(W)$ is not the empty word. If $\sigma(\rho(W)) = \rho(W)$, then it is clear that $\rho((\rho(W))^n) = (\rho(W))^n$ which is not the empty word. Further, since $[W]$ is not identity, it is not conjugate to identity, and hence, $[\sigma(\rho(W))]$ which is conjugate to $[\rho(W)]$ cannot define identity element. Clearly, $[\rho(W)]^n$ is conjugate to $[\sigma(\rho(W))]^n$ which is nonidentity. ♯

Now, we discuss isomorphism problem for presentations. Let $< X; \Re >$ be a presentation. We say that a word W in X is **derivable** from $\Re$ if it is reducible to the empty word after finite number of steps under the operation of deletions and insertions of members of $\Re \bigcup \Re^{-1}$, and also of trivial relators. The following four transformations on the presentation $< X; \Re >$ are called the **Tietze transformations**.

T_1. If a word W in X is derivable from $\Re$, then put W in the set $\Re$ of defining relators.

T_2. If a defining relator $P \in \Re$ is derivable from the rest of the relators in $\Re$, then delete it from the set $\Re$ of defining relators.

T_3. Let W be a word in X. Insert a symbol u in the set X of generating set and also insert $u^{-1}W$ or Wu^{-1} in the set $\Re$ of defining relators.

T_4. If there is a relator R in $\Re$ of the form $x^{-1}W$ or Wx^{-1}, where W is a word not involving the symbol x, then delete x from the generating set X, the relator R from the set $\Re$ of defining relations and also substitute W at the place of x in all the rest of the relators in $\Re - \{R\}$.

Theorem 10.4.8 *Let $< X; \Re >$ and $< Y; \Re' >$ be two presentations. They define isomorphic groups if and only if one can be obtained from the other by using Tietze transformations.*

Proof Suppose that $< Y; \Re' >$ is obtained from $< X; \Re >$ by using T_1 and T_2. Then, $X = Y$ and the normal subgroups of $F(X)$ generated by $\Re$ and $\Re'$ are same, and therefore, they define isomorphic groups. If $< Y; \Re' >$ is obtained from $< X; \Re >$ by T_3, then $X \subseteq Y$, and the map $[W] < \Re > \rightsquigarrow [W] < \Re' >$ defines an isomorphism. Since T_4 is the inverse transformation of T_3, it also does not change the group.

Conversely, suppose that the given two presentations define isomorphic groups. Let ϕ be the isomorphism from $F(X)/ < \Re >$ to $F(Y)/ < \Re' >$. Suppose that $\phi([x] < \Re >) = [W_x] < \Re' >$, where W_x is a word in Y, and $\phi^{-1}[y] < \Re' > = [W_y] < \Re >$, where W_y is a word in X. Then, the presentation $< X \bigcup Y; \Re \bigcup \Re' \bigcup \{x^{-1}W_x \mid x \in X\} \bigcup \{y^{-1}W_y \mid y \in Y\} >$ defines a group which is isomorphic to both the groups.   ♯

In general, there is no algorithm which decides in finite number of steps whether given two presentations define isomorphic groups. The philosophy of this part of group theory is to search some effective computable algebraic invariant of the group in terms of presentation. More precisely, given a class Σ of presentations, we try to have association f which associates to each presentation $P \in \Sigma$ a computable invariant $f(P)$. The discussion of these invariants is beyond the scope of this book.

We illustrate the use of Tietze transformations to solve isomorphism problem in some very simple cases by means of examples.

Example 10.4.9 The presentations

$$< a, b, c; b^2, (bc)^2 >$$

and

$$< x, y, z; y^2, z^2 >$$

define isomorphic groups.

Proof We show that the second presentation can be obtained from the first by means of Tietze transformations. Use T_3 to adjoin generators x, y, z and relators $x^{-1}a, y^{-1}b, z^{-1}bc$ in the first presentation. Thus, we arrive at the presentation $< a, b, c, x, y, z; x^{-1}a, y^{-1}b, z^{-1}bc, b^2, (bc)^2 >$. Now, use T_4 to remove the generators a, b, c from this presentation, and thus, we arrive at the presentation $< x, y, z; y^2, z^2 >$. It also follows that the group presented contains an element $[x]$ of infinite order. ♯

Example 10.4.10 The presentations

$$< a, b; a^4, b^4; a^2b^2a^{-1}b^{-1} >$$

and

$$< x, y; x^5, y^4, yxy^{-1}x^{-3} >$$

define isomorphic groups and it is of order 20.

Proof Adjoin the generators x, y with relators $x^{-1}ab, y^{-1}b$ in the first presentation and observe that $(ab)^5$ is derivable in the first presentation. The rest follows easily. ♯

We conclude the section by stating the following theorem of Nelson and Schreier without proof. For further study in the theory of presentation of groups, the reader is referred to combinatorial group theory by Magnus, Karras, and Solitor.

Theorem 10.4.11 *Every subgroup of a free group is free.*

Exercises

10.4.1 Show that the presentation $< x, y; x^3, y^4, (xy)^2 >$ describes the symmetric group of degree 4, and then show that the word problem is solvable for this presentation.

10.4.2 Show that the word and conjugacy problem are solvable for the presentation $< x_1, x_2, \ldots, x_n ; x_1^{r_1}, x_2^{r_2}, \ldots, x_n^{r_n} >$. Determine the number of conjugacy classes of elements of finite orders.

10.4.3 Let d be the g.c.d of m and $n^p - 1$. Show that the presentations

$$< x, y; x^m, y^p, y^{-1}xyx^{-n} >$$

and

$$< a, b; a^d, b^p, b^{-1}aba^{-n} >$$

define isomorphic groups.

10.4.4 Let $X = \{a_n \mid n \in N\}$ and $R = \{a_n(a_{n+1})^{-(n+1)}\}$. Solve the word problem for the presentation $< X; R >$. Show that it is a presentation of the additive group of rational numbers.

10.4.5 Find the order of the group having presentation

$$< a_1, a_2, \ldots, a_n; a_1a_2a_3^{-1}, a_2a_3a_4^{-1}, \ldots, a_{n-2}a_{n-1}a_n^{-1}, a_{n-1}a_na_1^{-1} >$$

for $n \leq 6$. Is it finite for $n \geq 7$?

10.4.6 Solve the word problem for the presentation $< a, b; a^{mn}, b^{mp}a^n b^{-nq} >$.

10.4.7 Identify the groups with following presentations.

(i) $< a, b; a^4, b^4, a^2b^2 >$.
(ii) $< a, b; a^2, b^2, (ab)^3 >$.

10.4.8 Find, if possible, the orders of the groups with following presentations.

(i) $< a, b; a^4, b^3, aba^3b >$.
(ii) $< a, b, c; a^4, b^4, c^3, a^2c^2, abc^2 >$.
(iii) $< a, b, c; a^2, b^3, c^5, abc >$.

10.4.9 Determine, if possible, the order of the commutator subgroups of the groups presented in Exercises 10.4.7 and 10.4.8.

10.4.10 Let $< X_1; \mathfrak{R}_1 >$ and $< X_2; \mathfrak{R}_2 >$ be presentations of a group G. Let $< X_1'; \mathfrak{R}_1' >$ and $< X_2'; \mathfrak{R}_2' >$ be presentations of a group G'. Assume that $X_1 \cap X_1' = \emptyset = X_2 \cap X_2'$ (note that there is no loss of generality in this assumption). Show that the presentations $< X_1 \cup X_1'; \mathfrak{R}_1 \cup \mathfrak{R}_1' >$ and $< X_2 \cup X_2'; \mathfrak{R}_2 \cup \mathfrak{R}_2' >$ define isomorphic groups. The group, thus obtained, is called the **free product** of G and G'. The free product of G and G' is denoted by $G \star G'$.
Hint. Use Theorem 10.4.8.

10.4.11 Show that there is an injective homomorphism i from G to $G \star G'$, and there is an injective homomorphism j from G' to $G \star G'$ such that if there is a group H together with a homomorphism μ from G to H, and a homomorphism ν from G' to H, then there is a unique homomorphism ϕ from $G \star G'$ to H such that $\phi o i = \mu$ and $\phi o j = \nu$.

Chapter 11
Arithmetic in Rings

This chapter is devoted to the study of rings in relation to their arithmetical properties.

11.1 Division in Rings

In this section, we introduce some arithmetical concepts in rings. Let R be a commutative integral domain with identity, and $R^\star$ denotes the set of nonzero elements of R. The multiplication in R is denoted by juxtaposition. Thus, the product of a and b is denoted by ab. An element $a \in R^\star$ is said to **divide** an element $b \in R^\star$, if there is an element $c \in R^\star$ such that $b = ac$. The notation a/b is used to say that a divides b. If a divides b, then we say that a is a **divisor** or a **factor** of b. We also say that b is a multiple of a.

Let u be a unit of R and $a \in R^\star$. Then $a = uu^{-1}a$. Since $u^{-1} \in R^\star$, u/a. Thus,

Proposition 11.1.1 *Every unit element of R divides every nonzero element of R.* ♯

Proposition 11.1.2 *Let $a, b \in R^\star$. The following two conditions are equivalent.*
(i) a/b and b/a.
(ii) a and b differ by a unit in the sense that there is a unit u such that $b = au$.

Proof $(i) \implies (ii)$. Assume (i). Then $b = ac$ and $a = bd$ for some $c, d \in R^\star$. But, then $b1 = b = bdc$ and $a1 = a = acd$. Since $a, b \in R^\star$, and R is an integral domain, we get $dc = 1 = cd$. Thus, a and b differ by a unit.

$(ii) \implies (i)$. Suppose that $b = au$, where u is a unit. Then clearly, a/b and also since $a = bu^{-1}$, b/a. ♯

We say that $a, b \in R^\star$ are **associates** to each other if they differ by a unit, or equivalently, a/b and b/a. The notation $a \sim b$ is used to say that a is an associate of b. It is easy to see that the relation $\sim$ of 'being associate to' is an equivalence

© Springer Nature Singapore Pte Ltd. 2017
R. Lal, *Algebra 1*, Infosys Science Foundation Series in Mathematical Sciences,
DOI 10.1007/978-981-10-4253-9_11

relation. Further, $a \sim b$ if and only if whenever a/c, b also divides c. Thus, while studying the arithmetic properties of rings, we need not distinguish associates.

Let $a \in R^*$. Then units of R and associates of a are always divisors of a. They are called **improper divisors**. Other divisors are called **proper divisors** of a.

A nonunit element $p \in R^*$ is called an **irreducible** element of R if it has no proper divisors. It is said to be a **prime** element of R if p/ab implies that p/a or p/b.

An element $d \in R^*$ is called a **greatest common divisor** (g.c.d) or a **greatest common factor** (g.c.f) of $a, b \in R^*$, if

(i) d/a and d/b, and

(ii) d'/a, d'/b implies that d'/d.

If d_1 and d_2 are greatest common divisors of $a, b \in R^*$, then d_1/d_2 and d_2/d_1. Thus, any two greatest common divisors of $a, b \in R^*$ are associates to each other. Therefore, if $a, b \in R^*$ has a greatest common divisor, then it is unique upto associate. A g.c.d of a and b will be denoted by (a, b).

An element $m \in R^*$ is called a **least common multiple** (l.c.m) of a and b, if

(i) a/m and b/m, and

(ii) a/m', b/m' implies that m/m'.

As in case of g.c.d, l.c.m, if exists, is unique up to associates. A l.c.m of a and b will be denoted by $[a, b]$.

The concepts defined above are well understood in the ring of integers right from our school days. The group $U(\mathbb{Z})$ of units of the ring of integers is the two element group $\{1, -1\}$. Here, irreducible and prime elements are same. Further, g.c.d and l.c.m exist, and there is an algorithm to find g.c.d and l.c.m of any two nonzero integers. Up to being associates, product of any two nonzero integers is same as the product of their g.c.d and l.c.m (see Chap. 3).

Example 11.1.3 The subset $\mathbb{Z}[i] = \{a + bi \mid a, b \in \mathbb{Z}\}$ of complex numbers under the usual addition and multiplication of complex numbers is a commutative integral domain with identity. The members of $\mathbb{Z}[i]$ are called Gaussian integers. The Gaussian integer $2 + 3i$ divides $-1 + 5i$, for $-1 + 5i = (2 + 3i)(1 + i)$. The Gaussian integer $2 + 3i$ does not divide $-1 + 4i$, for if $-1 + 4i = (2 + 3i)(a + bi)$, then taking the square of the moduli of each side, we get that $17 = 13(a^2 + b^2)$. This is impossible, for $(a^2 + b^2) \in \mathbb{N}$.

Let $a + bi$ be a unit of $\mathbb{Z}[i]$. Then, there is a $(c + di) \in \mathbb{Z}[i]$ such that $(a + bi)(c + di) = 1$. Again, taking the squares of the moduli of both sides, we get that $(a^2 + b^2)(c^2 + d^2) = 1$. Since $a^2 + b^2$ and $c^2 + d^2$ are natural numbers, $a^2 + b^2 = 1$. This implies that $a = \pm 1$ and $b = 0$ or $a = 0$ and $b = \pm 1$. Thus, the group $U(\mathbb{Z}[i])$ of units of $\mathbb{Z}[i]$ is the cyclic group $\{1, i, -1, -i\}$ of order 4. As such, the associates of $a + bi$ are $a + bi$, $-b + ai$, $-(a + bi)$, and $b - ai$.

$2 = 2 + 0i$ is not irreducible in $\mathbb{Z}[i]$, for $2 = (1 + i)(1 - i)$, where neither $1 + i$ nor $1 - i$ is a unit in $\mathbb{Z}[i]$. Observe that 2 is irreducible in the subring $\mathbb{Z}$ of $\mathbb{Z}[i]$.

The Gaussian integer $1 + i$ is irreducible in $\mathbb{Z}[i]$, for suppose that $(1 + i) = (a + bi)(c + di)$. Then, taking the squares of moduli of both sides, we get that $2 = (a^2 + b^2)(c^2 + d^2)$. Since $a^2 + b^2$ and $c^2 + d^2$ are natural numbers, $a^2 + b^2 = 1$ or $c^2 + d^2 = 1$. If $a^2 + b^2 = 1$, then $a = \pm 1$ and $b = 0$ or $a = 0$

and $b = \pm 1$. This means that $a + bi$ is unit. Similarly, in the other case, $c + di$ is a unit.

We shall show later in this chapter that irreducible elements and prime elements in $\mathbb{Z}[i]$ are same (as in the case of $\mathbb{Z}$). We will also show that any two nonzero elements of $\mathbb{Z}[i]$ have g.c.d as well as l.c.m and as in case of $\mathbb{Z}$, we have also algorithms to find them.

Example 11.1.4 (Arithmetic in $\mathbb{Z}[\sqrt{-5}]$). Consider the ring $\mathbb{Z}[\sqrt{-5}] = \{a + b\sqrt{-5} \mid a, b \in \mathbb{Z}\}$. Let $a + b\sqrt{-5}$ be a unit in $\mathbb{Z}[\sqrt{-5}]$. Then, there exists $c + d\sqrt{-5} \in \mathbb{Z}[\sqrt{-5}]$ such that $(a + b\sqrt{-5})(c + d\sqrt{-5}) = 1$. Equating the squares of moduli of each side, we get that $(a^2 + 5b^2)(c^2 + 5d^2) = 1$. Since $a^2 + 5b^2$ and $c^2 + 5d^2$ are natural numbers, $a^2 + 5b^2 = 1$. But, then $a = \pm 1$ and $b = 0$. This shows that the group $U(\mathbb{Z}[\sqrt{-5}])$ of units of $\mathbb{Z}[\sqrt{-5}]$ is $\{1, -1\}$. Thus, the associates of $a + b\sqrt{-5}$ in $\mathbb{Z}[\sqrt{-5}]$ are $\pm(a + b\sqrt{-5})$.

The element $2 = 2 + 0 \cdot \sqrt{-5}$ is irreducible in $\mathbb{Z}[\sqrt{-5}]$. For, suppose that $2 = (a + b\sqrt{-5})(c + d\sqrt{-5})$. Then, again equating the squares of moduli of each sides, we get that $4 = (a^2 + 5b^2)(c^2 + 5d^2)$. Since $a^2 + 5b^2$ and $c^2 + 5d^2$ are natural numbers, $a^2 + 5b^2 = 1$ or 2 or 4. Clearly, $a^2 + 5b^2$ can not be 2. If $a^2 + 5b^2 = 1$, then $a = \pm 1$ and $b = 0$ and in this case, $a + b\sqrt{-5} = \pm 1$ is a unit. If $a^2 + 5b^2 = 4$, then $c^2 + 5d^2 = 1$, and so in this case, $c + d\sqrt{-5}$ is a unit. This shows that 2 is irreducible. Using the same argument as above, it follows that $1 + \sqrt{-5}$, $1 - \sqrt{-5}$, and their associates are irreducible. Clearly, 2 does not divide $1 + \sqrt{-5}$, and it does not divide $1 - \sqrt{-5}$. But 2 divides $6 = (1 + \sqrt{-5})(1 - \sqrt{-5})$. This shows that 2 is not a prime element of $\mathbb{Z}[\sqrt{-5}]$. It also follows that $1 + \sqrt{-5}$, $1 - \sqrt{-5}$ and their associates are also not primes. Thus, contrary to the case in the domain $\mathbb{Z}$ of integers, an irreducible element in the integral domain $\mathbb{Z}[\sqrt{-5}]$ need not be a prime element.

Again, a pair of nonzero elements in $\mathbb{Z}[\sqrt{-5}]$ need not have a g.c.d. For example, consider $a = 6$ and $b = 2 \cdot (1 + \sqrt{-5})$. We show that a and b have no greatest common divisors. Suppose contrary. Let d be a greatest common divisor of a and b. Since $2/a$ and $2/b$, $2/d$. Let $d = 2 \cdot (m + n\sqrt{-5})$. Since d/b, $(m + n\sqrt{-5})/(1 + \sqrt{-5})$. Further, since $1 + \sqrt{-5}$ is irreducible, $m + n\sqrt{-5}$ is a unit or an associate of $1 + \sqrt{-5}$. This shows that $d \sim 2$ or $d \sim b$. Since $1 + \sqrt{-5}$ is a common divisor of a and b and $1 + \sqrt{-5}$ does not divide 2, d can not be an associate of 2. Again, since b does not divide a, d can not be an associate of b. Thus, such a d does not exist.

Finally, a pair of elements in $\mathbb{Z}[\sqrt{-5}]$ may have a g.c.d but no l.c.m. For example, consider the elements 2 and $1 + \sqrt{-5}$ of $\mathbb{Z}[\sqrt{-5}]$. Since 2 and $1 + \sqrt{-5}$ are both irreducible, and they are nonassociates, units and units are the only common divisors. Thus, 1 is a g.c.d of 2 and $1 + \sqrt{-5}$. We show that they do not have any l.c.m. Suppose contrary. Let m be a l.c.m of 2 and $1 + \sqrt{-5}$. Then, $m = 2(a + b\sqrt{-5})$ for some $a, b \in \mathbb{Z}$. Since 2 divides $2(1 + \sqrt{-5})$ and $1 + \sqrt{-5}$ also divides $2(1 + \sqrt{-5})$, it follows that $2(a + b\sqrt{-5})$ divides $2(1 + \sqrt{-5})$. But, then $a + b\sqrt{-5}$ divides $1 + \sqrt{-5}$. Since $1 + \sqrt{-5}$ is irreducible, $a + b\sqrt{-5}$ is a unit or an associate of $1 + \sqrt{-5}$. Hence, $m \sim 2$ or $m \sim 2(1 + \sqrt{-5})$. Since the common divisor $1 + \sqrt{-5}$ does not divide 2, m can not be an associate of 2. Also since $2(1 + \sqrt{-5})$ does not divide, the common multiple 6 of 2 and $1 + \sqrt{-5}$, it can not be an associate of $2(1 + \sqrt{-5})$. This proves the impossibility of the existence of l.c.m of 2 and $1 + \sqrt{-5}$.

Remark 11.1.5 It becomes evident from the above example that the arithmetic of $\mathbb{Z}[\sqrt{-5}]$ is significantly different from that of $\mathbb{Z}$ and $\mathbb{Z}[i]$. In general, an irreducible element need not be a prime element. Also g.c.d (or l.c.m) need not exist in general.

Proposition 11.1.6 *Every prime element is irreducible.*

Proof Let p be a prime element, and suppose that $p = ab$. Then, $p/(ab)$. Since p is assumed to be a prime element, p/a or p/b. Suppose that p/a. Then, $a = pc$ for some $c \in R^*$. Hence, $p = pcb$. By the restricted cancelation law in an integral domain, $cb = 1$. Since R is commutative, $bc = 1$ and b is a unit. Thus, $p \sim a$. Similarly, if p/b, then a is a unit and $p \sim b$. This shows that p has no proper divisors.   ♯

Proposition 11.1.7 *Let R be a commutative integral domain with identity. Let $a, b \in R^*$. Let d be a greatest common divisor and m a least common multiple of a and b. Then $ab \sim md$.*

Proof Since $a/(a \cdot b)$ and $b/(a \cdot b)$, $m/(a \cdot b)$. Suppose that $a \cdot b = m \cdot c$. Since a/m and b/m, c/a and also c/b. But, then c/d. Hence, $a \cdot b = m \cdot c$ divides $m \cdot d$. Now, we show that $m \cdot d$ divides $a \cdot b$. Since d is a g.c.d of a and b, $a = d \cdot u$ and $b = d \cdot v$ for some $u, v \in R^*$. But, then $a \cdot b = d \cdot d \cdot u \cdot v$. Clearly, $a/(d \cdot u \cdot v)$ and $b/(d \cdot u \cdot v)$. Hence, $m/(d \cdot u \cdot v)$. This shows that $m \cdot d$ divides $a \cdot b$.   ♯

Proposition 11.1.8 *Let R be a commutative integral domain with identity and $a, b, c \in R^*$. Then*
(i) $(c \cdot a, c \cdot b) \sim c \cdot (a, b)$,
and
(ii) $((a, b), c) \sim (a, (b, c))$
provided that both sides exist.

Proof (i) Since $c/(ca)$ and $c/(cb)$, $c/(ca, cb)$. Suppose that $(ca, cb) = cd$. Then, $(cd)/(ca)$ and $(cd)/(cb)$. Hence, d/a and d/b. Thus, cd divides $c(a, b)$, and so (ca, cb) divides $c(a, b)$. Also, since $(a, b)/a$ and $(a, b)/b$, $c(a, b)$ divides ca, and $c(a, b)$ divides cb. This shows that $c(a, b)$ divides (ca, cb). Thus, $c(a, b) \sim (ca, cb)$.
     The proof of (ii) is left as an exercise.   ♯

Corollary 11.1.9 *Let R be a commutative integral domain with identity. Suppose that g.c.d of any two nonzero elements of R exist in R. Then, every irreducible element of R is a prime element.*

Proof Suppose that g.c.d of any two nonzero elements of R exist. Let p be an irreducible element of R. Suppose that p does not divide a, and also p does not divide b. Since only divisors of p are units and associates of p, it follows that $(p, a) \sim 1 \sim (p, b)$. Now, using the above proposition, we have

$$(p, ab) \sim ((p, pa), ab) \sim (p, (pa, pb)) \sim (p, a(p, b)) \sim (p, a) \sim 1.$$

Thus, p does not divide ab. This shows that p is a prime element.   ♯

Arithmetic and Ideals.

Let R be a commutative integral domain with identity. Let $a \in R$. Then, the ideal of R generated by a is

$$Ra = \{xa \mid x \in R\}.$$

An ideal generated by a single element is called a **principal ideal**. Clearly, a is a unit if and only if $Ra = R$.

Proposition 11.1.10 *Let R be a commutative integral domain with identity. Then the following hold:*

(i) a/b if and only if $Rb \subseteq Ra$.

(ii) $a \sim b$ if and only if $Ra = Rb$.

(iii) a is a proper divisor of b if and only if Rb is properly contained in Ra and $Ra \neq R$.

(iv) m is l.c.m of a and b if and only if $Rm = Ra \cap Rb$.

(v) An element $d \in R$ is a g.c.d of a and b if and only if Rd is the smallest principal ideal containing the ideal $Ra + Rb$ generated by a and b.

Proof (i) Suppose that a/b. Then, $b = ca$ for some $c \in R^*$. But, then $Rb = Rca \subseteq Ra$. Conversely, if $Rb \subseteq Ra$, then since $b \in Rb$, $b \in Ra$. This implies that $b = ca$ for some $c \in R$.

(ii) We know that $a \sim b$ if and only if a/b and b/a. The result follows from (i).

(iii) a is a proper divisor of b if and only if a is nonunit divisor of b and b does not divide a. This is equivalent to say that $Ra \neq R$, $Ra \neq Rb$, and $Rb \subseteq Ra$.

(iv) Suppose that m is a l.c.m of a and b. Then, a/m and b/m and so $Rm \subseteq Ra \cap Rb$. Further, if $x \in Ra \cap Rb$, then a/x and b/x. Since m is l.c.m, m/x. But, then $x \in Rm$. This shows that $Rm = Ra \cap Rb$. Conversely, suppose that $Rm = Ra \cap Rb$. Then, $Rm \subseteq Ra$ and $Rm \subseteq Rb$, and hence, a/m and b/m. Also if a/n and b/n, then $n \in Ra \cap Rb = Rm$. But, then m/n. This shows that m is l.c.m of a and b.

(v) Suppose that d is a g.c.d of a and b. Then, d/a and d/b. Hence, d divides $xa + yb$ for all $x, y \in R$. This means that $Ra + Rb \subseteq Rd$. Next, if $Ra + Rb \subseteq Rd'$, then $Ra \subseteq Rd'$ and $Rb \subseteq Rd'$. This means that d'/a and d'/b. Since d is g.c.d of a and b, d'/d and so $Ra + Rb \subseteq Rd \subseteq Rd'$.

Conversely, let Rd be the smallest principal ideal containing $Ra + Rb$. Then, $Ra \subseteq Rd$ and $Rb \subset Rd$. This means that d/a and d/b. Further, if d'/a and d'/b. Then, $Ra \subseteq Rd'$, and $Rb \subseteq Rd'$. But, then $Ra + Rb \subseteq Rd'$. Since Rd is the smallest principal ideal containing $Ra + Rb$, $Rd \subseteq Rd'$. This means that d'/d, and so d is a g.c.d of a and b.   ♯

Remark 11.1.11 Clearly, $Ra + Rb = Rd$ implies that d is a g.c.d of a and b. However, d is a g.c.d of a, and b does not imply that $Ra + Rb = Rd$: For example, consider the polynomial ring $\mathbb{Z}[X]$. It is easy to see that 2 and X are irreducible elements of $\mathbb{Z}[X]$. Hence, 1 is g.c.d of 2 and X. But $\mathbb{Z}[X] \cdot 2 + \mathbb{Z}[X] \cdot X$ is the set of polynomials over $\mathbb{Z}$ whose constant terms are divisible by 2. Thus, $\mathbb{Z}[X] \cdot 2 + \mathbb{Z}[X] \cdot X \neq \mathbb{Z}[X] \cdot 1$.

Exercises

11.1.1 Find out the group of units of $\mathbb{Z}[\sqrt{-3}]$.

11.1.2 Show that 2 and $1 + \sqrt{-3}$ are irreducible in $\mathbb{Z}[\sqrt{-3}]$ but they are not primes. Do they have l.c.m?

11.1.3 Show that X is prime element in $\mathbb{Z}[X]$. Show that the ideal generated by X is a prime ideal but it is not maximal.

11.1.4 Let $a, b \in \mathbb{Z}^*$. Show that $a + bX$ is irreducible in $\mathbb{Z}[X]$ if and only if $(a, b) = 1$.

11.1.5 Show that $X^2 + 1$ is irreducible in $\mathbb{R}[X]$, where $\mathbb{R}$ is the field of real numbers. Show further that $aX^2 + bX + c$ is irreducible if and only if $b^2 - 4ac < 0$.

11.1.6 Show that $X^2 + \bar{a}$ is irreducible in $\mathbb{Z}_p[X]$ if and only if $(\frac{a}{p}) = -1$.

11.1.7 Show that every ideal in $\mathbb{Z}$ is a principal ideal. What are prime ideals of $\mathbb{Z}$? What are maximal ideals of $\mathbb{Z}$?

11.1.8 Show that $[ca, cb] \sim c[a, b]$ provided that both side exist.

11.1.9 Show that $[[a, b], c] \sim [a, [b, c]]$ provided that both side exist.

11.1.10 Find out the group of units of $\mathbb{Z}[\sqrt{-7}]$. Show that $3 + \sqrt{-7}$ is irreducible. Is it prime?

11.1.11 Show that $X^2 + \bar{1}$ is irreducible in $\mathbb{Z}_p[X]$ if and only if $p \equiv 1 (mod\ 4)$.

11.1.12 Show that $\mathbb{Z}[\sqrt{-5}]$ and $\mathbb{Z}[i]$ are not isomorphic.

11.1.13 Consider the integral domain $\mathbb{Z}[\omega] = \{a + b\omega \mid a, b \in \mathbb{Z}\}$, where ω is a primitive cube root of 1. Show that $a + b\omega$ is a unit in $\mathbb{Z}[\omega]$ if and only if $a^2 - ab + b^2 = 1$.

11.1.14 Find the group of units of $\mathbb{Z}[\omega]$.

11.1.15 Find the group of units of $\mathbb{Z}[\sqrt{-2}]$.

11.1.16 Show that $a + bi$ is irreducible in $\mathbb{Z}[i]$ provided that $a^2 + b^2$ is an irreducible integer.

11.1.17 Show that if $a^2 - ab + b^2$ is irreducible integer, then $a + b\omega$ is irreducible in $\mathbb{Z}[\omega]$.

11.1.18 Suppose that $a^2 + 2b^2$ is an irreducible integer. Show that $a + b\sqrt{-2}$ is an irreducible element of $\mathbb{Z}[\sqrt{-2}]$.

11.1.19 Let $a, b \in \mathbb{Z}$. Show that a/b in $\mathbb{Z}$ if and only if a/b in $\mathbb{Z}[i]$.

11.1.20 Let $a, b \in \mathbb{Z}$. Show that g.c.d of a and b in $\mathbb{Z}$ is same as a g.c.d of a and b in $\mathbb{Z}[i]$.

11.1.21 Determine the integral points on the circle $x^2 + y^2 = p$, and also the integral points on the ellipse $x^2 + 5y^2 = p$ for all primes $p \leq 19$.

11.2 Principal Ideal Domains

Definition 11.2.1 A commutative integral domain R with identity is called a **principal ideal domain** if every ideal of R is a principal ideal. If R is a principal ideal domain, then in short we express it by saying that R is a P.I.D.

Thus, a commutative integral domain with identity is a principal ideal domain if and only if every ideal of R is of the form Ra for some $a \in R$. The ring $\mathbb{Z}$ of integers is a principal ideal domain, for any ideal (indeed, any subgroup) of $\mathbb{Z}$ is of the $m\mathbb{Z}$ for some integer m. Every field is trivially a principal ideal domain, for the only ideals are $\{0\}$ and the field itself. A little later, we shall have more examples of principal ideal domains.

Proposition 11.2.2 *Greatest common divisor of any pair of nonzero elements in a P.I.D. exist. Further more, if d is a g.c.d of a and b in a P.I.D R, then $d = ua + vb$ for some $u, v \in R$.*

Proof Let R be a principal ideal domain and a, b be nonzero elements of R. Since R is a principal ideal domain, the ideal $Ra + Rb$ is a principal ideal. Hence, $Ra + Rb = Rd$ for some $d \in R$. From Proposition 11.1.10, d is the g.c.d of a and b. Since $d \in Rd = Ra + Rb$, $d = uv + vb$ for some $u, v \in R$. ♯

Corollary 11.2.3 *In a principal ideal domain, every irreducible element is prime element.*

Proof Follows from Proposition 11.2.2 and Corollary 11.1.9. ♯

Thus, $\mathbb{Z}[\sqrt{-5}]$ is not a P.I.D. Observe that a subring of a P.I.D need not be a P.I.D, for example, the field $\mathbb{C}$ of complex numbers is a P.I.D whereas the subring $\mathbb{Z}[\sqrt{-5}]$ is not a P.I.D. However,

Proposition 11.2.4 *If a homomorphic image (and hence, a difference ring) of a P.I.D is an integral domain, then it is also a P.I.D.*

Proof Let f be a surjective homomorphism from a P.I.D R to an integral domain R'. Clearly, R' is a commutative integral domain with identity. Let A be an ideal of R'. Then, $f^{-1}(A)$ is an ideal of R. Since R is a P.I.D, $f^{-1}(A)$ is generated by a single element a (say). Then, $A = f(f^{-1}(A))$ is generated by $f(a)$. Thus, every ideal of R' is a principal ideal.   ♯

Proposition 11.2.5 *Let R be a P.I.D and $p \in R$. Then the following conditions are equivalent.*
   (i) p is an irreducible element.
   (ii) p is a prime element.
   (iii) Rp is a maximal ideal of R.
   (iv) R/Rp is a field.

Proof Equivalence of (i) and (ii) follows from Proposition 11.1.6, and Corollary 11.2.3. Equivalence of (iii) and (iv) follows from the fact that M is a maximal ideal if and only if R/M is a field. We prove the equivalence of (i) and (iii).

   Assume (i) Let p be an irreducible element, and A an ideal containing Rp. Suppose that $A \neq Rp$. Then, there exists an element $a \in A - Rp$. Clearly, p does not divide a. Since p is irreducible, $(p, a) = 1$. From Proposition 11.2.2, $1 = xa + yp$ for some $x, y \in R$. Since A is an ideal containing a and p, $1 \in A$. Hence, $A = R$. Thus, Rp is a maximal ideal.

   Assume that Rp is a maximal ideal and a/p. Then, $Rp \subseteq Ra$. Since Rp is a maximal ideal, $Rp = Ra$ or $Ra = R$. Hence, $p \sim a$ or a is a unit. This shows that p is irreducible.   ♯

Theorem 11.2.6 *The polynomial ring $R[X]$ is a P.I.D if and only if R is a field.*

Proof Suppose that R is a field and A an ideal of $R[X]$. We have to show that A is a principal ideal. If $A = \{0\}$, then there is nothing to do. Suppose that $A \neq \{0\}$. Let $S = \{deg(f(X)) \mid f(X) \in A - \{0\}\}$. Clearly, S is a nonempty subset of $\mathbb{N} \bigcup \{0\}$. By the well-ordering principle in $\mathbb{N} \bigcup \{0\}$, S has the least element $deg(h(X))$ (say), where $h(X) \in A - \{0\}$. We show that $A = R[X]h(X)$. Clearly, $R[X]h(X) \subseteq A$. Let $f(X) \in A$. By the division algorithm in $R[X]$, there exist $q(X), r(X) \in R[X]$ such that

$$f(X) = q(X)h(X) + r(X),$$

where $r(X) = 0$ or else $deg(r(X)) < deg(h(X))$. Since $f(X), h(X) \in A$ and A is an ideal, $r(X) = f(X) - q(X)h(X)$ belongs to A. If $r(X) \neq 0$, then $deg(r(X)) \in S$ and $deg(r(X)) < deg(h(X))$. This is a contradiction to the choice of $h(X)$. Hence, $r(X) = 0$, and $f(X) = q(X)h(X) \in R[X]h(X)$.

   Conversely, suppose that $R[X]$ is a P.I.D. We first observe that X is irreducible in $R[X]$. Suppose that $X = f(X)g(X)$. Comparing the degrees, we may assume that $deg(f(X)) = 0$ and $deg(g(X)) = 1$. Suppose that $f(X) = u \in R$ and $g(X) = vX + w$. Comparing the coefficient of X, we see that $uv = 1$, and so $f(X) = u$ is a unit. Now, let $a \in R^{\star}$. Since $R[X]$ is assumed to be a P.I.D, a

and X will have a g.c.d. Since X is irreducible $(a, X) = 1$. By Proposition 11.2.2, $1 = af(X) + Xg(X)$ for some $f(X), g(X) \in R[X]$. Equating the constant term of both sides, we get that $1 = ab$ for some $b \in R$. This shows that a is invertible, and so R is a field.   ♯

Example 11.2.7 $\mathbb{Z}[X]$ is not a P.I.D, for $\mathbb{Z}$ is not a field. $\mathbb{R}[X]$ is a P.I.D, where $\mathbb{R}$ denotes the field of real numbers.

Proposition 11.2.8 *Let F be a field. Then, the following hold:*
(i) Any polynomial $aX + b$ of degree 1 is irreducible in $F[X]$.
(ii) Let $f(X)$ be a polynomial in $F[X]$ of degree greater than 1 which has a root in F. Then, $f(X)$ is reducible in $F[X]$
(iii) A polynomial $f(X)$ in $F[X]$ of degree 2 or degree 3 is irreducible if and only if it has no root in F.

Proof (i) Since the units of $F[X]$ are precisely those of F (Proposition 7.6.7), a polynomial $aX + b$ of degree 1 can not be a unit in $F[X]$. Again, since all nonunit elements of $F[X]$ are of positive degrees, and a polynomial of degree 1 can not be expressed as product of polynomials of positive degrees, it follows that any polynomial of degree 1 is irreducible.
(ii) Let $f(X)$ be a polynomial of degree at least 2, and let a be a root of $f(X)$. By the factor theorem (Corollary 7.6.17), it follows that $f(X) = (X - a)g(x)$ for some polynomial $g(X)$. Comparing the degrees, we see that $g(X)$ is of positive degree, and so it is not a unit. It follows that $f(X)$ is product of two nonunits.
(iii) Let $f(X)$ be a polynomial of degree 2 or 3. Suppose that $f(X) = g(X)h(X)$, where $g(X)$ and $h(X)$ are nonconstant polynomials. Comparing the degrees of both sides we see that $\deg g(X) = 1$ or $\deg h(X) = 1$. Suppose that $\deg g(X) = 1$. Then, $g(X) = aX + b$ with $a \neq 0$. But, then $-b/a$ is a root of $g(X)$, and hence, it is also a root of $f(X)$. This shows that $f(X)$ is reducible if and only if it has a root in F.   ♯

Example 11.2.9 $X^2 + 1$ is an irreducible element in $\mathbb{R}[X]$, for it has no root in $\mathbb{R}$. In general, a quadratic polynomial $aX^2 + bX + c$ is an irreducible element in $\mathbb{R}[X]$ if and only if $b^2 - 4ac < 0$. Since every cubic polynomial over $\mathbb{R}$ has a root in $\mathbb{R}$, it is reducible. Fundamental theorem of Algebra ensures that a polynomial with complex coefficients has all its roots in the field $\mathbb{C}$ of complex numbers. In particular, the irreducible elements in $\mathbb{C}[X]$ are precisely the linear polynomials. Further, it is evident that the conjugate of a complex root of a polynomial with real coefficients is again a root of the polynomial (this is because $f(\overline{z}) = \overline{f(z)}$ for all real polynomials $f(X)$). Again, for any complex number z, the polynomial $(X - z)(X - \overline{z}) = X^2 - (z + \overline{z}) + z\overline{z}$ is a real polynomial. In turn, it follows that any polynomial in $\mathbb{R}[X]$ of degree greater than 2 is reducible in $\mathbb{R}[X]$.

Let F be a field and $f(X)$ be an irreducible polynomial in $F[X]$ of degree n. Let $F[X]_{f(X)}$ denotes the set of polynomials of degrees less than n together with 0 (equivalently, the set of remainders obtained when polynomials are divided by

$f(X)$). The addition of polynomials induces the addition in $F[X]_{f(X)}$. Further, define a multiplication $\bullet$ in $F[X]_{f(X)}$ by defining $g(X) \bullet h(X)$ to be the remainder obtained when the usual product $g(X)h(X)$ of $g(X)$ and $h(X)$ is divided by $f(X)$.

Proposition 11.2.10 *The triple* $(F[X]_{f(X)}, +, \bullet)$ *is a field.*

Proof Clearly, $(F[X]_{f(X)}, +)$ is an abelian group. By the definition $r(X) \bullet (s(X) + t(X))$ is the remainder obtained when $r(X)(s(X) + t(X))$ is divided by $f(X)$. Further, the remainder obtained when the sum of the remainders of $r(X)s(X)$ and that of $r(X)t(X)$ is divided by $f(X)$ is same as the remainder obtained when $r(X)(s(X) + t(X))$ is divided by $f(X)$. Thus, $r(X) \bullet (s(X) + t(X)) = r(X) \bullet s(X) + r(X) \bullet t(X))$. Again, since $r(X)s(X) = s(X)r(X)$, it follows that $r(X) \bullet s(X) = s(X) \bullet r(X)$. The constant polynomial 1 is the identity element with respect to $\bullet$. Finally, we need to show that any non zero element in $F[X]_{f(X)}$ has a inverse with respect to the multiplication $\bullet$. Let $r(X)$ be a nonzero element in $F[X]_{f(X)}$. Since $deg\, r(X) < deg\, f(X)$ and $f(X)$ is irreducible, a greatest common divisor of $r(X)$ and $f(X)$ is 1. Since $F[X]$ is a P.I.D. (Theorem 11.2.6), by Proposition 11.2.2, there exist polynomials $u(X)$ and $v(X)$ in $F[X]$ such that

$$1 = u(X)r(X) + v(X)f(X).$$

It is evident that if we divide $u(X)r(X)$ by $f(X)$, the remainder obtained is 1. Let $s(X)$ be the remainder obtained when $u(X)$ is divided by $f(X)$. Then, $s(X) \in F[X]_{f(X)}$ and the remainder obtained when $s(X)r(X)$ is divided by $f(X)$ is 1. Thus, $s(X) \bullet r(X) = 1$. This shows that $(F[X]_{f(X)}, +, \bullet)$ is a field.   ♯

Corollary 11.2.11 *The field* $(F[X]_{f(X)}, +, \bullet)$ *is isomorphic to the quotient field* $F[X]/F[X]f(X)$. *Indeed, the map η from* $F[X]_{f(X)}$ *to* $F[X]/F[X]f(X)$ *given by* $\eta(r(X)) = r(X) + F[X]f(X)$ *is an isomorphism.*

Proof Let $g(X) + F[X]f(X)$ be an arbitrary element of $F[X]/F[X]f(X)$, where $g(X)$ is a polynomial in $F[X]$. Let $r(X)$ be the remainder obtained when $g(X)$ is divided by $f(X)$. Then, $g(X) - r(X)$ is a multiple of $f(X)$ and so it belongs to $F[X]f(X)$. Hence, $g(X) + F[X]f(X) = r(X) + F[X]f(X) = \eta(r(X))$. This shows that η is surjective map. Again, let $r(X)$ and $s(X)$ are polynomials in $F[X]_{f(X)}$ such that $\eta(r(X)) = \eta(s(X))$. Then, $r(X) - s(X) \in F[X]f(X)$. Since $r(X)$ and $s(X)$ are of degree less than that of $f(X)$, it follows that $r(X) = s(X)$. This shows that η is injective. Also, $\eta(r(X)) + \eta(s(X)) = (r(X) + F[X]f(X)) + (s(X) + F[X]f(X)) = (r(X) + s(X)) + F[X]f(X) = \eta(r(X) + s(X))$. Finally, since $r(X) \bullet s(X)$ is the remainder obtained when $r(X)s(X)$ is divided by $f(X)$, it follows that $\eta(r(X) \bullet s(X)) = (r(X) \bullet s(X)) + F[X]f(X) = r(X)s(X) + F[X]f(X) = (r(X) + F[X]f(X))(s(X) + F[X]f(X)) = \eta(r(X))\eta(s(X))$. This shows that η is a bijective homomorphism.   ♯

Example 11.2.12 Since $X^2 + 1$ has no root in $\mathbb{R}$, it follows that it is irreducible in $\mathbb{R}[X]$. Consider $\mathbb{R}[X]_{X^2+1} = \{a + bX \mid a, b \in \mathbb{R}\}$. The addition $+$ and the multiplication $\bullet$ in $\mathbb{R}[X]_{X^2+1}$ as described in the above proposition are given by

$$(a + bX) + (c + dX) = (a + c) + (b + dX)$$

and

$$(a + bX) \bullet (c + dX) = r + sX,$$

where $r + sX$ is the remainder obtained when $(a + bX)(c + dX)$ is divided by $X^2 + 1$. Now, $(a + bX)(c + dX) = bdX^2 + (ad + bc)X + ac$. Dividing it by $X^2 + 1$, we get the remainder $(bd - ac) + (ad + bc)X$. Thus,

$$(a + bX) \bullet (c + dX) = (bd - ac) + (ad + bc)X.$$

In particular, $X^2 = -1$. It is easily observed that the map $(a + bX) \longrightarrow (a + bi)$ defines an isomorphism from the field $\mathbb{R}[X]_{X^2+1}$ to the field $\mathbb{C}$ of complex numbers. In turn, the field $\mathbb{R}[X]_{X^2+1}$ can be called the field of complex numbers.

Example 11.2.13 Consider the field $\mathbb{Z}_2 = \{\bar{0}, \bar{1}\}$ of residue classes modulo 2. Clearly, the polynomial $X^2 + X + \bar{1}$ in $\mathbb{Z}_2[X]$ has no root in $\mathbb{Z}_2$, and as such it is irreducible in $\mathbb{Z}_2[X]$. For simplicity, let us denote $\mathbb{Z}_2[X]_{X^2 + X + \bar{1}}$ by F_4. Thus, F_4 is the set of polynomials of degree at most 1 together with $\bar{0}$. More explicitly, $F_4 = \{\bar{0}, \bar{1}, X, \bar{1} + X\}$. The addition and the multiplication $\bullet$ are given by the following table:

+	$\bar{0}$	$\bar{1}$	X	$\bar{1} + X$
$\bar{0}$	$\bar{0}$	$\bar{1}$	X	$\bar{1} + X$
$\bar{1}$	$\bar{1}$	$\bar{0}$	$\bar{1} + X$	X
X	X	$\bar{1} + X$	$\bar{0}$	$\bar{1}$
$\bar{1} + X$	$\bar{1} + X$	X	$\bar{1}$	$\bar{0}$

$\bullet$	$\bar{0}$	$\bar{1}$	X	$\bar{1} + X$
$\bar{0}$	$\bar{0}$	$\bar{0}$	$\bar{0}$	$\bar{0}$
$\bar{1}$	$\bar{0}$	$\bar{1}$	X	$\bar{1} + X$
X	$\bar{0}$	X	$\bar{1} + X$	$\bar{1}$
$\bar{1} + X$	$\bar{0}$	$\bar{1} + X$	$\bar{1}$	X

Since $X^2 + \bar{1}$ has no root in $\mathbb{Z}_3$, as in the above example we have the following:

Example 11.2.14 $X^2 + \bar{1}$ is irreducible in $\mathbb{Z}_3[X]$, and the field $\mathbb{Z}_3[X]_{X^2+\bar{1}}$ is a field of order 9.

Remark 11.2.15 In the chapter on Galois theory in the vol 2 of the book, we shall see that for every prime p and every natural number n, there is an irreducible polynomial of degree n in $\mathbb{Z}_p[X]$, and so also a field of order p^n. Indeed, we have a nice formula to find the number of irreducible polynomials of degree n in $\mathbb{Z}_p[X]$. All the fields of order p^n are isomorphic.

Exercises

11.2.1 Show that in a P.I.D., l.c.m of any pair of nonzero elements exist, and if m is l.c.m of a and b, then $mua + mvb = ab$ (or equivalently $v/a + u/a = 1/m$) for some u, v in the P.I.D.

11.2.2 Show that $X^2 + \bar{1}$ is irreducible in $\mathbb{Z}_3[X]$, and $\mathbb{Z}_3[X]_{X^2 + \bar{1}}$ is a field of order 9. Describe its elements, and also the operations of addition and multiplication in $\mathbb{Z}_3[X]_{X^2 + \bar{1}}$.

11.2.3 Show that $X^3 + X^2 + \bar{1}$ is irreducible in $\mathbb{Z}_2[X]$, and so is $X^3 + X + \bar{1}$. Describe the corresponding fields $\mathbb{Z}_2[X]_{X^3+X^2+\bar{1}}$ and $\mathbb{Z}_2[X]_{X^3+X+\bar{1}}$. Show that they are isomorphic.

11.2.4 Show that $X^3 + X^2 + X + \bar{1}$ is irreducible in $\mathbb{Z}_3[X]$. Describe the corresponding field which is of order 27. What are other irreducible polynomials of degree 3 in $\mathbb{Z}_3[X]$?

11.2.5 Let $f(X)$ be an irreducible polynomial in $F_q[X]$ of degree n, where F_q is a field of order q. Show that $F_q[X]/F_q[X](f(X))$ is a field of order q^n.

11.2.6 Show that $X^2 + \bar{a}$ is irreducible in $\mathbb{Z}_p[X]$, p an odd prime if and only if $(\frac{a}{p}) = -1$. In this case, $\mathbb{Z}_p[X]_{X^2+\bar{a}}$ is a field of order p^2.

11.2.7 Show that $X^2 + \overline{10}$ is irreducible in $\mathbb{Z}_p[X]$ if and only if the remainder obtained when the prime p is divided by 40 is different from 1, 3, 9, 13, 27, 31, 37 and 39. Thus, it is irreducible, for example, in $\mathbb{Z}_{43}[X]$.

11.2.8 Describe primes p for which $X^2 + \overline{15}$ is irreducible in $\mathbb{Z}_p[X]$.

11.2.9 Show that $X^2 + X + 1$ is irreducible in $\mathbb{R}[X]$. Describe the corresponding field. Exhibit an explicit isomorphism from this field to the field $\mathbf{C}$ of complex numbers.

11.2.10 Show that $\mathbb{C}[X, Y]/\mathbb{C}[X, Y](X^2 + Y^2 - 1)$ is a P.I.D. (assume that every polynomial over $\mathbb{C}$ has a root). Is this result true if we replace $\mathbb{C}$ by the field $\mathbb{R}$ of real numbers?

11.2.11 Show that $\mathbb{Q}[X]/\mathbb{Q}[X](X^3+3X^2-8)$ is a field. Is $\mathbb{R}[X]/\mathbb{R}[X](X^3+3X^2-8)$ a field?

11.2.12 What can we say about union of a chain of P.I.Ds? Is it always a P.I.D?

11.2.13 Let R be a P.I.D. Let $a, b \in R^\star$. Let d be a g.c.d and m a l.c.m of a and b. Show that Rm is an ideal of Ra, and Rb an ideal of Rd. Further show that $Ra/Rm \approx Rd/Rb$.

11.2.14 Show that $\mathbb{Z}[\sqrt{-5}]$ is not a PID.

11.3 Euclidean Domains

Definition 11.3.1 A pair (R, δ), where R is a commutative integral domain and δ a map from R^* to $\mathbb{N} \bigcup \{0\}$, is called an **Euclidean Domain**, if given $a, b \in R$, $b \neq 0$, there exist $q, r \in R$ such that

$$a = bq + r,$$

where $r = 0$ or else $\delta(r) < \delta(b)$.

Example 11.3.2 $(\mathbb{Z}, |\ |)$, where $|\ |$ is the absolute value map, is an Euclidean domain(follows from the division algorithm theorem). Further, $(\mathbb{Z}, \delta)$, where $\delta(a) = a^2$, is another Euclidean domain structure on $\mathbb{Z}$. Thus, on a commutative integral domain, there can be several Euclidean domain structures.

Example 11.3.3 Let F be a field. Then, $(F[X], deg)$ is an Euclidean domain (follows from the division algorithm in $F[X]$). $(F[X], \delta)$, where $\delta(f(X)) = 2^{deg(f(X))}$ is also an Euclidean domain (This fact also follows from the division algorithm).

Example 11.3.4 The pair $(\mathbb{Z}[i], \delta)$, where $\mathbb{Z}[i]$ is the ring of Gaussian integers, and $\delta(a + bi) = |a + bi|^2 = a^2 + b^2$, is an Euclidean domain.

Proof Let $n \in \mathbb{Z}$ and $m \in \mathbb{N}$. Then, by the division algorithm, there exist q and r in $\mathbb{Z}$ such that

$$n = mq + r, \ 0 \leq r < m, \text{ and so also}$$
$$n = m(q + 1) + r - m, \ 0 \leq r < m.$$

Clearly, $|r| \leq \frac{m}{2}$ or $|r - m| < \frac{m}{2}$. Thus, there exist integers u and v such that

$$n = mu + v, \ 0 \leq |v| \leq \frac{m}{2}.$$

Let $x = a + bi$ and $y = c + di$ be members of $\mathbb{Z}[i]$ with $y \neq 0$. Then, $|y|^2 = c^2 + d^2 \in \mathbb{N}$. From what we observed above, it follows that there exist integers α, β, u, and v such that

$$ac + bd = \alpha(c^2 + d^2) + \beta, \ 0 \leq |\beta| \leq \frac{c^2 + d^2}{2} \tag{11.1}$$

and

$$bc - ad = u(c^2 + d^2) + v, \ 0 \leq |v| \leq \frac{c^2 + d^2}{2}. \tag{11.2}$$

Thus,

$$xy = (a + bi)(c - di) = (ac + bd) + (bc - ad)i =$$
$$(\alpha + ui)(c^2 + d^2) + \beta + iv = (\alpha + ui)y\overline{y} + \beta + iv.$$

Hence,

$$x = (\alpha + ui)y + \frac{(\beta + iv)y}{c^2 + d^2}.$$

Put

$$\frac{(\beta + iv)y}{(c^2 + d^2)} = r.$$

Then, since x, y and $\alpha + ui \in \mathbb{Z}[i]$, $r \in \mathbb{Z}[i]$. Also $\delta(r) = \frac{\beta^2 + v^2}{c^2 + d^2}$. From Eqs. (11.1) and (11.2), we see that $\delta(r) < c^2 + d^2 = \delta(y)$. This shows that $(\mathbb{Z}[i], \delta)$ is an Euclidean domain.   ♯

Example 11.3.5 Let ω denotes a primitive cube root of unity. Thus, $\omega \neq 1$ and $\omega^3 = 1$. Clearly, ω^2 is also a primitive cube root of unity. We have $\overline{\omega} = \omega^2$, and $1 + \omega + \omega^2 = 0$. Consider the set $\mathbb{Z}[\omega] = \{a + b\omega \mid a, b \in \mathbb{Z}\}$. Then, $\mathbb{Z}[\omega]$ is a subring of the field $\mathbb{C}$ of complex numbers (verify). Thus, $\mathbb{Z}[\omega]$ is a commutative integral domain with identity. Define a map δ from $\mathbb{Z}[\omega]$ to $\mathbb{N} \bigcup \{0\}$ by

$$\delta(a + b\omega) = a^2 - ab + b^2 = (a + b\omega)\overline{(a + b\omega)}.$$

(Note that $\overline{(a + b\omega)} = a + b\omega^2$, and so $a^2 - ab + b^2$ is always a nonnegative integer). Then, $(\mathbb{Z}[\omega], \delta)$ is an Euclidean domain.

Proof Let $a + b\omega$ and $c + d\omega \neq 0$ be members of $\mathbb{Z}[\omega]$. As in the above example, we have integers α, β, u, and v such that

$$ac + bd - ad = \alpha(c^2 + d^2 - cd) + \beta, \text{ where } |\beta| \leq \frac{c^2 + d^2 - cd}{2} \cdots$$
$$(11.3)$$

and

$$bc - ad = u(c^2 + d^2 - cd) + v, \text{ where } |v| \leq \frac{c^2 + d^2 - cd}{2} \cdots. \quad (11.4)$$

Now, from Eqs. (11.3) and (11.4) we get

$$(a + b\omega)\overline{(c + d\omega)} = (ac + bd - ad) + (bc - ad)\omega =$$
$$(\alpha + u\omega)(c^2 + d^2 - cd) + \beta + v\omega$$

Since $c^2 + d^2 - cd = (c + d\omega)\overline{(c + d\omega)}$, we have

$$a + b\omega = (\alpha + u\omega)(c + d\omega) + \frac{\beta + v\omega}{c + d\omega} = (\alpha + u\omega)(c + d\omega) + \frac{(\beta + v\omega)(c + d\omega)}{c^2 + d^2 - cd}.$$

Take

$$r = \frac{(\beta + v\omega)(c + d\omega)}{c^2 + d^2 - cd}.$$

Then, $r = (a + b\omega) - (\alpha + u\omega)(c + d\omega)$ belongs to $\mathbb{Z}[\omega]$. Further,

$$\delta(r) = \frac{(\beta^2 + v^2 - \beta v)(c^2 + d^2 - cd)}{(c^2 + d^2 - cd)^2} = \frac{\beta^2 + v^2 - \beta v}{c^2 + d^2 - cd}.$$

Using the in equalities in (11.3) and (11.4), we get that

$$\delta(r) < \frac{3(c^2 + d^2 - cd)}{4} \leq c^2 + d^2 - cd = \delta(c + d\omega).$$

This shows that $(\mathbb{Z}[\omega], \delta)$ is an Euclidean domain.   ♯

Theorem 11.3.6 *Let (R, δ) be an Euclidean domain. Then, R is a principal ideal domain.*

Proof We first show that every ideal of R is of the form Ra for some $a \in R$. Let A be an ideal. If $A = \{0\}$, then $A = R \cdot 0$, and there is nothing to do. Suppose that $A \neq \{0\}$. Consider the set $S = \{\delta(a) \mid a \in A - \{0\}\}$. Clearly, S is a nonempty subset of $\mathbb{N} \bigcup \{0\}$. By the well-ordering principle in $\mathbb{N} \bigcup \{0\}$, there exists a $a_0 \in A - \{0\}$ such that $\delta(a_0)$ is the least element of S. We show that $A = Ra_0$. Since A is an ideal and $a_0 \in A$, $Ra_0 \subseteq A$. Let $b \in A$. Since $a_0 \neq 0$, there exist $q, r \in R$ such that

$$b = qa_0 + r,$$

where $r = 0$ or else $\delta(r) < \delta(a_0)$. Clearly, $r = b - a_0 q \in A$. Hence, $r = 0$, for otherwise $\delta(r) \in S$, and $\delta(r) < \delta(a_0)$, a contradiction to the choice of a_0. Thus, $b = qa_0 \in Ra_0$. This proves that every ideal of R is of the form Ra. It is sufficient, now, to show that R has identity. Since R is an ideal of R, it follows that $R = Ra_0$ for some $a_0 \in R$. Thus, there exists $e \in R$ such that $a_0 = ea_0$. Note that $a_0 \neq 0$, for $R \neq \{0\}$. Since R is commutative, $ea_0 = a_0 = a_0 e$. Let $b \in R$. Since $R = Ra_0$, there exists $c \in R$ such that $b = ca_0$. Now, $be = ca_0 e = ca_0 = b$. Since R is commutative, e is the identity of R.   ♯

Corollary 11.3.7 *In an Euclidean domain, g.c.d exists, and every irreducible element is prime.*

Proof Since every Euclidean domain is a principal ideal domain, the result follows from Proposition 11.2.2 and Corollary 11.2.3.   ♯

Corollary 11.3.8 $\mathbb{Z}[i]$ and $\mathbb{Z}[\omega]$ are principal ideal domains.   ♯

Example 11.3.9 Every field F is an Euclidean domain with respect to the δ defined by $\delta(a) = 1$ for all $a \neq 0$ (verify). In particular, the field $\mathbb{C}$ of complex numbers is an Euclidean domain. Clearly, $\mathbb{Z}[X]$ is not an Euclidean domain, for it is not even a principal ideal domain. $\mathbb{C}[X]$ is an Euclidean domain. Thus, subring of an Euclidean domain need not be an Euclidean domain.

Remark 11.3.10 Arithmetic properties of an Euclidean domain (R, δ) do not depend on a particular choice of δ. Some authors define an Euclidean domain to be a commutative integral domain for which such a δ exists.

Remark 11.3.11 There are principal ideal domains which are not Euclidean domains (see Euclidean Algorithms BAMS 1949, Vol 55).

Example 11.3.12 Let (R, δ) be an Euclidean domain. Let $a, b \in R^*$. Then, we can find a greatest common divisor d of a and b and also elements u and v of R such that $d = ua + vb$ as follows: Since (R, δ) is an Euclidean domain, there exist $q, r \in R$ such that

$$a = bq + r,$$

where $r = 0$ or else $\delta(r) < \delta(b)$. If $r = 0$, then b/a and $b = 0 \cdot a + 1 \cdot b$ is a g.c.d of a and b. Suppose that $r \neq 0$. The common divisors of a and b are same as those of b and r. Thus,

$$(a, b) \sim (b, r).$$

Again, since (R, δ) is an Euclidean domain, there exist q_1 and r_1 such that

$$b = q_1 r + r_1,$$

where $r_1 = 0$ or else $\delta(r_1) < \delta(r) < \delta(b)$. If $r_1 = 0$, then

$$(a, b) \sim (b, r) \sim r = 1a - qb$$

and we are through. Suppose that $r_1 \neq 0$. Then again common divisors of b and r are same as those of r and r_1. Thus,

$$(a, b) \sim (b, r) \sim (r, r_1),$$

where $r_1 = -q_1 r + b = -q_1 a + (q_1 q + 1)b$. Again, there exist q_2 and r_2 such that

$$r = q_2 r_1 + r_2,$$

where $r_2 = 0$ or else $\delta(r_2) < \delta(r_1) < \delta(r) < \delta(b)$. If $r_2 = 0$, then

$$(a, b) \sim (b, r) \sim (r, r_1) \sim r_1 = -q_1 a + (q_1 q + 1)b.$$

If $r_2 \neq 0$, proceed as above. This process stops after finitely many steps giving us

$$(a, b) \sim (b, r) \sim (r, r_1) \sim \cdots \sim (r_{i-1}, r_i) \sim r_i = u_i a + v_i b$$

for some $u_i, v_i \in R$.

Example 11.3.13 Using the above algorithm and the division algorithm in $\mathbb{Z}[i]$ as explained in Example 11.3.4, we find g.c.d and l.c.m of $8 + 6i$ and $5 + 15i$ in $\mathbb{Z}[i]$. This illustrates a method to find a g.c.d and l.c.m of any two Gaussian integers.

We observe that $100 = \delta(8+6i) < \delta(5+15i) = 250$. Now, $(5+15i)(8-6i) = 130 + 90i$. Further, $130 = 1 \times 100 + 30$ and $90 = 1 \times 100 + (-10)$. Thus,

$$(5 + 15i)(8 - 6i) = (1 + i) \times 100 + (30 - 10i) =$$
$$(1 + i)(8 + 6i)(8 - 6i) + (30 - 10i).$$

Hence,

$$(5 + 15i) = (1 + i)(8 + 6i) + \frac{30 - 6i}{8 - 6i} = (1 + i)(8 + 6i) + (3 + i).$$

Thus, the remainder obtained when $5 + 15i$ is divided by $8 + 6i$ is $3 + i$, and g.c.d of $5 + 15i$ and $8 + 6i$ is same as that of $8 + 6i$ and $3 + i$. Again consider

$$(8 + 6i)(3 - i) = 30 + 10i = (3 + i) \times 10 = (3 + i)(3 + i)(3 - i).$$

Hence, $8 + 6i = (3 + i)(3 + i)$. This shows that g.c.d of $8 + 6i$ and $3 + i$ is $3 + i$. Thus, g.c.d of $5 + 15i$ and $8 + 6i$ is $3 + i$, and also

$$3 + i = 1 \cdot (5 + 15i) + (-1 - i)(8 + 6i).$$

Further, l.c.m of $5 + 15i$ and $8 + 6i$ is $\frac{(5+15i)(8+6i)}{3+i} = (5 + 15i)(3 + i) = 50i$.

Exercises

11.3.1 Show that $(\mathbb{Z}[\sqrt{-2}], \delta)$, where $\delta(a + b\sqrt{-2}) = a^2 + 2b^2$ is an Euclidean domain.

11.3.2 Find g.c.d and l.c.m of the following pairs in $\mathbb{Z}[i]$. Also express g.c.d as linear sum of the corresponding pairs.
  (i) $3 + 4i$, $4 + 3i$.
  (ii) $4 + 2i$, $3 + 4i$.

11.3.3 Find g.c.d and l.c.m of the following pairs in $\mathbb{Z}[\omega]$. Also express g.c.d as linear sum of the corresponding pairs.

(i) $2 + 3\omega$, $6 + 8\omega$.

(ii) $5 + 11\omega$, $2 + 3\omega$.

11.3.4 Find g.c.d and l.c.m of the following pairs in $\mathbb{Z}[\sqrt{-2}]$.

(i) $5 + 7\sqrt{-2}$, $2 + \sqrt{-2}$.

(ii) $3 + \sqrt{-2}$, $4 + 3\sqrt{-2}$.

11.3.5 Generalize the Chinese remainder theorem of Chap. 6 to principal ideal domains and Euclidean domains.

11.3.6 Determine the field $\frac{\mathbb{Z}[i]}{\mathbb{Z}[i]\cdot(1+i)}$. What is its order?

11.3.7 Let p be a prime element of $\mathbb{Z}$, which is not a prime element of $\mathbb{Z}[i]$. Show that p is product of two complex conjugates in $\mathbb{Z}[i]$. Deduce that if a prime element of $\mathbb{Z}$ is not a prime element of $\mathbb{Z}[i]$, then it is sum of squares of two integers.

11.3.8 Let $a + bi$ be a prime element of $\mathbb{Z}[i]$. Show that $a^2 + b^2$ is prime element of $\mathbb{Z}$, or it is a square of a prime element of $\mathbb{Z}$.

11.3.9 Let $p \in \mathbb{Z}$ be a prime integer. Suppose that p is prime in $\mathbb{Z}[i]$. Show that $p \equiv 3 (mod\ 4)$.

11.3.10 Let p be a prime element of $\mathbb{Z}$ which is not prime element of $\mathbb{Z}[i]$. Show that $(\frac{-1}{p}) = 1$. Deduce that $p \equiv 1 (mod\ 4)$.

11.4 Chinese Remainder Theorem in Rings

In this section, we generalize the Chinese remainder theorem of Chap. 6 to arbitrary commutative rings with identities.

Definition 11.4.1 A pair (A, B) of ideals of R is said to be **co - maximal** if $A + B = R$.

Example 11.4.2 In a principal ideal domain, a pair (Ra, Rb) is co-maximal if and only if $Ra + Rb = R$. This is equivalent to say that $(a, b) \sim 1$.

Theorem 11.4.3 (Chinese Remainder Theorem). *Let R be a commutative ring with identity. Let $\{A_1, A_2, \ldots, A_r\}$ be a set of pairwise co-maximal ideals of R. Let $a_1, a_2, \ldots a_r$ be members of R. Then, there exists $x \in R$ such that $x - a_i \in A_i$ for all i.*

Proof We first prove the result for $r = 2$. Let $\{A_1, A_2\}$ be a pair of co-maximal ideals. Let $a_1, a_2 \in R$. Since $A_1 + A_2 = R$, $1 = u_1 + u_2$ for some $u_1 \in A_1$ and $u_2 \in A_2$. Take $x = a_2 u_1 + a_1 u_2$. Then, $x - a_1 = a_2 u_1 + a_1(u_2 - 1) =$

$a_2 u_1 + a_1(-u_1) = (a_2 - a_1)u_1 \in A_1$. Similarly, $x - a_2 \in A_2$. Assume that the result is true for r. Let $\{A_1, A_2, \ldots, A_r, A_{r+1}\}$ be a set of pairwise co-maximal ideals. Let $a_1, a_2, \ldots, a_r, a_{r+1}$ be elements of R. By the induction assumption, there exists a $y \in R$ such that $y - a_i \in A_i$ for all $i \le r$. We have

$$A_1 + A_{r+1} = R, \; A_2 + A_{r+1} = R, \ldots, A_r + A_{r+1} = R.$$

Multiplying the above equations, we get that

$$A_{r+1} + (A_1 \cap A_2 \cap \cdots \cap A_r) = R.$$

Put $(A_1 \cap A_2 \cap \cdots \cap A_r) = B$. Then, (A_{r+1}, B) is a pair of co-maximal ideals. From the case $r = 2$ (already proved), there exists an element $x \in R$ such that $x - a_{r+1} \in A_{r+1}$ and $x - y \in B$. But, then $x - a_i = (x - y) + (y - a_i)$ belongs to A_i for all $i \le r$. ♯

Corollary 11.4.4 *Let R be a principal ideal domain. Let $\{m_1, m_2, \ldots, m_r\}$ be a set of pairwise co-prime elements of R. Let $a_1, a_2, \ldots, a_r$ be elements of R. Then, there exists an element $x \in R$ such that*

$$x \equiv a_i \,(mod \; m_i) \; for \; all \; i \le r.$$

Proof Since R is a principal ideal domain,

$$(m_i, m_j) \sim 1 \; if \; and \; only \; if \; Rm_i + Rm_j = R \cdot 1 = R.$$

Thus, $\{Rm_1, Rm_2, \ldots, Rm_r\}$ is a set of pairwise co-maximal ideals of R. From the Chinese remainder theorem, there exists a $x \in R$ such that $x - a_i \in Rm_i$ for all i. This means that $x \equiv a_i \,(mod \; m_i)$ for all $i \le r$. ♯

Corollary 11.4.5 *Let R be a commutative ring with identity. Let $\{A_1, A_2, \ldots, A_r\}$ be a set of pairwise co-maximal ideals of R. Let $A = A_1 \cap A_2 \cap \cdots \cap A_r$. Then, R/A is isomorphic to $R/A_1 \times R/A_2 \times \cdots \times R/A_r$.*

Proof Define a map f from R to $R/A_1 \times R/A_2 \times \cdots \times R/A_r$ by

$$f(a) = (a + A_1, a + A_2, \ldots, a + A_r).$$

It is easy to see that f is a ring homomorphism. Let $(a_1 + A_1, a_2 + A_2, \ldots, a_r + A_r)$ be an arbitrary element of $R/A_1 \times R/A_2 \times \cdots \times R/A_r$. By the Chinese remainder theorem, there exists $x \in R$ such that $x - a_i \in A_i$ for all i. This means that $x + A_i = a_i + A_i$ for all i. Thus, f is surjective. Now,

$$ker \; f = \{x \in R \mid x + A_i = A_i \; \forall i\} = \{x \in R \mid x \in A_i \; \forall i\} = \bigcap_{i=1}^{r} A_i.$$

The result follows from the fundamental theorem of homomorphism. ♯

Corollary 11.4.6 *Let R be a principal ideal domain. Let $\{m_1, m_2, \ldots, m_r\}$ be a set of pairwise co-prime elements of R. Let $m = m_1 m_2 \cdots m_r$. Then, R/Rm is isomorphic to $R/Rm_1 \times R/Rm_2 \times \cdots \times R/Rm_r$.*

Proof The result follows, if we observe that $\{Rm_1, Rm_2, \ldots, Rm_r\}$ is a set of pairwise co-maximal ideals and $Rm = Rm_1 \cap Rm_2 \cap \cdots \cap Rm_r$.   ♯

Remark 11.4.7 Let A and B be ideals of a commutative ring R with identity. Recall that the ideal generated by the set $\{ab \mid a \in A \text{ and } b \in B\}$ is denoted by AB. Thus, $AB \subseteq A \cap B$. In general, AB need not be $A \cap B$. For example, $(4\mathbb{Z})(6\mathbb{Z}) = 24\mathbb{Z}$ where as $4\mathbb{Z} \cap 6\mathbb{Z} = 12\mathbb{Z}$. However, if $\{A_1, A_2, \ldots, A_r\}$ is a set of pairwise co-maximal ideals of R, then $A_1 A_2 \cdots A_r = A_1 \cap A_2 \cap \cdots \cap A_r$. The proof of this fact is by induction on r. Suppose that $r = 2$. Let (A_1, A_2) be a pair of co-maximal ideals. Then, $A_1 + A_2 = R$, and there is an element $u_1 \in A_1$ and an element $u_2 \in A_2$ such that $u_1 + u_2 = 1$. Let $x \in A_1 \cap A_2$. Then, $x = xu_1 + xu_2 \in A_1 A_2$, for $xu_1, xu_2 \in A_1 A_2$. This shows that $A_1 A_2 = A_1 \cap A_2$. Assume that the result is true for r. Let $\{A_1, A_2, \ldots, A_{r+1}\}$ be a set of pair wise co-maximal ideals of R. By the induction hypothesis, $A_1 A_2 \cdots A_r = A_1 \cap A_2 \cap \cdots \cap A_r$. As already observed in the proof of the Theorem 9.4.3, $(A_1 A_2 \cdots A_r, A_{r+1})$ is a pair of co-maximal ideals. Hence, $A_1 A_2 \cdots A_r A_{r+1} = (A_1 A_2 \cdots A_r) \cap A_{r+1} = A_1 \cap A_2 \cap \cdots \cap A_r \cap A_{r+1}$.

11.5 Unique Factorization Domain (U.F.D)

Definition 11.5.1 A commutative integral domain R with identity is called a **unique factorization domain** (**U.F.D**) if the following conditions are satisfied.

(i) Every nonzero nonunit element of R can be expressed as a product of irreducible elements of R,

and

(ii) the representation of any nonzero nonunit element a as product of irreducible elements is unique in the sense that if

$$a = p_1 p_2 \cdots p_r = p_1' p_2' \cdots p_s',$$

where p_i and p_j' are irreducible elements of R, then $r = s$, and there is a permutation σ of $\{1, 2, \ldots, r\}$ such that $p_i \sim p_{\sigma(i)}'$ for all i. A unique factorization domain is also termed as **Gaussian domain**.

Example 11.5.2 $\mathbb{Z}$ is a U.F.D (follows from the fundamental theorem of arithmetic).

Theorem 11.5.3 *Let R be a commutative integral domain with identity. Then, the following conditions are equivalent.*

1. R is a U.F.D.

2. (i) Given a sequence $a_1, a_2, \ldots, a_n, a_{n+1}, \ldots$ of elements of R such that a_{n+1}/a_n for all $n \in \mathbb{N}$, there exists a $n_0 \in \mathbb{N}$ such that $a_n \sim a_{n_0}$ for all $n \geq n_0$.

(ii) Greatest common divisors exist in R in the sense that any two nonzero elements of R have a g.c.d.

3. (i) Same as 2(i).

(ii) Every irreducible element is a prime element.

Proof 1 $\Longrightarrow$ 2. Suppose that R is a unique factorization domain. Let .

$$a_1, a_2, \ldots, a_n, a_{n+1}, \ldots$$

be a sequence of elements of R such that a_{n+1}/a_n for all n. Suppose that there is no m such that $a_n \sim a_m$ for all $n \geq m$, then no a_n is a unit (divisors of units are only units), and we can extract a subsequence

$$a_1, a_{n_1}, a_{n_2}, \ldots, a_{n_r}, a_{n_{r+1}}, \ldots$$

of the given sequence such that $a_{n_{r+1}}$ is a proper divisor of a_{n_r} for all r. Since a_1 is a nonunit, and R is a unique factorization domain,

$$a_1 = p_1 p_2 \cdots p_r,$$

where each p_i is irreducible element of R. Since a_{n_1} is a proper divisor of a_1, $a_1 = a_{n_1} b_{n_1}$, where a_{n_1} and b_{n_1} are nonunits. Since a_{n_2} is a proper divisor of a_{n_1}, $a_{n_1} = a_{n_2} b_{n_2}$, where a_{n_2} and b_{n_2} are again nonunits. But, then

$$a_1 = a_{n_2} b_{n_2} b_{n_1}.$$

Proceeding inductively, in this way, we get

$$a_1 = a_{n_r} b_{n_r} b_{n_{r-1}} \cdots b_{n_1},$$

where a_{n_r} is a nonunit and each b_{n_i} is a nonunit. Since R is a unique factorisation domain a_{n_r} and each b_{n_i} are products of irreducible elements of R. But, then a_1 will also be expressible as product of more that r irreducible elements of R. This is a contradiction to the supposition that R is a unique factorization domain. Thus, 1 implies 2(i).

Next, assuming 1, we show that g.c.d of any two nonzero elements exist. Let a, b be nonzero elements of R. If a is a unit, then a divides b, and so a is a g.c.d of a and b. Suppose that neither a nor b is a unit. We can write

$$a = u p_1^{\alpha_1} p_2^{\alpha_2} \cdots p_r^{\alpha_r}$$

and

$$b = p_1^{\beta_1} p_2^{\beta_2} \cdots p_r^{\beta_r},$$

where u is a unit $\alpha_i, \beta_i \geq 0$ for all i and $\{p_1, p_2, \ldots, p_r\}$ is a set of pairwise nonassociate irreducible elements of R. Take

$$d = p_1^{\mu_1} p_2^{\mu_2} \cdots p_r^{\mu_r},$$

where $\mu_i = min(\alpha_i, \beta_i)$. We show that d is a g.c.d of a and b. Clearly, d/a and d/b. Suppose that d'/a and d'/b. If d' is a unit, then it divides d. Suppose that d' is a nonunit. Suppose that p is an irreducible element which divides d'. Then, p divides a, and also p divides b. By the uniqueness of the factorization, $p \sim p_i$ for some i. Thus, we can write

$$d' = v p_1^{\delta_1} p_2^{\delta_2} \cdots p_r^{\delta_r},$$

where v is a unit. Since d'/a,

$$a = u p_1^{\alpha_1} p_2^{\alpha_2} \cdots p_r^{\alpha_r} = v p_1^{\delta_1} p_2^{\delta_2} \cdots p_r^{\delta_r} c.$$

for some c. Suppose that $\delta_1 > \alpha_1$. Then

$$u p_2^{\alpha_2} \cdots p_r^{\alpha_r} = v p_1^{\delta_1 - \alpha_1} p_2^{\alpha_2} \cdots p_r^{\alpha_r} c.$$

By the uniqueness of the factorization, $p_1 \sim p_j$ for some $j \geq 2$. This is a contradiction to the supposition that $\{p_1, p_2, \ldots, p_r\}$ is a set of pairwise nonassociate irreducible elements of R. Hence, $\delta_1 \leq \alpha_1$. Similarly $\delta_i \leq \alpha_i$ and $\delta_i \leq \beta_i$ for all i. Now, it is clear that d'/d. Hence, d is a g.c.d of a and b.

$2 \Longrightarrow 3$. Assume 2. 3(i) is same as 2(i). Again, since g.c.d of any two nonzero elements exist, it follows (Corollary 11.1.9) that every irreducible element is prime. Thus, $2 \Longrightarrow 3$.

$3 \Longrightarrow 1$. Assume 3. Using 3(i), we show that every nonzero nonunit element of R can be expressed as product of irreducible elements of R. Let a be a nonzero nonunit element of R. We first show that a has an irreducible factor. If a is irreducible, then there is nothing to do. If not, $a = a_1 b_1$, where a_1 is a proper divisor of a. If a_1 is irreducible, we get an irreducible factor of a. If not, $a_1 = a_2 b_2$, where a_2 is a proper divisor of a_1. If a_2 is irreducible, we get an irreducible factor of a_1, and so of a. If not, proceed. This process stops after finitely many steps giving us an irreducible factor p_1 of a, for otherwise we shall get a sequence

$$a_1, a_2, \ldots, a_n, a_{n+1}, \ldots$$

of elements of R such that each a_{i+1} is a proper divisor of a_i. This is a contradiction to 3(i). Let

$$a = p_1 c_1.$$

If c_1 is irreducible, we are through. If not, as before, there is an irreducible factor p_2 of c_1. Thus,

$$a = p_1 p_2 c_2,$$

where c_2 is a proper divisor of c_1. If c_2 is irreducible, we are through. If not, proceed. This process also stops after finitely many steps giving us a factorization

$$a = p_1 p_2 \cdots p_r$$

of a as product of irreducible elements of R, for otherwise we shall arrive at a sequence

$$c_1, c_2, \ldots, c_n, c_{n+1}, \ldots$$

of elements of R such that each c_{n+1} is a proper divisor of c_n (this is a contradiction to 3(i)).

Finally, we prove the uniqueness of the factorization. Let

$$a = p_1 p_2 \cdots p_r = p_1' p_2' \cdots p_s',$$

where p_i and p_j' are irreducible elements of R. Since every irreducible element is supposed to be prime, each p_i and p_j' are prime elements. Now, p_1 divides $a = p_1' p_2' \cdots p_s'$. Hence, p_1 divides p_j' for some j. After rearranging, we may assume that p_1 divides p_1'. Since p_1' is irreducible, and p_1 is not a unit (being irreducible), $p_1 \sim p_1'$. Suppose that $p_1' = u p_1$, where u is a unit. Then

$$p_1 p_2 \cdots p_r = u p_1 p_2' \cdots p_s'.$$

Canceling p_1, we get that

$$p_2 p_3 \cdots p_r = u p_2' p_3' \cdots p_s'.$$

As above $p_2 \sim u p_2' \sim p_2'$. Proceed inductively. In this process, $\{p_1, p_2, \ldots, p_r\}$ and $\{p_1', p_2', \ldots, p_s'\}$ both will exhaust simultaneously, for otherwise we shall arrive at a product of irreducible elements equal to identity. This is not possible, for an irreducible element can not be a unit. This ensures that $r = s$, and after some rearrangement $p_i \sim p_i'$ for all i. ♯

Corollary 11.5.4 *Every principal ideal domain is a unique factorization domain.*

Proof Let R be a principal ideal domain. By Proposition 11.2.2, g.c.d. of any two nonzero elements in R exists. Thus, it is sufficient to prove the condition 2(i) of the above theorem. Let

$$a_1, a_2, \ldots, a_n, a_{n+1}, \ldots$$

be a sequence of elements of R such that a_{n+1}/a_n for all n. Then, by Proposition 11.1.10, we get an ascending chain

$$Ra_1 \subseteq Ra_2 \subseteq \cdots Ra_n \subseteq Ra_{n+1} \subseteq \cdots$$

of principal ideals of R. Let

$$A = \bigcup_{n=1}^{\infty} Ra_n.$$

Then A, being union of ascending chain of ideals, is an ideal. Since R is a principal ideal domain, $A = Rd$ for some $d \in R$. Now, $d \in Rd = A = \bigcup_{n=1}^{\infty} Ra_n$. Hence, $d \in Ra_{n_0}$ for some $n_0 \in \mathbb{N}$. But, then

$$Rd \subseteq Ra_{n_0} \subseteq \bigcup_{n=1}^{\infty} Ra_n = A = Rd.$$

Hence,

$$Ra_{n_0} = A = \bigcup_{n=1}^{\infty} Ra_n.$$

This implies that $Ra_n = Ra_{n_0}$ for all $n \geq n_0$. By Proposition 11.1.10, $a_n \sim a_{n_0}$ for all $n \geq n_0$. This completes the proof. ♯

Since every Euclidean domain is a principal ideal domain, we have,

Corollary 11.5.5 *Every Euclidean domain is a unique factorization domain.* ♯

Example 11.5.6 It will follow soon (Gauss theorem) that $\mathbb{Z}[X]$ is a U.F.D. Clearly, then $\mathbb{Z}[X]$ is not a P.I.D for $\mathbb{Z}$ is not a field. Thus, a U.F.D need not be a P.I.D.

Example 11.5.7 Every field is a U.F.D. In particular, the field $\mathbb{C}$ of complex numbers is a U.F.D, whereas the subring $\mathbb{Z}[\sqrt{-5}]$ is not a U.F.D for 3 is irreducible in $\mathbb{Z}[\sqrt{-5}]$, but it is not prime. Thus, subring of a U.F.D need not be a U.F.D.

Example 11.5.8 A homomorphic image of a U.F.D need not be a U.F.D (even if it is an integral domain): For example, $\mathbb{Z}[X]$ is a U.F.D, the map $f(X) \rightsquigarrow f(\sqrt{-5})$ is a surjective homomorphism from $\mathbb{Z}[X]$ to $\mathbb{Z}[\sqrt{-5}]$ but, $\mathbb{Z}[\sqrt{-5}]$ is not a U.F.D.

Example 11.5.9 In a principal ideal domain, we noticed that if d is g.c.d of a and b, then there exist $u, v \in R$ such that $d = ua + vb$. This result is not true in a U.F.D. For example, $(2, X) \sim 1$ but $2f(X) + Xg(X)$ can never be 1.

Gauss Theorem

Our next aim is to prove the following theorem due to Gauss.

Theorem 11.5.10 (Gauss) *If R is a unique factorization domain, then $R[X]$ is also a unique factorization domain.*

To prove this result, we need some concepts and results.

Let R be a unique factorization domain. Let $f(X) \in R[X]$, $f(X) \neq 0$. A g.c.d of nonzero coefficients of powers of X in $f(X)$ is called a **content** of $f(X)$, and it is denoted by $c(f(X))$. Thus, a content of a polynomial is unique upto associates. Integers 2 and -2 are contents of the polynomial $2 + 10X^2 + 12X^4$ in $\mathbb{Z}[X]$. Gaussian integers $1 + i, -1 - i, 1 - i, and -1 + i$ are contents of the polynomial $2 + (1 + i)X + 2iX^2$ in $\mathbb{Z}[i][X]$ (prove it).

If $f(X)$ and $g(X)$ are associates in $R[X]$, then they differ by a unit in $R[X]$. Since units of $R[X]$ are those of R, $f(X) = ug(X)$ for some unit u of R. This shows that $c(f(X)) \sim c(g(X))$. Thus, polynomials which are associates have same contents. Nonassociate polynomials may also have same contents for example $2 + 10X^2$ and $2 + 10X^2 + 12X^4$ have same contents but, they are not associates.

A nonzero polynomial $f(X)$ is said to be a **primitive** polynomial if $c(f(X)) \sim 1$. Thus, $3 + 2X + 5X^2$ is a primitive polynomial in $\mathbb{Z}[X]$.

Let $f(X) \neq 0$ be a polynomial in $R[X]$. Then

$$f(X) = c(f(X))f_1(X),$$

where $f_1(X)$ is a primitive polynomial in $R[X]$. Also if

$$f(X) = af_1(X),$$

where $f_1(X)$ is primitive, then $c(f(X)) \sim a$.

Lemma 11.5.11 (Gauss) *Let R be a unique factorization domain. Then, the product of any two primitive polynomials in $R[X]$ is a primitive polynomial.*

Proof Let

$$f(X) = a_0 + a_1X + a_2X^2 + \cdots + a_nX^n$$

and

$$g(X) = b_0 + b_1X + b_2X^2 + \cdots + b_mX^m$$

be two primitive polynomials. Then, we have to show that

$$f(X)g(X) = c_0 + c_1X + c_2X^2 + \cdots + c_{n+m}X^{n+m},$$

where

$$c_r = \Sigma_{i+j=r}a_ib_j, \ r = 1, 2, \ldots, n+m$$

is a primitive polynomial. It is sufficient to show that no prime element of R divides each c_r. Let p be a prime element of R. Since $f(X)$ is primitive, p can not divide each a_i. Thus, there exists i_0 such that p/a_k for all $k < i_0$, and p does not divide a_{i_0}. Similarly, since $g(X)$ is also a primitive polynomial, there exists a j_0 such that p/b_l for all $l < j_0$, and p does not divide b_{j_0}. Consider

$$c_{i_0+j_0} = \Sigma_{i+j=i_0+j_0} a_i b_j \cdots . \tag{11.5}$$

If $i + j = i_0 + j_0$, $i \neq i_0$, and $j \neq j_0$, then $i < i_0$ or $j < j_0$. Hence, each term under summation in the right hand side of the Eq. (11.5) is divisible by p except $a_{i_0} b_{j_0}$ which is not divisible by p. This shows that p does not divide $c_{i_0+j_0}$, and hence, $f(X)g(X)$ is a primitive polynomial.   ♮

Corollary 11.5.12 *Let $f(X)$ and $g(X)$ be nonzero polynomials in $R[X]$, where R is a U.F.D. Then*

$$c(f(X)g(X)) \sim c(f(X))c(g(X)).$$

Proof $f(X) = c(f(X))f_1(X)$ and $g(X) = c(g(X))g_1(X)$, where $f_1(X)$ and $g_1(X)$ are primitive polynomials. Further,

$$f(X)g(X) = c(f(X))c(g(X))f_1(X)g_1(X),$$

where, by Gauss Lemma, $f_1(X)g_1(X)$ is primitive. It follows that $c(f(X))c(g(X))$ is a content of $f(X)g(X)$.   ♮

Lemma 11.5.13 *Every irreducible element of R is also an irreducible element of $R[X]$.*

Proof Let p be an irreducible element of R. Suppose that $p = f(X)g(X)$, where $f(X), g(X) \in R[X]$. Comparing the degrees, we find that $deg(f(X)) = 0 = deg(g(X))$. Thus, $f(X), g(X) \in R$. Since p is irreducible in R, $f(X)$ or $g(X)$ is a unit in R. Since units of R are also units of $R[X]$, the result follows.   ♮

Lemma 11.5.14 *Let R be a unique factorization domain, and F be its field of fractions. Let $f(X)$ and $g(X)$ be primitive polynomials in $R[X]$ which are associates in $F[X]$. Then $f(X)$ and $g(X)$ are associates in $R[X]$.*

Proof We know that the units of $F[X]$ are those of F. Thus, the units of $F[X]$ are precisely the nonzero elements of F. Let $f(X)$ and $g(X)$ be two primitive polynomials in $R[X]$ which are associates in $F[X]$. Then, $f(X) = ug(X)$, where u is a nonzero element of F. Since F is the field of fractions of R, $u = ab^{-1}$ for some $a, b \in R^\star$. Hence, $f(X) = ab^{-1}g(X)$. In turn, $bf(X) = ag(X)$ for some $a, b \in R^\star$. Since $f(X)$ and $g(X)$ are primitive in $R[X]$, comparing the contents we find that $b \sim a$ in R, and so $f(X)$ and $g(X)$ differ by a unit of R. Thus, $f(X) \sim g(X)$ in $R[X]$.   ♮

Lemma 11.5.15 *Let $f(X)$ be an irreducible element of $R[X]$ of positive degree. Then, $f(X)$ is irreducible in $F[X]$, where F is the field of fractions of R.*

Proof Suppose that $f(X)$ is an irreducible element of $R[X]$ of positive degree. Then, it is a primitive polynomial in $R[X]$, for otherwise a content of $f(X)$ would be a proper divisor of $f(X)$. Suppose that it is reducible in $F[X]$. Then, $f(X) = g(X)h(X)$, where $g(X)$ and $h(X)$ are polynomials in $F[X]$ of positive degrees (for nonzero elements of F are units of $F[X]$). Suppose that

$$g(X) = u_0 + u_1 X + u_2 X^2 + \cdots + u_n X^n$$

and

$$h(X) = v_0 + v_1 X + v_2 X^2 + \cdots + v_m X^m,$$

where u_i and v_j are members of F. If b is l.c.m of the denominators of nonzero coefficients of powers of X in $g(X)$, then there is an element $a \in R$ such that

$$g(X) = ab^{-1}g_1(X),$$

where $g_1(X)$ is a primitive polynomial in $R[X]$ of same degree as that of $g(X)$. Similarly,

$$h(X) = cd^{-1}h_1(X),$$

where $c, d \in R^*$ and $h_1(X)$ is a primitive polynomial in $R[X]$. Thus,

$$f(X) = ac(bd)^{-1}g_1(X)h_1(X)$$

or

$$bdf(X) = acg_1(X)h_1(X).$$

By the Gauss lemma, $g_1(X)h_1(X)$ is a primitive polynomial, and since $f(X)$ (being irreducible element of $R[X]$ of positive degree) is primitive, comparing the contents we obtain that

$$bd \sim ac \text{ and } f(X) \sim g_1(X)h_1(X) \text{ in } R[X].$$

Since $deg(g_1(X)) = deg(g(X)) > 0$ and $deg(h_1(X)) = deg(h(X)) > 0$, $g_1(X)$ and $h_1(X)$ are proper divisors of $f(X)$ in $R[X]$. But, then $f(X)$ would become reducible in $R[X]$. ♯

Proof of the Gauss theorem. We first show that every nonzero nonunit element $f(X)$ of $R[X]$ can be written as product of irreducible elements of $R[X]$. This, we prove by the induction on degree of $f(X)$. If $deg(f(X)) = 0$, then $f(X) \in R$. Since R is a U.F.D, $f(X)$ is a product of irreducible elements of R. By the Lemma 11.5.13, every irreducible element of R is also irreducible in $R[X]$, and hence, in this case $f(X)$ is product of irreducible elements of $R[X]$. Suppose that the assertion is true for all those polynomials whose degree is less than the $deg(f(X))$, where $deg(f(X)) \geq 1$. Now, $f(X) = c(f(X))f_1(X)$, where $f_1(X)$ is a primitive polynomial of degree same as that of $f(X)$. If $c(f(X))$ is a nonunit, it is expressible as product of irreducible elements of R, and so of $R[X]$. Hence, it is sufficient to show that $f_1(X)$ is product of irreducible elements of $R[X]$. If $f_1(X)$ is irreducible, nothing to do. Suppose not. Then, $f_1(X) = g(X)h(X)$, where neither $g(X)$ nor $h(X)$ is a unit. Since $f_1(X)$ is a primitive polynomial, $deg(g(X)) > 0$ and $deg(h(X)) > 0$. But, then $deg(g(X)) < deg(f_1(X)) = deg(f(X))$ and $deg(h_1(X)) < deg(f_1(X)) = deg(f(X))$. By the induction hypothesis, $g(X)$ and $h(X)$ are products of irreducible elements of $R[X]$. But, then $f_1(X)$ is also expressible as product of irreducible elements of $R[X]$.

Now, we prove the uniqueness of the factorization of $f(X)$. Suppose that

$$f(X) = p_1 p_2 \cdots p_r f_1(X) f_2(X) \cdots f_s(X) = p_1' p_2' \cdots p_t' f_1'(X) f_2'(X) \cdots f_u'(X),$$

where p_i, p_j' are irreducible elements of R, and $f_k(X)$ and $f_l'(X)$ are irreducible elements of $R[X]$ of positive degrees. Clearly, then $f_k(X)$ and $f_l'(X)$ are primitive polynomials of $R[X]$. By the Gauss lemma, $f_1(X) f_2(X) \cdots f_s(X)$ and $f_1'(X) f_2'(X) \cdots f_u'(X)$ are primitive polynomials. Comparing the contents,

$$p_1 p_2 \cdots p_r \sim p_1' p_2' \cdots p_t',$$

and

$$f_1(X) f_2(X) \cdots f_s(X) \sim f_1'(X) f_2'(X) \cdots f_u'(X).$$

Since R is a U.F.D, $r = t$, and after some rearrangement p_i is an associate of p_i' in R for all i. Hence, p_i is an associate of p_i' in $R[X]$ also. Next, since $f_k(X)$ and $f_l'(X)$ are irreducible elements of $R[X]$ of positive degrees, by Lemma 11.5.15, they are irreducible in $F[X]$, where F is the field of fractions of R. Since $F[X]$ (being a P.I.D) is a U.F.D, $s = u$, and after some rearrangement $f_k(X)$ is an associate of $f_k'(X)$ in $F[X]$ for all k. Since $f_k(X)$ and $f_k'(X)$ are primitive, by the Lemma 11.5.14, $f_k(X)$ is an associate of $f_k'(X)$ in $R[X]$ for all k. This shows that $R[X]$ is a unique factorization domain.   ♯

Corollary 11.5.16 *If R is a U.F.D, then $R[X_1, X_2, \ldots, X_n]$ is also a U.F.D.*

Proof Follows by Gauss theorem and induction on n.   ♯

Since $\mathbb{Z}$ is a U.F.D, we have the following.

Corollary 11.5.17 $\mathbb{Z}[X]$ *is a U.F.D.*   ♯

Following result is the converse of the Gauss theorem.

Theorem 11.5.18 *If $R[X]$ is a U.F.D, then R is also a U.F.D.*

Proof Suppose that $R[X]$ is a U.F.D. Let a be a nonzero nonunit element of R. Then, it is nonzero nonunit element of $R[X]$ also. Hence, a is product of irreducible elements of $R[X]$. Comparing degrees, we obtain that all irreducible factors are elements of R also. Next, suppose that

$$a = p_1 p_2 \cdots p_r = p'_1 p'_2 \cdots p'_s,$$

where p_i and p'_j are irreducible elements of R, and so of $R[X]$ also. Since $R[X]$ is a U.F.D, $r = s$, and after some rearrangement p_i and p'_i differ by units of $R[X]$. The result follows if we observe that the units of $R[X]$ are those of R. ♯

Criteria for Irreducibility of Polynomials

Let R_1 and R_2 be commutative integral domains with identities. Let σ be a nonzero ring homomorphism from R_1 to R_2. Let

$$f(X) = a_0 + a_1 X + a_2 X^2 + \cdots + a_n X^n$$

be a polynomial in $R_1[X]$. We have a polynomial $f^\sigma(X) \in R_2[X]$ given by

$$f^\sigma(X) = \sigma(a_0) + \sigma(a_1)X + \cdots + \sigma(a_n)X^n.$$

The map $f(X) \rightsquigarrow f^\sigma(X)$ is clearly a ring homomorphism which is surjective if and only if σ is surjective, and it is injective if and only if σ is injective.

Theorem 11.5.19 *Let R_1 and R_2 be commutative integral domains with identities, and σ a homomorphism from R_1 to R_2. Let F_2 be the field of fractions of R_2. Let $f(X) \in R_1[X]$. Suppose that $deg(f^\sigma(X)) = deg(f(X))$ (in particular $f(X)$ and $f^\sigma(X)$ are nonzero polynomials), and $f^\sigma(X)$ is irreducible in $F_2[X]$. Then $f(X)$ can not be expressed as product of two nonconstant polynomials in $R_1[X]$.*

Proof Suppose that $f(X) = g(X)h(X)$, where $g(X)$ and $h(X)$ are polynomials in $R_1[X]$. Then

$$f^\sigma(X) = g^\sigma(X)h^\sigma(X).$$

Since $deg(f(X)) = deg(f^\sigma(X))$, it follows that $deg(g(X)) = deg(g^\sigma(X))$ and $deg(h(X)) = deg(h^\sigma(X))$. Since $f^\sigma(X)$ is irreducible in $F_2[X]$, $g^\sigma(X)$ or $h^\sigma(X)$ is a unit of $F_2[X]$. But, then $deg(g(X)) = deg(g^\sigma(X)) = 0$ or $deg(h(X)) = deg(h^\sigma(X)) = 0$. Hence, $g(X)$ is constant or $h(X)$ is constant. ♯

Corollary 11.5.20 *Let R_1 be a U.F.D, and R_2 a commutative integral domain with identity. Let σ be a ring homomorphism from R_1 to R_2. Let $f(X)$ be a primitive polynomial in $R_1[X]$ with $deg(f(X)) = deg(f^\sigma(X))$. If $f^\sigma(X)$ is irreducible in $F_2[X]$, then $f(X)$ is irreducible in $R_1[X]$, and also in $F_1[X]$, where F_1 is the field of fractions of R_1.*

Proof Follows from the above theorem if we observe that a primitive polynomial in $R_1[X]$ is irreducible if and only if it can not be written as product of two nonconstant polynomials. Also a primitive polynomial in $R_1[X]$ is irreducible in $R_1[X]$ if and only if it is irreducible in $F_1[X]$. ♯

Corollary 11.5.21 *Let R be a U.F.D. Let p be a prime element of R and $\wp$ the prime ideal generated by p. Let*

$$f(X) = a_0 + a_1 X + a_2 X^2 + \cdots + a_n X^n$$

be a polynomial in $R[X]$ of positive degree such that p does not divide a_n. Suppose that

$$\overline{f}(X) = \overline{a_0} + \overline{a_1} X + \cdots + \overline{a_n} X^n,$$

where $\overline{a_i} = a_i + \wp \in R/\wp$. Suppose that $\overline{f}(X)$ is irreducible in $R/\wp[X]$. Then $f(X)$ is irreducible in $F[X]$, where F is the field of fractions of R.

Proof Let $f(X) = c(f(X)) f_1(X)$, where $f_1(X)$ is a primitive polynomial of same degree as that of $f(X)$. Also $f(X)$ is irreducible in $F[X]$ if and only if $f_1(X)$ is irreducible in $F[X]$. If p does not divide the leading coefficient of $f(X)$, then it does not divide the leading coefficient of $f_1(X)$ also. Next, if $\overline{f}(X)$ is irreducible in $R/\wp[X]$, then $\overline{f_1}(X)$ is irreducible in $R/\wp[X]$. Thus, we may assume that $f(X)$ is a primitive polynomial. Suppose that $f(X)$ is reducible in $F[X]$. Then, since $f(X)$ is primitive in $R[X]$,

$$f(X) = g(X)h(X),$$

where $g(X)$ and $h(X)$ are polynomials in $R[X]$ of positive degrees. But, then

$$\overline{f}(X) = \overline{g}(X)\overline{h}(X).$$

Since p does not divide a_n, $deg(f(X)) = deg(\overline{f}(X))$. Hence, $deg(\overline{g}(X) = deg(g(X))$ and $deg(\overline{h}(X)) = deg(h(X))$. Thus, $\overline{g}(X)$ and $\overline{h}(X)$ are of positive degrees. But, then $\overline{f}(X)$ would be reducible in $R/\wp[X]$. ♯

Corollary 11.5.22 *Let*

$$f(X) = a_0 + a_1 X + a_2 X^2 + \cdots + a_n X^n$$

be a polynomial in $\mathbb{Z}[X]$ such that p does not divide a_n for some prime p. Suppose that

$$\overline{f}(X) = \overline{a_0} + \overline{a_1} X + \overline{a_2} X^2 + \cdots + \overline{a_n} X^n$$

is irreducible in $\mathbb{Z}_p[X]$. Then, $f(X)$ is irreducible in $\mathbb{Q}[X]$.

Proof Since $\mathbb{Z}_p$ is a field, $\overline{f}(X)$ irreducible in $\mathbb{Z}_p[X]$ implies that $deg(\overline{f}(X)) = deg(f(X)) > 0$. The result follows from the above corollary. ♮

Remark 11.5.23 The above corollary gives us a sufficient condition for a polynomial in $\mathbb{Z}[X]$ to be irreducible in $\mathbb{Q}[X]$. Thus,

$$f(X) = a_0 + a_1 X + a_2 X^2 + \cdots + a_n X^n$$

in $\mathbb{Z}[X]$ is irreducible in $\mathbb{Q}[X]$ if for some prime p of $\mathbb{Z}$, p does not divide a_n and $\overline{f}(X)$ is irreducible in $\mathbb{Z}_p[X]$. This criteria is quite useful in checking the irreducibility of $f(X)$.

Example 11.5.24 We shall see in a later chapter that the polynomial $X^p - X + \overline{a}$ is irreducible in $\mathbb{Z}_p[X]$ whenever $\overline{a} \neq \overline{0}$. In particular, $X^p - X + \overline{1}$ is irreducible in $\mathbb{Z}_p[X]$ for every prime p. It follows from the above corollary that $X^p - X + 1$ is irreducible in $\mathbb{Q}[X]$. In fact $X^p - X + a$ is irreducible in $\mathbb{Q}[X]$ whenever p does not divide a. The polynomial $X^5 - 10X^3 - 6X - 1$ is also irreducible in $\mathbb{Q}[X]$, for in $\mathbb{Z}_5[X]$, and the polynomial $X^5 - \overline{10}X - \overline{6}X - \overline{1} = X^5 - X - \overline{1}$ is irreducible.

Theorem 11.5.25 (Eisenstein Irreducibility Criteria) *Let R be a unique factorization domain, and*

$$f(X) = a_0 + a_1 X + a_2 X^2 + \cdots + a_n X^n$$

be a polynomial in $R[X]$. Suppose that there is a prime element p in R such that
  (i) p/a_i for all $i < n$,
  (ii) p does not divide a_n, and
  (iii) p^2 does not divide a_0.
Then, $f(X)$ is irreducible in $F[X]$, where F is the field of fractions of R. Further, if $f(X)$ is primitive, then it is irreducible in $R[X]$ also.

Proof Consider the residue polynomial

$$\overline{f}(X) = \overline{a_0} + \overline{a_1}X + \overline{a_2}X^2 + \cdots + \overline{a_n}X^n = \overline{a_n}X^n$$

(for p/a_i for all $i < n$) of $f(X)$ in $R/\wp[X]$, where $\wp$ is the prime ideal generated by p. Note that $deg(f(X)) > 1$. Suppose that $f(X)$ is reducible in $F[X]$. Then, $f(X)$ is product of two polynomials in $R[X]$ of positive degrees. Suppose that

$$f(X) = g(X)h(X),$$

where $deg(g(X)) > 0$ and $deg(h(X)) > 0$. Then

$$\overline{f}(X) = \overline{a_n}X^n = \overline{g}(X)\overline{h}(X).$$

Clearly, $deg(\overline{g}(X)) = deg(g(X)) > 0$ and $deg(\overline{h}(X)) = deg(h(X)) > 0$. Hence

$$\overline{g}(X) = \overline{b_r}X^r \text{ and } \overline{h}(X) = \overline{c_s}X^s,$$

where $r, s \geq 1$ and $\overline{b_r} \neq \overline{0} \neq \overline{c_s}$. But, then $g(X)$ and $h(X)$ are polynomials in $R[X]$ whose constant terms are divisible by p. Since $f(X) = g(X)h(X)$, this would imply that the constant term a_0 of $f(X)$ is divisible by p^2. This is a contradiction to the hypothesis. Thus, $f(X)$ is irreducible in $F[X]$. In turn, if in addition to it, $f(X)$ is primitive also, then it is irreducible in $R[X]$ also.   ♯

Corollary 11.5.26 *Let p be a positive prime in $\mathbb{Z}$. Then the Cyclotomic polynomial*

$$f_p(X) = 1 + X + X^2 + \cdots + X^{p-1}$$

is irreducible in $\mathbb{Q}[X]$.

Proof Suppose that $f_p(X)$ is reducible in $\mathbb{Q}[X]$. Then $f_p(X + 1)$ is also reducible. Now,

$$f_p(X + 1) = (X + 1)^{p-1} + (X + 1)^{p-2} + \cdots + (X + 1) + 1 = \frac{(X + 1)^p - 1}{X + 1 - 1}.$$

Using the binomial theorem, we see that

$$f_p(X + 1) = X^{p-1} + {}^pC_1 X^{p-2} + \cdots + {}^pC_{p-1}.$$

Since ${}^pC_i$, $1 \leq i \leq p - 1$ is divisible by p and ${}^pC_{p-1} = p$ is not divisible by p^2, by the Eisenstein irreducibility criteria $f_p(X + 1)$ should be irreducible, a contradiction to the supposition. Hence, $f_p(X)$ is irreducible.   ♯

Corollary 11.5.27 *Let F be a field and $f(X, Y)$ an element of $F[X, Y]$, where*

$$f(X, Y) = a_0(X) + a_1(X)Y + a_2(X)Y^2 + \cdots + a_n(X)Y^n.$$

Suppose that $X/a_i(X)$ for all $i \leq n - 1$, X does not divide $a_n(X)$, and X^2 does not divide $a_0(X)$. Then, $f(X, Y)$ is irreducible in $F(X)[Y]$, where $F(X)$ is the field of fractions of $F[X]$. Further, if g.c.d of coefficients of powers of Y in $f(X, Y)$ is a unit in $F[X]$, then $f(X, Y)$ is irreducible in $F[X, Y]$.

Proof Follows from the Eisenstein irreducibility criteria if we note that $F[X]$ is a U.F.D, and X is a prime element of $F[X]$.   ♯

Example 11.5.28 The polynomial $X^6 + 6X^2 + 3X + 6$ is irreducible in $\mathbb{Q}[X]$ and also in $\mathbb{Z}[X]$, for 3 is a prime element of $\mathbb{Z}$ which satisfies the Eisenstein irreducibility criteria. The above polynomial is also primitive.

Exercises

11.5.1 Show that $\mathbb{Z}[\sqrt{-5}]$ is not a U.F.D but every nonzero nonunit element is expressible as product of irreducible elements. Also any two factorization contains the same number of irreducible factors.

11.5.2 Show that $3 + \sqrt{-7}$ is irreducible in $\mathbb{Z}[\sqrt{-7}]$. Is it prime? Is $\mathbb{Z}[\sqrt{-7}]$ a U.F.D?

11.5.3 Give an example of a commutative integral domain with identity in which not every nonzero nonunit element is expressible as product of irreducible elements.

11.5.4 Show by means of an example that homomorphic image of a U.F.D need not be a U.F.D even if it is an integral domain.

11.5.5 Show that the ring $\mathbb{R}[\![X]\!]$ of formal power series over the field $\mathbb{R}$ of real numbers is a U.F.D.

11.5.6 Prove that the following polynomials are irreducible in $\mathbb{Q}[X]$.
  (i) $X^5 - 3X^2 + 6X + 15$.
  (ii) $X^3 + 6X^2 + 7$.
  (iii) $X^{11} + 22X^9 + 11X^2 + 12X + 10$.

11.5.7 Show that the polynomial

$$(1+i) + 2X + (1-i)X^2 + X^3$$

is irreducible in $\mathbb{Z}[i][X]$.

11.5.8 Show that $1 + X + 2(1+X)Y + 3(1+X)^2Y^2 + Y^3$ is irreducible in $\mathbb{C}[X, Y]$, where $\mathbb{C}$ is the field of complex numbers.

11.5.9 Factor $X^4 + X^2 + 1$ in to irreducible factors of $\mathbb{Q}[X]$.

11.5.10 Show that $\mathbb{Z}[i]$ and $\mathbb{Z}[\sqrt{-5}]$ are not isomorphic.

11.5.11 Show that $\mathbb{Z}[X]$ and $\mathbb{Z}[i]$ are not isomorphic.

11.5.12 Show that $X^4 - 10X^2 - 1$ is irreducible in $\mathbb{Q}[X]$. Is it irreducible modulo 2, 3 or 5?

11.5.13 Are $\mathbb{Z}[i]$ and $\mathbb{Z}[\omega]$ isomorphic?

11.5.14 Show that if $p = a^2 + b^2$ is irreducible in $\mathbb{Z}$, then $a + bi$ is prime in $\mathbb{Z}[i]$.

11.5.15 Suppose that $f(X) \in \mathbb{Z}[X]$ is reducible. Show that there exists $n_0 \in \mathbb{N}$ such that for every prime $p \geq n_0$, the polynomial $\overline{f}(X)$ is reducible in $\mathbb{Z}_p[X]$.

11.5.16 Let p be a prime integer in $\mathbb{Z}$. Suppose that it is not prime in $\mathbb{Z}[i]$. Show that there is a prime element $z = a + bi$ in $\mathbb{Z}[i]$ such that $p = z\overline{z}$.

11.5.17 Let p be a prime integer in $\mathbb{Z}$. Show that p is prime in $\mathbb{Z}[i]$ if and only if $p \equiv 3 \pmod 4$.

Appendix

Category Theory

In this book, we introduced and studied several mathematical structures, viz. semigroups, groups, rings, and fields. We noticed some common features in the study of these structures. The category theory gives a unified general and abstract setting for all these and many more mathematical structures such as modules, vector spaces, and topological spaces. Quite often, in mathematics, the concrete results are expressed in the language of category theory. This appendix introduces the very basics in category theory for the purpose. The Gödel–Bernays axiomatic system for sets is the most suitable axiomatic system for category theory. As described in Chap. 2, class is a primitive term in this axiomatic system instead of sets. Indeed, sets are simply the members of classes. The classes which are not sets are termed as proper classes.

Definition A category Σ consists of the following:

1. A class $Obj\Sigma$ called the class of objects of Σ.
2. For each pair A, B in $Obj\Sigma$, we have a set $Mor_\Sigma(A, B)$ called the set of morphisms from the object A to the object B. Further,

$$Mor_\Sigma(A, B) \bigcap Mor_\Sigma(A', B') \neq \emptyset \text{ if and only if } A = A' \text{ and } B = B'.$$

3. For each triple A, B, C in $Obj\Sigma$, we have a map $\cdot$ from $Mor_\Sigma(B, C) \times Mor_\Sigma(A, B)$ to $Mor_\Sigma(A, C)$ called the law of composition. We denote the image $\cdot(g, f)$ by gf. Further, the law of composition is associative in the sense that if $f \in Mor_\Sigma(A, B)$, $g \in Mor_\Sigma(B, C)$ and $h \in Mor_\Sigma(C, D)$, then $(hg)f = h(gf)$.
4. For each $A \in Obj\Sigma$, there is an element I_A in $Mor_\Sigma(A, A)$ such that $fI_A = f$ for all morphisms from A, and $gI_A = g$ for all morphisms to A.

Clearly, for each object A of Σ, I_A is unique morphism, and it is called the identity morphism on A. The category Σ is called a **small category** if $Obj\Sigma$ is a set.

Examples 1. We have the category *SET* of sets whose objects are sets, and the morphisms from a set A to a set B are precisely the maps from A to B.

2. There is a category *GP* of groups whose objects are groups, and the morphisms from a group H to a group K are homomorphisms from H to K.

© Springer Nature Singapore Pte Ltd. 2017
R. Lal, *Algebra 1*, Infosys Science Foundation Series in Mathematical Sciences,
DOI 10.1007/978-981-10-4253-9

3. *SG* denotes the category of semigroups all of whose objects are semigroups, and morphisms are semigroup homomorphisms.

4. *AB* denotes the category of abelian groups whose objects are abelian groups, and morphisms are group homomorphisms.

5. *RING* denotes the category of rings whose objects are rings, and morphisms are ring homomorphisms.

6. *TOP* denotes the category of topological spaces whose objects are topological spaces, and the morphisms are continuous maps.

7. A group G can also be treated as a category having a single object G. The elements of the group can be taken as morphisms from G to G, and the composition of the morphisms is the binary operation of G.

Let Σ be a category. A morphism f from A to B is said to be a monomorphism (epimorphism) if it can be left (right) canceled in the sense that $fg = fh$ ($gf = hf$) implies that $g = h$. A morphism f from A to B is said to be an isomorphism if there is a morphism g from B to A such that $gf = I_A$ and $fg = I_B$. Clearly, such a morphism g is unique, and it is called the inverse of f. The inverse of f, if exists, is denoted by f^{-1}.

Remark In the category *SET* of sets, a morphism f from a set A to a set B is a monomorphism (epimorphism) if and only if it is an injective (surjective) map. Also in the category *GP*, a morphism f from a group H to a group K is a monomorphism if and only if f is injective (prove it). Clearly, an isomorphism is a monomorphism as well as an epimorphism. However, a morphism which is a monomorphism as well as epimorphism need not be an isomorphism. For example, consider the category *Haus* of Hausdorff topological spaces. The inclusion map from $\mathbb{Q}$ to $\mathbb{R}$ is a monomorphism as well as an epimorphism (it is an epimorphism because $\mathbb{Q}$ is dense in $\mathbb{R}$) but it is not an isomorphism.

Let Σ be a category and A be an object of Σ. Then, $Mor_\Sigma(A, A)$ is a monoid with respect to the composition of morphisms. The members of $Mor_\Sigma(A, A)$ are called the endomorphisms of A. The monoid $Mor_\Sigma(A, A)$ is denoted by $End(A)$. An isomorphism from A to A is called an automorphism of A. The set of all automorphisms of A is denoted by $Aut(A)$ which is a group under the composition of morphisms.

Let Σ be a category. We say that a category Γ is a subcategory of Σ if (i) $Obj\Gamma \subseteq \Sigma$, (ii) for each pair $A, B \in Obj\Gamma$, $Mor_\Gamma(A, B) \subseteq Mor_\Sigma(A, B)$, and (iii) the law of composition of morphisms in Γ is the restriction of the law of composition of morphisms in Σ to Γ. The subcategory Γ is said to be a full subcategory if $Mor_\Gamma(A, B) = Mor_\Sigma(A, B)$ for all $A, B \in Obj\Gamma$. *AB* is a full subcategory of *GP*.

Functors

Definition Let Σ and Γ be categories. A functor F from Σ to Γ is an association which associates with each member $A \in Obj\Sigma$, a member $F(A)$ of Γ, and to each morphism $f \in Mor_\Sigma(A, B)$, a morphism $F(f) \in Mor_\Gamma(F(A), F(B))$ such that the following two conditions hold:

(i) $F(gf) = F(g)F(f)$ whenever the composition gf is defined.
(ii) $F(I_A) = I_{F(A)}$ for all $A \in \Sigma$.

Let Σ be a category. Consider the category Σ^o whose objects are same as the objects of Σ, $Mor_\Sigma(A, B) = Mor_{\Sigma^o}(B, A)$, and the composition fg in Σ^o is same as gf in Σ. The category Σ^o is called the opposite category of Σ. A functor from Σ^o to the category Γ is called a contravariant functor from Σ to Γ.

If Σ is a category, then the identity map $I_{Obj\Sigma}$ from $Obj\Sigma$ to itself defines a functor called the identity functor. Composition of any two functors is again a functor.

Examples 1. Let H be a group. Denote its abelianizer $H/[H, K]$ by $Ab(H)$. Let f be a homomorphism from H to a group K. Then f induces a homomorphism $Ab(f)$ from $Ab(H)$ to $Ab(K)$ defined by $Ab(f)(h[H, H]) = f(h)[K, K]$. This defines a functor Ab from the category GP of groups to the category AB. This functor is called the abelianizer functor.

2. Let H be a group. Denote the commutator $[H, H]$ of H by $Comm(H)$. If f be a homomorphism from H to K, it induces a homomorphism $Comm(f)$ from $Comm(H)$ to $Comm(K)$ defined by $Comm(f)([a, b]) = [f(a), f(b)]$. This defines a functor $Comm$ from the category GP to itself. This functor is called the commutator functor.

3. We have a functor Ω from the category GP of groups to the category SET of sets which simply forgets the group structure and retains the set part of the group. More explicitly, $\Omega(G, o) = G$. Such a functor is called a forgetful functor. There is another such functor from the category $RING$ of rings to the category AB of abelian groups which forgets the ring structure, but retains the additive group part of the ring. There is still another forgetful functor from the category TOP to the category SET which forgets the topological structure, and retains the set part of the space.

4. Let f be a map from a set X to a set Y. Then, f induces a unique homomorphism $F(f)$ from the free group $F(X)$ to the free group $F(Y)$. This gives us the functor F from the category SET to the category GP. This functor is called the free group functor.

5. Let Σ be a category and A be an object of Σ. For each $B \in Obj\Sigma$, we put $Mor_\Sigma(A, -)(B) = Mor_\Sigma(A, B)$, and for each morphism f from B to C, we have a map $Mor_\Sigma(A, -)(f)$ from $Mor_\Sigma(A, B)$ to $Mor_\Sigma(A, C)$ defined by $Mor_\Sigma(A, -)(f)(g) = fg$. It is easily verified that $Mor_\Sigma(A, -)$ defined above is a functor from Σ to the category SET of sets. Similarly, we have a contravariant functor $Mor_\Sigma(-, A)$ from the category Σ to the category SET of sets.

Remark There is a useful important functor, viz, homotopy group functor π_1 from the category TOP^* of pointed topological spaces to the category GP of groups. It has tremendous application in geometry and topology. This functor will be discussed in Algebra 3.

Let F be a functor from a category Σ to a category Γ. The functor F is said to be faithful if for each pair $A, B \in Obj\Sigma$, the induced map $f \mapsto F(f)$ from $Mor_\Sigma(A, B)$ to $Mor_\Gamma(F(A), F(B))$ is injective. The functor F is said to be a full functor if these induced maps are surjective. The forgetful functor from GP to SET is faithful but it is not full. The abelianizer functor Ab is not faithful. A functor F is said to be an isomorphism from the category Σ to the category Γ if there is a functor G from Γ to Σ such that $GoF = I_\Sigma$ and $FoG = I_\Gamma$.

Natural Transformations

Definition Let F and G be functors from a category Σ to a category Γ. A natural transformation η from F to G is a family $\{\eta_A \in Mor_\Gamma(F(A), G(A)) \mid A \in Obj\Sigma\}$ of morphisms in Γ such that the diagram

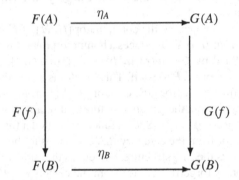

is commutative for all morphisms f in Σ.

Examples 1. Let ν_G denote the quotient homomorphism from G to $G/[G, G]$. Then, the family $\{\nu_G \mid G \in ObjGP\}$ defines a natural transformation ν from the identity functor I_{GP} to the abelianizer functor Ab.

2. For each set X, we have the inclusion map i_X from X to $\Omega(F(X))$ where F is the free group functor, and Ω is the forgetful functor. Evidently, the family $\{i_X \mid X \in ObjSET\}$ of maps defines a natural transformation from the identity functor I_{SET} to the functor ΩoF on the category SET.

Let F and G be two functors from a category Σ to a category Γ. A natural transformation η from F to G is called a natural equivalence if η_A is an isomorphism from $F(A)$ to $G(A)$ for all $A \in Obj\Sigma$. This is equivalent to say that there is a natural transformation ρ from G to F such that $\rho_A o \eta_A = I_{F(A)}$ and $\eta_A o \rho_A = I_{G(A)}$ for all objects A of Σ. A functor F from Σ to Γ is called an equivalence from Σ to Γ if there is a functor G from Γ to Σ such that FoG and GoF are naturally equivalent to the corresponding identity functors. Notice that there is a difference between isomorphism and equivalence between categories. An equivalence need not be an isomorphism.

Let Σ and Γ be categories. Then, $\Sigma \times \Gamma$ represents the category whose objects are pairs $(A, B) \in Obj\Sigma \times Obj\Gamma$, and a morphism from (A, B) to (C, D) is a pair (f, g), where f is a morphism from A to C in Σ, and g is a morphism from B to D in Γ. The composition law is coordinate-wise. This category is called the product category.

Let F be a functor from a category Σ to a category Γ. Let f be a morphism from C to A in Σ, and g be a morphism from B to D in Γ. This defines a map $Mor_\Gamma(F(f), g)$ from the set $Mor_\Gamma(F(A), B)$ to $Mor_\Gamma(F(C), D)$ given by $Mor_\Gamma(F(f), g)(h) = ghF(f)$. This defines a functor $Mor_\Gamma(F(-), -)$ from the product category $\Sigma^o \times \Gamma$ to the category SET of sets. Similarly, given a functor G from Γ to Σ, we have a functor $Mor_\Sigma(-, G(-))$ to the category SET of sets. We say that F is left adjoint to G, or G is right adjoint to F if there is a natural isomorphism η from the functor $Mor_\Gamma(F(-), -)$ to the functor $Mor_\Sigma(-, G(-))$. More explicitly, for each object A in Σ and each object B in Γ, we have a bijective map $\eta_{A,B}$ from $Mor_\Gamma(F(A), B)$ to the set $Mor_\Sigma(A, G(B))$ such that $Mor_\Sigma(f, G(g))o\eta_{A,B} = \eta_{C,D}oMor_\Gamma(F(f), g)$ for all morphisms (f, g) in $\Sigma^o \times \Gamma$ (look at the corresponding commutative diagram).

Examples 1. Consider the category SET of sets and the category GP of groups. We have the free group functor F from the category SET to the category GP. More explicitly, for each set X, we have the free group $F(X)$ on the set X. We also have the forgetful functor Ω from GP to SET. From the universal property of free group, every group homomorphism f from $F(X)$ to G determines and is uniquely determined by its restriction to X. This gives us a bijective map $\eta_{X,G}$ from $Hom(F(X), G)$ to $Map(X, \Omega(G))$. It is easy to observe (using the universal property of a free group) that η, thus obtained, is a natural equivalence. Hence, the free group functor F is left adjoint to the forgetful functor Ω.

2. We have the forgetful functor Ω from the category AB of abelian groups to the category GP of groups. We also have the abelianizer functor Ab from GP to AB. It can be easily verified that Ab is left adjoint to Ω.

Products and Coproducts in a Category

Definition Let A and B be objects in a category Σ. A product of A and B in Σ is a triple (P, f, g), where P is an object of the category Σ, f is morphisms from P to A, and g is a morphism from P to B such that given any such triple (P', f', g'), there is a unique morphism ϕ from P' to P such that $f\phi = f'$ and $g\phi = g'$.

It is easily observed from the definition that if (P, f, g) and (P', f', g') are two products of A and B, then there is an isomorphism ϕ from P' to P with $f\phi = f'$ and

$g\phi = g'$. Thus, if the product exists, then it is unique up to natural isomorphism. The product of A and B is usually denoted by $A \times B$.

In the category *SET* of sets, the Cartesian product $A \times B$ with the corresponding projection maps is the product in the category *SET*. Similarly, the direct product $H \times K$ of the groups H and K together with the corresponding projection maps is the product of H and K in the category *GP*.

Dually, we have the following:

Definition Let A and B be objects in a category Σ. A coproduct of A and B in Σ is a triple (U, f, g), where U is an object of the category Σ, f is morphisms from A to U, and g is a morphism from B to U such that given any such triple (U', f', g'), there is a unique morphism ϕ from U to U' such that $\phi f = f'$ and $\phi g = g'$.

It is easily observed from the definition that if (U, f, g) and (U', f', g') are two coproducts of A and B, then there is an isomorphism ϕ from U to U' with $\phi f = f'$ and $\phi g = g'$. Thus, if the co-product exists, then it is unique up to natural isomorphism. The coproduct of A and B is denoted by $A \coprod B$.

In the category *SET* of sets, the disjoint union $(A \times \{0\}) \bigcup (B \times \{1\})$ of A and B with the natural inclusion maps is the coproduct of A and B in the category *SET*. Similarly, the free product $H \star K$ of the groups H and K together with the natural inclusion maps is the coproduct of H and K in the category *GP*.

PullBack and Push-out Diagrams

Definition Let Σ be a category. Let $f \in Mor_\Sigma(A, C)$, and $g \in Mor_\Sigma(B, C)$. A commutative diagram

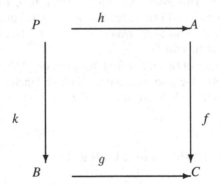

is said to be a pullback diagram if given any commutative diagram

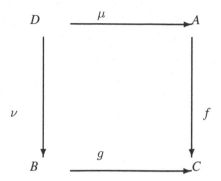

there exists a unique morphism ϕ from D to P such that $h\phi = \mu$ and $k\phi = \nu$.

Dually, a push-out diagram can be defined by reversing the arrows in the definition. The reader is advised to formulate the definition of push-out diagram.

In general, pullback and push out need not exist in a category. However, they exist in the category SET of sets and also in the category GP of groups: Let f be a morphism from A to C, and g be a morphism from B to C in the category SET/GP. Consider the product $A \times B$ in the category SET/GP. Let $P = \{(a, b) \in A \times B \mid f(a) = g(b)\}$. Let h denote first projection from P to A, and k denotes the second projection from P to B. This gives us a pullback diagram in SET/GP. Similarly, push out also exists in the category SET, and also it exists in the category GP.

Bibliography

1. Cohen, P.J.: Set Theory and Continuum Hypothesis. Dover Publication (2008)
2. Halmos, P.R.: Naive Set Theory. Springer (1914)
3. Kakkar, V.: Set Theory. Narosa Publishing House (2016)
4. Artin, M.: Algebra. Pearson Education (2008)
5. Herstein, I.N.: Topics in Algebra, 2nd edn. Wiley, New York (1975)
6. Jacobson, N.: Basic Algebra I. Freeman, San Francisco, II (1980)
7. Lang, S.: Algebra, 2nd edn. Addison-Wesley, MA (1965)
8. Birkoff, G., MacLane, S.: A survey of Modern Algebra, 3rd edn. Macmillan, New York (1965)
9. Hungerford, T.W.: Algebra, 8th edn. Springer, GTM (2003)
10. Rademacher, H.: Lectures on Elementary Number Theory. Krieger Publishing Co (1977)
11. Serre, J.P.: A Course in Arithmetic. Springer (1977)
12. Robinson, D.J.S.: A Course in The Theory of Groups, 2nd edn. Springer (1995)
13. Rotman, J.J.: An Introduction to the Theory of Groups, 4th edn. Springer, GTM (1999)
14. Suzuki, M.: Group Theory I and II. Springer (1980)
15. Curtis, M.L.: Matrix Groups. Springer (1984)
16. Magnus, W., Karrass, A., Solitar, D.: Combinatorial Group Theory. Wiley, Newyork (1966)
17. Rudin, W.: Principles of Mathematical Analysis, 3rd edn. McGraw Hill (1976)

Index

© Springer Nature Singapore Pte Ltd. 2017   429
R. Lal, *Algebra 1*, Infosys Science Foundation Series in Mathematical Sciences,
DOI 10.1007/978-981-10-4253-9

Printed in the United States
By Bookmasters